VOLCANOES

VOLCANOES

Editor

Berislav Vãduva

scitus
academics

Volcanoes

Edited by **Berislav Vãduva**

This Book is Printed on Acid-Free Paper. ♾

ISBN: 9781681178868
Library of Congress Control Number: 2017948913

SCITUS Academics LLC,
616, Corporate Way, Suite 2, 4766,
Valley Cottage, NY 10989

www.scitusacademics.com

Notice

Printed in United States of America.

Preface

A volcano is a vent or chimney which transfers molten rock known as magma from depth to the Earth's surface. Magma erupting from a volcano is called lava and is the material which builds up the cone surrounding the vent. A Volcanic hazard is the probability that a volcanic eruption or related geophysical event will occur in a given geographic area and within a specified window of time. The risk that can be associated with a volcanic hazard depends on the proximity and vulnerability of an asset or a population of people near to where a volcanic event might occur. Volcanoes emit gases during eruptions. Even when a volcano is not erupting, cracks in the ground allow gases to reach the surface through small openings called fumaroles. More than 90% of all gas emitted by volcanoes is water vapor (steam), most of which is heated ground water (underground water from rainfall and streams). Other common volcanic gases are carbon dioxide, sulfur dioxide, hydrogen sulfide, hydrogen, and fluorine. Sulfur dioxide gas can react with water droplets in the atmosphere to create acid rain, which causes corrosion and harms vegetation. Carbon dioxide is heavier than air and can be trapped in low areas in concentrations that are deadly to people and animals. Fluorine, which in high concentrations is toxic, can be adsorbed onto volcanic ash particles that later fall to the ground. The fluorine on the particles can poison livestock grazing on ash-coated grass and also contaminate domestic water supplies. Volcanic eruptions can drastically alter the geography of a region. Hot, liquid lava flowing downslope can force people living near volcanoes to flee their homes. Even more dangerous, however, are pyroclastic flows and lahars, in which volcanic rock fragments travel in fast-moving clouds and mudslides.

This book 'Volcanoes' provides the current trends in volcanological researches demonstrating clearly the research activity in volcanology from geophysical aspects of volcanic systems to their geological framework. It focuses on volcanological problems that could be used as valuable resource for both researchers and graduate research students. This book hence can likewise be valuable for students and researchers.

Table of Contents

CHAPTER 1

Monogenetic Basaltic Volcanoes: Genetic Classification, Growth, Geomorphology and Degradation

Gábor Kereszturi[1] and Károly Németh[1, 2]

[1] *Volcanic Risk Solutions, Massey University, Palmerston North, New Zealand*

[2] *King Abdulaziz University, Jeddah, Kingdom of Saudi Arabia*

1. INTRODUCTION

Plate motion and associated tectonics explain the location of magmatic systems along plate boundaries [1], however, they cannot give satisfactory explanations of the origin of intraplate volcanism. Intraplate magmatism such as that which created the Hawaiian Islands (Figure 1, hereafter for the location of geographical places the reader is referred to Figure 1) far from plate boundaries is conventionally explained as a result of a large, deep-sourced, mantle-plume [2-4]. Less volumetric magmatic-systems also occur far from plate margins in typical intraplate settings with no evidence of a mantle-plume [5-7]. Intraplate volcanic systems are characterized by small-volume volcanoes with dispersed magmatic plumbing systems that erupt predominantly basaltic magmas [8-10] derived usually from the mantle with just sufficient residence time in the crust to allow minor fractional crystallization or wall-rock assimilation to occur [e.g. 11]. However, there are some examples for monogenetic eruptions that have been fed by crustal contaminated or stalled magma from possible shallower depths [12-19]. The volumetric dimensions of such magmatic systems are often comparable with other, potentially smaller, focused magmatic systems feeding polygenetic volcanoes [20-21]. These volcanic fields occur in every known tectonic setting [1, 10, 22-28] and also on other planetary bodies such as Mars [29-33]. Due to the abundance of monogenetic volcanic fields in every tectonic environment, this form of volcanism represents a localized, unpredictable volcanic hazard to the increasing human populations of cities located close to these volcanic fields such as Auckland in New Zealand [34-35] or Mexico City in Mexico [36-37].

Importantly, research on monogenetic volcanoes and volcanic fields is focused on their "source to surface" nature, i.e. once the melt is extracted from the source it tends to ascend to the surface [11, 16-17, 38]. The rapid melt generation and short eruptive history of volcanoes fed by these magmas mean they can be used as 'probes' of various processes, particularly to detect short- and long-term changes occurring during emplacement of a single vent and/or a volcanic field. They also provide evidence of the evolution of magmatic systems that fed numerous individual small-volume volcanoes over time spans of millions of years [39-44]. This research has led to an understanding of the processes of melt extraction [17, 45-46], interactions in the lithospheric mantle [47-49], ascent within the lower to middle crust [16, 50] and in the shallow crust region [10, 51-54]. Other studies have elucidated plumbing and feeder systems of monogenetic volcanoes [8-9, 55-57], eruption mechanisms [58-61] and associated volcanic hazards [34, 62-67] as well as surface processes [68-71] and long-term landscape evolution [72-74].

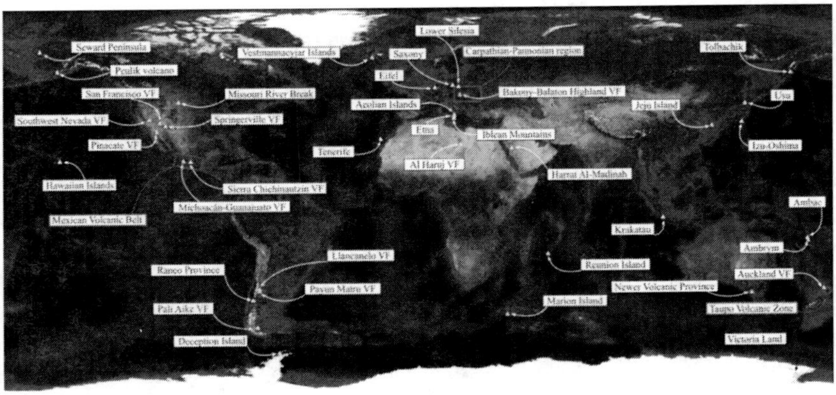

FIGURE 1. Overview map of the location of the volcanic field and zones mentioned in the text. The detailed location of specific volcanic edifices mentioned in the text can be downloaded as a Google Earth extension (.KMZ file format) from http://www.intechopen.com/.

Eruption of magma on the surface can be interpreted as the result of the dominance of magma pressure over lithostatic pressure [50, 75-76]. On the other hand, freezing of magma en route to the surface are commonly due to insufficient magma buoyancy, where the lithostatic pressure is larger than the magma pressure, or insufficient channelling/focusing of the magma [50, 76-78]. Once these small-volume magmas (0.001 to 0.1 km^3) intrude into the shallow-crust, they are vulnerable to external influences such as interactions with groundwater at shallow depth [79-82]. In many cases, the eruption style is not just determined by internal magma properties, but also by the external environmental conditions to which it has been exposed. Consequently, the eruption style becomes an actual balance between magmatic and environmental factors at a given time slice of the eruption. However, a combination of eruption styles is responsible for the formation of monogenetic volcanoes with wide range of morphologies, e.g. from conical-shaped to crater-shaped volcanoes. The morphology that results from the eruption is often connected to the dominant eruptive mechanisms, and therefore, it is an important criterion in volcano classifications. Diverse sources of information regarding eruption mechanism, edifice growth and hazards of monogenetic volcanism can be extracted during various stages of the degradation when the internal architecture of a volcano is exposed. Additionally, the rate and style of degradation may also help to understand the erosion and sedimentary processes acting on the flanks of a monogenetic volcano. The duration of the construction is of the orders of days to decades [83-84]. In contrast, complete degradation is several orders of magnitude slower process, from ka to Ma [68, 71, 73]. Every stage of degradation of a monogenetic volcano could uncover important information about external and internal processes operating at the time of the formation of the volcanic edifice. This information is usually extracted through stratigraphic, sedimentary, geomorphic and quantitative geometric data from erosion landforms. In this chapter, an overview is presented about the dominant eruption mechanism associated with subaerial monogenetic volcanism with the aim of understanding the syn- and post-eruptive geomorphic and morphometric development of monogenetic volcanoes from regional to local scales.

2. MONOGENETIC MAGMATIC SYSTEMS

Melt production from the source region in the mantle is triggered by global tectonic processes such as converging plate margins, e.g. Taupo Volcanic Zone in New Zealand [85-88] and in the Carpathian-Pannonian region in Central Europe [89-93] or diverging plate margins, for example sea-floor spreading along mid-oceanic ridges [94-95]. Melting also occurs in sensu stricto “convection plumes” or “hot spots” [2, 4, 96], which could alternatively result from small-scale, mantle wedge-driven convection cells [97]. This is often a passive effect of topographic differences between thick, cratonic and thin, oceanic lithosphere, as suspected by numerical modelling studies [97-101].

Typical ascent of the magma feeding eruptions through a monogenetic volcano starts in the source region by magma extraction from melt-rich bands. These melt-rich bands are commonly situated in a low angle (about 15–25°) to the plane of principal shear direction introduced by deformation of partially molten aggregates [95, 102-103]. The degree of efficiency of melt extraction is dependent on the interconnectivity, surface tension and capillary effect of the solid grain-like media in the mantle, which are commonly characterized by the dihedral angle between solid grains [104-105]. When deformation-induced strain takes place in a partially molten media, it increases the porosity between grains and triggers small-scale focusing and migration of the melt [104]. With the continuation of local shear in the mantle, the total volume of melt increases and enhances the magma pressure and buoyancy until it reaches the critical volume for ascent depending on favourable tectonic stress setting, depth of melt extraction and overlying rock (sediment) properties [16, 42]. The initiation of magma (crystals + melt) ascent starts as porous flow in deformable media and later transforms into channel flow (or a dyke) if the physical properties such as porosity/permeability of the host rock are high enough in elastic or brittle rocks in the crust [50, 75, 106-107]. The critical volume of melt essential for dyke injections is in the range of a few tens of m^3 [76], a volume which is several orders of magnitude less than magma batches feeding eruptions on the surface, usually ≥0.0001 km^3 [39, 108]. An increase in melt propagation distance is possible if small, pocket-fed initial dykes interact with each other [50, 76], which is strongly dependent on the direction of maximum (σ_1) and least principal stresses (σ_3), both in local and regional scales [109] and the vertical and horizontal separation of dykes [50, 76, 110]. These dykes move in the crust as self-propagating fractures controlled by the density contrast between the melt and the host rock from the over-pressured source zone [50]. The dykes could remain connected with the source region or propagate as a pocket of melt in the crust [111-112]. The geometry of such dykes is usually perpendicular to the least principal stress directions [108, 111]. The lateral migration of the magma en route is minimal in comparison with its vertical migration. This implies the vent location at the surface is a good approximation to the location of melt extraction at depth, i.e. the magma footprint [42, 54, 108]. The important implication of this behaviour is that interactions between magma and pre-existing structures are expected within the magma footprint area [54, 108]. Correlation between pre-existing faults and dykes are often recognized in volcanic fields [10, 53, 108, 113-115]. The likelihood of channelization of magma by a pre-existing fracture such as a fault, is preferable in the case of high-angle faults, i.e. 70–80°, and shallow depths [53] when the magma pressure is less than the tectonic strain taken up by faulting [42, 53].

These monogenetic eruptions have a wide variation in eruptive volumes. Volumetrically, two end-members types of volcanoes have been recognized [5, 109, 116]. Large-volume (≥1 km^3 or polygenetic) volcanoes are formed by multiple ascent of magmas that use more or less the same conduit system over a long period of time usually ka to Ma and have complex phases of construction and destruction [86, 117-119]. The spatial concentration of melt ascents, and temporally the longevity of such systems are usually caused by the formation of magma storage systems at various levels of the crust beneath the volcanic edifices [120-122]. In this magma chamber stalled magma can evolve by differentiation and crystallization in ka time scales [123]. On the other hand, a small-volume (≤1 km^3 or monogenetic) volcano is referred to as "[it] erupts only once" [e.g. 116]. The relationship between large and small volume magmatic systems and their volcanoes is poorly understood [1, 5, 109, 124-127]. Nevertheless, there is a wide volumetric spectrum between small and large (monogenetic and polygenetic) volcanoes and these two end-members naturally offer the potential for transition types of volcanoes to exist. An ascent event is not always associated with a single batch of magma, but commonly involves multiple tapping events (i.e. multiple magma batches), creating a diverse geochemical evolution over even a single eruption [9, 11, 16-17, 45, 128]. Multiple melt batches involved in a single event may be derived from the mantle directly or from some stalling magma ponds around high density contrast zones in the lithosphere such as the upper-mantle/crust boundary [9, 128] and/or around the ductile/brittle boundary zone in the crust [16].

A volcanic eruption on the surface is considered to be a result of a successful coupling mechanism between internal processes, such as melt extraction rate and dyke interaction en-route to the surface [50, 76, 110], and external processes, such as local and regional stress fields in the crust [42, 109]. Therefore, the spatial and temporal location of a volcanic event represents the configuration of the magmatic system at the time of the eruption. However, mantle-derived, usually primitive magmas feeding monogenetic magmatic systems are uncommon and rarely erupt individually. They tend to concentrate in space forming groups of individual volcanoes or clusters [7, 24, 129-130], and in time constitute volcanic cycles [39, 42, 131-132]. The spatial component of volcanism is dependent on the susceptibility of magma to be captured by pre-existing structures such as faults [10, 53-54], and the regional stress field at the time of the melt ascent [7, 43, 109, 133]. Temporal controls are also significantly influenced by internal and external forces. The monogenetic magmatic systems can be classified into two groups [131, 134]. The volume-predictable systems

[134-135] are internally-controlled, i.e. it is magmatically-controlled [42]. In this system, an eruption on the surface is a direct result of successful separation of melt from a heterogeneous mantle, which is independent from the tectonics. Therefore, the total volume of magma erupting at the surface is usually a function of magma production rates of the system and repose time since the previous eruption [42]. These magmatic systems are usually characterized by high magma flux, promoting frequent dyke injections and high magmatic contribution to local extensional strain accumulation. These could trigger earthquakes, faulting and surface deformations, such as ruptures, associated with the high rates of magma intrusions [111, 136] similar to the intrusion at tensional rift zones [e.g. 137-138]. Magma ascent is often dominated by the regional-scale direction of stress rather than the location of pre-existing faults and topography [111]. In contrast, the time-predictable magmatic system [131, 139] is a passive by-product of tectonic shear-triggered melt extraction [42, 95, 103, 131]. Without tectonic forces, the melt would not be able to be extracted from partially molten aggregates [42]. Consequently, this magma generation process is externally- or tectonically-controlled [42]. The overall magma supply of these volcanic fields is generally low. Magmatic pressure generated by the magma injections are commonly suppressed by lithostatic pressure, resulting in a greater chance of interaction between magma and pre-existing structures in the shallow crust [53, 111, 140]. Dyke capturing commonly takes place if the orientation of the dyke plane is not parallel with the direction of maximum principal stress, causing vent alignments and fissure orientation to not always be perpendicular with the least principal stress direction [42, 54].

Restriction of magma ascent to a small area usually results in monogenetic volcanoes forming volcanic fields in a well-defined geographic area. These eruptions normally take place from hours to decades resulting in the accumulation of small-volume eruptive products on the surface predominantly from basaltic magmas. However, a monogenetic volcanic field could experience monogenetic eruptions over time scales of Ma [5, 39, 141-142] and the lifespan is characterized by waxing and waning stages of volcanism and cyclic behaviour [39, 108]. In a single monogenetic volcanic field, tens to thousands of individual volcanoes may occur [143] with predominantly low SiO_2 content eruptive products ranging from ca. 40 wt% up to 60 wt% [16, 40, 128, 144-146]. However, monogenetic volcanism does not depend on the chemical composition because there are similar small-volume monogenetic volcanoes that have been erupted from predominantly silica-rich melt such as Tepexitl tuff ring, Serdán-Oriental Basin, Mexican Volcanic Belt, Mexico [147] or the Puketarata tuff ring, Taupo Volcanic Zone, New Zealand [148].

3. CONSTRUCTION OF MONOGENETIC VOLCANOES

The ascent of magma from source to surface usually involves thousands of interactions between external and internal processes, thus the pre-eruptive phase works like an open system. Once single or multiple batch(es) of magma start their ascent to the surface, there is continuous degassing and interactions with the environment at various levels en route. On the surface, the ascending magma ascent can feed a volcanic eruption that can be explosive or effusive. Important characteristics of the volcanic explosion are determined at shallow depth (≤1–2 km) by the balance between external and internal factors such as chemical composition or availability of external water. The volcanic eruptions are usually characterised by discrete eruptive and sedimentary processes that are important entities of the formation and emplacement of a monogenetic vent itself.

3.1. Internal Versus External-Driven Eruptive Styles

The current classification of volcanic eruptions is based mainly on characteristics such as magma composition, magma/water mass ratio, volcanic edifice size and geometry, tephra dispersal, dominant grain-size of pyroclasts and (usually eye-witnessed) column height [e.g. 149]. If the ascending melt or batches of melts reach the near-surface or surface region, it will either behave explosively or intrusively/effusively. Explosive magma fragmentation is triggered either by the dissolved magmatic volatile-content [150] or by the thermal energy to kinetic energy conversion and expansion during magma/water interactions [151-152], producing distinctive eruption styles. These eruption styles can be classified on the basis of the dominance of internal or external processes.

Internally-driven eruptions are promoted by dissolved volatiles within the melt that exsolve into a gas-phase during decompression of magma [153-155]. The volatiles are mainly H_2O with minor CO_2, the latter exsolving at higher pressure and therefore greater depths than H_2O [e.g. 156]. Expansion of these exsolved gases to form bubbles in the magma suddenly lowers the density of the rising fluid, causing rapid upward magma acceleration and eventually fragmentation along bubble margins [150, 155, 157-159]. The growth of gas bubbles by diffusion and decompression in the melt occurs during magma rise, until the volume faction exceeds 70–80% of the melt, at which point magma fragmentation occurs [160-161]. Magmas with low SiO_2 contents, such as basalts and undersaturated magmas have low viscosity, allowing bubbles to expand easily in comparison to andesitic and rhyolitic magmas. Thus these low-silica magmas generate mild to moderate explosive types of eruptions such as Hawaiian [e.g. 162], Strombolian [e.g. 153], violent Strombolian [e.g. 163] and in very rare instances sub-Plinian types [e.g. 164, 165]. There is a conceptual difference between Hawaiian and Strombolian-style eruptions because in the former case magmatic gases rise together with the melt [154], whereas in Strombolian-style eruptions an essentially stagnant magma has gas slugs that rise and bubble through it – generating large gas slug bursts and foam-collapse at the boundary of the conduit [153, 166]. According to the rise speed-dependent model, bubbles form during magma ascent [150], while in the case of the foam collapse model, bubbles up to 2 m in diameter are generated deeper, in the upper part of a shallow magma chamber, based on acoustic measurements at the persistently active Stromboli volcano in the Aeolian Islands, Italy [153].

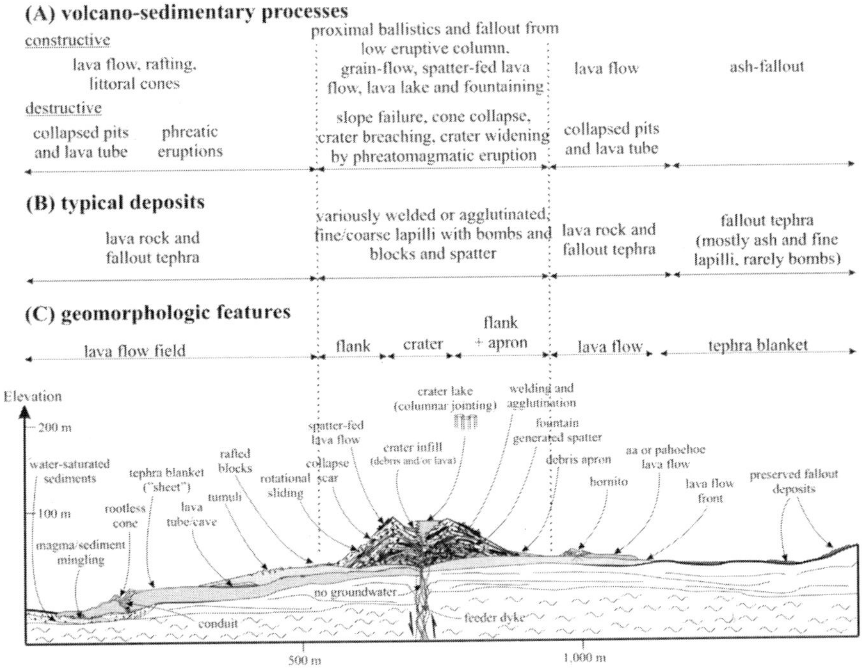

FIGURE 2. Schematic cross-section through a spatter cone showing the typical volcano-sedimentary processes and geomorphologic features.

A Hawaiian eruption results from one of the lowest energy magma fragmentation that are driven mostly by the dissolved gas content of the melt, which produces lava fountaining along fissures or focussed fountains up to 500 m in height [150, 162, 167-168]. The lava fountaining activity ejects highly deformable lava 'rags' at about 200–300 m/s exit velocity with an exit angle that typically ranges between 30–45° from vertical [169-170].

The nature and the distribution of the deposits associated with lava fountaining depend on the magma flux and the magma volatile content [162, 171-172]. Magmatic discharge rates during lava fountain activity range typically between 10 and 10^5 kg/s [162, 166]. The duration of typical lava fountaining activity may last only days or up to decades. An example for the former is Kilauea Iki, which erupted in 1959 [167, 170], while an example for the latter is Pu'u 'O'o-Kupaianaha, which began to erupt in 1983 [173]. Both are located on the Kilauea volcano in the Big Island of Hawaii, USA. Pyroclasts generated by lava fountaining are coarsely fragmented clots of magma which do not travel far from and above the vent [170-171]. They commonly land close to the vent and weld (i.e. mechanical compaction of fluid pyroclasts due to overburden pressure), agglutinate (i.e. flattening and deformation of fluid pyroclasts) or coalesce (i.e. homogenously mixed melt formed by individual fluidal clots) due to the high emplacement temperature of fragmented lava lumps on the depositional surface and/or the fast burial of lava fragments, which can retain heat effectively for a long time [171-172, 174]. The degree of welding and agglutinating of lava spatter is dependent on the [170-172, 174-175]:

1. accumulation rate and thickness of the deposit,
2. duration of eruption,
3. lava fountain height,
4. initial temperature determined by the magma composition,
5. heat loss rate, as well as
6. the grain size.

As a result of the limited energy involved in this type of magma fragmentation, the coarsely fragmented lava clots are transported ballistically, while the fines are transported by a low eruption column, as in the case of Plinian eruptions [162, 176]. The fragments tend to accumulate in proximal position, forming a cone-shaped pile, a spatter cone (Figure 2), which is built up by alternation of lava spatter and lava fountain-fed flows <100 m in diameter and a few tens of meters in height [170-172, 177-179].

Based on the grain size and the limited areal dispersion of tephra associated with typical Strombolian-style eruptions, it is considered a result of a mild magma fragmentation [149, 155]. However, larger volumes of tephra are produced than Hawaiian-style eruptions [159, 180]. Tephra production is derived from relatively low, non-sustained eruption columns [111, 153, 158, 181]. Individual explosions last <1 min and eject 0.01 to 100 m^3 of pyroclasts to <200 m in height with an exit velocity of particles of 3–100 m/s [180]. The magma discharge rate of 10^3 to 10^5 kg/s is based on historical examples of volcanoes erupted from water-rich, subduction-related magma [156]. The near surface fragmentation mechanism and limited energy released in a single eruption results in coarse lapilli-to-block-sized pyroclasts, predominantly between 1 and 10 cm in diameter, accumulating in close proximity to the vent [84, 182-183]. The exit velocity and angle of ballistic trajectories of particles of 20–25° determines the maximum height of the edifice and produces a limited size range of clasts in these volcanic edifices [184-185]. The repetition of eruptions produces individual, moderately-to-highly vesicular pyroclasts, called scoria or cinder, that do not agglutinate in most situations, but tend to avalanche downward forming talus deposits on the flanks of the growing cone [185-187]. Due to the mildly explosive nature of the eruptions, and the relatively stable pyroclast exit angles and velocity, a well-defined, conical-shaped volcano is constructed and is commonly referred to as a scoria or cinder cone (Figure 3). These cones have a typical basal diameter of 0.3 to 2.5 km, and they are up to 200 m high [153, 179, 182, 185, 188-189].

A more energetic magma fragmentation than is normally associated with Strombolian-activity cause violent Strombolian eruptions [163, 190]. In the 'normal' Strombolian-style eruptions, the magma is separated by gas pockets, which rise periodically in the magma through the conduit forming a coalescence of gas pockets, or a slug flow regime [153]. When the gas segregation increases, the eruptions become more explosive due to episodic rupture of liquid films of large bubbles, causing alternation of the flow regime from slug flow to churn flow, which is a typical characteristic of the violent Strombolian activity [163]. Based on numerical simulations, the increases in the gas flux, which creates the "churn flow", is caused by factors such as an increased length of conduit, the change in magma flux from 10^4 to 10^5 kg/s, the gas content, and/or the ascent speed variations that allow magma to vesiculate variably within the conduit [156, 163, 191]. Larger energy release during more explosive eruptions produces a higher degree of fragmentation, and hence finer-grained, ash-lapilli dominated beds [191], as well as higher eruption columns (<10 km) that disperse tephra efficiently over longer distances [83, 163].

Externally-driven fragmentation occurs when the melt interacts with external water leading to phreatomagmatic or Surtseyan-style eruptions [152, 192-195]. These explosive interactions take place when magma is in contact with porous- or fracture-controlled groundwater aquifers or surface water [151, 194, 196-202]. In special cases when explosive interactions take place between lava and lake, sea water or water-saturated sediments, littoral cone [203-204] and rootless cone [205-208] are generated. Processes and eruption mechanisms associated with these eruptions are not discussed in the present chapter. The evidences of the role of water in the formation of tuff rings and maar have been proofed by many studies [e.g. 152, 201, 209]. However, there are similar eruptive processes and eruption styles have been described from eruptions of silica-undersaturated magmas (e.g. foidite, melilitite and carbonatite) in environments, where the role of external water on the eruptive style is limited [e.g. 210-212].

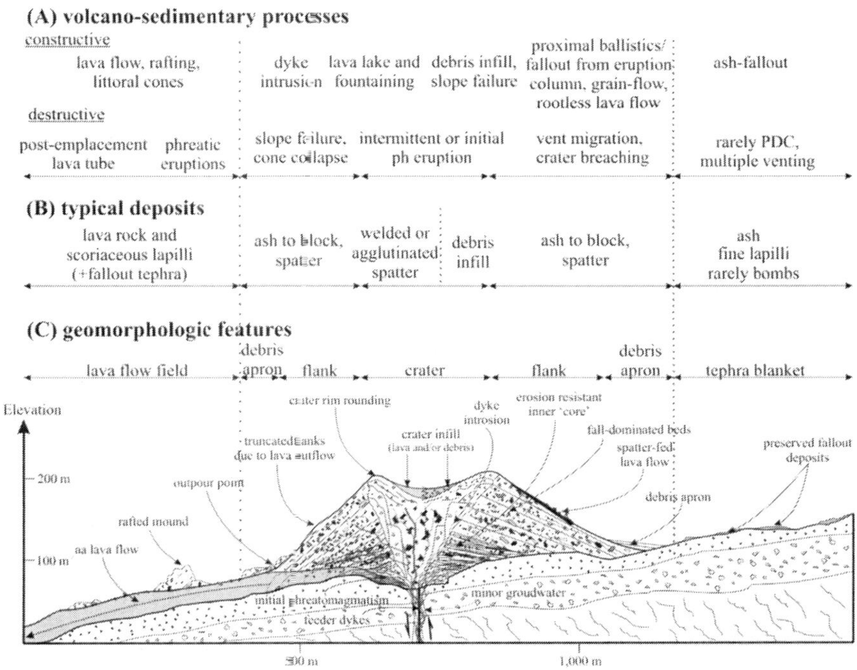

FIGURE 3. Schematic cross-section through a typical scoria cone showing the typical volcano-sedimentary processes and geomorphologic features. Abbreviations: PDC – pyroclastic density current, ph – phreatomagmatic eruption.

Phreatomagmatic eruptions (rarely called Taalian eruptions) are defined by some as being in subaerial environments [194]. These eruptions may produce a series of volcanic craters which vary in size between 0.1 km and 2.5 km in diameter [213]. The largest ones are very likely to be generated by multiple eruptions forming amalgamated craters such as Lake Coragulac maar, Newer Volcanics Province, south-eastern Australia [214] and/or formed in specific environment such as Devil Mountain maar, Seward Peninsula, Alaska [215]. The fragmentation itself is triggered by a molten fuel-coolant interaction (MFCI) processes requiring conversion of magmatic heat to mechanical energy [151, 193-194, 216-220]. The MFCI proceeds as follows [151, 220]:

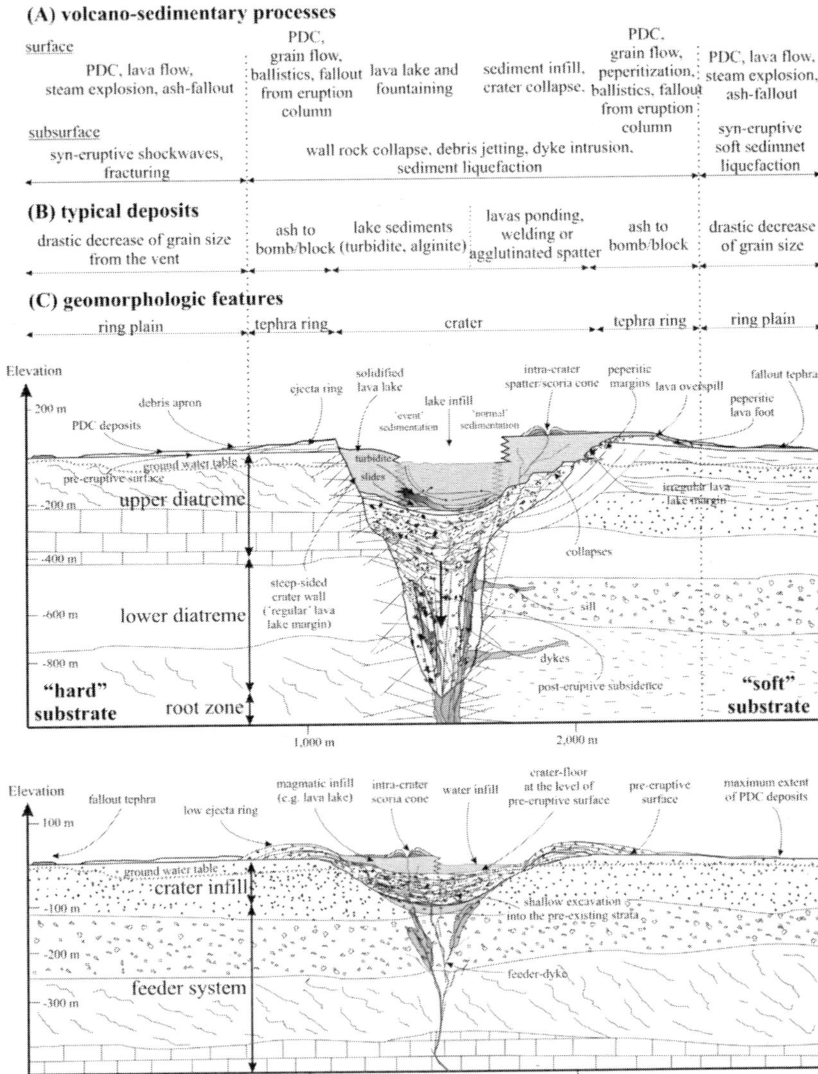

FIGURE 4. Schematic cross-sections through a maar-diatreme (top Figure) and a tuff ring (bottom Figure) showing the typical volcano-sedimentary processes and geomorphologic features. Note that the left-hand side represents the characteristics of a maar-diatreme volcano formed in a hard-substrate environment, while the right-hand side is the soft rock environment. Abbreviations: PDC – pyroclastic density current.

1. coarse premixing of magma and water producing a vapour film between fuel and coolant,
2. collapse of the vapour film, generating fragmentation of magma and producing shock waves,
3. rapid expansion of superheated steam to generate thermohydraulic explosions, as well as
4. post-eruption (re)fragmentation of molten particles.

In some cases the MFCI process is self-driven and, after the initial interactions, the fragmentation does not involve any other processes [194, 196, 217, 221-222]. The series of eruptions may excavate a crater that cuts into the pre-existing topography, forming a hole-in-the-ground structure called a maar (Figure 4) [152]. If the explosion locus stays at shallow depths, the resulting volcano is tuff ring, which has a crater floor normally near the pre-eruptive surface (Figure 4) [223]. Both eruptions result in a surface accumulation of tephra by fallout and pyroclastic density currents, mostly base surges, forming a usually circular ejecta ring around the crater [51, 58, 195, 200, 202, 224-228]. These eruptions produce pyroclastic deposits that have a diversity of juvenile pyroclasts (e.g. various shape, grain-size, vesicularity and microlite content) and variety of accidental lithic clasts derived from the underlying strata [52, 79, 229-231]. Pyroclastic successions of phreatomagmatic volcanoes can form coarse grained, chaotic breccias related to vent construction, conduit wall collapse or migration, as well as well-stratified, lapilli and ash-dominated beds with various degrees of sorting and large ballistically ejected, fluidal-shaped juvenile bombs or angular to heavily milled accidental lithic blocks [41, 59, 81, 223, 232-237]. Due to density current transportation of pyroclasts, the accumulating deposits are stratified and are commonly cross- or dune-bedded [81, 229-230, 238-240]. The craters of most of these phreatomagmatic volcanoes are filled by either post-maar eruptive products such as solidified lava lakes commonly showing columnar jointing and/or intra-crater scoria/spatter cones [81, 241-244], or non-volcanic sediments, such as lacustrine alginate, volcaniclastic turbidite deposits [195, 245-250].

Surtseyan-style eruptions occur when the external water is 'technically' unlimited during the course of the eruption when eruptions occur through a lake or the sea [222, 251-254]. In contrast with phreatomagmatic eruptions, Surtseyan-style eruptions require a sustained bulk mixing of melt and coolant, which generates more abrupt and periodic eruptions [194, 196]. During Surtseyan-style eruptions, water is flashed to steam which tears apart large fragments of the rising magma tip [222, 252, 254-256]. This process is far less efficient than self-sustained typical MFCI and causes a near continuous ejection of tephra [194]. This tephra feeds subaqueous pyroclastic density currents, which build up a subaqueous volcanic pile that may emerge to become an island in the course of the eruption [222, 253-254, 257-259], as was the case during the well-documented eruption of Surtsey tuff cone, Vestmannaeyjar Islands, Iceland in 1963–1967 AD [260-261]. After emergence, a conical volcano can cap the edifice and build a typical steep-sided tuff cone (Figure 5). The tuff cone gradually grows by rapidly expelled and frequent (every few seconds) tephra-laden jets that eject muddy, water-rich debris, which may initiate mass flows later on in the inner-crater wall and on the outer, steepening flank of the growing cone [253, 260, 262-265]. These shallow explosions eventually produce a cone form, although it often has irregular geometry with a breached or filled crater by late-stage lava flows or asymmetric crater rim [223, 234, 259, 264]. The diameters of craters of these tuff cones are comparable to the tuff rings and maars, but the elevation of the crater rims are higher, reaching up to 300 m [223]. Monogenetic volcanoes that formed by Surtseyan-eruptions typically have no diatreme below their crater, however, some recent research suggested that diatremes may exist beneath a few tuff cones, such as Saefell tuff cone, south Iceland [266] or Costa Giardini diatreme, Iblean Mountains, Sicily [267].

3.2. Spectrum Of Basaltic Monogenetic Volcanoes

As documented above, five types of monogenetic volcanoes are conventionally recognized [177, 179, 223]:

1. lava spatter cones,
2. scoria or cinder cones,
3. maars or maar-diatremes,
4. tuff rings and
5. tuff cones.

This classification is primarily based on the morphological aspects and dominant eruption styles of these volcanoes. Furthermore, there is a strong suggestion that a given eruption style results in a given type of volcanic edifice, e.g. Strombolian-style eruptions create scoria cones [e.g. 111, 182]. The conventional classification also fails to account for the widely recognized diversity or transitions in eruption styles that may form 'hybrid' edifices, e.g. intra-maar scoria cones with lava flows or scoria cones truncation by late stage phreatomagmatism [229, 244, 268-271]. The variability in the way a monogenetic volcano could be constructed also means that the conventional classification hides important details of complexity that may be important from volcanic hazard perspective (e.g. a volcano built up by initial phreatomagmatic eruptions and later less dangerous Strombolian eruptions). The diversity of pyroclastic successions relates to fluctuation of eruption styles that may be triggered by changing conduit conditions, such as geometry, compositional change, and variations in both magma and/or ground water supply [41, 52, 150, 163, 272-273]. Due to the abundance of intermediate volcanoes, a classification scheme is needed, where the entire eruptive history can be parameterized numerically.

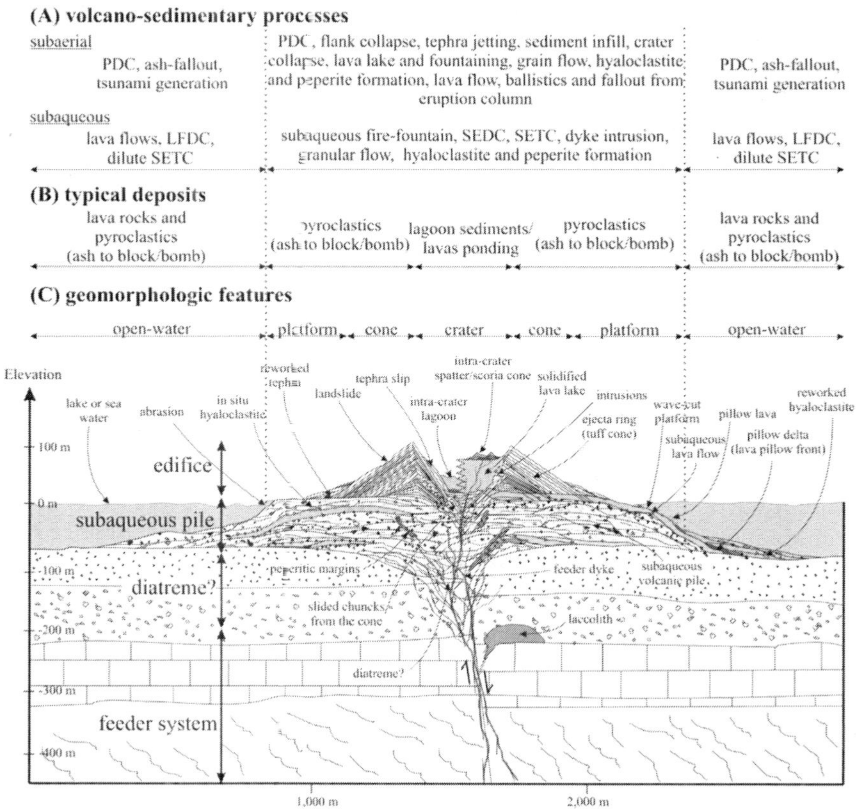

FIGURE 5. Schematic cross-section through a tuff cone showing the typical volcano-sedimentary processes and geomorphologic features. Abbreviations: PDC – pyroclastic density current, SEDC – subaqueous eruption-fed density current, SETC – subaqueous eruption-fed turbidity current, LFDC – lava flow-fed density current

In the present study, the construction of a small-volume volcano is based on two physical properties (Figure 6):

1. eruption style and associated sedimentary environment during an eruption and
2. number of eruption phases.

A given eruption style is a complex interplay between internal and external controlling parameters at the time of magma fragmentation. The internally-driven eruption styles are, for example, controlled by the ascent speed, composition, crystallization, magma degassing, number of magma batches involved, rate of cooling, dyke and conduit wall interactions, depth of gas segregation and volatile content such as H_2O, CO_2 or S [9, 11, 17, 111, 128, 150, 154-156, 163, 188, 191, 274-276]. These processes give rise to eruption styles in basaltic magmas that are equivalent to the Hawaiian, Strombolian and violent Strombolian eruption styles. However, due to the small-volume of the ascending melt, the controls on magma fragmentation are dominated by external parameters, including conduit geometry, substrate geology, vent stability/migration, climatic settings, and the physical characteristics of the underlying aquifers [39, 82, 234, 277-279]. Another important parameter in the construction of a monogenetic volcanic edifice is the number of eruptive phases contributing to its eruption history (Figure 6). The complexity of a monogenetic landform increases with increasing number or combination of eruptive phases. These can be described as "single", "compound" and "complex" volcanic edifices or landforms [280-281]. In this classification, the volcano is the outcome of combinations of eruption styles repeated by *m* phases. For example, a one-phase volcano requires only one dominant eruption mechanism during its construction. Due the single eruption style, the resulting volcano is considered to be a simple landform with possibly simple morphology. However, monogenetic volcanoes tend to involve two or multiple phases (Figure 6). Their construction requires two or more eruption styles and the result is a compound or complex landforms respectively, e.g. maar-like scoria cones truncated by late stage phreatomagmatic eruptions [e.g. 82, 270, 282] or a tuff cone with late-stage intra-crater scoria cone(s) [e.g. 265, 283]. These phases may occur at many scales from a single explosion (e.g. a few m^3) to an eruptive unit comprising products of multiple explosions from the same eruption style.

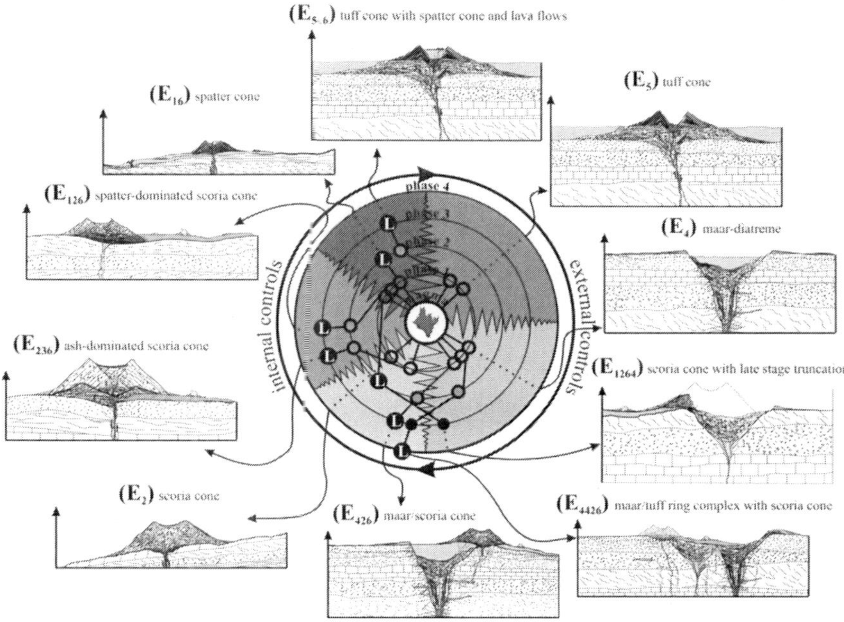

FIGURE 6. Eruption history (E) defined by a spectrum of eruptive processes determined by internal and external parameters at a given time. The initial magma (in the centre of the graph in red) is fragmented by the help of internal and external processes which determine the magma fragmentation mechanism and eruption style (phase 1). If a change (e.g. sudden or gradual exhaustion of groundwater, shift in vent position or arrival of new magma batch) occurs, it will trigger a new phase (phase 2, 3, 4,..., n); moving away from the pole of the diagram) until the eruption ceases. Note that black circles with white "L" mean lava effusion. If the eruption magma is dominantly basaltic in composition, the colours correspond to Surtseyan (dark blue), phreatomagmatic (light blue), Strombolian (light orange), violent Strombolian (dark orange) and Hawaiian (red) eruption styles.

To put this into a quantitative context, this genetic diversity can be expressed as set of matrices, similar to Bishop [280]. In Bishop [280], the quantitative taxonomy is represented by matrices of volcanic landforms that were based on surface morphologic complexity and eruption sequences. The role of geomorphic signatures is reduced due to the fact that in the case of an eruption centre built up by multiple styles of eruptions, not all eruption styles contribute to the geomorphology. For example, a scoria cone constructed by tephra from a Strombolian-style eruption, might be destroyed by a late stage phreatomagmatic eruption, as documented from Pinacate volcanic field in Sonora, Mexico [82, 284] and Al Haruj in Libya [282]. In these cases, the final geomorphologies resemble to maar craters, but the formation such volcanoes are more complex than a classical, simple maar volcano. In the proposed classification scheme, the smallest genetic entity (i.e. eruption style and their order) was considered to define the eruption history of a monogenetic volcano quantitatively. Considering only the typical, primitive basaltic composition range ($SiO_2 \leq 52\%$ w.t.), internally-driven eruption styles are the Hawaiian, Strombolian and violent Strombolian eruption styles [111]. At the other end of the spectrum, externally-driven eruption styles are the phreatomagmatic and Surtseyan-types [81, 285]. In addition, the effusive activity can also be involved in this genetic classification. The abovementioned eruption/effusion styles can build up a volcano in the following combination: 6×1, 6×6, 6×6^2 or 6×m (or n×m) matrices, depending on the number of volcanic phases involved in the course of the eruption. This means that an eruption history (E) of a simple volcano (E_{simple}) could be written as:

$$E_{simple} = [1 \quad 2 \quad 3 \quad 4 \quad 5 \quad 6] \tag{1}$$

where the elements 1, 2, 3, 4 and 5 corresponds to explosive eruptions such as Hawaiian, Strombolian, violent Strombolian, Taalian (or phreatomagmatic) and Surtseyan-type eruptions, respectively, while the 6 is the effusive eruption. For a more complex eruption history involving two ($E_{compound}$) and multiple ($E_{complex}$) eruption styles can be written as:

$$E_{compound} = \begin{bmatrix} 11 & 12 & 13 & 14 & 15 & 16 \\ 21 & 22 & 23 & 24 & 25 & 26 \\ 31 & 32 & 33 & 34 & 35 & 36 \\ 41 & 42 & 43 & 44 & 45 & 46 \\ 51 & 52 & 53 & 54 & 55 & 56 \\ 61 & 62 & 63 & 64 & 65 & 66 \end{bmatrix} \tag{1}$$

For instance, a monogenetic volcano with an eruption history of fire-fountain activity associated with a Hawaiian-type eruption and effusive activity could be described as having a compound eruption history (or E_{16} in Figure 6). While an example of a monogenetic volcano with a complex eruption history could be a volcanic edifice with a wide, 'maar-crater-like' morphology, but built up from variously welded or agglutinated scoriaceous pyroclastic rock units (e.g. E_{1264} in Figure 6), similar to Crater Elegante in Pinacate volcanic field, Sonora, Mexico [284]. In some cases, gaps, paucity of eruptions or opening of a new vent site after vent migration between eruptive phases is observed/expected based on reconstructed stratigraphy [e.g. 52, 279] and geochemistry [e.g. 9, 286]. In this classification system, the recently recognized polymagmatic or polycyclic behaviour of monogenetic volcanoes, e.g. an eruption fed by more than one batch of magma with distinct geochemical signatures [17, 23, 287], can be integrated. For example, the volcano could be E_{44} if the controls on eruption style remained the same or E_{42} if that chemical change is associated with changes in eruption style. The number of rows and columns in these matrices could be increased until all types of eruption style are described numerically, thus an n×m matrix is created. Increasing the number of volcanic phases will increase the range of volcanoes that could possibly be created. Of course, the likelihood of various eruptive combinations described by these matrices is not the same because there are 'unlikely' (e.g. E_{665}) and 'common' eruptive scenarios (e.g. E_{412}).

$$E_{complex} = \begin{bmatrix} 111 & 112 & 113 & 114 & 115 & 116 \\ . & & & & & . \\ . & & & & & . \\ . & & & & & . \\ 661 & 662 & 663 & 664 & 665 & 666 \end{bmatrix} \quad (3)$$

In summary, a volcano from monogenetic to polygenetic can be described as a matrix with elements corresponding to discrete volcanic phases occurring through its evolution. The major advantage of these matrices is that their size is infinite (n×m), thus an infinite number of combinations of eruption styles could be described (Figure 6). In this system each volcano has a unique eruptive history, in other words, each volcano is a unique combination of *n* number of eruption styles through *m* number of volcanic phases (Figure 6). This matrix-based classification scheme helps to solve terminological problems and to describe volcanic landforms numerically. For example, the diversity of scoria cones from spatter-dominated to ash-dominated end-members [68, 288] cannot be easily expressed within the previous classification scheme. This completely quantitative coding of volcanic eruption styles into matrices could be used for numerical modelling or volcanic hazard models, e.g. spatial intensity of a given eruption style.

4. GEOMORPHOLOGY OF MONOGENETIC VOLCANOES

4.1. Historical Perspective

The combination of eruption styles (listed above) and related sedimentary processes are often considered to be the major controlling conditions on a monogenetic volcano's geomorphic evolution [84]. Thus, the quantitative topographic parameterization of volcanoes is an important source of information that helps to reveal details about their growth, eruptive processes and associated volcanic hazards and its applicable to both conical [119, 190, 289-292] and non-conical volcanoes [199, 293]. These methods are commonly applied to both polygenetic [119, 289, 294-297] and monogenetic volcanic landforms [68-69, 298-299]. Morphometric measurements on monogenetic volcanoes began with the pioneering work of Colton [300], who noticed a systematic change in the morphology of volcanic edifices over time due to erosional processes such as surface wash and gullying. A surge of research in volcanic morphometry, focused mostly on scoria cones, occurred from the 1970s to 1990s, when the majority of morphometric formulae were established and tested [70-71, 84, 179, 185, 199, 213, 223, 293, 301-304]. This intense period of research was initiated by National Aeronautics and Space Administration (NASA) in the 1960s and 1970s due to an increasing interest in extraterrestrial surfaces that could be expected to be encountered during landings on extraterrestrial bodies such as the Moon or Mars [e.g. 33]. Additional interests were to understand magma ascent, the lithospheric settings of extraterrestrial bodies, the evolution of volcanic eruptions, the geometry of volcanoes in different atmospheric conditions, surface processes and seeking H_2O in extraterrestrial bodies [30, 33, 305-308]. Given the lack of field data from extraterrestrial bodies, many parameters that were able to be measured remotely, such as edifice height (H_{co}), basal (W_{co}) and crater diameters (W_{cr}) were introduced. There were measured manually from images captured by Mariner and Viking orbiter missions [e.g. 33] and Luna or Apollo missions for the Moon [e.g 309] in order to compare these data with the geometry of volcanic landforms on the Earth [e.g. 179, 293]. Dimensions, such as crater diameter, were measured directly from these images, while the elevation of the volcanic edifices was estimated from photoclinometry (i.e. from shadow dimensions of the studied landform) [33, 309-310]. Because elevation measurements were indirect, the horizontal dimensions such as W_{co} and W_{cr} were preferred in the first morphometric parameterization studies [179]. The increased need for Earth analogues led to intense and systematic study of terrestrial small-volume volcanoes [179, 185, 189, 293]. The terrestrial input sources, such as topographic/geologic maps and field measurements, were more accurate than the extraterrestrial input resources; however, they were still below the accuracy required (i.e. the contour line intervals of ≥20 m were not dense enough to capture the topography of a monogenetic volcano having an average size of ≤1500–2000 m horizontally and of ≤100–150 m vertically). The extensive research on monogenetic volcanoes identified general trends regarding edifice growth, eruption mechanism and subsequent degradation [71, 84, 185, 293]. In addition, morphometric signatures were recognized that associated a certain type of monogenetic volcanic landform with the discrete eruption style that formed it. The morphometric signatures of Earth examples were then widely used to describe and identify monogenetic volcanoes on extraterrestrial bodies such as the Moon and Mars [179]. Basic morphometric parameters were calculated and geometrically averaged to get morphometric signatures for four types of terrestrial, monogenetic volcanoes [179], including spatter cones (W_{co} = 0.08 km, W_{cr}/W_{co} = 0.36 km and H_{co}/W_{co} = 0.22 km), scoria cones (W_{co} = 0.8 km, W_{cr}/W_{co} = 0.4 km and H_{co}/W_{co} = 0.18 km), as well as maars and tuff rings (W_{co} = 1.38 km, W_{cr}/W_{co} = 0.6 km and H_{co}/W_{co} = 0.02 km). These morphometric signatures are still used in landform recognition [e.g. 31, 311].

In terrestrial settings, the morphometric studies of monogenetic volcanoes in volcanic fields and on polygenetic volcanoes have targeted

1. the characterization of the long-term and short-term evolution of magmatic systems [39, 42, 131, 134],
2. understanding of eruption mechanisms and processes [84, 182, 185, 312-314],
3. expression of tectonic influences on edifice growth [298, 315-316],
4. dating of conical landforms such as scoria or cinder cones [189, 304, 317-320],
5. examination of erosion processes [69-70, 290, 299, 321-327] and landscape evolution [39, 68, 73, 328],
6. reconstruction of the original size, geometry and facies architecture of polygenetic [329-332] and monogenetic edifices [291, 333-336], as well as
7. detection of climate and climate change influences on degradation [71, 324-325, 337].

Morphology quantified via morphometric parameters could be a useful tool to address some of these questions in volcanology, geology and geomorphology. The morphology of a volcanic edifice contains useful information from every stage of its evolution, including eruptive processes, edifice growth and degradation phases. However, the geomorphic information extracted through morphometric parameters often show bi- or even multi-modality, i.e. the morphometry is a mixture of primary and secondary attributes [e.g. 338]. The following section explores the dominant volcanological processes that influence the geomorphology of a monogenetic volcano.

4.2. Syn-Eruptive Process-Control On Morphology

The eruption styles shaping the volcanic edifices may undergo many changes during the eruption history of a monogenetic volcano (Figure 6). A given volcano's morphology and the grain size distribution of its eruptive products are generally viewed as the primary indicator of the eruption style that forms a well-definable volcanic edifice (i.e. "Strombolian-type scoria cones"). This oversimplification of monogenetic volcanoes, together with the widely used definition that "they erupts only once" [116], suggest a simplicity in terms of magma generation, eruption mechanism and sedimentary architecture. This supposedly simple and homogenous inner architecture of each classical volcanic edifice, such as spatter cones, scoria cones, tuff rings and maars, led to the identification of a "morphometric signature". The morphometric signature of monogenetic volcanoes was used in the terrestrial environment, e.g. to ascribe a relationship between morphometry and "geodynamic setting" [337], as well as extraterrestrial environments, e.g. for volcanic edifice recognition [31, 179, 339-341]. Certain types of volcanoes could be discriminated from each other based on their morphometric signature, but some general assumptions need to be made. For example,

1. the morphometric signature concept is entirely based on the assumption that a volcanic landform directly relates to a certain well-defined eruption style,
2. thus the pyroclast diversity within the edifice is minimal (i.e. homogenous), as well as
3. the resultant volcanic landform is emplaced in a closed-system with no transitions between eruption styles, especially from externally to internally-driven eruption styles and vice versa.

Consequently, the edifice studied was believed to have a relatively simple eruption history, which is a classical definition of a "monogenetic volcano". As demonstrated above, monogenetic volcanoes develop in an open-system. This section explores the volcanological/geological constrains of geomorphic processes responsible for the final volcanic edifice and the morphometric development of two end-member types of monogenetic volcanoes such as crater-type (4.2.1.) and cone-type edifices (4.2.2).

4.2.1. Crater-type monogenetic volcanoes

Crater-type monogenetic volcanoes such as tuff rings and maar volcanoes (Figure 4), are characterized by a wide crater with the floor above or below the syn-eruptive surface, respectively [81, 152, 223, 234]. Their primary morphometric signature parameters are major/minor crater diameter and depth, crater elongation and breaching direction, volume of ejecta ring, and crater or slope angle of the crater wall [313, 342-345]. Of these morphometric parameters the crater diameters were used widely for interpreting crater growth during the formation of a phreatomagmatic volcano. For the genetic integration of crater growth and, consequently, the interpretation of crater diameter values of terrestrial, dominantly phreatomagmatic volcanoes, there are fundamentally two end-member models.

The first model is the incremental growth model (Figure 7A). In this model, the crater's formation is related to many small-volume eruptions and subsequent mass wasting, shaping the crater and underlying diatreme [81, 151, 199-200, 209, 221, 285, 346-347]. Growth initiates when the magma first interacts with external water, possibly groundwater along the margin of the dyke intrusions, triggering molten-fuel-coolant interactions (MFCI) [192-193, 220, 348]. These initial interactions excavate a crater on the surface, while the explosion loci along the dyke gradually deepen the conduit beneath the volcano towards the water source, resulting in a widening crater diameter [199]. This excavation mechanism initiates some gravitational instability of the conduit walls, triggering slumping and wall rock wasting, contributing to the growing crater [81, 195, 199, 223, 229-230, 349]. This classical model suggests that

1. crater evolution is related to diatreme growth underneath, and
2. the crater's growth is primarily a function of the deep-seated eruption at the root zone.

However, it is more likely that the pyroclastic succession created at the rim of the crater preserves only a certain stage of the evolution of whole volcanic edifice. For instance, the possibility of juvenile and lithic fragments being erupted and deposited within the ejecta ring from a deep explosion (i.e. at the depth of a typical diatreme, about 2 km) is highly unlikely. Rather than being dominated by the deep-seated eruptions, explosions can occur at variable depths within the diatreme [347]. The individual phreatomagmatic eruptions from various levels of the volcanic conduit create debris jets (solids + liquid + magmatic gases and steam), which are responsible for the transportation of tephra [285, 347, 350]. Every small-volume explosion causes upward transportation of fragmented sediment in the debris jet, giving rise to small and continuous subsidence/deepening of the crater floor [198-199]. This is in agreement with stratigraphic evidence from eroded diatremes, such as Coombs Hills, Victoria Land, Antarctica [285, 350] and Black Butte diatreme, Missouri River Breaks, Montana [351].

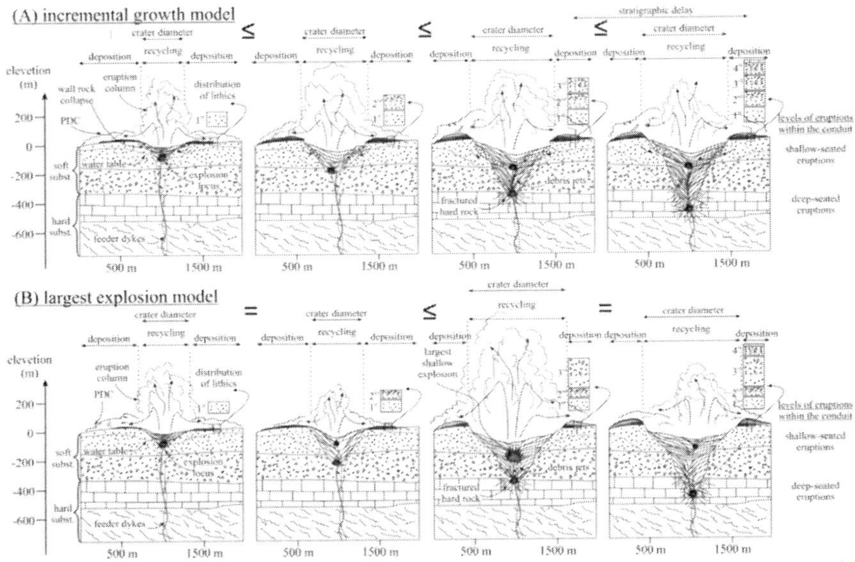

FIGURE 7. Crater growth envisaged by the incremental growth (A) and largest explosion models (B). In the top Figure , the crater grows by each explosion and the subsequent mass wasting processes (e.g. slumping of wall rock). This results in the gradual growth of the crater and the underlying diatreme (if any). On the other hand, in the bottom Figure the crater can reach its final size in the middle of the eruption history by the largest near-surface explosion and the further explosions make no contribution to its size, geometry and morphology. Abbreviations: PDC – pyroclastic density current.

The second model is where the crater geomorphology is dominated by largest explosion event during the eruption sequence. Thus, the crater size directly represents the 'peak' (or maximum) energy released during the largest possible shallow explosion [202, 313, 343-344, 352]. This model of crater growth (Figure 7B) for phreatomagmatic volcanoes is proposed on the basis of analogues from phreatic eruption, such as Uso craters, Hokkaido, Japan in 2000 [352], and experiments on chemical and nuclear explosions [344]. In this model, the crater diameter (D) is a function of the total amount of ejected tephra (V_{ejecta}) [313]:

$$D = 0.97V_{ejecta}^{0.36} \tag{4}$$

which can further be converted into explosion energy (E) as:

$$E = 4.45 \times 10^6 D^{3.05} \tag{5}$$

which approximates the largest energy released during the eruptions. This relationship between crater size and ejected volume was based on historical examples of phreatomagmatic eruptions [313]. These historical eruptions are, however, associated with usually polygenetic volcanoes, such as Taupo or Krakatau, and not with classical monogenetic volcanoes, except the Ukinrek maars, near Feulik volcano, Alaska. In this model, the largest phreatomagmatic explosion governs the final morphology of the crater, so the crater size correlates with the peak energy of the maar-forming eruption directly [343]. Scaled experiments showed that there is a correlation between the energy and the crater depth and diameter [353] if the explosions take place on the surface [344]. Most of the explosions modelled in Goto et al. [344] were single explosions, only a few cratering experiments involved multiple explosions at the same point [344], which is more realistic for monogenetic eruptions. In these multiple explosions, the crater did not grow by subsequent smaller explosions, possibly because the blast pressure was lower than the rock strength when it reached the previously formed crater rim [344]. As noted by Goto et al. [344], such experimental explosions on cratering are not applicable to underground eruptions; therefore, they do not express the energy released by deep-seated eruptions generating three-phase (solid, gas and fluid) debris jets during diatreme formation [e.g. 350]. Morphologically, these deep-seated eruptions have a minor effect on crater morphology and diameter, and their deposits rarely appear within the ejecta ring around the crater.

Theoretically, both emplacement models are possible because both mechanisms can contribute significantly to the morphology of the resulting landform. The incremental growth model is based on statigraphy, eye-witnessed historical eruptions and experiments [81, 152, 198-199, 346-347], while the largest explosion model is based on analogues of chemical or nuclear explosion experiments, phreatic eruptions and impact cratering [313, 344, 353-354]. Based on eye-witnessed eruptions and geological records, the crater diameter as a morphometric signature for maar-diatreme and tuff ring volcanoes is the result of complex interplay between the eruptions and the substrate. The dominant processes, such as many, small-volume explosions with various energies migrating within the conduit system vertically and horizontally, as well as gradual mass wasting depending on the physical properties of rock strength, are what control the final crater diameter. On the other hand, the substrate beneath the volcano also plays an important role in defining crater morphology, as highlighted for terrestrial volcanic craters [209, 229, 355], as well as extraterrestrial impact craters [356]. In different substrates, different types of processes are responsible for the mass wasting. For example, an unconsolidated substrate tends to be less stable due to explosion shock waves that may liquefy water-rich sediments, and induce grain flow and slumping, enlarging the crater [229]. On the other hand, in a hard rock environment, the explosions and associated shock waves tend to fracture the country rock, depending on its strength, leading to rock falls and sliding of large chucks from the crater rim [229]. The crater walls in these two contrasting environments show different slope angles [229, 345]. These differences in the behaviour of the substrate in volcanic explosions may cause some morphological variations in the ejecta distribution and the final morphology of the crater (Figure 8).

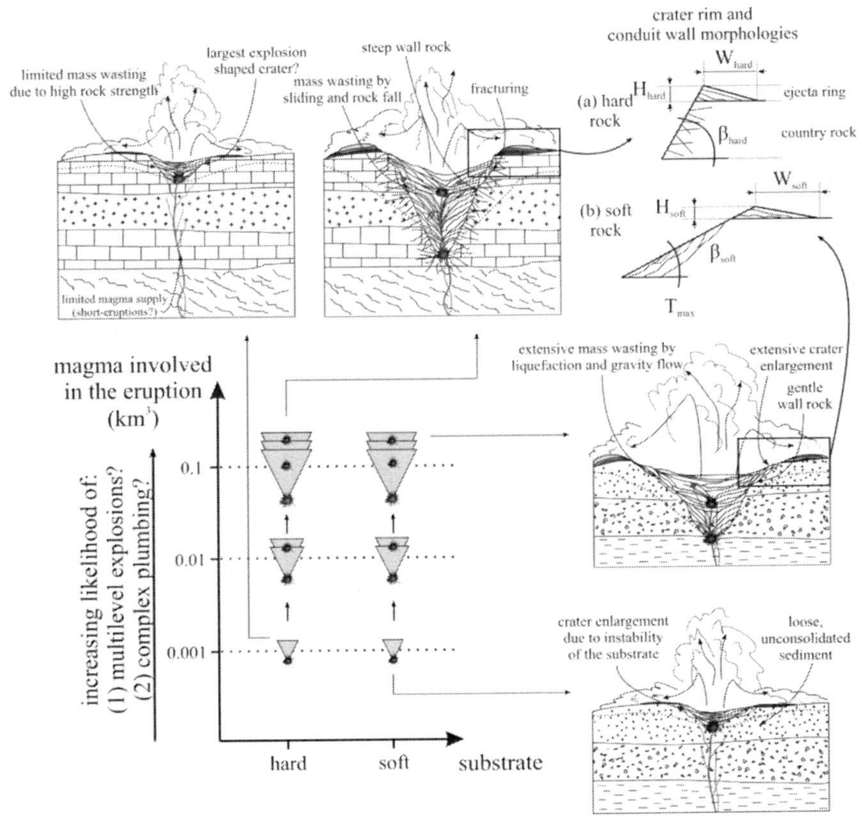

FIGURE 8. Morphology of volcanic crater and diatreme generated by phreatomagmatic eruptions in soft and hard substrate environments. The parameters (β – wall rock angle, H – ejecta ring height over the surrounding topography, W – flank width of the ejecta ring) for crater morphologies have the following possible relationships: $\beta_{hard} > \beta_{soft}$; while the $H_{hard} > H_{soft}$ (?) and $W_{hard} > W_{soft}$ (?).

The crater diameter is an important morphometric parameter in volcanic landform recognition; however, the final value of the crater diameter is the result of a complex series of processes, usually involving syn-eruptive, mass wasting processes of the crater walls, e.g. the 1977 formation of the Ukinrek maars in Alaska [357-358]. This makes the direct interpretation of crater diameter values more complicated than predicted by simple chemical and nuclear cratering experiments [313, 344]. Thus, the incremental growth (multiple eruptions + mass wasting) model seems to be a better explanation of the growth of a crater during phreatomagmatic eruptions [198-200, 267, 346, 359] and kimberlite volcanism [360-361]. Thus, the morphometric data of a fresh maar or tuff ring volcano contain cumulative information about the eruption energy, the location and depth of (shallow) explosion loci, as well as the stability of the country rock and associated mass wasting. The largest eruption dominated model may only be suitable to express energy relationships without the effects of mass wasting from the crater walls. This probably exists in only a few limited sites. For example, these eruptions should take place from a small-volume of magma, limiting the duration of volcanic activity and reducing the possibility of development of a diatreme underneath (Figure 8). Additionally, these eruptions should be in a consolidated hard rock environment, with high rock strength and stability. The following model for the interpretation of crater diameter and morphology data can be applied for phreatomagmatic volcanoes (Figure 8). This model integrates both conceptual models for crater and edifice growth, but the majority of craters experience complex development instead of the dominance of the largest explosion event. The likelihood of the largest explosion event dominated morphology is limited to simple, short-lived eruptions due to limited magma supply or vent migration in a hard rock environment (Figure 8). If the magma supply lasts, the development of a diatreme underneath starts that has a further effect on the size, geometry and morphology of the crater of the resultant volcano.

The crater diameter is often a function of basal edifice diameter (W_{co}) or height (H_{co}), creating ratios which are commonly used in landform recognition in extraterrestrial environments [177, 179, 340]. $W_{co}$is often difficult to measure because of the subjectivity in boundary delimitation [293]. This high uncertainty in delimitation of the crater boundary is a result of the gradual thinning of tephra with distance from the crater, with usually a lack of a distinct break in slope between the ejecta ring's flanks and the surrounding tephra sheet, e.g. the Ukinrek maars, Alaska [357-358]. Any break in slope could also be smoothed away by post-eruptive erosional processes. The crater height estimates vary greatly for maar-diatreme and tuff ring volcanoes, but they are usually ≤50 m [199, 223, 293]. This small elevation difference from the surrounding landscape gives rise to some accuracy issues, particularly regarding the establishment of the edifice height.

To demonstrate the limitations (e.g. input data accuracy, data type, and genetic oversimplification) of morphometric signature parameters on phreatomagmatic volcanoes, two examples (Pukaki and Crater Hill) were selected from the Quaternary Auckland volcanic field in New Zealand. Both volcanoes above were used to establish the average morphometric signature of an Earth analogue phreatomagmatic volcano [e.g. 293]. The early morphometric parameters were measured from topographic maps having coarse contour line intervals (e.g. 20–30 m), which cannot capture the details of the topography accurately. Some cross-checks were made on the basic morphometric parameters established from topographic maps and Digital Elevation Models (DEMs) derived from airborne Light Detection And Ranging (LiDAR) survey. The results showed that the differences in each parameter could be as high as ±40%. In addition, both the Pukaki and the Crater Hill volcanoes from Auckland were listed as "tuff rings" [293], due to the oversimplified view of monogenetic volcanism in the 1970s and 80s. Their eruption history, including volume, facies architecture and morphology, are completely different. The present crater floor of Pukaki volcano is well under the syn-eruptive surface, thus it is a maar volcano sensu stricto following Lorenz [199]. This was formed by a magma-water interaction driven phreatomagmatic eruption from a small volume of magma of 0.01 km^3 estimated from a DEM and corrected to Dense Rock Equivalent (DRE) volume [362-363]. The present facies architecture of this volcano seems quite simple (e.g. like an E_4 volcano in Figure 6). On the other hand, Crater Hill has a larger eruptive volume of 0.03 km^3 [362-363] and experienced multiple stages of phreatomagmatism (at least 3) and multiple stages of magmatic eruptions (at least 5) with many transitional layers between them, forming an initial tuff ring and an intra-crater scoria cone [80]. Later eruption formed an additional scoria cone and associated lava flow that filled the crater with lava up to 120 m in thickness [80, 364]. Consequently, Crater Hill is an architecturally complex volcano with complex eruption history, i.e. at least an E_{4226}. The important implication of the examples above and the usual complex pattern and processed involved on the establishment of the final geomorphology (e.g. incremental crater growth) are that morphometric signature (if it exists) can be only used for phreatomagmatic volcanoes if the eruption history of the volcano is reconstructed. In other words, parameters to express morphometric signature cannot be compared between volcanoes with different eruption histories (i.e. they are characterized by different phases with different eruption styles). Comparison without knowledge of the detailed eruption history could be misleading. Furthermore, the morphometric signature properties used in extraterrestrial volcano recognition for crater-type volcanoes should be reviewed, using a volcanological constraint on the reference volcano selection.

4.2.2. Cone-type monogenetic volcanoes

Cone-type monogenetic volcanoes, such as spatter- (Figure 2), scoria (or cinder; Figure 3) and tuff cones (Figure 5), are typically built up by proximal accumulation of tephra from low to medium (0.1–10 km in height) eruption columns and associated turbulent jets, as well as block/bombs that follow ballistic trajectories [188, 223, 288, 365-367]. Deposition from localized pyroclastic density currents is possible, mostly in the case of tuff cones [223, 265, 283] and rarely in the case of scoria cones from violent Strombolian eruptions [188]. The primary morphology of cone-type monogenetic volcanoes could be expressed by various morphometric parameters, including height (H_{co}), basal (W_{co}) and crater diameter (W_{cr}) and their ratios (H_{co}/W_{co} or W_{cr}/W_{co}), inner and outer slope angle or elongation. On a fresh edifice, where no post-eruptive surface modification has taken place, these morphometric parameters are related to the primary attributes of eruption dynamics and syn-eruptive sedimentary processes. However, there are potentially two valid models to explain their dominant construction mechanisms, including a ballistic emplacement with drag forces and fallout from turbulent, momentum-driven jets at the gas-thrust region [84, 182, 185, 368-369]. In both models, the angle of repose requires loose, dry media. This criterion is rarely fulfilled in the case of a tuff cone [e.g. 223, 234] and littoral cones that form during explosive interactions between lava and water [e.g. 203, 204]. In these cases the ejected fragments have high water-contents that block the free avalanching of particles upon landing [223, 252, 259, 283, 286]. This is inconsistent with other magmatic cone-type volcanoes; the growth processes of tuff and littoral cones are not discussed in further detail here.

The ballistics model with and without drag for scoria cone growth was proposed as a result of eye-witness accounts of eruptions of the NE crater at Mt. Etna in Sicily, Italy [185]. This model is based on the assumption that the majority of the ejecta of a volcanic cone is coarse lapilli and block/bombs (≥8–10 cm in diameter), thus they follow a (near) ballistic trajectory after exiting the vent (Figure 9A). Consequently, the particle transport is momentum-driven, as documented for the bomb/block fraction during bursting of large bubbles in the upper conduit during Strombolian style explosive eruptions [180, 183, 370]. The particle velocity of such bomb/blocks was up to 70–80 m/s for a sensu stricto Strombolian style eruption measured from photoballistic data [371-372]. However, recent studies found that the typical exit velocities are about 100–120 m/s [180, 183] and they could reach as high as 400 m/s [373]. These studies also showed that the typical particle diameter is cm-scale or less instead of dm-scale [180, 183], which cannot be derived purely from impact breakage of clasts upon landing [182, 368]. Especially during paroxysmal activity at Stromboli [374] or more energetic violent Strombolian activity [83, 163], the dominance of fine particles in the depositional records contradicts the ballistics emplacement model for the cones.

To solve the debate about cone growth, the jet fallout model was proposed [182], based on the fact that there is a considerably high proportion of fines in cone-building pyroclastic deposits [60, 182, 288]. This proposed behaviour is similar to the proximal sedimentation from convective plumes, forming cones with similar geometry to scoria cones, e.g. the 1886 eruption of Tarawera, Taupo Volcanic Zone, New Zealand [375] or the 1986 eruption of Izu-Oshima volcano, Japan [368]. The fines content does not fulfil the criteria of pure ballistic trajectory, thus turbulent, momentum-driven eruption jets (Figure 9A) should be part of the cone growth mechanisms [182, 368, 376]. As documented above, scoria cones demonstrate a wider range of granulometric characteristics [182] than previously thought [185]. These slight or abrupt changes of grain size within an edifice imply that the term "scoria cone" is not as narrow and well-defined as proposed in earlier studies [71, 185, 189, 301]. Consequently, there should be a spectrum of characteristics within the "scoria cones" indicating the existence of spatter-, lapilli- and ash-dominated varieties [68, 288]. Such switching from classical lapilli-dominated to ash- or block-dominated cone architectures reflects syn-eruptive reorganization of conduit-scale processes, including

1. multiple particle recycling and re-fragmentation during conduit cleaning [180, 377] or
2. changes in magma ascent velocity (i.e. increase or decrease in the efficiency of gas segregation) that in turn effect the viscosity of the magma [150, 163, 378].

The latter case may or may not cause a change in eruption style (e.g. a shift from normal Strombolian-style to violent Strombolian-style eruption) that could effectively lead to changes in the grain size distribution of the ejecta by possible skewing towards finer fractions (≤1–2 cm). This switching has significant consequences on pyroclast transport as well. The higher efficiency of magma fragmentation and production of finer pyroclasts (e.g. ash) causes more effective pyroclast-to-gas heat transfer in the gas-thrust region. A buoyant eruption column is created when the time of heat transfer is shorter than the residence time of fragments in the lowermost gas-thrust region [111, 182, 379]. Thus, particle transport shifts from momentum- to buoyancy-driven modes [182-183]. Based on numerical simulations, these changes in the way pyroclasts are transported are consistent with modelled sedimentation trends from jet fallout as a function of vent distance [182]. Once the eruption has produced medium ($Md\phi \leq 10$–20 mm) and coarse fragments ($Md\phi \leq 50$–100 mm), pyroclasts show an exponential decrease in sedimentation rates away from the vent [182]. This is in agreement with the trend predicted by the ballistic emplacement model [185]. However, once the overall fragment size is dominated by fines (Mdϕ ≤ 2–3 mm), the maximum sedimentation rate departs further towards the crater rim [182]. The threshold particle launching velocity is about ≥50 m/s [182]. The fragment diameter of Mdϕ ≤ 2–3 mm is consistent with the calculated threshold for formation of buoyant, eruptive columns during violent Strombolian activity [111], but is significantly finer than fragments (Mdϕ ≤10–12 mm) generated during some paroxysmal events recorded at Stromboli in Italy [374].

In the case of scoria cone growth the final morphology is not only dependent on the mode of pyroclast transportation via air, but also on significant post-emplacement redeposition. If particles are sufficiently molten and hot, their post-emplacement sedimentation processes usually involve some degree of welding/agglutination and rootless lava flows may form [171-172]. In this case, high irregularity in the flank morphology is expected due to the variously coalescent large lava clots and spatter [171-172, 380]. If the particles are brittle and cool, as well as having enough kinetic energy to keep moving, they tend to avalanche on the inclined syn-eruptive depositional surface [84, 141, 172, 185, 234, 288]. The avalanching grain flows often give rise to inversely-graded horizons or segregated lenses within the overall homogenous, clast-supported successions of the accumulating pyroclast pile, while the hot particles cause spatter-horizons in the statigraphy (Figure 10). In the earlier case, the criterion to sustain efficient grain flow processes on the initial flank of a pyroclastic construct is that the particles have to be granular media (i.e. loose and sufficiently chilled). The properties such as grain size, shape and surface roughness determine the angle of repose, which is a material constant [381]. These are all together responsible for the formation of usually smooth cone flank morphologies.

Classically, scoria cones are referred to as being formed by Strombolian-style eruptions, in spite of the fact that Stromboli volcano, Aeolian Islands, is not a scoria cone. In reality, scoria cones are formed by “scoria cone-forming” eruptions. Thus, the term scoria cone includes every sort of small-volume volcano with a conical shape and basaltic to andesitic composition. Additionally, during scoria cone growth, three major styles of internally-driven eruption types can be distinguished, Hawaiian, Strombolian and violent Strombolian, and an additional externally-driven eruption style, such as phreatomagmatism-dominated, is also expected (e.g. Figure 6). From the eruption styles listed above, at least the first three could individually form a “scoria cone”, or similar looking volcano, which is rarely or never taken into account during interpretation of geomorphic data of a monogenetic volcano. The cone growth mechanism is, here, considered to be a complex interplay between many contrasting modes of sedimentation of primary pyroclastic materials, including transport via air (by turbulent jets and as ballistics) and subsequent redistribution by particle avalanching (Figure 9A). It is also important to note that cone growth is not only a constructive process; there could also be destructive phases. These processes (e.g. flank failure or crater breaching) alter the morphology in a short period of time. Consequently, the edifice growth is not a straightforward process (e.g. a simple piling up of pyroclastic fragments close to the vent), but rather a combination of constructive and destructive phases at various scales. The spatial and temporal contexts of such constructive and destructive processes are important factors from a morphometric stand point. In this chapter, two modes of cone growth mechanisms are recognized (Figure s 9B and C) cones formed by:

1. a distinct and stable eruption style (e.g. E_{simple}) and by
2. various magma eruption styles with transitions between them (e.g. $E_{compound}$ or $E_{complex}$), during the eruption histories.

The simple cone growth model is applicable to cone growth from a single and stable eruption style, e.g. Strombolian, Hawaiian or violent Strombolian styles only (Figure 9B). Theoretically, if an edifice is formed by a repetition of one of these well-defined and stable eruption style without fluctuation of in efficiency or any changes, the crater excavation and diameter, as well as the mode of pyroclast transport, should vary in a narrow range, i.e. fragmentation mechanism 'constant' (Figure 9B). The first explosions when the gas bubbles manage to escape from the magma leading to the explosive fragmentation of the melt, usually take place on the pre-eruptive surface or at a few tens of meters deep [57, 84, 185, 382-383]. After the first eruption, there is a time involved to either excavate the crater or pile up ejecta around the vent, which is in turn dependent on the eruption style and the tephra accumulation rate. Once the threshold crater rim height is reached, the height and its position are attached to properties of certain eruption dynamics (Figure 9B). In other words, the eruption style and related pyroclast transport distribute tephra to limited vertical and horizontal directions. Due to the steadiness of eruption style, particle fallout from the near-vent, dilute jets at the gas-thrust region and ballistics have an 'average' vertical distance that they travel. This 'average' will determine the width and relative offset of the crater rim above the crater floor, which grows rapidly during the initial establishment of the crater morphology (first cartoon in Figure 9B) and then slows down (second and third cartoons in Figure 9B). Of course the location, morphology of the crater rim and floor are not just dependent on the efficiency of the magma fragmentation, but the subsequent wall rock failure, as documented by Gutmann [369], similar to the development of maar-diatreme volcanoes (Figure 7). Due to the stability of the conduit and a single eruption style, the role of such failures in the control of morphology is minimal in comparison with complex modes of edifice growth (see later).

This growth model is applicable to simple eruptions, with steady eruption styles and possibly steady magma discharge rates, such as the violent-Strombolian eruptions during the Great Tolbachik fissure eruptions in Kamchatka, Russia [84, 382] or the Strombolian-style eruption during the growth of the NE crater at Mt. Etna [185]. During the Tolbachik fissure eruptions, the rim-to-rim crater diameter of Cone 1 grew rapidly from 56 m to 127 m during the first 5 days, and later slowed and stayed in a narrow range around 230–280 m during the rest of the eruptions [84, 382-385]. This is similar to certain stages of growth of the NE crater at Mt. Etna [84, 185]. This means the crater widens initially until a threshold width is reached, which corresponds to the maximum strength of the pyroclastic pile and the limits of the eruption style (Figure 9B). This trend seems to be consistent with an exponential growth of crater width over time until the occurrence of lava flows, as documented at Cone 1 and 2 of the Tolbachik fissure eruptions [e.g. 383]. The maximum range of the crater width appeared to be reached once the lava outflowed from the foot of the cones. This can be interpreted as an actual decrease in magma flux fuelling the explosion, and therefore the pyroclast supply for flank growth. Assuming that the magma is torn apart into small particles and are launched with sufficiently high initial velocity to the air, e.g. Strombolian eruptions, the particles have enough time to cool down, thus upon landing they initiate avalanching due to their kinetic energy. These processes will smooth the syn-eruptive surface to the angle of repose of the ejected pyroclasts if the pyroclast-supply is high enough to cover the entire flank of the growing edifice. If the angle of repose of the tephra, θ, depending on granulometric characteristics, and the height of the crater rim, H, are known at every stage of the eruption (i = 1,2,..., n), the flank width (W_i) would indicate at a certain stage of growth (Figure 9B):

$$W_i = H_i / \tan \theta_i \qquad (6)$$

When the eruption style is constant during the eruption and produces pyroclasts of the same characteristics, the pyroclasts behave as granular media (i.e. they are controlled by the angle of repose); the aspect ratio between the height and flank width should be in a narrow range until the 'tandem' relationship is established between the crater rim and explosion loci. Once the eruption is in progress, the explosion locus stay either at the same depth (e.g. at the pre-eruptive ground level) or migrates upwards if enough material is piled up within the crater resulting in a relative up-migration of the crater floor over time (Figure 9B). Consequently, this upward migration of the explosion locus should result in an elevation increase of the crater rim, if the eruption style remains the same. Once the crater rim rises, the majority of the pyroclasts avalanche downward from higher position, which creates a wider flank and increases the overall basal

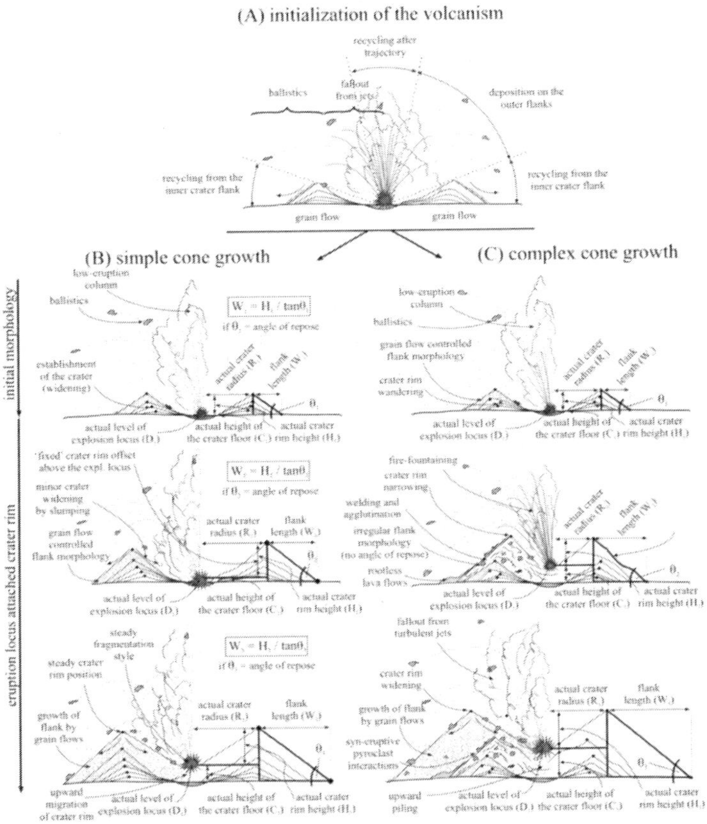

FIGURE 9. Growth of cone-type volcanoes (e.g. scoria cones) through ballistics' emplacement and jet fallout models. After the initialization of the volcanism (A), there are two different types of cone growth: simple and complex. The simple cone growth model (B) supposes a steady fragmentation mechanism and associated eruptive style and sedimentary processes, thus the angle of repose is near constant, $\theta_1=\theta_2=\theta_3$ over the eruption history. The variation of the relative height of the crater rim (H) above the location of the explosion locus and the radius of the crater (R) is 'fixed' or varies in a narrow range ($H_1 \leq H_2=H_3$ and $R_1 \leq R_2=R_3$), after the initial rapid growth of rim height and crater width. The simple cone growth model implies that constructional and flank morphologies are the results of a major pyroclast-transport mechanism (e.g. grain flows in the case of a scoria cone or welding, formation of rootless lava flows in the case of spatter cones). On the other hand, the complex cone growth model (C) involves gradual or abrupt changes in eruption style, triggering multiple modes of pyroclast transport, and possible changes in the relative height and diameters of the crater rim ($H_1 \neq H_2 \neq H_3$). The granulometric diversity allows for post-emplacement pyroclast interactions which permits or blocks the free-avalanching of the particle on the outer flanks. Therefore, the angle of repose is constant and may not always be reached ($\theta_1 \neq \theta_2 \neq \theta_3$), especially when clast accumulation rate and temperature is higher, causing welded and agglutinated spatter horizons.

width of the edifice. The repetition of such crater rim growth and flank formation takes place until the magma supply is exhausted. Due to the dominance of loose and brittle scoria in the edifice, the H_{co}/W_{co} ratio is in a narrow range in accordance with Eq. 3. Although, there is a slight difference between H_{co}/W_{co} and H_i/W_i, because the former contains the crater. The crater is possibly the most sensitive volcanic feature of a cone that could be modified easily (e.g. shifting in eruption style in the course of the eruptions and/or vent migration). Finally, this growing process is in agreement with the earlier documented narrow ranges of the H_{co}/W_{co} ratio that are governed by the angle of repose of the ejecta [84, 301].

The complex cone growth model assumes the cone is the result of many distinctive eruption styles and changes between them, which trigger a complex cone growth mechanism from various eruptive and sedimentary processes (Figure 9C). Such changes usually relate to changes from one eruption style to another, which have consequences for the morphological evolution of the growing cone. The switching in efficiency of magma fragmentation can be triggered by the relative influence of externally and internally-governed processes or reorganization of internal or external controls without shifting from one to the other. An example for the first change is a gradual alteration in the abundance of ground water and consequent shift from phreatomagmatism to magmatic eruption styles. On the other hand, the reorganization of processes in either in the internally- or externally-driven eruption styles could be related to the changes in degree of vesiculation and efficiency of gas segregation in the conduit system. Each of these changes could modify the dominant eruption style that determines the grain size, pyroclast transport and in turn edifice growth processes. An internal gas-driven magma fragmentation leading to a Hawaiian eruption produces larger (up to a few meters) magma clots that are emplaced ballistically, while fines are deposited from turbulent jets and a low-eruption column, in agreement with the processes observed at Kilauea Iki in Hawaii [e.g. 176]. In this eruption style, the dominance of coarse particles (e.g. lava clots up to 1–2 m in diameter) are common. These large lava clots cannot solidify during their ballistic transport, and therefore after landing they could deform plastically, weld and/or agglutinate together or with other smaller pyroclasts [171, 380], depending on the accumulation rate and the clast temperature [172]. As a result of the efficient welding processes, the landing is not usually followed by free-avalanching, unlike loose, sufficiently cooled, brittle particles from other eruption styles, e.g. normal Strombolian styles. At the other end of the spectrum, energetic eruptions styles, such as violent Strombolian or phreatomagmatic eruptions, tend to generate localized sedimentation from pyroclastic density currents such as base surges. Similar to spatter generation, pyroclasts from pyroclastic density currents do not conform to the angle of repose. In contrast, the pyroclasts from sensu stricto

Strombolian-eruptions are sufficiently fragmented to suffer rapid cooling during transport either as ballistics or fallout from turbulent jets and eruption columns (Figure 9C). Thus, the particle with sufficient kinetic energy post-emplacement can fuel grain avalanches on the outer flanks of a growing pyroclastic pile. The active grain flows deposit pyroclasts on flanks in accordance with the syn-eruptive depositional surface properties and granulometric characteristics that provide the kinetic and static angle of repose of the granular pile. This is sustained until the ejected material is characterized by the same granulometric characteristics. With increasing degree of magma fragmentation, the dominant grain size of the tephra decreases. The increased efficiency of the magma fragmentation (e.g. violent Strombolian eruptions) commonly results in a higher eruption column, and therefore broader dispersion of tephra. The high variability and fluctuating eruption styles form a wide range of textural varieties of edifice such as Lathrop Wells in Southwest Nevada Volcanic Field, Nevada [60, 188], Pelagatos in Sierra Chichinautzin, Mexico [14, 378] or Los Morados, Payún Matru, Argentina [273]. Due to the variability in eruption styles, particles have different surfaces or granulometric characteristics. These differences induce some fine-scale post-emplacement pyroclast interactions. The effect of these syn-eruptive pyroclast interactions could prevent effective grain flow processes, 'reset' or delay (previous) sedimentary processes on the flanks of a growing conical volcano. Thus it could be a key control to determine the flank morphology of the resultant volcanic edifice. Examples for syn-eruptive granulometric differences can be found in the pyroclastic succession of the Holocene Rangitoto scoria cone in Auckland, New Zealand (Figure 10). During deposition of beds with contrasting dominant particle sizes, the angle of repose is not always a function of granulometric properties, but can be a result of the mode of pyroclast interactions with the syn-eruptive depositional surface. There is wide range of combinations of pyroclast interactions among ash, lapilli, block and spatter (Figure 10). Such transitions in eruption style and resulting pyroclast characteristics are important due to their blocking of freshly landed granular particles conforming (immediately) to the angle of repose expected if they were circular, dry and hard grains. For instance, the deposition of an ash horizon on a lapilli-dominated syn-eruptive surface must fill the inter-particle void, causing ash 'intrusions' into the lapilli media (Figure 10). These pyroclast interactions possibly slow the important cone growing mechanisms (e.g. grain flow efficiency) down, creating a sedimentary delay when the angle of repose is established on the syn-eruptive surface. In the complex cone growth model, the changes of eruptions style have an effect on the relative position of the crater rim shifting the position of the maximum sedimentation and the mode of pyroclast transport [182]. For example, switching from normal Strombolian to violent Strombolian style eruptions could increase the initial exit velocity from the usual 60–80 m/s [180, 183, 372, 386] to higher values, ≥150–200 m/s [111, 382, 387-388]. This shift in eruption style causes further decrease in the grain size of the ejecta [163, 367, 389]. This change could altogether cause migration of the location of the maximum sedimentation point further towards the crater rim if the tephra is transported by jet instead of pure ballistic trajectories [182]. Such changes in eruption styles introduce significant horizontal and vertical crater rim wandering during the eruption history of a monogenetic volcano (Figure 9C). This wandering modifies, in turn, the sedimentary environment leading the formation of cones with complex inner facies architectures, which may or may not be reflected by the morphology. Consequently, the major control in this cone growth model is on the eruption styles and their fluctuations over the eruption history.

In both cone growth models, lava flows can occur as a passive by-product of the explosive eruptions 'draining' the degassed magma away from the vent. Once the magma reaches the crater without major fragmentation it can either form lava lakes and/or (later) intrude into the flanks, increasing the stress [111, 298, 369]. When the magma pressure exceeds the strength of the crater walls, the crater wall may collapse or be rafted outwards [111, 369]. Sometimes, the magma flows out from foot of the cone, possibly fed from dykes [188, 273, 367, 369, 388, 390-391]. In this case the flank collapse is initiated by the inflation of the lava flow by discrete pulsation of magma injected beneath the cooler dyke margins [392-393] beneath a certain flank sector of a cone [111]. If the lava yield strength is reached and overtakes the pressure generated by the total weight of a certain flanks sector, flank collapse and subsequent rafting of remnants are common [111, 273, 390], leading to complex morphologies with breached craters and overall horse-shoe shape [322, 394]. The direction of crater breaching of scoria cones may not always be the consequence of effusion activity, but may coincide with regional/local principal stress orientations or fault directions [298, 315, 395-396]. If the magma supply is sufficient, the edifice that has been (partially) truncated by slope failure could be (partially) rebuilt or 'reheal' as documented from Los Morados scoria cone in Payun Matru, Argentina [273] or Red Mountain, San Francisco Volcanic Field, Arizona [397]. The direction of lava outflow from a cone commonly overlaps with the overall direction of background syn-eruptive surface inclination. This overall terrain tilt could cause differences in the tension in the downhill flank sector [82, 298, 391, 398]. Any kinds of changes during basaltic monogenetic eruptions could cause sudden decompression of the conduit system, leading to a change in the eruption style [273, 399]. Consequently, these destructive processes are more likely to occur during complex edifice growth, instead of simple cone growth. These changes could account for the fine-scale morphometric variability and architectural diversity observed in granular pile experiments and field observations [68, 182, 298, 314, 398]. Evidence of fine-scale morphometric variability due to lava outflow and crater breaching is observed through systematic morphometric analysis on young (≤4 ka) scoria cones in Tenerife [398]. Two types of morphometric variability were found: intra-cone and inter-cone variability. Intra-cone variability was characterised among individual flank facets. The slope angle variability was calculated to be as high as 12° between flanks sectors along (±45°) and perpendicular (±45°) to the main axis of the tilt direction of the syn-eruptive terrain [398]. This is about a third of the entire range of the natural spectrum of angles of repose of loose, granular material, i.e. scoria-dominated flanks [84, 182]. Inter-cone variability was detected on cones of the same age. According to Wood [71], these fresh cones should be in a narrow morphometric range (e.g. slope angle of 30.8±3.9°) based on fresh and pristine scoria cones analysed from the San Francisco Volcanic Field, Arizona [71, 324]. In contrast to this expected high value, the average slope angles of the studied cones from Tenerife turned out to vary from 22° to 30°, which has significant impact on many traditionally used interpretations of morphometric data including morphometric-based dating [e.g. 338], the morphometric signature concepts and erosion rate calculations. Both inter- and intra-cone variability were interpreted as a sign of differences in syn-eruptive processes coupled between internal, such as changes in efficiency of fragmentation, magma flux, effusive activity and associated crater breaching [398], and external controlling factors, such as interaction between pre-existing topography and the eruption processes [398]. All of these diverse eruptive and sedimentary processes are somehow integrated into the fresh morphology that is the subject of morphometric parameterization. A few of these processes could be detected while others could not, using morphometric parameters at one or multiple scales. For instance, some of this morphometric variability is usually undetectable using topographic maps and manual geomorphic analysis. This narrow variability, possibly associated with syn-eruptive differences in cone growth rates and trends, is in some instances in the range of the accuracy of the morphometric parameterization technique. For example, a manual calculation of slope angle from a 1:50 000 topographic maps with contour intervals of 20 m is ±5° [e.g. 312]. The fine-scale morphometric variability cannot be assessed accurately with this high analytical error range. On a DEM (either contour-based or airborne-based) with high vertical and horizontal accuracies (e.g. Root Mean Square Error under a few meters), this small-scale variability can be detected [e.g. 398]. An important consequence of this variability is that the initial geometry of the cone-type volcanoes, such as scoria cones, is not in a narrow range as previously expected [e.g. 71]. In other words, the morphometric signature of cone-type volcanoes are wider than described before, limiting the possibility of morphometric comparisons of individual edifices (especially eroded edifices) due to the lack of control on their initial geometries. It seems on the basis of the presented eruptive diversity, comparative morphologic

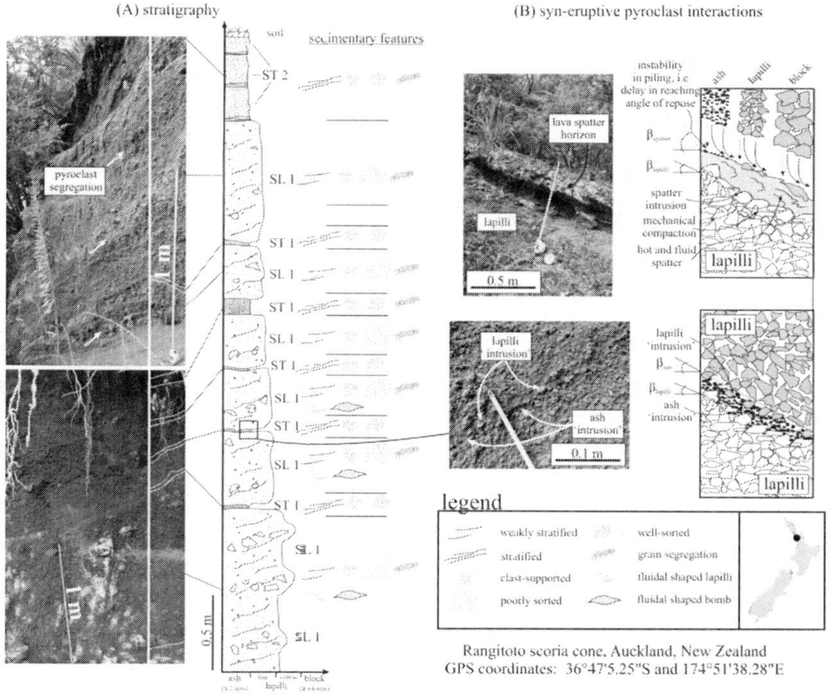

FIGURE 10. Stratigraphic log (A) of Rangitoto scoria cone, Auckland Volcanic Field, New Zealand showing the typical alteration of scoriaceous lapilli (SL1) and scoriaceous ash (ST1-2). Some examples are for syn-eruptive pyroclast interactions during growth of cone-type volcanoes such as scoria cones (B).

studies should be focused on comparing cones that have similar processes involved in their formation (i.e. E_{simple}, $E_{compound}$ or $E_{complex}$) and limited post-eruptive surface modifications (i.e. younger than a few ka in age). The morphology of a fresh (≤ a few ka) cone-type volcano is the result of primary eruptive processes; therefore, the morphometric parameters should be interpreted as the numerical integration of such eruptive diversity and mode of edifice growth. As stated by Wood [84], only fresh cones must be used for detecting causes and consequences of changes in morphology. When cone-type volcanoes from a larger age spectrum, e.g. up to a few Ma [e.g. 312, 337], are studied, the primary, volcanic morphometric signatures are modified by post-eruptive processes. Thus, they contain a mixture of syn- and post-eruptive morphometric signatures. Hence, interpretation of large morphometric datasets should be handled with care. Furthermore, it is also evident that not all geomorphic changes experienced by the edifice during the eruption history are preserved in the final volcano morphology. This could be due to, for instance, rehealing of the edifice after a collapse event, or changing eruption style, reflecting the complex nature of cone growth. Future research should focus on finding the link between the eruptive processes and morphology, as well as finding out how syn-eruptive constructive and destructive processes can be discriminated from each other on an 'unmodified', fresh cone.

5. DEGRADATION OF MONOGENETIC VOLCANOES

Once the eruption ceases, a bare volcanic surface is created with all of the primary morphologic attributes that have been determined by the temporal and spatial organization of the internally- and externally-controlled eruptive and sedimentary processes during the eruption history (Figure 6). The fresh surface is usually 'unstabilized' and highly permeable due to the unconsolidated pyroclasts, but there is often some degree of welding/agglutination, the presence of compacted ash, or lava flow cover. The degradation processes of a volcanic landform have a significant effect on the alteration of primary, volcanic geomorphic attributes (e.g. lowering of H_{co}/W_{co} or slope angle values). The significant transition from primary (pristine) volcanic to erosion landforms fundamentally starts when, for example, soil is formed, vegetation succession is developed or the surface is dissected over the primary eruptive products. However, modification of the pristine, eruption-controlled morphology could happen by non-erosion processes. For example, rapid, post-eruptive subsidence of the crater of a phreatomagmatic volcano due to diagenetic compaction, or lithification of the underlying diatreme infill during and immediately after the eruptions [400-401]. This usually leads to deepening of the crater or thickening of the sediments accumulated within the crater. Some compaction and post-eruptive surface fracturing, due to the gradual cooling down of the conduit and fissure system, is also expected at cone-type volcanoes, such as after the formation of Laghetto scoria cones at Mt. Etna, Italy [184] or Pu'u 'O'o spatter/scoria cone in Hawaii [173]. These processes could cause some geomorphic modification that may affect the morphometric parameters. On the other hand, the long-term surface modification of a monogenetic volcanic landform is related to degradation and aggradation processes over the erosion history. The structure of the degradation processes that operate on volcanic surfaces can be classified into two groups based on their frequency of occurrence and efficiency:

1. long-term (ka to Ma), slow mass movements, called 'normal degradation', as well as

2. short-term (hours to days), rapid mass movements, called 'event degradation'.

In the following section a few common degradation and aggradation processes are discussed briefly.

5.1. Long-Term, Normal Degradation Of Monogenetic Volcanoes

Normal degradation is a long-term (ka to Ma) mass wasting process that occurs by a combination of various sediment transport mechanisms and erosion processes such as rill and gully erosion, raindrop splash erosion, abrasion or deflation. Normal degradation requires initiation of the erosion agent that is usually the 'product' of the actual balance between many internal and external degradation controls at various levels, such as the climate or inner architecture (Figure 11). The external environment (e.g. annual precipitation, temperature or dominant wind direction etc.) is recognized as a major control on degradation [68, 71, 324-325, 337-338], influencing the chemical weathering and rates of CO_2consumption [e.g. 402]. However, there is no single control on chemical weathering rates; the actual weathering rates are often a function of many controlling conditions [e.g. 403]. Important controls on weathering rate could be the climatic settings (e.g. surface runoff, moisture availability, temperature, atmospheric CO_2 level, rates of evaporation [e.g. 404]), the tectonics (e.g. the post-orogenic increase in chemical weathering that decreases the atmospheric CO_2 concentration [e.g. 405]), the geomorphology (e.g. age of the surface, surface drainage system, rates of sediment transport, relief, soil cover, sediment composition [e.g. 406]) or the biology (e.g. microorganism, plant cover, animal activity [e.g. 407]).

A combination of the abovementioned controls and processes on chemical weathering interacts in many ways depending on the internal composition and characteristics of the volcanics exposed to the environment (Figure 11). The importance of internal controls on degradation seems to be neglected by earlier studies on monogenetic volcanic edifices [e.g. 71] in contrast with recent studies [e.g. 68]. The facies architecture and granulometric characteristics of a volcanic surface govern how the edifice reacts to the environmental impacts, e.g. the flanks drain the rainwater 'overground' leading to the formation of rills and gullies or allow infiltration [71, 408-412]. The pyroclast-scale properties are determined by fluctuation of eruption styles during the eruption history, leading to accumulation of pyroclasts with contrasting geochemical, textural and granulometric characteristics. This pyroclast diversity will be responsible for the various rates of chemical weathering. Additionally, this diversity has an effect on the mode and efficiency of sediment transport during the course of degradation. For instance, the 'stability' (or amount of loose particles on the flanks) causes slight differences in rates, styles and susceptibility for erosion. The stability could increase with the formation of mature/immature soil, thick accumulation of weathering products, denudation of a lava-spatter horizon and/or heavy vegetation cover, which altogether help to stabilize the landscape. These changes on the mineral- to pyroclast-scales lead to transitions from 'unstabilized' to 'stabilized' stages. The duration of the transition depends on many factors (e.g. Figure s 12A and B), such as the initial surface morphology, granulometric characteristics, volcanic environment and climatic settings [408-410, 413-417]. The transition could be as short as a couple of years if the volcanic surface is characterized by the dominance of fines, e.g. ejecta ring around a tuff ring, and typically exposed to a humid, tropical climate [408]. In arid climates the lag time between soil formation is significantly longer (if it takes place at all), up to 0.1–0.2 Ma [188, 413]. There are extreme environments where the soil/vegetation cover can barely be developed due to the high rates of volcanic degassing and acid rain, e.g. the intra-caldera environment in Ambrym, Vanuatu [e.g. 408] or cold polar regions, e.g. Deception Island, Antarctica [e.g. 418, 419]. Changes in surface stability could be governed by gradual denudation of inner, texturally compacted (e.g. welded or agglutinated spatter horizons or zones) pyroclastic units. This leads to rock selective erosion and higher preservation potential of an edifice in the long-term [e.g. 68]. Consequently, the degradation processes cannot be separated from the architecture of the degrading volcanic edifice, and therefore the erosion history is strongly attached to the eruption history. In this respect, the erosion history and rates seem to be governed (at least on one hand) by the time-lagged denudation of pyroclastic beds with varying susceptibility to erosion. In other words, the rate and style of degradation are theoretically the 'inverse' of the eruption history if the external controls are steady over the erosion history.

The actual balance between the internal and external controls determines the dominant rates and mode of sediment transport mechanism at a given point on the flanks of a monogenetic volcano (Figure 11). The mode and style of erosion of monogenetic volcanic landforms can be subdivided into 'overground' and 'underground' erosion. The long-term, overground degradation of volcanic surfaces can be accounted for through water-gravity (including rainfall, sea or freshwater, underground water or ice or various lateral movements of sediment/soil cover due to gravitation and water), and wind erosion agents. The sediment transport fluxes of such erosion agents and their time-scales are highly variable. The most significant, normal degradation processes on bare unstable (volcanic) surfaces are possibly triggered by the presence of water. Rainfall erosion

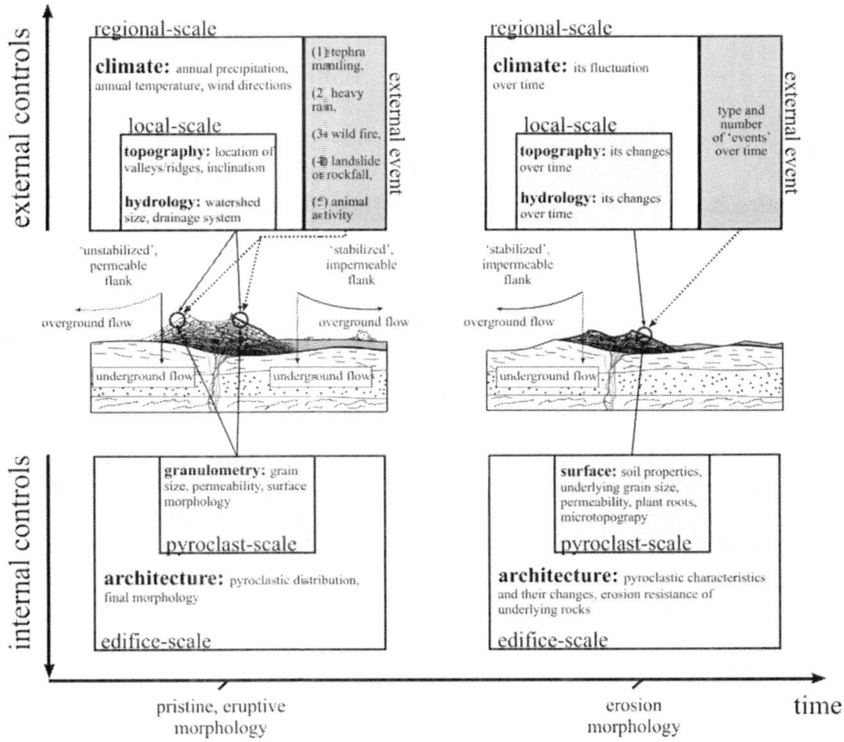

FIGURE 11. Conceptualized model for the configuration of internal and external degradation controls on determining erosion agents and sediment transport processes at a given point on 'unstabilized' and 'stabilized' volcanic surfaces over the erosion history. Degradation of a monogenetic volcano takes place by both long-term (ka to Ma; dark and light yellow boxes) and short-term, event-degradation processes (hours to years; green boxes) with different rates over the erosion history. The effects of such degradation processes take place both 'overground' and 'underground'.

causes small-scale movement of particles up to a couple of cm in diameter or chunks of soil when the rain drops impact on the surface [e.g. 420]. Besides the impact-triggered (or raindrop splash) sediment detachments, there are associated processes such as minor raindrop-induced surface wash [420-421]. On volcanic surfaces, the raindrop introduced and/or wind-driven rain splash erosion and related downward sediment transport on steep flanks (20–25°) is about 15.8 cm/yr for pyroclasts with diameter of 1.7 cm [323]. The regional-scale lowering of the landscape (over a 100 m length with 10° of slope angle) associated with rain-splash could be in the range of 0.1–1 mm/yr [e.g. 422]. This equivalent to a long-term erosion rates of 0.1 t/km/yr, calculating with 2.5 mm median rain drop diameter on a flank with 10° of slope angle [422]. If rainfall intensity exceeds the soil's infiltration capacity at any time during a rainfall event, overground flow, such as unchannelized sheet flow can be generated [e.g. 423]. The erosion capacity of sheet flow is higher than the rain-splash, but it is significantly lower than mass-wasting once the rill and gully network is developed (e.g. Figure s 12E and F). Drainage system development on the flanks of monogenetic volcanoes is found to correlate with the age of the volcanic edifice [70, 321, 324, 424]. Although, the required time for their formation could be as short as a couple of months or years and it could develop on the gentle flanks (e.g. a tuff ring with slope angle of 5–10° in Figure s 12E and F) which anomalously short period of time could introduce error in the relative morphology-based dating [425]. The fluvial erosion could remove sediment in a range of 10–100 000 t/km^2/yr [e.g. 426]. These overground surface processes, such as rill and gullies, and various soil/sediment creep [70-71, 323, 408, 425] have been accounted as a major mass wasting mechanism over the degradation history of a monogenetic volcanoes. The effect of these overground degradation processes were modelled mostly on scoria cones [324-325, 327, 337].

All varieties of soil and sediment creep and solifluction processes, such as soil and frost creep and gelifluction, are usually slow processes [e.g. 427], in comparison with surface runoff. Thus, these processes modify the volcano flanks' morphology constantly and over a longer period of time. The rates of erosion vary depending on the topography (e.g. slope gradient), sediment/soil properties (e.g. proportion of fines, moist content) and predominant climate (e.g. amount and type of precipitation, annual temperature), but rarely exceed the downhill movement rates of 1 m/yr and the volumetric velocity of between 1×10^{-10}–1×10^{-8} km^3/km/yr [e.g. 427]. The sediment transport rates introduced by solifluction are many orders of magnitude smaller than the erosion loss of a volcano by fluvial processes. In cold semi-arid and arid regions, the ice plays the major role on the sediment transport [e.g. 428]. Consequently, the surface modification and movements are related to diurnal and annual frost-activity, such as ground freezing and thawing cycles. On scoria cones at the periglacial Marion Island, South Indian Ocean, dominant sediment transportation processes on scoria cone flanks are the needle-ice-induced frost creep related to the diurnal and possibly annual frost cycles [428]. The frost creep rates are 53.2 cm/yr for ash (≥70% of grains ≤2 mm), 16.1 cm/yr for lapilli (≥30% of grains between 2–60 mm), and 2.6 cm/yr for bomb/blocks (≥70% of grains ≤60 mm), based on measurements on painted rocks [428]. The rates are primarily controlled by the predominant grain-size of the sediment, the slope angle of the underlying terrain and altitude [428]. Based on the transportation rates, the processes are the same or an order of magnitude faster than rain-splash-induced pyroclast transport in a semi-arid environment, such as San Francisco volcanic field in Arizona [323].

Probably the most effective overground degradation process on a freshly created volcanic surface is the wave-cut erosion. In rocky, coastal regions, the wave cut notch moving back and sideward removes mass by hydraulic action and abrasion [e.g. 429]. Wave-cut erosion mostly affects tuff cones that are located on or offshore (Figure s 12C, D, E and F). In this environment abrasion is a common syn-eruptive [430] or post-eruptive erosion process, e.g. Surtsey [431-432] or the early formation Jeju Island, Korea [433]. For instance, post-eruptive coastal modification through abrasion generated an area-loss of about 0.2 km^2 between 1975 and 1980 on Surtsey island in Iceland [431]. The rate of volume-loss of non-volcanic coastal regions is in the range of 10 000 to 100 000 t/km/yr [e.g. 434]. This enhanced rate of mass-removal is in agreement with advanced states of erosion on a recently formed tuff cone within Lake Vui in the caldera of Ambae volcano, Vanuatu, which formed in 2005 [263]. The initial surface has been intensively modified by wave-cut erosion and slumping from the crater walls, leading to an enormous enlargement of the crater and crater breaching over 10 months (Figure s 12C and D).

The effect of wind deflation is often limited, especially in humid climates. However, there are examples in volcanic environments where, in spite of the high annual precipitation, the sediment transport rates are still significantly high due to wind action, e.g. in some parts of south Iceland, around 600 t/km/yr [435]. Expressing the long-term effect and rate of sediment transport by wind is complicated due the high variability of wind intensities (i.e. storm events versus normal background intensities) and directions [e.g. 436]. Sediment transport fluxes can vary in a wide range as a function of wind energy and surface characteristics (e.g. sediment availability, vegetation cover or water saturation). Long-term sediment transport by wind in volcanic areas (e.g. Iceland) is in the range of 100 to 1000 t/km/yr [435-436]. However, this sediment transport rate is in relation, but it is not equivalent to the erosion rates. Furthermore, efficient wind transportation as bed load by creep and saltation, and suspended load, is limited to particles generally ≤8 mm in diameter [e.g. 436]. This granulometric limit is crucial for the long-term erosion of volcanic landforms that built up from coarser pyroclasts, such as coarse lapilli-dominated scoria cones. Significant increase is observed during storm events, when these sediment transport rates could reach as high as a couple of percentage of the annual fluxes within an hour [435]. Thus, the long-term approximation of the sediment transport rates could be interpreted as cumulative values of normal, background and increased, storm/related erosion rates [e.g. 435]. In addition, there are a few examples such as Surtsey, where the wind deflation is considerable. In Surtsey, the strong wind is responsible for the polishing of palagonitized tephra surfaces and transporting and redistributing unconsolidated tephra on the freshly created island [431]. Direct observation of short-term volumetric change of a young scoria cones (e.g. Laghetto or Monte Barbagallo, ca. 2700–2800 m asl) is through deflation by wind in the summit region of Mt. Etna, Sicily [326]. Surface modification is inferred to occur on the windward side of the Monte Barbagallo cones [326]. The wind likely induces some minor pyroclast disequilibrium on the flanks that may lead to minor rock fall events or initiate grain flows [326]. In real semi-arid areas, wind-erosion is an important transport agent and surface modificator over unstabilized volcanic surfaces such as the Carapacho tuff ring in the Llancanelo Volcanic Field, Mendoza, Argentina (Figure 12A). The layer-by-layer stripping of the volcanic edifice is completely visible on the windward side of the erosion remnant facing the Andes. On the other hand, the wind-blown sediments can accumulate over time leading to sometimes expressible aggradation on volcanic surfaces (Figure 13A). Accumulation of aeolian addition could significantly contribute to the soil formation by gaining excess material, e.g. quartz or mica [415, 437]. Due to the generally high roughness of pyroclast- or lava rock-dominated surfaces (e.g. highly vesicular scoria or a'a lava flow), the wind slows down, leading to sedimentation and later accumulation of wind-transported particles [415-416, 437-438]. The wind-induced aggradation helps to reduce the transition time between an 'unstabilized' to a 'stabilized' surface by developing desert pavement in semi-arid/arid desert environments [415-416].

The previously mentioned, generally long-term overground degradation processes often account for most of the volumetric loss and surface modification of monogenetic volcanoes [e.g. 71]. In the case of the underground degradation, the surface water leaves the system through the groundwater if the actual soil infiltration capacity exceeds the rainfall intensity [423]. This underground water can remove weathering products (e.g. leaching of cations from the regolith) as dissolved sediment fluxes [e.g. 439]. The rate of chemical weathering could be extremely high in humid [e.g. 439] and lower in moderate climates [e.g. 440], based on chemical and solute-derived weathering data from rivers draining mafic to intermediate igneous rocks. The rates of underground chemical weathering could be in the range of 10 to 1000 $t/km^2/year$ in a humid climate [e.g. 439], which is in the same order of magnitude as the erosion rate by surface runoff. This efficient removal of weathering products by infiltrating rainwater and subsequent underground flow has an important, previously unaccounted for effect on degradation of monogenetic volcanic landforms.

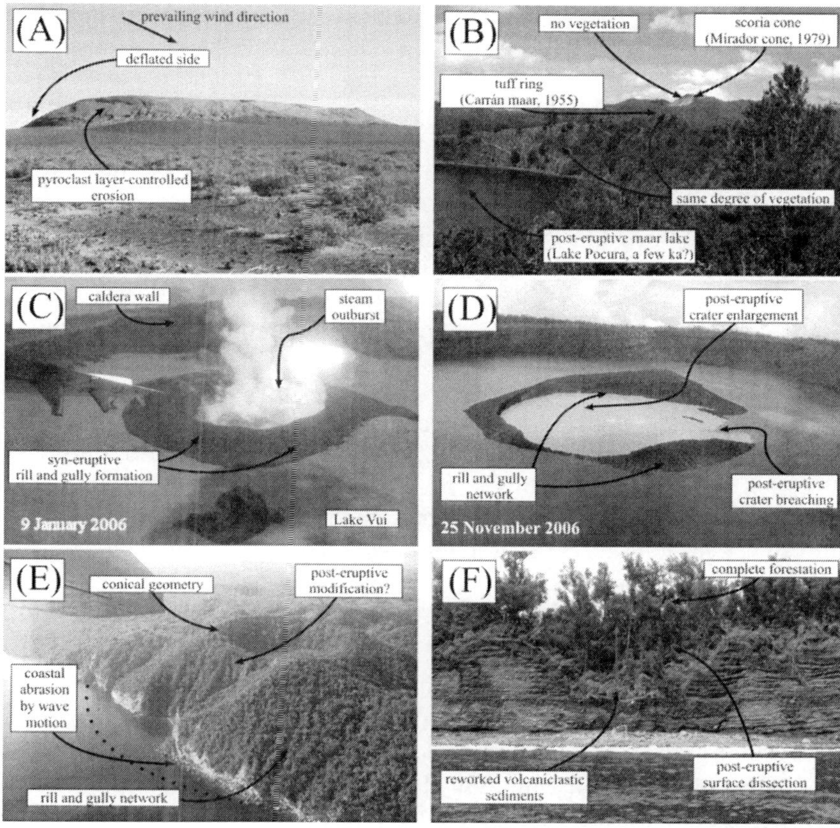

FIGURE 12. A) Contrasting rates of wind erosion on Pleistocene Carapacho tuff ring [**61**] under a semi-arid climate in the Llancanelo Volcanic Field, Mendoza, Argentina. It is interesting to note that the wind-erosion is strongly controlled by the resistance of individual beds to wind deflation. Lack of vegetation cover helps to maintain the long-term, slow erosion on the windward side of the volcanic edifice. (B) Contrasting style and rates of revegetation of volcanic surfaces on the flanks of maar/tuff ring and scoria cones. Lake Pocura (Ranco Province, Chile) is a few ka old maar crater and is characterised by the same degree of vegetation cover as the recently formed Carran maar (1955). The Mirador scoria cone (1979) lacks of vegetation cover. In comparison with Carapacho tuff ring (in A), and, the surface stabilization in semi-arid/arid climates, the time for vegetation to develop is much shorter, in a few decade time-scales. (C and D) Rapid syn- and post-eruptive erosion is observed on a freshly formed tuff cone in the caldera of Ambae, Vanuatu in 2006. (E) A tuff cone (≤2 ka old) located along the coastal region in Ambrym, Vanuatu. Note the geomorphic similarities of this tuff cone with scoria cones. (F) Cross-section through a post-eruptive, well-developed, gully exposed by intensive wave-cut erosion since 1913. The gully developed on the gentle-flanks of a tuff ring formed in the phreatomagmatic eruption in Ambrym, Vanuatu [**425**].

It is also worth noting that volcanoes with conical geometry are suspected to have different weathering regimes in accordance with microclimatic and topographic conditions [e.g. 437]. Taking the Rangitoto scoria cone (Auckland, New Zealand) as an example, the area of the cone is about 0.41 km^2, while the bulk DEM-based volume is about 0.022 km^3 above 140 m asl [363, 441]. The edifice has a basal diameter of about 600 m and a crater diameter of 200 m. Considering the abovementioned ranges of subsurface weathering and erosion rates, the time scales of complete erosion of the scoria cone can be calculated. Assuming a 1200 kg/m^3 average density for moderately to highly vesiculated scoria deposits (i.e. 3.4 to 341.6 m^3/yr volume loss rates), the time scale of complete degradation would be between 6.4 Ma and 0.06 Ma, using the constant degradation rates mentioned above. Of course, the rates of chemical weathering tend to slow down when the soil coverage becomes thicker and the weatherable parental rocks are reduced [e.g. 442]. Apart from this, the underground erosion of volcanic edifices by infiltrating surface water (e.g. initial stages of scoria cone degradation) and groundwater flow could be very important and effective long-term degradation process that should have some influence on the morphology of monogenetic volcanoes, especially for volcanic areas with strong chemical weathering rates (e.g. humid, tropical areas with high annual temperature).

In summary, the degradation of a monogenetic volcano is many orders of magnitude longer (≥100 ka to ≤ 50 000 ka) than their formation (≤0.01 ka). For example, the degradation of a small-volume (≤0.1 km^3) volcanic edifice usually takes place in a couple or 10s of Ma for welded and/or spatter-dominated edifices, such as in the Bakony-Balaton Highland Volcanic Field in Hungary [68] or in Sośnica hill volcano is Lower Silesia, Poland [335]. Phreatomagmatic volcanoes, especially those with diatremes, could degrade over a longer period of time due to their significant vertical extent, e.g. the Oligocene Kleinsaubernitz maar–diatreme volcano in Eastern Saxony, Germany [401]. During such a long degradation time, the rates and style of post-eruptive surface modification of monogenetic volcanic landforms are generally vulnerable to changes in the configuration and balance between internal and external degradation controls (Figure 11). These could be triggered internally, e.g. denudation of a spatter-dominated or a fine ash horizon (e.g. Figure 13B), or externally, such as long-term climate change or climate oscillation [39], initializing a gradual shift in the dominant mode of sediment transport. Each of these gradual changes (e.g. soil formation, granulometric and climatic changes etc.) causes a partial or complete reorganization of the controls on degradation. This adjustment of erosion settings could result in a change in erosion agent that may or may not increase or decrease sediment yield on the flanks of a volcano. All of these changes over the long erosion history open systemically new potential ‘pathways’ for erosion, leading to diverse erosion scenarios. The long-term degradation seems to be an iterative process, repeating a constant erosion agent adjustment that is triggered by many gradual changes over the erosion history of a volcano. In many previous erosion studies, the edifices are usually treated as individuals sharing the same internal (i.e. configuration of pyroclastic successions) and initial geometry [e.g. 71] and degrading in accordance with the climate of the volcanic field [324-325, 337]. Of course the climate is in general a important control on degradation, but the climatic forces are in continuous interactions with the volcanic surface, promoting the importance of architecture and granulometric characteristics of the exposed pyroclasts and lava rocks in the volcanic edifice. In extreme cases such as Pukeonake scoria cone (having typical monogenetic edifice dimensions of 150 m in height and 900 m in basal width) in the Tongariro Volcanic Complex in New Zealand, there is an unusual wide granulometric contrast within the pyroclastic succession (Figure 13B). Additionally, the trends and processes in degradation are in close relationship with the exposed pyroclast characteristics which determine the rates of chemical weathering, soil characteristics, surface permeability and, in turn, the mode of sediment transport on the flanks (e.g. Figure s 13C and D).

5.2. Short-Term, Event Degradation Of Monogenetic Volcanoes

The event degradation processes take place in a short time frame (hours to days), but they could cause sudden disequilibrium in the degradation and sedimentary system. A monogenetic magmatic system tends to operate inhomogeneously both spatially, forming volcanic clusters, and temporally, forming volcanic cycles. Additionally, there are monogenetic volcanoes that can be found as parasitic or satellite vents on the flanks of larger, polygenetic volcanoes, such as Mt. Etna in Sicily [317] or Mauna Kea in Hawaii [314]. The spatial and temporal closeness of volcanic events, however, pose a generally overlooked problem related to the degradation of monogenetic volcanic landforms, such as tephra mantling or geomorphic truncation by eruptive processes of a surrounding volcano. Tephra mantling is considered to be an important process for the degradation of volcanic edifices as stated by Wood [71] and White [412]. The average distance between neighbouring volcanoes in intraplate settings, such as Auckland in New Zealand, is about 1340 m (i.e. 5.6 km^2), while on the flank of a polygenetic volcanic/volcanic island, such as Tenerife in Canary Islands, that average is about 970 m (i.e. 2.9 km^2). On the other hand, the typical area of a tephra blanket 1–2 cm thick ranges from 10 km^2 for Hawaiian eruptions [162] to 10^3 km^2 for violent Strombolian eruptions [163, 443-444] and for phreatomagmatic eruptions [358]. Consequently, the individual edifices commonly overlap each other's eruption footprint (i.e.

area affected by the primary sedimentation from the eruptions), showing the importance of tephra mantling. Furthermore, there are monogenetic volcanic edifices that are developed on flanks of larger polygenetic volcanoes where the mantling by tephra could be more frequent and more significant than in intraplate volcanic fields (e.g. Mt. Roja in the southern edge of Tenerife in the Canary Islands, Figure 13E). A few cm thick tephra cover could cause complete or partial damage to the vegetation canopy [411, 445-447]. Mantling could reset all dominant surface processes, including sediment transport systems, erosion agents, vegetation cover or soil formation processes. This leads to similar reorganization of the degradation controls to those seen with the long-term gradual changes of external or internal factors during normal degradation, but in much shorter time-scales (hours to years). The sedimentary responses to mantling could occur instantly or with a slight delay. Increased erosion rates of older cones were documented instantly after the tephra mantling by fine/coarse ash from Paricutin, Michoacán-Guanajuato, Mexico between 1943 and 1952 [411]. The tephra that mantled the topography was fine ($Md\phi$ = 0.1–0.5 mm) and relatively impermeable, which led to the formation of new, extensive incisions by rill channels and significant deepening of older gullies by the increased sediment yield [411]. In contrast, the sedimentary response for the Tarawera eruption in New Zealand was delayed by the well-sorted, coarse and high permeability of the tephra accumulated over the landscape [448]. The mantling may have an effect on vegetation coverage (e.g. cover or burn the vegetation) and the erosional agent responsible for shaping the morphology of the volcano. The long-term effect of this may be the longer preservation of the landform, or increased dissection which temporarily enhances the overall rates of erosion. These changes will have an important influence on the majority of the morphometric parameters and their pattern of changes over the erosion history.

Surface modification of an already formed monogenetic volcanic edifice could also be triggered by the formation of another monogenetic vent close by [52, 412, 449]. The amalgamated or nested volcanic complexes that have some time delay between their formation are common in volcanic fields, for example Tihany in the Bakony-Balaton Highland, Hungary [41], Rockeskyllerkopf volcanic complex in the Eifel, Germany [450] or Songaksan in Jeju Island, Korea [9, 278]. The eruption of nearby volcano(es) seems to be a common process that may lead to the minor truncation of surfaces, bomb/block-dominated horizons and discordances in the stratigraphic log. This type of 'event' degradation by monogenetic eruption could modify the previously formed topography instantly.

On pyroclast surfaces with limited permeability (e.g. fine ash, lava spatter horizon or lava flows), the rain water tends to simply runoff depending on the actual infiltration rate and the rain fall intensity rate [409-410, 447, 451-452]. On this fine ash surface, the infiltration rates are an order of magnitude lower than on a loose, lapilli-covered flank. This is visible on the flank of La Fossa cone in Vulcano, Aeolian Islands (Figure 13C). The La Fossa cone is not a typical monogenetic volcano, but it has similar geometry and size to a typical monogenetic volcanic edifice. Erosion on the La Fossa cone is characterized by surface runoff on the upper steeper flanks (≥30°) built up by fine indurated ash (Mdϕ = 100 μm), while the erosion of the lower flanks (≥28° and Mdϕ = 1–2 mm) is usually due to debris flows forming levees and terminal lobes [410]. The strikingly different style of mass wasting mechanism is interpreted to be the result of the lack of vegetation cover and strong contrast in permeability and induration of the underlying pyroclastic deposits [410]. Erosion by debris flows forms deep and wide gullies even on a flank built up by permeable rocks, e.g. La Fossa in Vulcano [410] or Benbow tuff cone in Ambrym, Vanuatu [408]. The triggering mechanism for a volcaniclastic debris flow is limited to a period of intense, heavy rainfall [408, 410, 451, 453]. Thus it operates infrequently and it tends to typically redistribute a pocket of a few tens of m^3 of sediments [410].

Similarly to the volcaniclastic debris flows, landslides could also be part of the event degradation processes, especially on steep flanks (e.g. cone-type morphology). The susceptibility for landslides that remove large chunks from the original volcanic edifice, increases by diversity of the pyroclast in the succession. In other words, the layer-cake, usually bedded, inner architecture of either the ejecta ring around a phreatomagmatic volcano or a scoria cone, is extremely susceptible to landsliding triggered, for instance by heavy rain, earthquake, animal activity or surface instability of freshly deposited mantling tephra [e.g. 411].

Another event degradation process is the wild fire that is responsible for the temporal increase (by 100 000 times the 'background' sediment yield) in erosion rates and sediment yields on steep flanks [e.g. 456, 457]. The major effects on a surface by a wild-fire include accumulation of ash, partial or complete damage of vegetation, and organic matter in the soil, and modification of soil structure, if any, and its nutrient content [e.g. 456]. These changes of the surface properties lead to modifications of porosity, bulk density and infiltration rates of the surface, promoting overground flow which is able to carry the increased sediment yield [e.g. 456]. The overground flow removes the fine sediment (e.g. volcanic ash and lapilli and non-volcanic ash) and the topsoil, causing enrichment of coarser sediment on the surface.

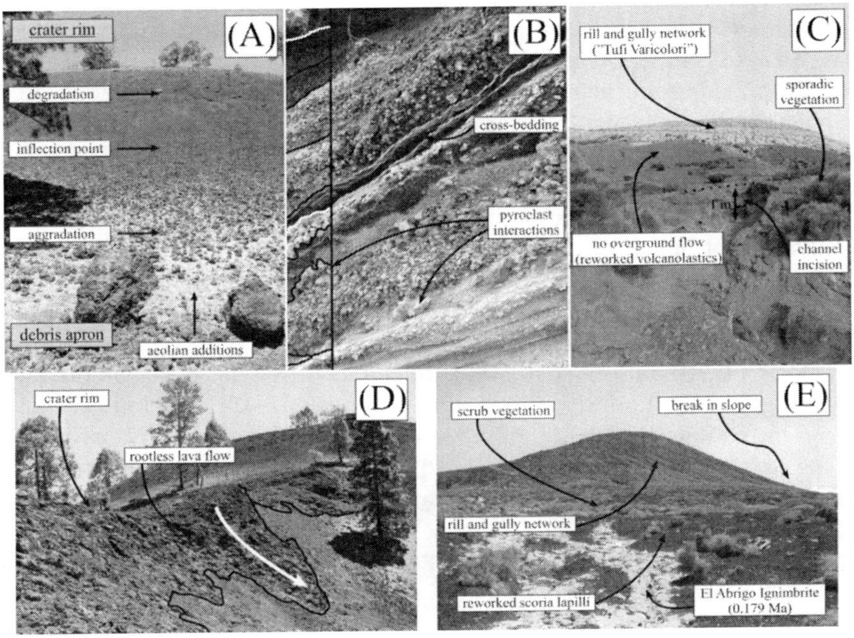

FIGURE 13. A) Degradation/aggradation on the outer flanks of Mt. Cascajo volcano (a few ka old?) in the NW rift zone, Tenerife, Canary Island. (B) Textural and granulometric inhomogeneity of the Pukeonake scoria cone, at the foot of the Tongariro Volcanic Complex in New Zealand. It is important to note that each of these units has different erosion resistance that may have an effect on the trends and patterns in the erosion history. (C) Contrasting granulometry and permeability caused difference in erosion surface modification on the flanks of La Fossa cone at Vulcano, Aeolian Islands. (D) Spatter accumulated on the crater rim feeds a small-volume rootless lava flow at the Mt. Cascajo volcano in Tenerife, Canary Island. The spatter is relatively impermeable in comparison with its environment (e.g. loose, scoria lapilli), thus will have an influence on the subsequent erosion patterns. (E) View of Mt. Roja in the southern edge of Tenerife, Canary Island. The Mt. Roja edifice was partly or completely covered by the El Ambrigo Ignimbrite (0.18 Ma) sourced from the former Las Cañadas edifice [**454-455**]. This mantling resulted in development of a rill and gully system on the flanks facing toward north.

Animal activity is a commonly recognized erosion type due to its effect on the compaction of the uppermost soil [e.g. 458, 459] and/or linear dissection of the surface by trampling [e.g. 460, 461]. As a result of compaction, the rainwater cannot penetrate through the soil cover easily, leading to overground flow that increases the erosion rate and sediment yield [e.g. 458]. Wild animal tramping could be a source of rill and gully formation on flanks, mostly in semi-arid and arid environments [e.g. 460], particularly in, the crater-type volcanoes that commonly host post-eruptive lakes within the crater basins, such as Laguna Potrok Aike in Pali Aike Volcanic Field, Patagonia, Argentina [462] or Pula maar, Bakony-Balaton Highland, Western Hungary [195, 463]. These maar lakes create special habitats that could increase animal activity, creating more opportunity for animal activity-induced erosion.

The event degradation processes, such as heavy rain-induced debris flow, tephra mantling, landslide, post-wild fire runoff or animal activity, could individually trigger rapid geomorphic modifications that affect the long-term degradation rates of the volcanic edifices (Figure 11). The rates of sediment yield could be a thousand times larger in response to event degradation than in the case of normal degradation. These events are usually randomly or inhomogeneously distributed over the erosion history of a volcanic landform, making the quantification to the total erosion-loss complicated. Due to significant surface modification, their effect on the morphometric parameters could be large and difficult to quantify.

5.3. Post-Eruptive Erosion Of Monogenetic Volcanoes By Normal And Event Degradation Processes

The geomorphic state of (monogenetic) volcanoes is commonly expressed by various morphometric parameters, including edifice height, slope angle or H_{co}/W_{co} ratios. The values of these morphometric parameters usually show a decreasing trend over the course of the erosion history [70-71, 290, 299, 317, 320, 464-465]. Consequently, the morphometric parameters show strong time-dependence. This systematic change in morphology was observed mostly on scoria cones in many classical volcanic fields, such as San Francisco volcanic field in Arizona [71] or Cima Volcanic Field in California [70]. This recognition led to the most obvious interpretation: the morphology is dependent on the degree of erosion, which is a function of time and the climate [e.g. 71]. Therefore, morphometric parameters may be used as a dating tool for volcanic edifices if the final geometry and internal architecture are similar among the volcanic edifices being compared [e.g. 71, 324]. These fundamental assumptions (regardless if stated or not) are valid for all comparative morphometric studies targeted volcanoes.

The concept outlined above is, however, sometimes oversimplified and the assumptions are not always fulfilled. The concern about the classical interpretation of morphometric parameters and their change over time is derived from various sources.

- The architecture of a single monogenetic edifice is commonly inhomogeneous in terms of internal facies characteristics. This architectural diversity usually results in diversity in erosion-resistance and susceptibility to chemical weathering of the pyroclastic rocks exposed to the external environment (e.g. Figure s 10 and 13B or [e.g. 68]. Due to the continuous denudation of internal beds, the internal architectural irregularities could cause different rates of weathering and erosion leading to hardly predictable trends in degradation.
- The previous concept of monogenetic volcanism implied that the morphology of a volcanic landform is linked only to the specific eruption styles (i.e. Strombolian-type scoria cone). However, this is an oversimplification and belies the complex pattern in edifice growth (e.g. Figure s 6–9).
- The final, pristine edifice morphology is mostly controlled by syn-eruptive processes (e.g. explosion energy, substrate stability, mass wasting and mode of pyroclast transport, e.g. Figure s 7–9). Any change in either the internal or external controls during the course of an eruption could modify the final morphology of the edifice partially or dramatically. This is in agreement with the measured high variability of slope angle [e.g. 398] or aspect ratios [e.g. 314] on relatively fresh edifices. This supposedly eruptive process-related morphometric variability is observed on both the intra-edifice scale (e.g. between various parts of an edifice) and inter-edifice scale [398].
- There is a large difference in the rates and mode of chemical weathering and sediment transport operating on different types of pyroclastic deposits or lava rock surfaces (e.g. Figure s 12 and 13). For instance, there is a contrast between mass wasting rates by under- and overground flow processes, e.g. spatter or a higher degree of welding/agglutination could cause asymmetric patterns in the permeability and, therefore, the subsequent initialization of the erosion on a freshly created surface.
- One theoretical concern about morphometric parameters, such as edifice height, aspect ratio or slope angle, is that they are intra-edifice, 'static' descriptors. Thus, they only express the current geomorphic state of the volcano. In contrast, they are often used to reveal and describe 'dynamic' processes, such as erosion patterns over time. It is obvious that trends in erosion processes cannot be seen based on these intra-edifice, 'static' parameters unless they are compared with other edifice parameters, or measure direct geomorphic modification by erosion processes over short periods of time (e.g. Figure s 12C and D or [e.g. 464]). The first seems an 'extrapolation' due to the assumption of

that all volcanic edifices in the comparison have similar eruptive- and erosion histories. The second is more suitable, because it is the direct observation of erosion loss and surface modification [e.g. 326].

- Comparative morphometric studies often lack or have limited age constraints (e.g. a few % of the total population of the studied edifices are dated) on the morphology, or inversely in special cases, the dating is the purpose of the comparison. There are just a few studies with complete age constraint [e.g. 68, 70, 326].
- The long-term surface modification is often believed as a result of the climate forces and climate-induced erosion processes. Wood [71] stated the importance of tephra mantling as a possible source of acceleration of erosion rates, but such event-degradation (e.g. tephra mantling, edifice truncation by eruption nearby, landsliding, wild fire, animal activity etc.), are usually neglected. They occur infrequently, but they could cause rapid and significant modification that may influence the patterns of future degradation.

Due to the concerns and arguments listed above, the morphometric parameters and their classical interpretations should be revised. Referring back to the complexity of construction of monogenetic volcanoes (Figure 6) and their primary geomorphic development (Figure s 7–9), it is obvious that on a fresh volcanic landform the geomorphic feature is determined by syn-eruptive processes, which in turn are governed by the internally and externally-driven processes during the eruption history. Once the eruption ceases, the 'input' configuration of a monogenetic volcano in terms of architecture, pyroclast granulometric characteristics, geometry and geomorphology is given. The erosion agents at the start of the erosion history are determined by the interactions between the internal (e.g. pyroclastic rocks on the surface) and external processes (e.g. climate; Figure 11). The results of such series of interactions between these properties lead to surface and subsurface weathering, soil formation and development of vegetation succession over time. Each of these developments on the flanks of a monogenetic volcano has a feedback to the original controls modifying the actual balance towards one side. This leads to disequilibrium in the system and subsequent adjusting mechanism. These processes are called normal degradation, operating at a longer-time scale (ka to Ma). However, the degradation mechanism sometimes does not function as 'normal'. During the erosion history of a monogenetic volcano, there are some environmental effects called 'events' such as tephra mantling or heavy-rain-induced grain flows. These 'events' are documented to cause orders of magnitude larger surface modification and possibly initialize new rates and trends in the dominant sediment-transport system and increase the sediment yield [e.g. 411-412, 447-448, 457]. Consequently, the erosion history of a monogenetic volcano comprises both normal ('background') and event degradation processes (Figure 11). The cumulative result of many interactions, reorganization of erosion agents, and effects of event degradation processes over the erosion history, are integrated into the geomorphic state at the time of examination.

The degradation of the volcanic edifice leads to aggradation at the foot of the edifice and the development of a debris apron (Figure 13A). Based on the behaviour and changes of intensity of the abovementioned major sediment transport processes, it is evident that the individual contribution of such erosion processes is not constant over time. It is more likely that they are enhanced or eased by each other at certain stages of the degradation. The gradual changes in style, rate and mode of sediment transport on the flank of a monogenetic volcanic edifice are likely triggered by the shifting of dominant external (e.g. climate change) and internal environments (e.g. variability of erosion resistant layer within the edifice as observed in Figure 14A). Consequently, the degradation of the monogenetic edifice as a whole cannot be linear (or maybe just certain parts of the erosion history) and must erode faster at the beginning and slower at the end of the degradation [71, 324] in accordance with the wide range of rates and time-scales of sediment transport processes. Consequently, a single geomorphic agent cannot account for a volcano's degradation. Instead, it seems to be the result from the overall contribution of all processes with complex temporal distribution. Without event degradation processes, given the fact that the erosion history lasts at least over a time-scale of a couple of ka for a typical monogenetic volcanoes, this increases the likelihood of some changes in the external environment that could modify the degradation trend.

These surface modifications and degradation processes should be in correlation with the values of morphometric parameters, but their interpretation is possibly not a straightforward process. The pristine unmodified geomorphic stage of a monogenetic volcano is predominantly controlled by the processes that occurred in the eruption history. Once the degradation proceeds (e.g. erosion surface modification, soil formation or development of vegetation cover), these primary geomorphic attributes are gradually replaced by excess 'signatures' of the various post-eruptive processes. This will result in 'noise' in the original syn-eruptive state of morphometric parameters extracted from the topographic attributes. The soil cover on the surface creates a buffer zone between the pyroclastic deposits and the environment. In this buffer zone, most of the weathering and erosion processes take place (e.g. overground flow). During the degradation the actual erosion surface, regardless of whether it is 'unstabilized' or 'stabilized', could contain pyroclasts with contrasting granulometric and textural characteristics (e.g. Figure 13B). For instance, the rates of weathering, weathering product transport and soil formation could be different at the base of a volcanic cone than at the crater rim, due to the differences in flank morphology, aspect or microclimate. These differences are demonstrated for various sectors of a cone-type volcano by the variation in microclimatic setting, e.g. insolation, freeze-thaw cycles or snow cover [437]. If there are a couple of meter difference in sediment accumulation/loss, chemical weathering and soil formations, it could cause a variation of a few degrees in the slope angle values. In extreme cases, these differences could cause misinterpretation of the morphometric parameters, thus these should be taken into account or stated as an assumption of the interpretation. The increase of post-eruptive 'noise' of the morphometric parameters will possibly increase over the erosion history, and possibly the largest in the late stage degradation of the edifice (e.g. Ma after its formation). In the case of older scoria cones, the architectural control could increase, as the well-compacted and welded units are exposed, leading to rock-selective erosion styles and longer preservation potential for a volcanic landform. This is found to be important to the good preservation of the Pliocene (2.5–3.8 Ma) scoria cones such as Agár-tető or Bondoró at the Bakony-Balaton Highland in Hungary [68]. For instance, these scoria cones are old but they resemble considerably younger cone morphologies, due to their higher morphometric values (e.g. height about 40–80 m, slope angle of 10–15°). These parameters could be similar to the degradation signatures of a much younger cone, e.g. Early Pleistocene cones (slope angle of 13±3.8°) from Springerville volcanic field, Arizona [324].

Many lines of evidence suggest that the neglected internal architecture, initial variability in geomorphic state or effect of 'event' degradation processes play an important role on edifice degradation rates and trends. Once the degradation histories for various edifices are characterized by

1. different 'input' morphometric conditions (e.g. Figure 14B), and
2. large variability of rates and trends in mass wasting processes in accordance with the susceptibility of chemical weathering of the underlying volcanic rocks and the total capacity of sediment transport, it is possible that the same geomorphic state can be reached not only by ageing of the edifice, but via a combination of other processes.

This further implies that the monogenetic volcanic edifice has a unique eruptive (e.g. Figure 6) and erosion history (e.g. Figure s 12–14). As a result of the eruptive diversity, the erosion history is not independent from the eruption history (i.e. the complexity of the monogenetic edifices). In this interpretation, there is a chance to have edifices showing the same 'geomorphic state' (in terms of the basic geometric parameters) reached through different 'degradation paths'. An example for this could be the case

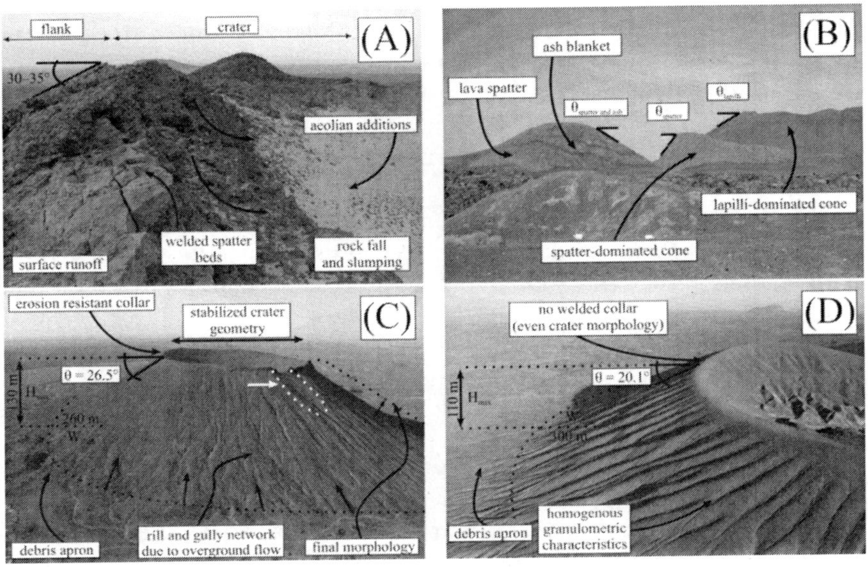

FIGURE 14. A) Difference in mode of erosion (rock fall or surface runoff) due to spatter accumulation on the crater rim of a 1–3 Ma old scoria cone in the Al Haruj Volcanic Field in Libya [**288, 466**]. (B) Variability in slope angle on the flanks of spatter-dominated and lapilli-dominated cones (1256 AD) in the last eruptions at the Harrat Al-Madinah Volcanic Field, Saudi Arabia [**467**]. Due to the young ages, these differences could be the results of differences in syn-eruptive processes (e.g. fragmentation mechanism, degree of welding and granulometric properties). These different 'input' geomorphic states alone can also lead to the large variability of degradation paths of monogenetic volcanic landforms. (C and D) Architecturally-controlled erosion pattern on Pleistocene scoria cones in the Harrat Al-Madinah Volcanic Field, Saudi Arabia. The ages are between 1.2 and 0.9 Ma for the cone in Figure C, and only a couple of ka for the cone in Figure D [**467**]. The geomorphic contrast between the edifices is striking in the slope angles, θ, calculated as $\theta = \arctan(H_{max}/W_{flank})$ from basic morphometric data. The erosion resistant collar on the crater rim changes the erosion patterns by keeping the crater rim at the same level over even Ma. This results in the 'undermining' of the flanks (small black arrows at the foot of the cones represented by a dashed line in Figure D) leading to a gradual increase of the slope angles in contrast to all previously proposed erosion models for cone-type monogenetic volcanoes. The white arrow near the rim (Figure C) indicates the significant surface modification by event degradation (e.g. mass wasting of the erosion-resistant spatter collar). It is speculative, but the consequence of this irreversible and possibly 'random' event may have initialized the formation of a deeper gully (white dashed lines) leading to crater breaching over a longer time-scale.

of the two scoria cones in Figure s 14C and D. In Figure 14C, the geometry of the edifice is strongly attached to the erosion-resistance and to the position of the spatter-dominated collar along the crater rim. In this eruptive history and subsequent erosion, the slope angle can be increased due to the undermining of the flanks. Consequently, the morphology of the cone is becoming 'younger' over time, that is the slope angle or H_{co}/W_{co} ratio will increase rather than decrease. On the other hand, a classical-looking cone (Figure 14D) that has a homogenous inner architecture, experiences different rates and degrees of erosion over different time scales. Therefore, both cones degrade through different patterns and rates. To confidently say that the decreasing trend in morphometric parameters is associated with age, it is important to reconstruct the likely environment where the edifice degradation has taken place, including the number of 'event' and major changes in the degradation controls. This includes understanding the combination and diversity of facies architecture [68, 468], the stratigraphic position of the edifice within the stratigraphic record of the volcanic field [412, 469], the approximate likelihood of aggradation by, for example, tephra mantling [411], and spatial and temporal combination and fluctuation of 'normal' and 'event' degradation processes over the erosion history.

6. CONCLUSIONS: TOWARDS UNDERSTANDING THE COMPLEXITY OF MONOGENETIC VOLCANOES

A typical monogenetic volcanic event begins at the magma source region, usually in the mantle, and ends when the volcanics have been fully removed by, for example, erosion processes. Within this conceptualized life cycle of a monogenetic volcano, there is an active stage (e.g. propagation of the magma towards the surface feeding a monogenetic eruption; Figure 15) and a passive stage (e.g. post-eruptive degradation until the feeder system is exposed; Figure 15). The active stage of evolution is dependent on many interactions between internally- or externally-driven factors. The magma (left hand side on Figure 15) intrudes into shallow parts of the crust that can be fragmented in accordance with the actual balance between the magmatic and external conditions at the time of the fragmentations. This could result in 6 varieties of volcanic eruptions if the composition is dominantly basaltic, which is responsible for the construction of a monogenetic volcanic edifice with a simple eruption history (E_{simple}, that is 6^1 combinations of eruption styles). Once there is some disequilibrium in the system during the course of the eruption that will result in changing eruption styles adjusting the balance in the system, forming compound

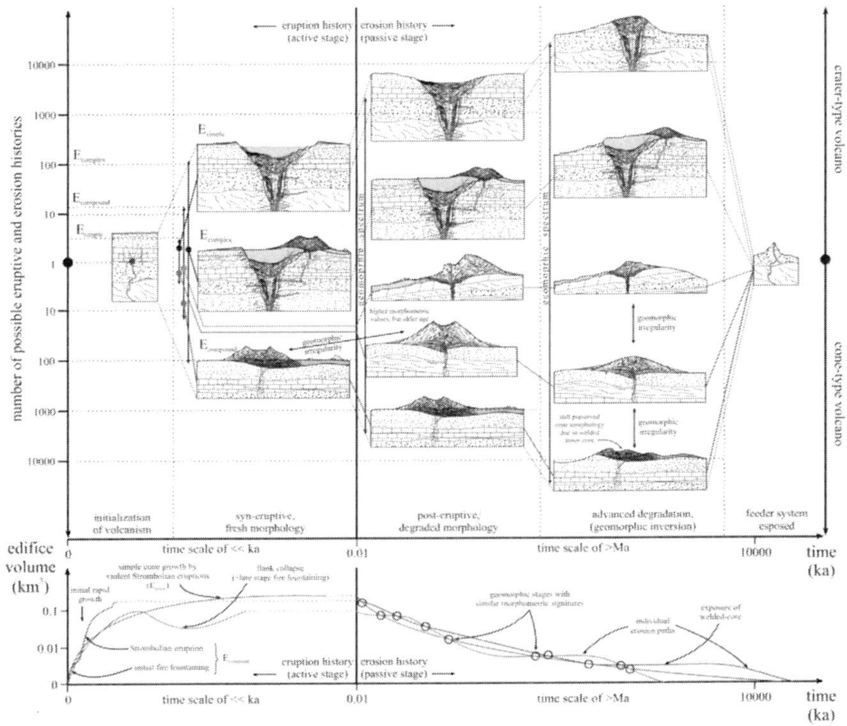

FIGURE 15. Conceptualized model to understand causes and consequences in the life cycle of crater- and cone-type monogenetic volcanoes. See the text for a detailed explanation. Note that the colour coding in the bottom graph corresponds the conceptualized degradation paths for the three cone-type volcanoes in the top graph. The double-headed arrows (top graph) and black circles (bottom graph) show the time slices when geomorphic states of the edifices are similar, expressed by the morphometric parameters.

eruption histories ($E_{compound}$ that is 6^2 combinations of eruption styles). Each number of shifts in dominant eruption style opens a new phase of edifice growth and therefore increases the complexity of the eruption history towards $E_{complex}$ (that is 6^3 or more combinations of eruption styles). There could be even thousands of theoretical combinations of eruption styles if the volcano is built up by more than 4 phases with different eruption styles, until the magma supply is completely exhausted or new vent is established by migration of the magma focus. With increasing complexity of the eruption history, the complexity of the facies architecture of the volcanic edifice increases. Conceptually, these eruption histories can be numerically described by matrices, based on spatial and temporal characteristics of eruption styles (e.g. Figure 6). The coding of eruption styles could be

1. Hawaiian,
2. Strombolian,
3. violent Strombolian,
4. phreatomagmatic,
5. Surtseyan and
6. effusive activity

, if the erupting melt is characterised by basaltic to basaltic andesitic in composition. This systems can be modified by adding further eruption styles such as sub-Plinian. The syn-eruptive geomorphology of a volcano is, however, not only the result of the eruption style and associated pyroclast transport mechanism, but there are stages of destructive processes, such as flank collapse during scoria cone growth (e.g. Figure 3) or wall rock mass wasting during excavation of a maar crater (e.g. Figure 4). These common syn-eruptive processes (constructive and destructive phases during the eruption history) have an important role on the resulting morphology, but they are not always visible/detectable in the morphology of the edifice.

On the other hand, after the eruption ceases a passive stage of surface modification takes place (right hand side on Figure 15). In the passive stage, the erosion history is also governed significantly by a series of interactions between the exposed pyroclastic deposits and lava rocks (and their textural and granulometric characteristics determining permeability) and external influences such as climate, location or hydrology of the area. These interactions determine the long-term (ka to Ma) degradation processes and rates. However, it is important to note that the erosion history is often a function of 'normal' and 'event' degradation. The effect of event degradation is expected to be larger, in some cases, than the cumulative surface modification by normal degradation processes. The relationship between event and normal degradation should be a subject for future studies. Due to the large number of combinations of eruption styles that can generate edifices with different pyroclastic successions and different initial geometries (at least a broader range than previously thought), volcanoes can have very different susceptibilities for erosion. This implies that degradation trends and patterns of monogenetic volcanoes should be individual volcano-specific (right hand side on Figure 15). In addition, the combination of erosion path of individual monogenetic volcanoes is an order of magnitude larger than during the eruption history due to the larger number of controlling factors (6 eruption styles versus varieties of 'normal' and 'event' mass wasting processes) and the longer time-scale of degradation (<<ka versus >Ma). This has an important practical conclusion: there are certain stages during the degradation when some morphometric irregularity occurs if two or more volcanic edifices are compared. The morphometric irregularity refers to the state when two volcanoes appear similar through morphometric parameters such as H_{co}/W_{co} ratio or slope angle, but they have different absolute ages (black double-headed arrows of the top graph and black circles on the bottom graph in Figure 15). An important practical application of the volcano-specific degradation is that the correlation between the morphology of the edifice is not always a function of the time elapsed since formation of the volcanic edifice. As a consequence of the diverse active and passive evolution of a volcanic edifice, age grouping based on geomorphic parameters, such as H_{co}/W_{co} ratio or average or maximum cone slope angle, should be avoided. In terms of interpretation of the morphometric data, the post-eruptive surface modification causes unfortunate 'noise' in the primary morphometric signatures, which can be only reduced by using edifices with absolute age constraints. Due to the long-lived evolution of monogenetic volcanic fields (Ma-scale), there are usually volcanoes that are freshly formed, sometimes close to volcanoes with no primary morphological features at the time of the examination. The large contrasting and dynamic geological environment of such monogenetic volcanoes makes the interpretation of available topographic information more complicated than previously thought. Future studies should target this particular issue and define the meaning of morphology of these monogenetic volcanic edifices at many scales.

7. ACKNOWLEDGEMENTS

Constructive discussion about various aspects of monogenetic volcanism and remote sensing with Shane Cronin, Jon Procter, Jan Lindsay, Javier Agustín-Flores, Marco Brenna, Natalia Pardo, Nicola Le Corvec, Mark Bebbington and Mike Tuohy is acknowledged. Remarks and suggestion by Bob Stewart, Adrian Pittari, Marco Brenna, Kate Arentsen and Javier Agustín-Flores and an anonymous reviewer are gratefully appreciated. This study was supported by the NZ MSI-IIOF Grant "Facing the challenge of Auckland's Volcanism" (MAUX0808). The authors would like to thank the Auckland City Council for the LiDAR dataset and the Department of Conservation (Te Papa Atawhai) for the logistical help during field trips.

REFERENCES

1. E. Canon-Tapia, G. P. L. Walker, 2004 Global aspects of volcanism: the perspectives of "plate tectonics" and "volcanic systems Earth-Science Reviews 66 163 82
2. W. J. Morgan, 1971 Convection Plumes in the Lower Mantle. Nature 230 42 3
3. J. Tarduno, H. Bunge-P, N. Sleep, U. Hansen, 2009 The Bent Hawaiian-Emperor Hotspot Track: Inheriting the Mantle Wind Science 324 50 3
4. D. J. De Paolo, M. Manga, 2003 Deep Origin of Hotspots- The Mantle Plume Model. Science 300 920 1
5. K. Németh, 2010 Monogenetic volcanic fields: Origin, sedimentary record, and relationship with polygenetic volcanism The Geological Society of America Special Paper 470 43 66
6. CP Conrad, T. A. Bianco, E. I. Smith, P. Wessel, 2011 Patterns of intraplate volcanism controlled by asthenospheric shear Nat Geosci 4 317 21
7. M. S. Bebbington, S. J. Cronin, 2011 Spatio-temporal hazard estimation in the Auckland Volcanic Field, New Zealand, with a new event-order model. Bull Volcanol 73 1 55 72
8. G. N. Keating, G. A. Valentine, D. J. Krier, F. V. Perry, 2008 Shallow plumbing systems for small-volume basaltic volcanoes. Bull Volcanol 70 563 82
9. M. Brenna, S. J. Cronin, K. Németh, I. E. M. Smith, Y. K. Sohn, 2011 The influence of magma plumbing complexity on monogenetic eruptions, Jeju Island, Korea Terra Nova 23 70 5
10. C. B. Connor, J. A. Stamatakos, D. A. Ferrill, B. E. Hill, G. Ofoegbu, F. M. Conway, et al. 2000 Geologic factors controlling patterns of small-volume basaltic volcanism: Application to a volcanic hazards assessment at Yucca Mountain, Nevada. J Geophys Res 105 1 417 32
11. I. E. M. Smith, S. Blake, C. J. N. Wilson, B. F. Houghton, 2008 Deep-seated fractionation during the rise of a small-volume basalt magma batch: Crater Hill, Auckland, New Zealand Contrib Mineral Petrol 155 4 511 27
12. G. Gencalioglu-Kuscu, 2011 Geochemical characterization of a Quaternary monogenetic volcano in Erciyes Volcanic Complex: Cora Maar (Central Anatolian Volcanic Province, Turkey). Int J Earth Sci 100 8 1967 85
13. B. Browne, M. Bursik, J. Deming, M. Louros, A. Martos, S. Stine, 2010 Eruption chronology and petrologic reconstruction of the ca. 8500 yr B.P. eruption of Red Cones, southern Inyo chain, California Geol Soc Am Bull 122 9-10 1401 1422
14. J. Agustín-Flores, C. Siebe, M. Guilbaud-N, 2011 Geology and geochemistry of Pelagatos, Cerro del Agua, and Dos Cerros monogenetic volcanoes in the Sierra Chichinautzin Volcanic Field, south of México City J Volcanol Geotherm Res 201 1-4 143 162
15. M. Strong, J. Wolff, 2003 Compositional variations within scoria cones Geology 31 2 143 6
16. K. Németh, J. D. L. White, A. Reay, U. Martin, 2003 Compositional variation during monogenetic volcano growth and its implication for magma supply to continantal volcanic fields. J Geol Soc London 160 523 30
17. M. Brenna, S. J. Cronin, I. E. M. Smith, Y. K. Sohn, K. Németh, 2010 Mechanisms driving polymagmatic activity at a monogenetic volcano, Udo, Jeju Island, South Korea Contrib Mineral Petrol 160 931 50
18. J. M. Cebriá, BM Martiny, J. López -Ruiz, D. J. Morán-Zenteno, 2011 The Parícutin calc-alkaline lavas: New geochemical and petrogenetic modelling constraints on the crustal assimilation process J Volcanol Geotherm Res 201 1-4 113 125
19. C. S. J. Shaw, 2004 The temporal evolution of three magmatic systems in the West Eifel volcanic field, Germany. J Volcanol Geotherm Res 131 213 40

20. T. Hasenaka, M. Ban, H. D. Granados, 1994 Contrasting volcanism in the Michoacán-Guanajuato Volcanic Field, central Mexico: Shield volcanoes vs. cinder cones. Geofisica Internacional 33 1 125 38

21. C. B. Connor, F. M. Conway, 2000 Basaltic volcanic fields. Sigurdsson H, Houghton BF, McNutt SR, Rymer H, Stix J, editors. Encyclopedia of Volcanoes. San Diego: Academic Press 331 343

22. K. Hoernle, J. D. L. White, P. van den Bogaard, F. Hauff, DS Coombs, R. Werner, et al. 2006 Cenozoic intraplate volcanism on New Zealand: Upwelling induced by lithospheric removal. Earth Planet Sci Lett 248 350 67

23. K. Németh, S. J. Cronin, MJ Haller, M. Brenna, G. Csillag, 2010 Modern analogues for Miocene to Pleistocene alkali basaltic phreatomagmatic fields in the Pannonian Basin: "soft-substrate" to "combined" aquifer controlled phreatomagmatism in intraplate volcanic fields. Cent Eur J Geosci 2 3 339 61

24. C. D. Condit, C. B. Connor, 1996 Recurrence rates of volcanism in basaltic volcanic fields: An example from the Springerville volcanic field, Arizona. Geol Soc Am Bull 108 1225 41

25. P. V. Kshirsagar, H. C. Sheth, B. Shaikh, 2011 Mafic alkalic magmatism in central Kachchh, India: a monogenetic volcanic field in the northwestern Deccan Traps Bull Volcanol 73 5 595 612

26. H. C. Sheth, G. Mathew, K. Pande, S. Mallick, B. Jena, 2004 Cones and craters on Mount Pavagadh, Deccan Traps: Rootless cones? Earth and Planetary Sciences (Proceedings of the Indian Academy of Sciences) 113 4 831 8

27. A. Márquez, S. P. Verma, F. Anguita, R. Oyarzun, J. L. Brandle, 1999 Tectonics and volcanism of Sierra Chichinautzin: extension at the front of the Central Trans-Mexican Volcanic belt J Volcanol Geotherm Res 93 1-2 125 150

28. A. Germa, X. Quidelleur, P. Y. Gillot, P. Tchilinguirian, 2010 Volcanic evolution of the back-arc Pleistocene Payun Matru volcanic field (Argentina) J S Am Earth Sci 29 3 717 30

29. S. M. Baloga, L. S. Glaze, B. C. Bruno, 2007 Nearest-neighbor analysis of small features on Mars: applications to tumuli and rootless cones. J Geophys Res 112 E03002 10.1029/2005JE002652

30. M. A. Bishop, 2007 Higher-order neighbor analysis of the Tartarus Colles cone groups, Mars: The application of geographical indices to the understanding of cone pattern evolution Icarus 197 1 73 83

31. P. Broz, E. Hauber, 2012 A unique volcanic field in Tharsis, Mars: Pyroclastic cones as evidence for explosive eruptions Icarus 218 88 99

32. L. P. Keszthelyi, W. L. Jaeger, C. M. Dundas, S. Martínez-Alonso, AS Mc Ewen, M. P. Milazzo, 2010 Hydrovolcanic features on Mars: Preliminary observations from the first Mars year of HiRISE imaging Icarus 205 211 29

33. L. Wilson, J. W. Head, 1994 Mars: Review and analysis of volcanic eruption theory and relationships to observed landforms. Rev Geophys 32 3 221 63

34. L. Sandri, G. Jolly, J. Lindsay, T. Howe, W. Marzocchi, 2011 Combining long- and short-term probabilistic volcanic hazard assessment with cost-benefit analysis to support decision making in a volcanic crisis from the Auckland Volcanic Field, New Zealand Bull Volcanol 74 3 705 23

35. J. Lindsay, W. Marzocchi, G. Jolly, R. Constantinescu, J. Selva, L. Sandri, 2010 Towards real-time eruption forecasting in the Auckland Volcanic Field: application of BET_EF during the New Zealand National Disaster Exercise 'Ruaumoko' Bull Volcanol. 72 2 185 204

36. C. Siebe, V. Rodriguez-Lara, P. Schaaf, M. Abrams, 2004 Radiocarbon ages of Holocene Pelado, Guespalapa, and Chichinautzin scoria cones, south of Mexico City: implications for archaeology and future hazards Bull Volcanol 66 203 25

37. H. Delgado, R. Molinero, P. Cervantes, J. Nieto-Obregón, Cruz. R. Lozaro-Santa, H. L. Macías-González, et al. 1998 Geology of Xitle volcano in southern Mexico City-a 2000- year-old monogenetic volcano in an urban area. Rev Mex Cienc Geol 15 2 115 31

38. G. A. Valentine, N. Hirano, 2010 Mechanisms of low-flux intraplate volcanic fields-Basin and Range (North America) and northwest Pacific Ocean Geology 38 1 55 8
39. G. Kereszturi, K. Németh, G. Csillag, K. Balogh, J. Kovács, 2011 The role of external environmental factors in changing eruption styles of monogenetic volcanoes in a Mio/Pleistocene continental volcanic field in western Hungary J Volcanol Geotherm Res 201 1-4 227 240
40. Brenna, M., Cronin, S. J., Smith, I. E. M., & Maas, R., Sohn, Y. K. (2012). How Small-volume Basaltic Magmatic Systems Develop: a Case Study from the Jeju Island Volcanic Field, Korea. *J Petrol*, 53(5), 985-1018.
41. K. Németh, U. Martin, S. Harangi, 2001 Miocene phreatomagmatic volcanism at Tihany (Pannonian Basin, Hungary) J Volcanol Geotherm Res 111 1-4 111 35
42. G. A. Valentine, F. V. Perry, 2007 Tectonically controlled, time-predictable basaltic volcanism from a lithospheric mantle source (central Basin and Range Province, USA). Earth Planet Sci Lett 261 1-2 201 16
43. A. Geyer, J. Martí, 2010 The distribution of basaltic volcanism on Tenerife, Canary Islands: Implications on the origin and dynamics of the rift systems Tectonophysics 483 3-4 310 326
44. K. Németh, 2012 An Overview of the Monogenetic Volcanic Fields of the Western Pannonian Basin: Their Field Characteristics and Outlook for Future Research from a Global Perspective. Stoppa F, editors. Updates in Volcanology- A Comprehensive Approach to Volcanological Problems. In-Tech 27 52 http://www. intechopen. com/books/updates-in-volcanology-a-comprehensive-approach-to-volcanological-problems/an-overview-of-the-monogenetic-volcanic-fields-of-the-western-pannonian-basin-their-field-characteri
45. M. É. Jankovics, S. Harangi, B. Kiss, T. Ntaflos, 2012 Open-system evolution of the Füzes-tó alkaline basaltic magma, western Pannonian Basin: Constraints from mineral textures and compositions Lithos 140-141 25 37
46. E. Bali, A. Zanetti, C. Szabó, D. W. Peate, T. E. Waight, 2008 A micro-scale investigation of melt production and extraction in the upper mantle based on silicate melt pockets in ultramafic xenoliths from the Bakony-Balaton Highland Volcanic Field (Western Hungary) Contrib Mineral Petrol 155 165 79
47. I. Kovács, G. Falus, G Stuart, K. Hidas, C. Szabó, M. F. J. Flower, et al. 2012 Seismic anisotropy and deformation patterns in upper mantle xenoliths from the central Carpathian-Pannonian region: Asthenospheric flow as a driving force for Cenozoic extension and extrusion? Tectonophysics 514-517 168 79
48. K. Yang, K. Hidas, G. Falus, C. Szabó, B. Nam, I. Kovács, et al. 2010 Relation between mantle shear zone deformation and metasomatism in spinel peridotite xenoliths of Jeju Island (South Korea): Evidence from olivine CPO and trace elements J Geodyn 50 5 424 40
49. K. Hidas, G. Falus, C. Szabó, P. J. Szabó, I. Kovács, T. Földes, 2007 Geodynamic implications of flattened tabular equigranular textured peridotites from the Bakony-Balaton Highland Volcanic Field (Western Hungary) J Geodyn 43 484 503
50. A. M. Rubin, 1995 Propagation of Magma-Filled Cracks. Annual Review Of Earth And Planetary Sciences 23 287 336
51. K. Németh, S. J. Cronin, 2011 Drivers of explosivity and elevated hazard in basaltic fissure eruptions: The 1913 eruption of Ambrym Volcano, Vanuatu (SW-Pacific). J Volcanol Geotherm Res 201 1-4 194 209
52. G. Kereszturi, K. Németh, 2011 Shallow-seated controls on the evolution of the Upper Pliocene Kopasz-hegy nested monogenetic volcanic chain in the Western Pannonian Basin, Hungary. Geol Carpath 62 6 535 46
53. E. S. Gaffney, B. Damjanac, G. A. Valentine, 2007 Localization of volcanic activity: 2. Effects of pre-existing structure Earth Planet Sci Lett 263 323 38
54. G. A. Valentine, K. E. C. Krogh, 2006 Emplacement of shallow dikes and sills beneath a small basaltic volcanic center- The role of pre-existing structure (Paiute Ridge, southern Nevada, USA). Earth Planet Sci Lett 246 3-4 217 30

55. Kiyosugi, K., Connor, C. B., Wetmore, P. H., Ferwerda, B. P., Germa, A. M., Connor, L. J., et al. (2012). Relationship between dike and volcanic conduit distribution in a highly eroded monogenetic volcanic field: San Rafael, Utah, USA. *Geology*, 40(8), 695-698.
56. N. S. Lefebvre, J. D. L. White, B. A. Kjarsgaard, 2012 Spatter-dike reveals subterranean magma diversions: Consequences for small multivent basaltic eruptions. Geology 40 5 423 6
57. A. R. Hintz, G. A. Valentine, 2012 Complex plumbing of monogenetic scoria cones: New insights from the Lunar Crater Volcanic Field (Nevada, USA). J Volcanol Geotherm Res 239-240 0 19 32
58. R. Sulpizio, R. De Rosa, P. Donato, 2008 The influence of variable topography on the depositional behaviour of pyroclastic density currents: The examples of the Upper Pollara eruption (Salina Island, southern Italy) J Volcanol Geotherm Res 175 3 367 85
59. G. Kereszturi, G. Csillag, K. Németh, K. Sebe, K. Balogh, V. Jáger, 2010 Volcanic architecture, eruption mechanism and landform evolution of a Pliocene intracontinental basaltic polycyclic monogenetic volcano from the Bakony-Balaton Highland Volcanic Field, Hungary. Cent Eur J Geosci 2 3 362 84
60. G. A. Valentine, D. Krier, F. V. Perry, G. Heiken, 2005 Scoria cone construction mechanisms, Lathrop Wells volcano, southern Nevada, USA Geology 33 8 629 32
61. C. Risso, K. Németh, A. M. Combina, F. Nullo, M. Drosina, 2008 The role of phreatomagmatism in a Plio-Pleisotcene high-density scoria cone field: Llancanelo Volcanic Field, Argentina. J Volcanol Geotherm Res 168 61 86
62. C. Magill, R. Blong, 2005 Volcanic risk ranking for Auckland, New Zealand. II: Hazard consequences and risk calculation Bull Volcanol 67 4 340 9
63. L. J. Connor, C. B. Connor, K. Meliksetian, I. Savov, 2012 Probabilistic approach to modeling lava flow inundation: a lava flow hazard assessment for a nuclear facility in Armenia. Journal of Applied Volcanology 1 3 http://www.appliedvolc.com/1//3
64. A. J. L. Harris, M. Favalli, R. Wright, H. Garbeil, 2011 Hazard assessment at Mount Etna using a hybrid lava flow inundation model and satellite-based land classification Nat Hazards 58 3 1001 27
65. K. Németh, S. J. Cronin, 2009 Phreatomagmatic volcanic hazards where rift-systems meet the sea, a study from Ambae Island, Vanuatu J Volcanol Geotherm Res 180 2-4 246 258
66. Cronin, S. J., Bebbington, M., & Lai, C. D. (2001). A probabilistic assessment of eruption recurrence on Taveuni volcano, Fiji. *Bull Volcanol*, 63, 274-89.
67. C. Siebe, L. Arana-Salinasa, M. Abrams, 2005 Geology and radiocarbon ages of Tláloc, Tlacotenco, Cuauhtzin, Hijo del Cuauhtzin, Teuhtli, and Ocusacayo monogenetic volcanoes in the central part of the Sierra Chichinautzin, México J Volcanol Geotherm Res 141 225 43
68. G. Kereszturi, K. Németh, 2012 Structural and morphometric irregularities of eroded Pliocene scoria cones at the Bakony-Balaton Highland Volcanic Field, Hungary Geomorphology 136 1 45 58
69. A. Rodriguez-Gonzalez, J. L. Fernandez-Turiel, F. J. Perez-Torrado, M. Aulinas, J. C. Carracedo, D. Gimeno, et al. 2011 GIS methods applied to the degradation of monogenetic volcanic fields: A case study of the Holocene volcanism of Gran Canaria (Canary Islands, Spain) Geomorphology 134 3-4 249 259
70. J. C. Dohrenwend, S. G. Wells, B. D. Turrin, 1986 Degradation of Quaternary cinder cones in the Cima volcanic field, Mojave Desert, California. Geol Soc Am Bull 97 4 421 7
71. C. A. Wood, 1980 Morphometric analysis of cinder cone degradation. J Volcanol Geotherm Res 8 2-4 137 60
72. K. Németh, 2004 Calculation of long-term erosion in Central Otago, New Zealand, based on erosional remnants of maar/tuff rings Z Geomorphol 47 1 29 49
73. J. Büchner, O. Tietz, 2012 Reconstruction of the Landeskrone Scoria Cone in the Lusatian Volcanic Field, Eastern Germany- Long-term degradation of volcanic edifices and implications for landscape evolution. Geomorphology 151-152 175 187

74. K. Németh, U. Martin, 1999 Late Miocene paleo-geomorphology of the Bakony-Balaton Highland Volcanic Field (Hungary) using physical volcanology data. Zeitschrift fr Geomorphologie 43 4 417 38
75. J. R. Lister, R. C. Kerr, 1991 Fluid-Mechanical Models of Crack Propagation and Their Application to Magma Transport in Dykes. J Geophys Res 96 B6 10049 77
76. C. O'Neill, M. Spiegelman, 2010 Formulations for Simulating the Multiscale Physics of Magma Ascent. Dosseto A, Turner SP, van Orman JA, editors. Timescales of Magmatic Processes: From Core to Atmosphere. Chichester, UK. : John Wiley & Sons, Ltd 87 101
77. M. Gardine, M. West, T. Cox, 2011 Dike emplacement near Parícutin volcano, Mexico in 2006. Bulletin of Volcanology 73 2 123 32
78. S. de la Cruz-Reyna, I. Yokoyama, 2011 A geophysical characterization of monogenetic volcanism Geofisica Internacional 50 4 465 84
79. K. Németh, 2010 Volcanic glass textures, shape characteristics and compositions of phreatomagmatic rock units from the Western Hungarian monogenetic volcanic fields and their implications for magma fragmentation Cent Eur J Geosci 2 3 399 419
80. B. F. Houghton, C. J. N Wilson, M. D. Rosenberg, I. E. M. Smith, R. J. Parker, 1996 Mixed deposits of complex magmatic and phreatomagmatic volcanism: an example from Crater Hill, Auckland, New Zealand. Bull Volcanol 58 59 66
81. V. Lorenz, 2007 Syn- and posteruptive hazards of maar-diatreme volcanoes J Volcanol Geotherm Res 159 285 312
82. J. T. Gutmann, 2002 Strombolian and effusive activity as precursors to phreatomagmatism: eruptive sequence at maars of the Pinacate volcanic fields, Sonora, Mexico. J Volcanol Geotherm Res 113 1-2 345 56
83. W. F. Foshag, J. R. Gonzalez, 1956 Birth and development of Paricutin volcano, Mexico Geological Survey Bulletin. 965-D 355 487
84. C. A. Wood, 1980 Morphometric evolution of cinder cones. J Volcanol Geotherm Res 7 3-4 387 413
85. F. J. Davey, 2010 Crustal seismic reflection measurements across the northern extension of the Taupo Volcanic Zone, North Island, New Zealand. J Volcanol Geotherm Res 190 1-2 75 81
86. C. J. N. Wilson, B. F. Houghton, M. O. Mc Williams, M. A. Lanphere, S. D. Weaver, R. M. Briggs, 1995 Volcanic and structural evolution of Taupo Volcanic Zone, New Zealand: a review. J Volcanol Geotherm Res 68 1-3 1 28
87. B. F. Houghton, C. J. N. Wilson, M. O. Mc Williams, MA Lanphere, S. D. Weaver, R. M. Briggs, et al. 1995 Chronology and dynamics of a large silicic magmatic system: Central Taupo Volcanic Zone, New Zealand Geology 23 13 6
88. Wallace, L. M., Beavan, J., Mc Caffrey, R., & Darby, D. (2004). Subduction zone coupling and tectonic block rotations in the North Island, New Zealand. *J Geophys Res*, 109(B12), B12406. 10.1029/2004jb003241.
89. I. Seghedi, H. Downes, 2011 Geochemistry and tectonic development of Cenozoic magmatism in the Carpathian-Pannonian region Gondwana Res 20 655 72
90. J. Lexa, I. Seghedi, K. Németh, A. Szakács, V. Konečný, Z. Pécskay, et al. 2010 Neogene-Quaternary Volcanic forms in the Carpathian-Pannonian Region: a review. Cent Eur J Geosci 2 3 207 70
91. Z. Pécskay, J. Lexa, A. Szakács, I. Seghedi, K. Balogh, V. Konecny, et al. 2006 Geochronlogy of Neogene magmatism in the Carpathian arc and intra-Carpathian area. Geol Carpath 57 6 511 30
92. S. Harangi, L. Lenkey, 2007 Genesis of the Neogene to Quaternary volcanism in the Carpathian-Pannonian region: Role of subduction, extension, and mantle plume Beccaluva L, Bianchini G, Wilson M, editors. Cenozoic Volcanism in the Mediterranean Area. London: Geological Society of America 67 92
93. C. Szabó, G. Falus, Z. Zajacz, I. Kovács, E. Bali, 2004 Composition and evolution of lithosphere beneath the Carpathian-Pannonian region: a review. Tectonophysics 393 1-4 119 37

94. R. D. Müller, M. Sdrolias, C. Gaina, W. R. Roest, 2008 Age, spreading rates and spreading symmetry of the world's ocean crust. Geochem Geophys Geosyst 9 Q04006
95. R. F. Katz, M. Spiegelman, B. Holtzman, 2006 The dynamics of melt and shear localization in partially molten aggregates. Nature 442 676 9
96. V. Courtillot, A. Davaille, J. Besse, J. Stock, 2006 Three distinct types of hotspots in the Earth's mantle Earth Planet Sci Lett 205 295 308
97. S. D. King, D. L. Anderson, 1998 Edge-driven convection. Earth Planet Sci Lett 160 289 96
98. MD Ballmer, G. Ito, J. van Hunen, P. J. Tackley, 2011 Spatial and temporal variability in Hawaiian hotspot volcanism induced by small-scale convection Nat Geosci 4 457 60
99. M. D. Ballmer, J. van Hunen, G. Ito, P. J. Tackley, T. A. Bianco, 2007 Non-hotspot volcano chains originating from small-scale sublithospheric convection. Geophysical Rerearch Letters 43 L23310 10.1029/2007GL031636
100. S. D. King, J. Ritsema, 2000 African Hot Spot Volcanism: Small-Scale Convection in the Upper Mantle Beneath Cratons. Science 290 1137 40
101. G. R. Foulger, J. H. Natland, 2003 Is "Hotspot" Volcanism a Consequence of Plate Tectonics? Science 300 921 2
102. D. S. King, M. E. Zimmerman, D. L. Kohlstedt, 2010 Stress-driven Melt Segregation in Partially Molten Olivine-rich Rocks Deformed in Torsion. J Petrol 51 1-2 21 42
103. B. K. Holtzman, N. J. Groebner, M. E. Zimmerman, S. B. Ginsberg, D. L. Kohlstedt, 2003 Stress-driven melt segregation in partially molten rocks. Geochem Geophys Geosyst 4 5 8607
104. D. Mc Kenzie, 1984 The Generation and Compaction of Partially Molten Rock. J Petrol 25 3 713 65
105. Laporte, D., & Watson, E. B. (1995). Experimental and theoretical constraints on melt distribution in crustal sources: the effect of crystalline anisotropy on melt interconnectivity. *Chem Geol*, 124(3-4), 161-184.
106. P. B. Kelemen, G. Hirth, N. Shimizu, M. Spiegelman, H. J. Dick, 1997 A review of melt migration processes in the adiabatically upwelling mantle beneath oceanic spreading ridges. Philosophical Transactions of the Royal Society A: Mathematical Physical & Engineering Sciences 355 1723 283 318
107. N. H. Sleep, 1988 Tapping of Melt by Veins and Dikes. J Geophys Res 93 B9 10255 72
108. G. A. Valentine, F. V. Perry, 2006 Decreasing magmatic footprints of individual volcanoes in a waning basaltic field. Geophys Res Lett 33 L14305 10.1029/2006GL026743.
109. Takada, A. (1994). The influence of regional stress and magmatic input on styles of monogenetic and polygenetic volcanism. *J Geophys Res*, 99(B7), 10.1029/94JB00494.
110. G. Ito, S. J. Martel, 2002 Focusing of magma in the upper mantle through dike interaction. J Geophys Res 107 B10 10.1029/2001JB000251.
111. G. A. Valentine, T. K. P. Gregg, 2008 Continental basaltic volcanoes- Processes and problems. J Volcanol Geotherm Res 177 4 857 73
112. T. Dahm, 2000 On the shape and velocity of fluid-filled fractures in the Earth Geophys J Int 142 1 181 92
113. P. T. Delaney, D. D. Pollard, J. I. Ziony, E. H. Mc Kee, 1986 Field Relations Between Dikes and Joints: Emplacement Processes and Paleostress Analysis. J Geophys Res 91 B5 4920 38
114. F. Mazzarini, M. D'Orazio, 2003 Spatial distribution of cones and satellite-detected lineaments in the Pali Aike Volcanic Field (southernmost Patagonia): insights into the tectonic settings of a Neogene rift system J Volcanol Geotherm Res 125 291 305
115. L. E. Lara, A. Lavenu, J. Cembrano, C. Rodríguez, 2006 Structural controls of volcanism in transversal chains: Resheared faults and neotectonics in the Cordón Caulle-Puyehue area (40.5°S), Southern Andes. J Vol canol Geotherm Res 158 1-2 70

116. G. P. L. Walker, 2000 Basaltic volcanoes and volcanic systems. Sigurdsson H, Houghton BF, McNutt SR, Rymer H, Stix J, editors. Encyclopedia of Volcanoes. San Diego: Academic Press 283 290

117. A. V. Zernack, J. N. Procter, S. J. Cronin, 2009 Sedimentary signatures of cyclic growth and destruction of stratovolcanoes: A case study from Mt. Taranaki, New Zealand. Sediment Geol 220 288 305

118. J. C. Carracedo, Badiola. E. Rodríguez, H. Guillou, M. Paterne, S. Scaillet, F. J. Pérez Torrado, et al. 2007 Eruptive and structural history of Teide Volcano and rift zones of Tenerife, Canary Islands Geol Soc Am Bull 119 9 1027 51

119. P. Grosse, B. van Wyk de Vries, I. A. Petrinovic, P. A. Euillades, G. E. Alvarado, 2009 Morphometry and evolution of arc volcanoes. Geology 37 7 651 4

120. R. W. Buck, P. Einarsson B. Brandsdóttir, 2006 Tectonic stress and magma chamber size as controls on dike propagation: Constraints from the 1975-1984 Krafla rifting episode. J Geophys Res 111 B12404 10.1029/2005JB003879

121. B. D. Marsh, 1989 Magma Chambers. Annual Review of Earth and Planetary Sciences 17 439 74

122. E. Burov, C. Jaupart, L. Guillou-Frottier, 2003 Ascent and emplacement of buoyant magma bodies in brittle-ductile upper crust. J Geophys Res 108 2177 10.1029/2002JB001904

123. C. J. Hawkesworth, S. Blake, P. Evans, R. Hughes, R. Macdonald, L. E. Thomas, et al. 2000 Time scales of crystal fractionation in magma chambers- Integrating physical, isotopic and geochemical perspectives. J Petrol 41 7 991 1006

124. S. A. Alaniz-Alvarez, A. F. Nieto-Samaniego, L. Ferrari, 1998 Effect of strain rate in the distribution of monogenetic and polygenetic volcanism in the Transmexican volcanic belt Geology 26 7 591 4

125. S. Fedotov, 1981 Magma rates in feeding conduits of different volcanic centres J Volcanol Geotherm Res 9 4 379 94

126. M. Suter, J. Contreras, A. Gómez-Tuena, C. Siebe, O. Quintero-Legorreta, A. García-Palomo, et al. 1999 Effect of strain rate in the distribution of monogenetic and polygenetic volcanism in the Transmexican volcanic belt: Comments and Reply Geology 27 6 571 5

127. H. Mattsson, Á. Höskuldsson, 2003 Geology of the Heimaey volcanic centre, south Iceland: early evolution of a central volcano in a propagating rift? J Volcanol Geotherm Res 127 1-2 55

128. L. E. Mc Gee, C. Beier, I E. M. Smith, S. P. Turner, 2011 Dynamics of melting beneath a small-scale basaltic system: a U-Th-Ra study from Rangitoto volcano, Auckland volcanic field, New Zealand Contrib Mineral Petrol 162 3 547 63

129. K. Kiyosugi, C. B. Connor, D. Zhao, L. J. Connor, K. Tanaka, 2010 Relationships between volcano distribution, crustal structure, and P-wave tomography: an example from the Abu Monogenetic Volcano Group, SW Japan Bull Volcanol 72 3 331 40

130. V. Toprak, 1998 Vent distribution and its relation to regional tectonics, Cappadocian Volcanics, Turkey J Volcanol Geotherm Res 85 1-4 55

131. C. R. Bacon, 1982 Time-predictable bimodal volcanism in the Coso Range, California. Geology 10 2 65 9

132. H. M. Frey, R. A. Lange, C. M. Hall, H. Delgado-Granados, 2004 Magma eruption rates constrained by 40Ar/39Ar chronology and GIS for the Ceboruco-San Pedro volcanic field, western Mexico Geol Soc Am Bull 116 3-4 259

133. K. Nakamura, 1977 Volcanoes as possible indicators of tectonic stress orientation-principal and proposal. J Volcanol Geotherm Res 2 1 1 16

134. M. A. Kuntz, D. E. Champion, E. C. Spiker, R. H. Lefebvre, 1986 Contrasting magma types and steady-state, volume-predictable, basaltic volcanism along the Great Rift, Idaho. Geol Soc Am Bull 97 5 579 94

135. C. King-Y, 1989 Volume predictability of historical eruptions at Kilauea and Mauna Loa volcanoes. J Volcanol Geotherm Res 38 281 5

136. T. Parsons, G. A. Thompson, 1991 The role of magma overpressure in suppressing earthquakes and topography: worldwide examples. Science 253 5026 1399 402
137. C. J. Ebinger, D. Keir, A. Ayele, E. Calais, T. J. Wright, M. Belachew, et al. 2008 Capturing magma intrusion and faulting processes during continental rupture: seismicity of the Dabbahu (Afar) rift Geophys J Int 174 3 1138 52
138. A. Gudmundsson, I. F. Loetveit, 2005 Dyke emplacement in a layered and faulted rift zone J Volcanol Geotherm Res 144 1-4 311
139. M. Bebbington, 2008 Incorporating the eruptive history in a stochastic model for volcanic eruptions J Volcanol Geotherm Res 175 325 33
140. J. J. Aranda-Gómez, J. F. Luhr, T. B. Housh, C. B. Connor, T. Becker, C. D. Henry, 2003 Synextensional Pliocene-Pleistocene eruptive activity in the Camargo volcanic field, Chihuahua, México Geol Soc Am Bull 115 3 298 313
141. M. Guilbaud-N, C. Siebe, P. Layer, S. Salinas, R. Castro-Govea, V. H. Garduño-Monroy, et al. 2011 Geology, geochronology, and tectonic setting of the Jorullo Volcano region, Michoacán, México. J Volcanol Geotherm Res 201 1-4 97 112
142. J. Wijbrans, K. Németh, U. Martin, K. Balogh, 2007 40Ar/39Ar geochronology of Neogene phreatomagmatic volcanism in the western Pannonian Basin, Hungary J Volcanol Geotherm Res 164 4 193 204
143. C. B. Connor, 1990 Cinder Cone Clustering in the TransMexican Volcanic Belt: Implications for Structural and Petrologic Models. J Geophys Res 95 B12 19 395-19,405
144. T. Hasenaka, I. S. E. Carmichael, 1987 The Cinder Cones of Michoacán-Guanajuato, Central Mexico: Petrology and Chemistry. J Petrol 28 2 241 69
145. Y. Huang, C. Hawkesworth, P. van Calsteren, I. Smith, P. Black, 1997 Melt generation models for the Auckland volcanic field, New Zealand: constraints from U-Th isotopes Earth Planet Sci Lett 149
146. K. D. Putirka, MA Kuntz, D. M. Unruh, N. Vaid, 2009 Magma Evolution and Ascent at the Craters of the Moon and Neighboring Volcanic Fields, Southern Idaho, USA: Implications for the Evolution of Polygenetic and Monogenetic Volcanic Fields J Petrol 50 9 1639 65
147. A. Austin-Erickson, M. H. Ort, G. Carrasco-Núñez, 2011 Rhyolitic phreatomagmatism explored: Tepexitl tuff ring (Eastern Mexican volcanic belt) J Volcanol Geotherm Res 201 1-4 325 41
148. M. R. Brooker, B. F. Houghton, C. J. N. Wilson, J. A. Gamble, 1993 Pyroclastic phases of a rhyolitic dome-building eruption: Puketarata tuff ring, Taupo Volcanic Zone, New Zealand. Bull Volcanol 55 6 395 406
149. G. P. L. Walker, 1973 Explosive volcanic eruptions- a new classification scheme. Geol Rundsch 62 2 431 46
150. E. A. Parfitt, 2004 A discussion of the mechanisms of explosive basaltic eruptions. J Volcanol Geotherm Res 134 1-2 77 107
151. K. H. Wohletz, 1986 Explosive magma-water interactions: Thermodynamics, explosion mechanism and field studies. Bull Volcanol 48 5 245 64
152. V. Lorenz, 1985 Maars and diatremes of phreatomagmatic origin: a review. S Afr J Geol 88 2 459 70
153. S. Vergniolle, G. Brandeis, J. Mareschal-C, 1996 Strombolian explosions: Eruption dynamics determined from acoustic measurements. J Geophys Res 101 20449 66
154. E. A. Parfitt, L. Wilson, 1995 Explosive volcanic eruptions- IX. The transition between Hawaiian-style lava fountaining and Strombolian explosive activity. Geophys J Int 121 1 226 32
155. E. A. Blackburn, R. S. J. Sparks, 1976 Mechanism and dynamics of Strombolian activity. J Geol Soc 132 429 40
156. L. Pioli, B. J. Azzopardi, K. V. Cashman, 2009 Controls on the explosivity of scoria cone eruptions: Magma segregation at conduit junctions J Volcanol Geotherm Res 186 3-4 407 415
157. E. Johnson, P. Wallace, K. Chashman, H. D. Granados, A. Kent, 2008 Magmatic volatile contents and degassing-induced crystallization at Volcán Jorullo, Mexico:

Implications for melt evolution and the plumbing systems of monogenetic volcanoes. Earth Planet Sci Lett 269 478 87
158. G. P. L. Walker, R. Croasdale, 1972 Characteristics of some basaltic pyroclastics Bull Volcanol 35 2 303 17
159. B. F. Houghton, H. M. Gonnermann, 2008 Basaltic explosive volcanism: Constraints from deposits and models. Chem Erde 68 117 40
160. L. Wilson, J. W. Head, 1981 Ascent and eruption of basaltic magma on the Earth and Moon J Geophys Res 86 2971 3001
161. R. S. J. Sparks, 1978 The dynamics of bubble formation and growth in magmas: A review and analysis. J Volcanol Geotherm Res 3 1-2 1 37
162. E. A. Parfitt, 1998 A study of clast size distribution, ash deposition and fragmentation in a Hawaiian-style volcanic eruption J Volcanol Geotherm Res 84 197 208
163. L. Pioli, E. Erlund, E. Johnson, N. K. Cashman, P. Wallace, M. Rosi, et al. 2008 Explosive dynamics of violent Strombolian eruptions: The eruption of Parícutin Volcano 1943-1952 (Mexico). Earth Planet Sci Lett 271 1-4 359 68
164. M. Coltelli, P. Del Carlo L. Vezzoli, 1998 Discovery of a Plinian basaltic eruption of Roman age at Etna volcano, Italy Geology 26 12 1095 8
165. B. F. Houghton, C. J. N. Wilson, P. Del Carlo, M. Coltelli, J. E. Sable, R. Carey, 2004 The influence of conduit processes on changes in style of basaltic Plinian eruptions: Tarawera 1886 and Etna 122 BC. J Volcanol Geotherm Res 137 1 14
166. S. Vergniolle, M. T. Mangan, 2000 Hawaiian and Strombolian eruptions. Sigurdsson H, Houghton B, McNutt SR, Rymer H, Stix J, editors. Encyclopedia of VolcanoesSan Diego: Academic Press 447 461
167. W. K. Stovall, B. F. Houghton, H. Gonnermann, S. A. Fagents, D. A. Swanson, 2011 Eruption dynamics of Hawaiian-style fountains: the case study of episode 1 of the Kilauea Iki 1959 eruption. Bull Volcanol 73 5 511 29
168. M. Polacci, R. A. Corsaro, D. Andronico, 2006 Coupled textural and compositional characterization of basaltic scoria: Insights into the transition from Strombolian to fire fountain activity at Mount Etna Italy. Geology 34 3 201 4
169. W. A. Duffield, R. L. Christiansen, R. Koyanagi, D. W. Peterson, 1982 Storage, migration, and eruption of magma at Kilauea volcano, Hawaii, 1971-1972. J Volcanol Geotherm Res 13 273 307
170. J. W. Head, L. Wilson, 1987 Lava fountain heights at Pu'u'O'o, Kilauea, Hawaii: Indicators of amount and variations of exsolved magma volatiles. J Geophys Res 92 B13 13715
171. J. Sumner, S. Blake, R. Matela, J. Wolff, 2005 Spatter. J Volcanol Geotherm Res 142 49 65
172. J. W. Head, L. Wilson, 1989 Basaltic pyroclastic eruptions: Influence of gas-release patterns and volume fluxes on fountain structure, and the formation of cinder cones, spatter cones, rootless flows, lava ponds and lava flows J Volcanol Geotherm Res 37 3-4 261 271
173. C. Heliker, J. Kauahikaua, D. R. Sherrod, M. Lisowski, P. F. Cervelli, 2003 The Rise and Fall of Pu'u'Ö'ö Cone, 1983-2002. US Geological Survey Professional Paper 1676 29 52
174. J. A. Wolff, J. M. Sumner, 2000 Lava fountains and their products. Sigurdsson H, Houghton BF, McNutt SR, Rymer H, Stix J, editors. Encyclopedia of Volcanoes. San Diego: Academic Press 321 329
175. B. Capaccioni, F. Cuccoli, 2005 Spatter and welded air fall deposits generated by fire-fountaining eruptions : Cooling of pyroclasts during transport and deposition J Volcanol Geotherm Res 145 3-4 263 280
176. E. A. Parfitt, L. Wilson, 1999 A Plinian treatment of fallout from Hawaiian lava fountains. J Volcanol Geotherm Res 88 1-2 67 75
177. J. W. Head, W. B. Bryan, R. Greeley, J. E. Guest, P. H. Schultz, R. S. J. Sparks, et al. 1981 Distribution and morphology of basalt deposits on planets. Project BVS, editors. Basaltic Volcanism on the Terrestrial Planets. New York: Pergamon Press 701 800

178. J. Kauahikaua, D. R. Sherrod, K. V. Cashman, C. Heliker, K. Hon, T. N. Mattox, et al. 2003 Hawaiian Lava-Flow Dynamics During the Pu'u 'Ō'ō-Kūpaianaha Eruption: A Tale of Two Decades. US Geological Survey Professional Paper 1676 63 88
179. C. A. Wood, 1979 Monogenetic volcanoes in terrestrial planets. Proceedings of the 10th Lunar and Planetary Science Conference
180. M. R. Patrick, A. J. L. Harris, M. Ripepe, J. Dehn, D. A. Rothery, S. Calvari, 2007 Strombolian explosive styles and source conditions: insights from thermal (FLIR) video Bull Volcanol 69 769 84
181. D. Vespermann, H. Schmincke-U, 2000 Scoria cones and tuff rings. Sigurdsson H, Houghton BF, McNutt SR, Rymer H, Stix J, editors. Encyclopedia of VolcanoesSan Diego: Academic Press 683 694
182. C. Riedel, G. G. J. Ernst, M. Riley, 2003 Controls on the growth and geometry of pyroclastic constructs J Volcanol Geotherm Res 127 1-2 121 52
183. A. J. L. Harris, M. Ripepe, E. A. Hughes, 2012 Detailed analysis of particle launch velocities, size distributions and gas densities during normal explosions at Stromboli J Volcanol Geotherm Res 132 109 131
184. S. Calvari, H. Pinkerton, 2004 Birth, growth and morphologic evolution of the'Laghetto' cinder cone during the 2001 Etna eruption. J Volcanol Geotherm Res 132 225 39
185. T. R. Mc Getchin, M. Settle, BA Chouet, 1974 Cinder cone growth modeled after Northeast Crater, Mount Etna, Sicily J Geophys Res 79 3257 72
186. B. F. Houghton, C. J. N. Wilson, 1989 A vesicularity index for pyroclastic deposits. Bull Volcanol 51 6 451 62
187. M. T. Mangan, K. V. Cashman, 1996 The structure of basaltic scoria and reticulite and inferences for vesiculation, foam formation, and fragmentation in lava fountains. J Volcanol Geotherm Res 73 1 18
188. G. A. Valentine, D. J. Krier, F. V. Perry, G. Heiken, 2007 Eruptive and geomorphic processes at the Lathrop Wells scoria cone volcano J Volcanol Geotherm Res 161 1-2 57
189. M. Settle, 1979 The structure and emplacement of cinder cone fields Am J Sci 279 10 1089 107
190. Donald. G. A. Mac, 1972 Volcanoes. Englewood Cliffs: Prentice-Hall
191. E. J. Erlund, K. V. Cashman, P. J. Wallace, L. Pioli, M. Rosi, E. Johnson, et al. 2010 Compositional evolution of magma from Parícutin Volcano, Mexico: The tephra record J Volcanol Geotherm Res 197 1-4 167 87
192. B. Zimanowski, R. Büttner, 2002 Dynamic mingling of magma and liquefied sediments J Volcanol Geotherm Res 114 37 44
193. B. Zimanowski, R. Büttner, V. Lorenz, 1997 Premixing of magma and water in MFCI experiments. Bull Volcanol 58 6 491 5
194. J. D. L. White, 1996 Impure coolants and interaction dynamics of phreatomagmatic eruptions. J Volcanol Geotherm Res 74 3-4 155 70
195. K. Németh, K. Goth, U. Martin, G. Csillag, P. Suhr, 2008 Reconstructing paleoenvironment, eruption mechanism and paleomorphology of the Pliocene Pula maar, (Hungary) J Volcanol Geotherm Res 177 2 441 56
196. P. Kokelaar, 1986 Magma-water interactions in subaqueous and emergent basaltic volcanism. Bull Volcanol 48 275 89
197. B. Zimanowski, K. H. Wohletz, 2000 Physics of phreatomagmatism I. Terra Nostra 6 515 23
198. V. Lorenz, S. Kurszlaukis, 2007 Root zone processes in the phreatomagmatic pipe emplacement model and consequences for the evolution of maar-diatreme volcanoes J Volcanol Geotherm Res 159 4 32
199. V. Lorenz, 1986 On the growth of maar and diatremes and its relevance to the formation of tuff rings. Bull Volcanol 48 5 265 74
200. J. D. L. White, P. Ross-S, 2011 Maar-diatreme volcanoes: A review. J Volcanol Geotherm Res 201 1-4 1 29

201. R. Büttner, P. Dellino, B. Zimanowski, 1999 Identifying modes of magma/water interaction from the surface features of ash particles. Nature 401 688 90
202. G. Sottili, D. Palladino, M. Gaeta, M. Masotta, 2012 Origins and energetics of maar volcanoes: examples from the ultrapotassic Sabatini Volcanic District (Roman Province, Central Italy) Bull Volcanol 74 1 163 86
203. T. N. Mattox, M. T. Mangan, 1997 Littoral hydrovolcanic explosions: a case study of lava-seawater interaction at Kilauea volcano. J Volcanol Geotherm Res 75 1-2 1 17
204. Z. Jurado-Chichay, S. K. Rowland, G. P. L. Walker, 1996 The formation of circular littoral cone from tube-fed pahoehoe: Mauna Loa, Hawaii Bull Volcanol 57 7 471 82
205. S. Thorarinsson, 1953 The crater groups in Iceland Bull Volcanol 14 3 44
206. C. W. Hamilton, T. Thordarson, S. A. Fagents, 2010 Explosive lava-water interactions I: architecture and emplacement chronology of volcanic rootless cone groups in the 1783-1784 Laki lava flow, Iceland Bull Volcanol 72 449 67
207. C. W. Hamilton, S. A. Fagents, T. Thordarson, 2010 Explosive lava-water interactions II: self-organization processes among volcanic rootless eruption sites in the 1783-1784 Laki lava flow, Iceland Bull Volcanol 72 469 85
208. S. A. Fagents, T. Thordarson, 2007 Rootless volcanic cones in Iceland and on Mars. Chapman MG, editors. The geology of Mars: evidence from Earth-based analogs.New York, USA: Cambridge University Press 151 177
209. V. Lorenz, 2003 Maar-Diatreme Volcanoes, their Formation, and their Setting in Hard-rock or Soft-rock environments. Geolines 15 72 83
210. H. B. Mattsson, B. A. Tripoli, 2011 Depositional characteristics and volcanic landforms in the Lake Natron-Engaruka monogenetic field, northern Tanzania. J Volcanol Geotherm Res 203 23 34
211. F. Stoppa, G. Rosatelli M. Schiazza, A. Tranquilli, 2012 Hydrovolcanic vs Magmatic Processes in Forming Maars and Associated Pyroclasts: The Calatrava-Spain- Case History. Stoppa F, editors. Updates in Volcanology- A Comprehensive Approach to Volcanological Problems. InTech 3 26http://www.intechopen.com/books/updates-in-volcanology-a-comprehensive-approach-to-volcanological-problems/hydrovolcanic-vs-magmatic-processes-in-forming-maars-and-associated-pyroclasts-the-calatrava-spain-c
212. F. Stoppa, 1996 The San Venanzo maar and tuff ring, Umbria, Italy: eruptive behaviour of a carbonatite-melilitite volcano Bull Volcanol 57 7 563 77
213. V. Lorenz, A. R. Mc Birney, H. Williams, 1970 An investigation of volcanic depressions. Part III. Maars, tuff-rings, tuff-cones, and diatremes. Houston, Texas
214. T. N. Blaikie, L. Ailleres, R. A. F. Cas, P. G. Betts, 2012 Three-dimensional potential field modelling of a multi-vent maar-diatreme- The Lake Coragulac maar, Newer Volcanics Province, southeastern Australia. J Volcanol Geotherm Res 235-236 0 70 83
215. J. E. Beget, D. M. Hopkins, S. D. Charron, 1996 Largest known maars (Espenberg Maars) on earth, Seward Peninsula, northwest Alaska. Arctic 49 1 62 9
216. U. Grunewald, B. Zimanowski, R. Büttner, L. F. Phillips, K. Heide, G. Büchel, 2007 MFCI experiments on the influence of NaCl-saturated water on phreatomagmatic explosions J Volcanol Geotherm Res 159 1-3 126
217. M. F. Sheridan, K. H. Wohletz, 1981 Hydrovolcanic Explosions: The Systematics of Water-Pyroclast Equilibration. Science 212 1387 9
218. K. H. Wohletz, G. A. Valentine, 1990 Computer simulations of explosive volcanic eruptions. Ryan MP, editors. Magma transport and storage 114 134
219. Büttner, R., Dellino, P., La Volpe, L., Lorenz, V., & Zimanowski, B. (2002). Thermohydraulic explosions in phreatomagmatic eruptions as evidenced by the comparison between pyroclasts and products from Molten Fuel Coolant Interaction experiments. *J Geophys Res*, 107(B11), 10.1029/2001JB000511.
220. R. Büttner, B. Zimanowski, 1998 Physics of thermohydraulic explosions. Physical Review E 57 5 5726 9

221. K. H. Wohletz, R. G. Mc Queen, 1984 Experimental Studies of Hydrovolcanic Explosions. editors. Explosive volcanism: Inception, evolution and hazards. Washington D.C., USA: National Academy Press 158 169
222. B. P. Kokelaar, 1983 The mechanism of Surtseyan volcanism. J Geol Soc 140 939 44
223. K. H. Wohletz, M. F. Sheridan, 1983 Hydrovolcanic explosions II. Evolution of basaltic tuff rings and tuff cones. Am J Sci 283 385 413
224. G. Gençalioğlu-Kuşcu, C. Atilla, R. A. F. Cas, K. Ilkay, 2007 Base surge deposits, eruption history, and depositional processes of a wet phreatomagmatic volcano in Central Anatolia (Cora Maar). J Volcanol Geotherm Res 159 1-3 198 209
225. R. Funiciello, G. Giordano, D. De Rita, 2003 The Albano maar lake (Colli Albani Volcano, Italy): recent volcanic activity and evidence of pre-Roman Age catastrophic lahar events. J Volcanol Geotherm Res 123 43 61
226. J. A. Vazquez, M. H. Ort, 2006 Facies variation of eruption units produced by the passage of single pyroclastic surge currents, Hopi Buttes volcanic field, USA J Volcanol Geotherm Res 154 3-4 222 36
227. R. Sulpizio, P. Dellino, 2008 Chapter 2 Sedimentology, Depositional Mechanisms and Pulsating Behaviour of Pyroclastic Density Currents Joachim G, Joan M, iacute, editors. Developments in Volcanology. Elsevier. 57 96
228. G. Carrasco-Núñez, M. H. Ort, C. Romero, 2007 Evolution and hydrological conditions of a maar volcano (Atexcac crater, Eastern Mexico) J Volcanol Geotherm Res 159 179 97
229. A. Auer, U. Martin, K. Németh, 2007 The Fekete-hegy (Balaton Highland Hungary) „soft-substrate" and „hard-substrate" maar volcanoes in an aligned volcanic complex- Implications for vent geometry, subsurface stratigraphy and the paleoenvironmental setting. J Volcanol Geotherm Res 159 1-3 225 45
230. N. Pardo, J. L. Macias, G. Giordano, P. Cianfarra, D. R. Avellán, F. Bellatreccia, 2009 The ~1245 yr BP Asososca maar eruption: The youngest event along the Nejapa-Miraflores volcanic fault, Western Managua, Nicaragua J Volcanol Geotherm Res 184 3-4 292 312
231. H. Raue, 2004 A new model for the fracture energy budget of phreatomagmatic explosions J Volcanol Geotherm Res 129 1-3 99
232. J. D. L. White, H. Schmincke-U, 1999 Phreatomagmatic eruptive and depositional processes during the '49 eruption on La Palma (Canary Islands). J Volcanol Geotherm Res 94 1-4 283 304
233. U. Martin, K. Németh, 2005 Eruptive and depositional history of a Pliocene tuff ring hat developed in a fluvio-lacustrine basin: Kissomlyó volcano (western Hungary). J Volcanol Geotherm Res 147 342 56
234. Y. K. Sohn, 1996 Hydrovolcanic processes forming basaltic tuff rings and cones on Cheju Island, Korea. Geol Soc Am Bull 108 10 1199 211
235. S. K. Chough, Y. K. Sohn, 1990 Depositional mechanics and sequences of base surges, Songaksan tuff ring, Cheju Island, Korea. Sedimentology 37 1115 35
236. J. J. Aranda-Gómez, J. F. Luhr, 1996 Origin of the Joya Honda maar, San Luis Potosí, Mexico. J Volcanol Geotherm Res 74 1 18
237. V. Lorenz, 1974 Vesiculated tuffs and associated features Sedimentology 21 2 273 91
238. G. E. Alvarado, G. J. Soto, F. M. Salani, P. Ruiz, L. H. de Mendoza, 2011 The formation and evolution of Hule and Río Cuarto maars, Costa Rica J Volcanol Geotherm Res 201 1-4 342 356
239. C. N. Ngwa, C. E. Suh, C. W. Devey, 2010 Phreatomagmatic deposits and stratigraphic reconstruction at Debunscha Maar (Mt Cameroon volcano). J Volcanol Geotherm Res 192 3-4 201 11
240. Dellino, P., Isaia, R., La Volpe, L., & Orsi, G. (2004). Interaction between particles transported by fallout and surge in the deposits of the Agnano-Monte Spina eruption (Campi Flegrei, Southern Italy). *J Volcanol Geotherm Res*, 133(1-4), 193-210.

241. K. Németh, C. M. White, 2009 Intra-vent peperites related to the phreatomagmatic 71 Gulch Volcano, western Snake River Plain volcanic field, Idaho (USA) J Volcanol Geotherm Res 183 1-2 30 41
242. G. Hetényi, B. Taisne, F. Garel, É. Médard, S. Bosshard, H. Mattsson, 2012 Scales of columnar jointing in igneous rocks: field measurements and controlling factors Bull Volcanol 74 2 457 82
243. J. Martí, L. Planagumà, A. Geyer, E. Canal, D. Pedrazzi, 2011 Complex interaction between Strombolian and phreatomagmatic eruptions in the Quaternary monogenetic volcanism of the Catalan Volcanic Zone (NE of Spain) J Volcanol Geotherm Res 201 1-4 178 93
244. J. J. Aranda-Gómez, J. F. Luhr, G. Pier, 1992 The La Brena- El Jagüey Maar Complex, Durango, México: I. Geological evolution. Bull Volcanol 54 5 393 404
245. V. Lorenz, 2003 Syn- and post-eruptive processes of maar-diatreme volcanoes and their relevance to the accumulation of port-eruptive maar creater sediments. Földtani Kutatás 11 1-2 13 22
246. C. Molloy, P. Shane, P. Augustinus, 2009 Eruption recurrence rates in a basaltic volcanic field based on tephra layers in maar sediments: Implications for hazards in the Auckland volcanic field Geol Soc Am Bull 121 11-12 1666 77
247. M. Pirrung, C. Fischer, G. Büchel, R. Gaupp, H. Lutz, F. Neuffer-O, 2003 Lithofacies succession of maar crater deposits in the Eifel area (Germany) Terra Nova 15 2 125 32
248. K. Németh, 2001 Deltaic density currents and turbidity deposits related to maar crater rims and their importance for paleogeographic reconstructionof the Bakony-Balaton Highland Volcanic Field, Hungary. Kneller B, McCaffrey B, Peakall J, Druitt T, editors. Sediment transport and deposition by particulate gravity currents. Oxford: Blackwell Sciences 261 277
249. J. Pastre-F, A. Gauthier, S. Nomade, P. Orth, A. Andrieu, F. Goupille, et al. 2007 The Alleret maar (Massif Central, France): A new lacustrine sequence of the early Middle Pleistocene in western Europe CR Geosci 339 987 97
250. B. Zolitschka, 1993 Palaeoecological implications from the sedimentary record of a subtropical maar lake (Eocene Eckfelder Maar; Germany). Negendank JW, Zolitschka B, editors. Paleolimnology of European Maar Lakes. Springer Berlin Heidelberg 477 484
251. V. Lorenz, 1974 Studies of the Surtsey tephra deposits. Surtsey Research Progress Report VII 72 9
252. K. Kano, 1998 A shallow-marine alkali-basalt tuff cone in the Middle Miocene Jinzai Formation, Izumo, SW Japan J Volcanol Geotherm Res 87 1-4 173
253. J. D. L. White, 2000 Subaqueous eruption-fed density currents and their deposits. Precambrian Res 101 2-4 87 109
254. J. D. L. White, 1996 Pre-emergent construction of a lacustrine basaltic volcano, Pahvant Butte, Utah (USA). Bull Volcanol 58 4 249 62
255. R. A. F. Cas, C. A. Landis, R. E. Fordyce, 1989 A monogenetic, Surtla-type, Surtseyan volcano from the Eocene-Oligocene Waiareka-Deborah volcanics, Otago, New Zealand: A model. Bulletin of Volcanology 51 4 281 98
256. U. Martin, C. Breitkreuz, S. Egenhoff, P. Enos, L. Jansa, 2004 Shallow-marine phreatomagmatic eruptions through a semi-solidified carbonate platform (ODP Leg 144, Site 878, Early Cretaceous, MIT Guyot, West Pacific) Mar Geol 204 3-4 251
257. J. B. Garvin, R. S. Williams, J. J. Frawley, W. B. Krabill, 2000 Volumetric evolution of Surtsey, Iceland, from topographic maps and scanning airborne laser altimetry. Surtsey Research 11 127 34
258. R. G. Vaughan, P. W. Webley, 2010 Satellite observations of a surtseyan eruption: Hunga Ha'apai, Tonga. J Volcanol Geotherm Res 198 1-2 177 86
259. V. Zanon, J. Pacheco, A. Pimentel, 2009 Growth and evolution of an emergent tuff cone: Considerations from structural geology, geomorphology and facies analysis of São Roque volcano São Miguel (Azores). J Volcanol Geotherm Res 180 2-4 277 291

260. B. P. Kokelaar, G. P. Durant, 1983 The submarine eruption and erosion of Surtla (Surtsey), Iceland. J Volcanol Geotherm Res 19 3-4 239 46
261. S. Thorarinsson, 1965 Surtsey eruption course of events and the developement of the new island. Surtsey Research Progress Report I 51 5
262. H. B. Mattsson, 2010 Textural variation in juvenile pyroclasts from an emergent, Surtseyan-type, volcanic eruption: The Capelas tuff cone, São Miguel (Azores). J Volcanol Geotherm Res 189 1-2 81 91
263. K. Németh, S. J. Cronin, D. Charley, M. Harrison, E. Garae, 2006 Exploding lakes in Vanuatu- "Surtseyan-style" eruption witnessed on Ambae Island. Episodes 29 2 87 92
264. B. D. Brand, A. B. Clarke, 2009 The architecture, eruptive history, and evolution of the Table Rock Complex, Oregon: From a Surtseyan to an energetic maar eruption. J Volcanol Geotherm Res 180 2-4 203 24
265. Y. K. Sohn, S. K. Chough, 1993 The Udo tuff cone, Cheju Island, South Korea: transformation of pyroclastic fall into debris fall and grain flow on a steep volcanic cone slope. Sedimentology 40 769 86
266. H. B. Mattsson, A. Höskuldsson, S. Hand, 2005 Crustal xenoliths in the 6220 BP Saefell tuff-cone, south Iceland: Evidence for a deep, diatreme-forming, Surtseyan eruption J Volcanol Geotherm Res 145 3-4 234 248
267. I. Suiting, H. Schmincke-U, 2009 Internal vs. external forcing in shallow marine diatreme formation: A case study from the Iblean Mountains (SE-Sicily, Central Mediterranean) J Volcanol Geotherm Res 186 3-4 361 378
268. U. Martin, K. Németh, 2004 Mio/Pliocene Phreatomagmatic Volcanism in the Western Pannonian Basin. Budapest
269. J. Cassidy, S. J. France, CA Locke, 2007 Gravity and magnetic investigation of maar volcanoes, Auckland volcanic field, New Zealand J Volcanol Geotherm Res 159 1-3 153
270. Y. Weinstein, 2007 A transition from strombolian to phreatomagmatic activity induced by a lava flow damming water in a valley J Volcanol Geotherm Res 159 267 84
271. G. E. Alvarado, G. J. Soto, F. M. Salani, P. Ruiz, L. H. de Mendoza, 2011 The formation and evolution of Hule and Río Cuarto maars, Costa Rica Journal of Volcanology and Geothermal Research 201 1-4 342 356
272. J. F. Luhr, 2001 Glass inclusions and melt volatile contents at Paricutin Volcano, Mexico. Contrib Mineral Petrol 142 3 261 83
273. K. Németh, C. Risso, F. Nullo, G. Kereszturi, 2011 The role of collapsing and rafting of scoria cones on eruption style changes and final cone morphology: Los Morados scoria cone, Mendoza Argentina. Cent Eur J Geosci 3 2 102 18
274. A. J. Needham, J. M. Lindsay, I. E. M. Smith, P. Augustinus, P. A. Shane, 2011 Sequential Eruption of Alkaline and Sub-Alkaline Magmas from a small Monogenetic volcano in The Auckland Volcanic Field, New Zealand J Volcanol Geotherm Res 201 1-4 126 42
275. P. Cervantes, P. Wallace, 2003 Magma degassing and basaltic eruption styles: a case study of 2000 year BP Xitle volcano in central Mexico. J Volcanol Geotherm Res 120 3-4 249 70
276. E. R. Johnson, P. J. Wallace, K. V. Cashman, Granados. H. Delgado, 2010 Degassing of volatiles (H2O, CO2, S, Cl) during ascent, crystallization, and eruption of basaltic magmas. J Volcanol Geotherm Res 197 1-4 225 38
277. B. F. Houghton, H. Schmincke-U, 1986 Mixed deposits of simultaneous strombolian and phreatomagmatic volcanism: Rothenberg volcano, East Eifel Volcanic Field. J Volcanol Geotherm Res 30 117 30
278. Y. K. Sohn, K. H. Park, 2005 Composite tuff ring/cone complexes in Jeju Island, Korea: possible consequences of substrate collapse and vent migration. J Volcanol Geotherm Res 141 157 75

279. M. H. Ort, G. Carrasco-Núñez, 2009 Lateral vent migration during phreatomagmatic and magmatic eruptions at Tecuitlapa Maar, east-central Mexico J Volcanol Geotherm Res 181 1-2 67 77
280. M. A. Bishop, 2009 A generic classification for the morphological and spatial complexity of volcanic (and other) landforms. Geomorphology 111 1-2 104 9
281. G. Walker, 1971 Compound and simple lava flows and flood basalts Bull Volcanol 35 3 579 90
282. K. Németh, 2004 The morphology and origin of wide craters at Al Haruj al Abyad, Libya: maars and phreatomagmatism in a large intracontinental flood lava field? Z Geomorphol 48 4 417 39
283. H. Solgevik, H. Mattsson, O. Hennelin, 2007 Growth of an emergent tuff cone: Fragmentation and depositional processes recorded in the Capelas tuff cone, São Miguel, Azores J Volcanol Geotherm Res 159 246 66
284. J. T. Gutmann, 1976 Geology of Crater Elegante, Sonora, Mexico. Geol Soc Am Bull 87 12 1718 29
285. P. Ross-S, J. D. L. White, 2006 Debris jets in continental phreatomagmatic volcanoes: a field study of their subterranean deposits in the Coombs Hills vent complex, Antarctica. Volcanol Geotherm Res 149 62 84
286. Sohn, Y. K., Cronin, S. J., Brenna, M., Smith, I. E. M., Németh, K., White, J. D. L., et al. (2012). Ilchulbong tuff cone, Jeju Island, Korea, revisited: A compound monogenetic volcano involving multiple magma pulses, shifting vents, and discrete eruptive phases. *Geol Soc Am Bull*, 10.1130/B30447.1.
287. C. Shaw, A. Woodland, 2012 The role of magma mixing in the petrogenesis of mafic alkaline lavas, Rockeskyllerkopf Volcanic Complex, West Eifel, Germany Bull Volcanol 74 2 359 76
288. U. Martin, K. Németh, 2006 How Strombolian is a "Strombolian" scoria cone? Some irregularities in scoria cone architecture from the Transmexican Volcanic Belt, near Volcán Ceboruco (Mexico), and Al Haruj (Libya). J Volcanol Geotherm Res 155 1-2 104 118
289. D. Karátson, 1996 Rates and factors of stratovolcano degradation in a continental climate: a complex morphometric analysis for nineteen Neogene/Quaternary crater remnants in the Carpathians J Volcanol Geotherm Res 73 1-2 65
290. M. Inbar, M. Gilichinsky, I. Melekestsev, D. Melnikov, N. Zaretskaya, 2011 Morphometric and morphological development of Holocene cinder cones: A field and remote sensing study in the Tolbachik volcanic field, Kamchatka J Volcanol Geotherm Res 201 1-4 301 311
291. A. Rodriguez-Gonzalez, J. L. Fernandez-Turiel, F. J. Perez-Torrado, D. Gimeno, M. Aulinas, 2009 Geomorphological reconstruction and morphometric modelling applied to past volcanism Int J Earth Sci 99 3 645 60
292. D. Karátson, T. Telbisz, G. Wörner, 2012 Erosion rates and erosion patterns of Neogene to Quaternary stratovolcanoes in the Western Cordillera of the Central Andes: An SRTM DEM based analysis Geomorphology 139-140 122 135
293. R. J. Pike, 1978 Volcanoes on the inner planets: Some preliminary comparisons of gross topography. Proceedings of the 10th Lunar and Planetary Science Conference 3 3239 73
294. D. Karátson, M. Favalli, S. Tarquini, A. Fornaciai, G. Wörner, 2010 The regular shape of stratovolcanoes: A DEM-based morphometrical approach J Volcanol Geotherm Res 193 3-4 171 181
295. P. Grosse, B. van Wyk de Vries, P. A. Euillades, M. Kervyn, I. Petrinovic, 2012 Systematic morphometric characterization of volcanic edifices using digital elevation models. Geomorphology 136 114 31
296. J. N. Procter, S. J. Cronin, T. Platz, A. Patra, K. Dalbey, M. Sheridan, et al. 2010 Mapping block-and-ash flow hazards based on Titan 2D simulations: a case study from Mt. Taranaki, NZ. Nat Hazards 53 483 501

297. J. Procter, S. J. Cronin, I. C. Fuller, G. Lube, V. Manville, 2010 Quantifying the geomorphic impacts of a lake-breakout lahar, Mount Ruapehu, New Zealand. Geology 38 1 67 70
298. A. Tibaldi, 1995 Morphology of pyroclastic cones and tectonics. J Geophys Res 100 B12 24521 35
299. M. Inbar, C. Risso, 2001 A morphological and morphometric analysis of a high density cinder cone volcanic field.- Payun Matru, south-central Andes, Argentina. Z Geomorphol 45 3 321 43
300. H. S. Colton, 1937 The basaltic cinder cones and lava flows of the San Francisco Mountain Volcanic Field. Museum of Northern Arizona Bulletin 10 1 58
301. S. C. Porter, 1972 Distribution, Morphology, and Size Frequency of Cinder Cones on Mauna Kea Volcano, Hawaii. Geol Soc Am Bull 83 12 3607 12
302. I. Moriya, 1986 Morphometry of Pyroclastic Cones in Japan. The geographical reports of Kanazawa University 3 58 76
303. T. Hasenaka, I. S. E. Carmichael, 1985 A compilation of location, size, and geomophological parameters of volcanoes of the Michoacan-Guanajuato volcanic field, central Mexico. Geofisica Internacional 24 4 577 607
304. T. Hasenaka, I. S. E. Carmichael, 1985 The cinder cones of Michoacán-Guanajuato central Mexico: their age, volume and distribution, and magma discharge rate. J Volcanol Geotherm Res 25 1-2 105 24
305. J. E. Bleacher, L. S. Glaze, R. Greeley, E. Hauber, S. M. Baloga, S. E. H. Sakimoto, et al. 2009 Spatial and alignment analyses for a field of small volcanic vents south of Pavonis Mons and implications for the Tharsis province, Mars. J Volcanol Geotherm Res 185 96 102
306. S. C. Mest, D. A. Crown, 2005 Millochau crater, Mars: Infilling and erosion of an ancient highland impact crater. Icarus 175 335 59
307. L. S. Glaze, S. W. Anderson, E. R. Stofan, S. E. Smrekar, 2005 Statistical distribution of tumuli on pahoehoe flow surfaces: analysis of examples in Hawaii and Iceland and potential applications to lava flows on Mars. J Geophys Res 110 B08202 10.1029/2004JB003564
308. J. C. Armstrong, C. B. Leovy, 2005 Long term wind erosion on Mars. Icarus 176 57 74
309. R. J. Pike, 1976 Crater dimensions from Apollo data and supplemental sources. The Moon 15 463 77
310. R. J. Pike, 1977 Apparent depth/apparent diameter relation for lunar craters. Proceedings of the 8th Lunar Science Conference 3 3427 36
311. R. R. Ghent, S. W. Anderson, T. M. Pithawala, 2012 The formation of small cones in Isidis Planitia, Mars through mobilization of pyroclastic surge deposits. Icarus 217 1 169 83
312. K. Bemis, J. Walker, A. Borgia, B. Turrin, M. Neri, III C. Swisher, 2011 The growth and erosion of cinder cones in Guatemala and El Salvador: models and statistics J Volcanol Geotherm Res 201 1-4 39 52
313. H. Sato, H. Taniguchi, 1997 Relationship between crater size and ejecta volume of recent magmatic and phreato-magmatic eruptions: Implications for energy partitioning. Geophys Res Lett 24 3 205 8
314. M. Kervyn, G. G. J. Ernst, J. Carracedo-C, P. Jacobs, 2012 Geomorphometric variability of "monogenetic" volcanic cones: Evidence from Mauna Kea, Lanzarote and experimental cones. Geomorphology 136 1 59 75
315. C. Corazzato, A. Tibaldi, 2006 Fracture control on type, morphology and distribution of parasitic volcanic cones: an example from Mt. Etna, Italy. J Volcanol Geotherm Res 158 1-2 177 94
316. A. Tibaldi, 2003 Influence of cone morphology on dykes, Stromboli, Italy. J Volcanol Geotherm Res 126 1-2 79 95
317. M. Favalli, D. Karátson, F. Mazzarini, M. T. Pareschi, E. Boschi, 2009 Morphometry of scoria cones located on a volcano flank: A case study from Mt. Etna (Italy), based on high-resolution LiDAR data. J Volcanol Geotherm Res 186 3-4 320 30

318. F. M. Conway, C. B. Connor, B. E. Hill, C. D. Condit, K. Mullaney, C. M. Hall, 1998 Recurrence rates of basaltic volcanism in SP cluster, San Francisco volcanic field, Arizona. Geology 26 7 655 8

319. S. R. Rodríguez, W. Morales-Barrera, P. Layer, E. González-Mercado, 2010 A quaternary monogenetic volcanic field in the Xalapa region, eastern Trans-Mexican volcanic belt: Geology, distribution and morphology of the volcanic vents. J Volcanol Geotherm Res 197 1-4 149 66

320. G. J. Aguirre-Díaz, M. C. Jaimes-Viera, J. Nieto-Obregón, 2006 The Valle de Bravo Volcanic Field: Geology and geomorphometric parameters of a Quaternary monogenetic field at the front of the Mexican Volcanic Belt. Geological Society of America Special Papers 402 139 54

321. J. Dóniz, C. Romero, J. Carmona, A. García, 2011 Erosion of cinder cones in Tenerife by gully formation, Canary Islands, Spain. Physical Geography 32 2 139 60

322. J. Doniz, C. Romero, E. Coello, C. Guillen, N. Sanchez, L. Garcia-Cacho, et al. 2008 Morphological and statistical characterisation of recent mafic volcanism on Tenerife (Canary Islands, Spain). J Volcanol Geotherm Res 173 3-4 185 95

323. D. M. Hooper, 1999 Cinder movement experiments on scoria cone slopes: Rates and direction of transport. Landform Analysis 2 5 18

324. D. M. Hooper, M. F. Sheridan, 1998 Computer-simulation models of scoria cone degradation. J Volcanol Geotherm Res 83 3-4 241 67

325. D. M. Hooper, 1995 Computer-simulation models of scoria cone degradation in the Colima and Michoacán-Guanajuato volcanic fields, Mexico. Geofisica Internacional 34 3 321 40

326. A. Fornaciai, B. Behncke, M. Favalli, M. Neri, S. Tarquini, E. Boschi, 2011 Detecting short-term evolution of Etnean scoria cones: a LIDAR-based approach. Bull Volcanol 72 10 1209 22

327. J. D. Pelletier, M. L. Cline, 2007 Nonlinear slope-dependent sediment transport in cinder cone evolution Geology 35 12 1067 70

328. J. Thouret-C, 1999 Volcanic geomorphology- an overview. Earth-Science Reviews 47 95 131

329. D. Karátson, T. Telbisz, BS Singer, 2010 Late-stage volcano geomorphic evolution of the Pleistocene San Francisco Mountain, Arizona (USA), based on high-resolution DEM analysis and 40Ar/39Ar chronology. Bull Volcanol 72 7 833 46

330. B. Székely, D. Karátson, 2004 DEM-based morpometry as a tool for reconstructing primary volcanic landforms: examples from the Börzsöny Mountains, Hungary. Geomorphology 63 25 37

331. P. Lahitte, A. Samper, X. Quidelleur, 2012 DEM-based reconstruction of southern Basse-Terre volcanoes (Guadeloupe archipelago, FWI): Contribution to the Lesser Antilles Arc construction rates and magma production. Geomorphology 136 1 148 64

332. B. Csatho, T. Schenk, P. Kyle, T. Wilson, W. B. Krabill, 2008 Airborne laser swath mapping of the summit of Erebus volcano, Antarctica: Applications to geological mapping of a volcano. J Volcanol Geotherm Res 177 3 531 48

333. J. Parrot-F, 2007 Study of Volcanic Cinder Cone Evolution by Means of High Resolution DEMs. MODSIM 2007 International Congress on Modelling and Simulation; 2007: Modelling and Simulation Society of Australia and New Zealand

334. Parrot-F, J. (2007) Tri-dimensional parameterisation: an automated treatment to study the evolution of volcanic cones. *Géomorphologie: relief, processus, environnement,* 2007(3), 247-57.

335. M. Awdankiewicz, 2005 Reconstructing an eroded scoria cone: the Miocene Sosnica Hill volcano (Lower Silesia, SW Poland). Geological Quarterly 49 439 48

336. V. Rapprich, V. Cajz, M. Kostak, Z. Pécskay, T. Ridkosil, P. Raska, et al. 2007 Reconstruction of eroded monogenetic Strombolian cones of Miocene age: A case study on character of volcanic activity of the Jicin Volcanic Field (NE Bohemia) and subsequent erosional rates estimation. Journal of Geoscience 52 3-4 169 80

337. Fornaciai, A., Behncke, B., Favalli, M., Neri, M., Tarquini, S., & Boschi, E. (2012). Morphometry of scoria cones, and their relation to geodynamic setting: A DEM-based analysis. *J Volcanol Geotherm Res*, 217-218, 56-72.

338. Kereszturi, G., Geyer, A., Martí, J., Németh, K., & Dóniz-Páez, J. F. (in press) Evaluation of morphometry-based dating of monogenetic volcanoes – A case study from Bandas del Sur, Tenerife (Canary Islands). *Bull Volcanol.*

339. J. K. Lanz, R. Wagner, U. Wolf, J. Kröchert, G. Neukum, 2010 Rift zone volcanism and associated cinder cone field in Utopia Planitia, Mars. J Geophys Res 115 E12 E12019

340. R. J. Pike, 1974 Craters on Earth, Moon, and Mars: Multivariate classification and mode of origin. Earth Planet Sci Lett 22 3 245 55

341. H. Frey, M. Jarosewich, 1982 Subkilometer Martian Volcanoes: Properties and Possible Terrestrial Analogs. J Geophys Res 87 B12 9867 79

342. Avellán, D. R., Macías, J. L., Pardo, N., Scolamacchia, T., & Rodriguez, D. (2012). Stratigraphy, geomorphology, geochemistry and hazard implications of the Nejapa Volcanic Field, western Managua, Nicaragua. *J Volcanol Geotherm Res*, 213-214, 51-71.

343. J. Taddeucci, G. Sottili, D. M. Palladino, G. Ventura, P. Scarlato, 2009 A note on maar eruption energetics: current models and their application. Bull Volcanol 72 1 75 83

344. A. Goto, H. Taniguchi, M. Yoshida, T. Ohba, H. Oshima, 2001 Effects of explosion energy and depth to the formation of blast wave and crater: Field Explosion Experiment for the understanding of volcanic explosion. Geophys Res Lett 28 22 4287 90

345. A. Martín-Serrano, J. Vegas, A. García-Cortés, L. Galán, J. L. Gallardo-Millán, S. Martín-Alfageme, et al. 2009 Morphotectonic setting of maar lakes in the Campo de Calatrava Volcanic Field (Central Spain, SW Europe). Sediment Geol 222 1-2 52 63

346. G. A. Valentine, N. L. Shufelt, A. R. L. Hintz, 2011 Models of maar volcanoes, Lunar Crater (Nevada, USA). Bull Volcanol 73 6 753 65

347. G. A. Valentine, 2012 Shallow plumbing systems for small-volume basaltic volcanoes, 2: Evidence from crustal xenoliths at scoria cones and maars. J Volcanol Geotherm Res 223-224 47 63

348. B. Zimanowski, 1998 Phreatomagmatic explosions. Freundt A, Rosi M, editors. From Magma to Tephra: Developments in Volcanology 4. Amsterdam: Elsevier 25 54

349. P. T. Leat, R. N. Thompson, 1988 Miocene hydrovolcanism in NW Colorado, USA, fuelled by explosive mixing of basic magma and wet unconsolidated sediment. Bull Volcanol 50 4 229 43

350. M. Mc Clintock, J. White, 2006 Large phreatomagmatic vent complex at Coombs Hills, Antarctica: Wet, explosive initiation of flood basalt volcanism in the Ferrar-Karoo LIP. Bull Volcanol 68 3 215 39

351. B. C. Hearn, 1968 Diatremes with Kimberlitic Affinities in North-Central Montana. Science 159 3815 622 5

352. Yokoo, A., Taniguchi, H., Goto, A., & Oshima, H. (2002). Energy and depth of Usu 2000 phreatic explosions. *Geophys Res Lett*, 29(24), 10.1029/2002GL015928.

353. M. D. Nordyke, 1962 An Analysis of Cratering Data from Desert Alluvium. J Geophys Res 67 5 1965 74

354. A. J. Chabai, 1965 On Scaling Dimensions of Craters Produced by Buried Explosives. J Geophys Res 70 20 5075 98

355. P. Ross-S, S. Delpit, MJ Haller, K. Németh, H. Corbella, 2011 Influence of the substrate on maar-diatreme volcanoes- an example of a mixed setting from the Pali Aike volcanic field, Argentina. J Volcanol Geotherm Res 201 1-4 253 71

356. D. G. Jankowski, S. W. Squyres, 1992 The Topography of Impact Craters in "Softened" Terrain. Icarus 100 26 39

357. S. Self, J. Kienle, J. Hout-P, 1980 Ukinrek maar, Alaska, II Deposits and formation of the 1977 craters. J Volcanol Geotherm Res 7 39 65

358. J. Kienle, P. R. Kyle, S. Self, R. J. Motyka, V. Lorenz, 1980 Ukinrek maar, Alaska, I April 1977 Eruption sequence, petrology and tectonic setting. J Volcanol Geotherm Res 7 11 37

359. N. Geshi, K. Németh, T. Oikawa, 2011 Growth of phreatomagmatic explosion craters: A model inferred from Suoana crater in Miyakejima Volcano, Japan. J Volcanol Geotherm Res 201 1-4 30 8
360. R. S. J. Sparks, L. Baker, R. J. Brown, M. Field, J. Schumacher, G. Stripp, et al. 2006 Dynamical constraints on kimberlite volcanism. J Volcanol Geotherm Res 155 1-2 18 48
361. A. L. Walters, J. C. Phillips, R. J. Brown, M. Field, T. Gernon, G. Stripp, et al. 2006 The role of fluidisation in the formation of volcaniclastic kimberlite: Grain size observations and experimental investigation. J Volcanol Geotherm Res 155 1-2 119 37
362. S. R. Allen, I. E. M. Smith, 1994 Eruption styles and volcanic hazard in the Auckland Volcanic Field, New Zealand. Geoscience Reports of Shizuoka University 20 5 14
363. Kereszturi, G., Németh, K., Cronin, J. S., Smith, I. E. M., Agustin-Flores, J., Lindsay, J., (in preparation). Eruptive magmatic volumes of the Quaternary Auckland Volcanic Field (New Zealand).
364. D. J. Rout, J. Cassidy, CA Locke, I. E. M. Smith, 1993 Geophysical evidence for temporal and structural relationships within the monogenetic basalt volcanoes of the Auckland volcanic field, northern New Zealand J Volcanol Geotherm Res 57 1-2 71 83
365. R. M. Murtagh, J. D. L. White, Y. K. Sohn, 2011 Pyroclast textures of the Ilchulbong 'wet' tuff cone, Jeju Island, South Korea. J Volcanol Geotherm Res 201 1-4 385 96
366. P. D. Cole, J. E. Guest, A. M. Duncan, J. Pacheco-M, 2001 Capelinhos 1957-1958, Faial, Azores: deposits formed by an emergent surtseyan eruption. Bull Volcanol 63 2-3 204 20
367. J. F. Luhr, T. Simkin, 1993 Parícutin: the volcano born in a Mexican cornfield Geoscience Press
368. Mannen, K., & Ito, T. (2007). Formation of scoria cone during explosive eruption at Izu-Oshima volcano, Japan. *Geophys Res Lett*, 34(L18302), 10.1029/2007GL030874.
369. J. T. Gutmann, 1979 Structure and eruptive cycle of cinder cones in the Pinacate volcanic field and controls of strombolian activity. J Geol 87 448 54
370. M. Ripepe, A. J. L. Harris, 2008 Dynamics of the 5 April 2003 explosive paroxysm observed at Stromboli by a near-vent thermal, seismic and infrasonic array. Geophys Res Lett 35 7
371. B. Chouet, N. Hamisevicz, T. R. Mc Getchin, 1974 Photoballistics of Volcanic Jet Activity at Stromboli, Italy. J Geophys Res 79 32 4961 76
372. M. Ripepe, M. Rossi, G. Saccorotti, 1993 Image processing of explosive activity at Stromboli. J Volcanol Geotherm Res 54 3-4 335 51
373. Taddeucci, J., Scarlato P., Capponi, A., Del Bello, E., Cimarelli, C., Palladino, D. M., et al. (2012). High-speed imaging of Strombolian explosions: The ejection velocity of pyroclasts. *Geophys Res Lett*, 39(2), L02301, 10.1029/2011GL050404.
374. M. Rosi, A. Bertagnini, A. J. L. Harris, L. Pioli, M. Pistolesi, M. Ripepe, 2006 A case history of paroxysmal explosion at Stromboli: Timing and dynamics of the April 5, 2003 event. Earth Planet Sci Lett 243 594 606
375. J. Sable, B. Houghton, C. Wilson, R. Carey, 2006 Complex proximal sedimentation from Plinian plumes: the example of Tarawera 1886. Bull Volcanol 69 1 89 103
376. M. Kervyn, G. Ernst, J. Keller, R. Vaughan, J. Klaudius, E. Pradal, et al. 2010 Fundamental changes in the activity of the natrocarbonatite volcano Oldoinyo Lengai, Tanzania. Bull Volcanol 72 8 913 31
377. Patrick, M. R. (2007). Dynamics of Strombolian ash plumes from thermal video: Motion, morphology, and air entrainment. *J Geophys Res*, 112(B6), B06202-40, 10.1029/2006JB004387.
378. Guilbaud-N, M., Siebe, C., & Agustín-Flores, J. (2009). Eruptive style of the young high-Mg basaltic-andesite Pelagatos scoria cone, southeast of México City. *Bull Volcanol*, 71(8), 859-880.
379. G. G. J. Ernst, R. S. J. Sparks, S. N. Carey, M. I. Bursik, 1996 Sedimentation from turbulent jets and plumes. J Geophys Res 101 B3 5575 89

380. G. W. Bertotto, E. A. Bjerg, C. A. Cingolani, 2006 Hawaiian and Strombolian style monogenetic volcanism in the extra-Andean domain of central-west Argentina. J Volcanol Geotherm Res 158 3-4 430 44
381. A. Van Burkalow, 1945 Angle of repose and angle of sliding friction: an experimental study. Geol Soc Am Bull 56 6 669 707
382. P. Tokarev, 1978 Prediction and characteristics of the 1975 eruption of Tolbachik volcano, Kamchatka. Bull Volcanol 41 3 251 8
383. S. Fedotov, V. Enman, Y. Nikitenko, M. Maguskin, V. Levin, S. Enman, 1980 Crustal deformations related to the formation of new tolbachik volcanoes in 1975-1976, Kamchatka. Bull Volcanol 43 1 35 45
384. P. I. Tokarev, 1983 Calculation of the magma discharge, growth in the height of the cone and dimensions of the feeder channel of Crater I in the Great Tolbachik Fissure Eruption, July 1975. Fedotov SA, Markhinin YK, editors. The Great Tolbachik Fissure Eruption: Geological and Geophysical Data, 1975-1976 Cambridge: Cambridge University Press 27 35
385. S. Fedotov, A. Chirkov, N. Gusev, G. Kovalev, Y. Slezin, 1980 The large fissure eruption in the region of Plosky Tolbachik volcano in Kamchatka, 1975-1976. Bull Volcanol 43 1 47 60
386. M. Ripepe, S. Ciliberto, M. Della Schiava, 2001 Time constraints for modeling source dynamics of volcanic explosions at Stromboli. J Geophys Res 106 B5 8713 27
387. J. Taddeucci, M. Pompilio, P. Scarlato, 2004 Conduit processes during the July-August 2001 explosive activity of Mt. Etna (Italy): inferences from glass chemistry and crystal size distribution of ash particles. J Volcanol Geotherm Res 137 1-3 33 54
388. S. Self, R. S. J. Sparks, B. Booth, G. P. L. Walker, 1974 The 1973 Heimaey Strombolian Scoria deposit, Iceland. Geol Mag 111 539 48
389. P. Doubik, B. E. Hill, 1999 Magmatic and hydromagmatic conduit development during the 1975 Tolbachik Eruption, Kamchatka, with implications for hazards assessment at Yucca Mountain, NV. J Volcanol Geotherm Res 91 1 43 64
390. R. F. Holm, 1987 Significance of agglutinate mounds on lava flows associated with monogenetic cones: An example at Sunset Crater, northern Arizona. Geol Soc Am Bull 99 3 319 24
391. G. A. Valentine, F. V. Perry, D. Krier, G. N. Keating, R. E. Kelley, A. H. Cogbil, 2006 Small-volume basaltic volcanoes: Eruptive products and processes, and posteruptive geomorphic evolution in Crater Flat (Pleistocene), southern Nevada. Geol Soc Am Bull 118 11-12 1313 30
392. S. Self, L. Keszthelyi, T. Thordarson, 1998 The importance of pahoehoe. Annual Review of Earth and Planetary Sciences 26 81 110
393. S. Calvari, H. Pinkerton, 1999 Lava tube morphology on Etna and evidence for lava flow emplacement mechanisms. J Volcanol Geotherm Res 90 3-4 263 80
394. M. Inbar, A. R. Enriquez, J. H. G. Graniel, 2001 Morphological changes and erosion processes following the 1982 eruption of El Chichón volcano, Chiapas, Mexico. Géomorphologie: relief, processus, environnement 2001 3 175 83
395. T. M. Lutz, J. T. Gutmann, 1995 An improved method of determining alignments of point-like features and its implications for the Pinacate volcanic field, Mexico. J Geophys Res 100 B9 17659 70
396. T. S. Paulsen, T. J. Wilson, 2010 New criteria for systematic mapping and reliability assessment of monogenetic volcanic vent alignments and elongate volcanic vents for crustal stress analyses. Tectonophysics 482 1-4 16 28
397. N. R. Riggs, W. A. Duffield, 2008 Record of complex scoria cone eruptive activity at Red Mountain, Arizona, USA, and implications for monogenetic mafic volcanoes. J Volcanol Geotherm Res 178 763 76
398. Kereszturi, G., Jordan, G., Németh, K., & Dóniz-Páez, J. F. (2012). Syn-eruptive morphometric variability of monogenetic scoria cones. *Bull Volcanol*, 74(9), 2171-2185.
399. F. Di Traglia, C. Cimarelli, D. de Rita, Torrente. D. Gimeno, 2009 Changing eruptive styles in basaltic explosive volcanism: Examples from Croscat complex scoria cone,

Garrotxa Volcanic Field (NE Iberain Peninsula). J Volcanol Geotherm Res 180 2-4 89 109

400. Suhr, P., Goth, K., Lorenz, V., & Suhr, S. (2006). Long lasting subsidence and deformation in and above maar-diatreme volcanoes- a never ending story. *Zeitschrift der Deutschen Gesellschaft für Geowissenschaften*, 157(3), 491-511.
401. Lorenz, V., Suhr, P., & Goth, K. (2003). Maar-Diatrem-Vulkanismus- Ursachen und Folgen. Die Guttauer Vulkangruppe in Ostsachsen als Beispiel für die komplexen Zusammenhänge. *Zeitschrift für Geologische Wissenschaften*, 31(4-6), 267-312.
402. C. Dessert, B. Dupré, J. Gaillardet, L. M. François, C. J. Allègre, 2003 Basalt weathering laws and the impact of basalt weathering on the global carbon cycle. Chem Geol 202 3-4 257 73
403. A. J. West, A. Galy, M Bickle, 2005 Tectonic and climatic controls on silicate weathering. Earth Planet Sci Lett 235 1-2 211 28
404. P. V. Brady, S. A. Carroll, 1994 Direct effects of CO2 and temperature on silicate weathering: Possible implications for climate control. Geochim Cosmochim Acta 58 7 1853 6
405. M. E. Raymo, W. F. Ruddiman, 1992 Tectonic forcing of late Cenozoic climate. Nature 359 6391 117 22
406. C. S. Riebe, J. W. Kirchner, R. C. Finkel, 2004 Erosional and climatic effects on long-term chemical weathering rates in granitic landscapes spanning diverse climate regimes. Earth Planet Sci Lett 224 3-4 547 62
407. P. Hinsinger, Barros. O N. Fernandes, M. F. Benedetti, Y. Noack, G. Callot, 2001 Plant-induced weathering of a basaltic rock: experimental evidence. Geochim Cosmochim Acta 65 1 137 52
408. K. Németh, S. J. Cronin, R. B. Stewart, D. Charley, 2009 Intra- and extra-caldera volcaniclastic facies and geomorphic characteristics of a frequently active mafic island-arc volcano, Ambrym Island, Vanuatu. Sediment Geol 220 3-4 256 70
409. J. J. Major, T. Yamakoshi, 2005 Decadal-scale change of infiltration characteristics of a tephra-mantled hillslope at Mount St Helens, Washington. Hydrol Processes 19 18 3621 30
410. M. Ferrucci, S. Pertusati, R. Sulpizio, G. Zanchetta, M. T. Pareschi, R. Santacroce, 2005 Volcaniclastic debris flows at La Fossa Volcano (Vulcano Island, southern Italy): Insights for erosion behaviour of loose pyroclastic material on steep slopes. J Volcanol Geotherm Res 145 3-4 173 91
411. K. Segerstrom, 1950 Erosion Studies at Paricutin, State of Michoacán, Mexico Geological Survey Bulletin 965-A 1 163
412. J. D. L. White, 1991 The depositional record of small, monogenetic volcanoes within terrestrial basins. Fisher EV, Smith GA, editors. Sedimentation in Volcanic Settings 155 171
413. G. A. Valentine, C. D. Harrington, 2006 Clast size controls and longevity of Pleistocene desert pavements at Lathrop Wells and Red Cone volcanoes, southern Nevada. Geology 34 7 533 6
414. L. D. Mc Fadden, E. V. Mc Donald, S. G. Wells, K. Anderson, J. Quade, S. L. Forman, 1998 The vesicular layer and carbonate collars of desert soils and pavements: formation, age and relation to climate change. Geomorphology 24 2-3 101 45
415. L. D. Mc Fadden, S. G. Wells, M. J. Jercinovich, 1987 Influences of eolian and pedogenic processes on the origin and evolution of desert pavements. Geology 15 6 504 8
416. S. G. Wells, J. C. Dohrenwend, L. D. Mc Fadden, B. D. Turrin, K. D. Mahrer, 1985 Late Cenozoic landscape evolution on lava flow surfaces of the Cima volcanic field, Mojave Desert, California. Geol Soc Am Bull 96 12 1518 29
417. T. Kato, T. Kamijo, T. Hatta, K. Tamura, T. Higashi, 2005 Initial Soil Formation Processes of Volcanogenous Regosols (Scoriacious) from Miyake-jima Island, Japan. Soil Science & Plant Nutrition 51 2 291 301

418. R. Melo, G. Vieira, A. Caselli, M. Ramos, 2012 Susceptibility modelling of hummocky terrain distribution using the information value method (Deception Island, Antarctic Peninsula). Geomorphology 155-156 0 88 95

419. J. López-Martínez, E. Serrano, T. Schmid, S. Mink, C. Linés, 2012 Periglacial processes and landforms in the South Shetland Islands (northern Antarctic Peninsula region). Geomorphology 155-156 0 62 79

420. P. I. A. Kinnell, 2005 Raindrop-impact-induced erosion processes and prediction: a review. Hydrol Processes 19 14 2815 44

421. A. I. J. M. Van Dijk, L. A. Bruijnzeel, S. E. Wiegman, 2003 Measurements of rain splash on bench terraces in a humid tropical steepland environment. Hydrol Processes 17 3 513 35

422. T. Dunne, D. V. Malmon, S. M. Mudd, 2010 A rain splash transport equation assimilating field and laboratory measurements. J Geophys Res 115 F1 F01001

423. R. E. Horton, 1945 Erosional development of streams and their drainage basins; hydrophysical approach to quantitative morphology. Geol Soc Am Bull 56 3 275 370

424. S. J. Cronin, V. E. Neall, 2001 Holocene volcanic geology, volcanic hazard and risk on Taveuni, Fiji New Zeal J Geol Geophys 44 417 37

425. K. Németh, S. J. Cronin, 2007 Syn- and post-eruptive erosion, gully formation, and morphological evolution of a tephra ring in tropical climate erupted in 1913 in West Ambrym, Vanuatu. Geomorphology 86 115 30

426. J. Poesen, J. Nachtergaele, G. Verstraeten, C. Valentin, 2003 Gully erosion and environmental change: importance and research needs. Catena 50 2-4 91 133

427. N. Matsuoka, 2001 Solifluction rates, processes and landforms: a global review. Earth-Science Reviews 55 1-2 107 34

428. S. D. Holness, 2004 Sediment movement rates and processes on cinder cones in the maritime Subantarctic (Marion Island). Earth Surf Processes Landforms 29 1 91 103

429. C. D. Woodroffe, 2002 Coasts: Form, Process and Evolution. Cambridge: Cambridge University Press

430. Y. K. Sohn, S. Yoon-H, 2010 Shallow-marine records of pyroclastic surges and fallouts over water in Jeju Island, Korea, and their stratigraphic implications. Geology 38 8 763 6

431. B. Calles, K. Lindé, J. O. Norrman, 1982 The geomorphology of Surtsey island in 1980. Surtsey Research Progress Report IX 117 32

432. J. O. Norrman, B. Calles, R. A. Larsson, 1974 The geomorphology of Surtsey Island in 1972. Surtsey Research Progress Report VII 61 71

433. Y. K. Sohn, K. H. Park, S. Yoon-H, 2008 Primary versus secondary and subaerial versus submarine hydrovolcanic deposits in the subsurface of Jeju Island, Korea. Sedimentology 55 4 899 924

434. V. Rachold, M. N. Grigoriev, F. E. Are, S. Solomon, E. Reimnitz, H. Kassens, et al. 2000 Coastal erosion vs riverine sediment discharge in the Arctic Shelf seas. Int J Earth Sci 89 3 450 60

435. Arnalds, O., Gisladottir, F. O., & Orradottir, B. (2012). Determination of aeolian transport rates of volcanic soils in Iceland. *Geomorphology*, 167–168, 4–12.

436. E. F. Thorarinsdottir, O. Arnalds, 2012 Wind erosion of volcanic materials in the Hekla area, South Iceland. Aeolian Research 4 39 50

437. J. A. Rech, R. W. Reeves, D. M. Hendricks, 2001 The influence of slope aspect on soil weathering processes in the Springerville volcanic field, Arizona. Catena 43 1 49 62

438. J. C. Dohrenwend, L. D. Mc Fadden, B. D. Turrin, S. G. Wells, 1984 K-Ar dating of the Cima volcanic field, eastern Mojave Desert, California: Late Cenozoic volcanic history and landscape evolution. Geology 12 3 163 7

439. S. D. Rad, C. J. Allègre, P. Louvat, 2007 Hidden erosion on volcanic islands. Earth Planet Sci Lett 262 1-2 109 24

440. J. Gaillardet, R. Millot, B. Dupré, 2003 Chemical denudation rates of the western Canadian orogenic belt: the Stikine terrane. Chem Geol 201 3-4 257 79

441. Kereszturi, G., Procter, J., Cronin, J. S., Németh, K., Bebbington, M., & Lindsay, J. (2012). LiDAR-based quantification of lava flow hazard in the City of Auckland (New Zealand). *Remote Sens Environ*, 125, 198-213.
442. P. Oliva, J. Viers, B. Dupré, 2003 Chemical weathering in granitic environments. Chem Geol 202 3-4 225 56
443. M. H. Ort, M. D. Elson, K. C. Anderson, W. A. Duffield, T. L. Samples, 2008 Variable effects of cinder-cone eruptions on prehistoric agrarian human populations in the American southwest. J Volcanol Geotherm Res 176 3 363 76
444. S. K. Rowland, Z. Jurado-Chichay, G. J. Ernst, 2009 Pyroclastic deposits and lava flows from the 1759-1774 eruption of El Jorullo, México: Aspects of "violent strombolian" activity and comparison with Parícutin. Hoskuldsson A, Thordarson T, Larsen G, Self S, Rowland S, editors. The Legacy of George PL Walker, Special Publications of IAVCEI 2. London, UK: Geological Society of London 105 128
445. S. J. Cronin, M. J. Hedley, V. E. Neall, R. G. Smith, 1998 Agronomic impact of tephra fallout from the 1995 and 1996 Ruapehu Volcano eruptions, New Zealand. Environ Geol 34 1 21 30
446. S. Grishin, R. del Moral, 1996 Dynamics of forests after catastrophic eruptions of Kamchatka's volcanoes. Turner IM, Diong CH, Lim SSL, Ng PKL, editors. Biodiversity and Dynamics of Ecosystems. DIWPA Series 1 133 146
447. T. Chinen, A. Riviere, 1990 Post-eruption erosion processes and plant recovery in the summit atrio of Mt. Usu, Japan. Catena 17 3 305 14
448. J. D. L. White, B. F. Houghton, K. A. Hodgson, C. J. N. Wilson, 1997 Delayed sedimentary response to the A.D. 1886 eruption of Tarawera, New Zealand. Geology 25 5 459 62
449. P. Kralj, 2011 Eruptive and sedimentary evolution of the Pliocene Grad Volcanic Field, North-east Slovenia J Volcanol Geotherm Res 201 1-4 272 84
450. S. J. Shaw, A. B. Woodland, J. Hopp, N. D. Trenholm, 2010 Structure and evolution of the Rockeskyllerkopf Volcanic Complex, West Eifel Volcanic Field, Germany. Bull Volcanol 72 971 90
451. F. M. Bullard, 1947 Studies on Paricutin volcano, Michoacan, Mexico. Geol Soc Am Bull 58 433 50
452. T. Chinen, 1986 Surface erosion associated with tephra deposition on Mt. Usu and other volcanoes. Environmental Science, Hokkaido 9 1 137 49
453. G. A. Valentine, D. M. Palladino, E. Agosta, J. Taddeucci, R. Trigila, 1998 Volcaniclastic aggradation in a semiarid environment, northwestern Vulcano Island, Italy. Geol Soc Am Bull 110 5 630 43
454. S. E. Bryan, J. Marti, R. A. F. Cas, 1998 Stratigraphy of the Bandas del Sur Formation; an extracaldera record of Quaternary phonolitic explosive eruptions from the Las Canadas edifice Tenerife (Canary Islands) Geol Mag 135 5 605 36
455. J. Martí, J. Mitjavila, V. Araña, 1994 Stratigraphy, structure and geochronology of the Las Cañadas caldera (Tenerife, Canary Island). Geol Mag 131 6 715 27
456. R. A. Shakesby, 2011 Post-wildfire soil erosion in the Mediterranean: Review and future research directions. Earth-Science Reviews 105 3-4 71 100
457. M. Inbar, M. Tamir, L. Wittenberg, 1998 Runoff and erosion processes after a forest fire in Mount Carmel, a Mediterranean area. Geomorphology 24 1 17 33
458. S. W. Trimble, A. C. Mendel, 1995 The cow as a geomorphic agent- A critical review. Geomorphology 13 1-4 233 53
459. D. R. Butler, 1995 Zoogeomorphology- Animals as Geomorphic Agents. Cambridge, UK: Cambridge University Press
460. J. Boelhouwers, T. Scheepers, 2004 The role of antelope trampling on scarp erosion in a hyper-arid environment, Skeleton Coast, Namibia. J Arid Environ 58 4 545 57
461. D. R. Butler, 1993 The Impact of Mountain Goat Migration on Unconsolidated Slopes in Glacier National Park, Montana. The Geographical Bulletin 35 2 98 106
462. B. Zolitschka, F. Schäbitz, A. Lücke, H. Corbella, B. Ercolano, M. Fey, et al. 2006 Crater lakes of the Pali Aike Volcanic Field as key sites for paleoclimatic and

paleoecological reconstructions in southern Patagonia, Argentina. J S Am Earth Sci 21 3 294 309

463. L. Hably, Z. Kvaček, 1998 Pliocene mesophytic forests surrounding crater lakes in western Hungary. Review of Palaeobotany and Palynology 101 1 257 69
464. M. Inbar, J. L. Hubp, L. V. Ruiz, 1994 The geomorphological evolution of the Paricutin cone and lava flows, Mexico, 1943-1990. Geomorphology 9 57 76
465. I. G. B. E. Sucipta, I. Takashima, H. Muraoka, 2006 Morphometric age and petrological characteristic of volcanic rocks from the Bajawa cinder cone complex, Flores, Indonesia. J Mineral Petrol Sci 101 2 48 68
466. K. Németh, S. K. Suwesi, Z. Peregi, Z. Gulácsi, J. Ujszászi, 2003 Plio/Pleistocene Flood Basalt Related Scoria and Spatter Cones, Rootless Lava Flows, and Pit Craters, Al Haruj Al Abiyad, Libya. Geolines 15 98 103
467. M. R. Moufti, A. M. Moghazi, K. A. Ali, 2012 Geochemistry and Sr-Nd-Pb isotopic composition of the Harrat Al-Madinah Volcanic Field, Saudi Arabia. Gondwana Res 21 2-3 670 89
468. A. P. Cabrera, P. J. Caffe, 2009 The Cerro Morado Andesites: Volcanic history and eruptive styles of a mafic volcanic field from northern Puna, Argentina. J S Am Earth Sci 28 2 113 31
469. V. Manville, K. Németh, K. Kano, 2009 Source to sink: A review of three decades of progress in the understanding of volcaniclastic processes, deposits, and hazards. Sediment Geol 220 3-4 136 61

CHAPTER 2

An Overview of Mud Volcanoes Associated to Gas Hydrate System

Umberta Tinivella[1] and Michela Giustiniani

[1] OGS – National Institute of Oceanography and Experimental Geophysics, Borgo Grotta Gigante 42C, 34010, Trieste, Italy

1. INTRODUCTION

Humankind has been aware of volcanic activity in this planet since ancient times. This can be inferred from the remains of human settlements giving evidence of destruction by volcanic activity, and by the many myths around the world describing events that can be interpreted in relation to volcanic eruptions. Furthermore, such occurrences are evidenced by the special words that some human groups created to designate the special cases of "fire mountains", thus distinguishing these from other (non-volcanic) mountains found in the same region. More recently, as suggested by [1] Cañón-Tapia and Szakács, films and television shows devoted to exploring volcanoes have become very common, making it easier for the general public to gain access to "firsthand" experiences concerning this type of natural phenomenon. Consequently, it is only fair to say that at present almost everyone has an "intuitive" knowledge of what a volcano is [1]. In last years, many authors have devoted efforts to provide some additional theoretical aspects about volcanism, and one of the most recent publications is the book title "What is a volcano?" [1], addressing also a phylosophical question about this topic.

The definition of volcano is often missed or just defined as "opening-in-the-ground" in the textbooks. Starting from the classical Aristotelian requirements of a definition, it is shown that only a definition that is part of a hierarchically organized system of definitions can be accepted. Thus, conceptual constructs should reflect the same type of makeup as nature's processes, which are hierarchically organized. Such a line of reasoning implies that a volcano should be defined by making an explicit mention of the hierarchy of systems to which it belongs. Therefore, volcano can be defined as either a subsystem (i.e., the eruptive subsystem) of the broader igneous system or as a particular type of igneous system (i.e., one reaching the surface of Earth). A volcano, viewed as a volcanic system, is composed of a magma-generation subsystem, a magma transport subsystem, magma storage subsystem(s), and an eruptive subsystem. The accurate definition and identification of each subsystem should allow distinction between individual volcanoes in both space and time. Minimal conventional requirements need to be agreed upon by volcanologists to identify and recognize a particular volcano from other volcanoes, including those partially occupying the same space but separated in time, or those partially overlapping in both space and time. Conceptual volcanology can be envisaged as addressing the issue of accurate definition of basic terms and concepts, besides nomenclature and systematics, aiming at reaching the conceptualization level of more basic sciences. In contrast to volcanologists, sedimentologists are not only interested in "classical" volcanoes, but also in a second type, sedimentary volcanoes. This type of volcano is helpful for sedimentologists in understanding the processes that occur in the commonly unconsolidated subsoil, even after deep burial. Sedimentary volcanoes can be grouped in three classes: mud volcanoes, sand volcanoes, and associated structures such as water-escape and gas-escape structures. Sedimentary volcanoes have several characteristics in common with "classical" volcanoes, including their shapes and the processes that contribute to their genesis, as well described by [2] and here reported. These two type of volcanoes have largely similar morphologies, they overlap each other in size (large sedimentary volcanoes reach size that compete with those of many classical volcanoes), and they have a genesis that is comparable in many respect. For instance, comparison of the material flowing out from classical and sedimentary volcanoes shows that the sedimentary outflows resemble basaltic lava in the case of mud volcanoes, and acid lava in the case of sand volcanoes; this results in sedimentary volcanoes that morphologically resemble shield volcanoes and stratovolcanoes respectively.

It could be established that sedimentary volcanoes have, like the "classical" one, some kind of magma chamber in the form of a gas-and/or liquid-bearing layer with increasing pressure (either continuously, for instance, from the weight of the ever thickening sedimentary overburden, or incidentally, for instance, from an earthquake-induced shock wave). If the pressure exceeds a threshold, the pressured gas and/or liquid (commonly pore water with dissolved air or hydrocarbons) breaks upward through the overlying sediment, often following an already existing zone of weakness, to finally flow out at the sedimentary surface, either subaerially or subaqueously. In fact, the material that rises up through the connection between the source in the subsoil and the sedimentary surface behaves in several respects as magma on its journey from the magma chamber to the vent or crater opening: Magma becomes more fluid and develops gas bubbles while rising owing to decreasing pressure, and in sedimentary volcanoes with a deep-seated source the formation of bubbles also take place. These similarities should be promoted a fruitful collaboration between the researchers that can be mutually beneficial.

Among the sedimentary volcanoes, the mud volcanoes are the most interesting form many points of view, as described in this chapter. Mud volcanoes and mud volcanism are some of nature's most anonymous, mysterious and undiscussed geological features, which have been studied for millennia. This is strong interested and is explained considering a number of facts. First of all, thousands of mud volcanoes exist worldwide, defining and affecting the habitat and the daily lives of the millions of people living amongst them. Secondly, mud volcanism and mud volcano distribution is strongly connected to the formation and the distribution of the world's petroleum assets, thus serving as an indicator for valuable natural resources [3]. Thirdly, mud volcanoes offer an insight into otherwise hidden deep structural and diagenetic processes, such as the formation of gas hydrates, mineral dissolution and transformation, degradation of organic material and high pressure/temperature-reactions [4]. Lastly, mud volcanism generally involves voluminous generation and emission of both methane and carbon dioxide whereby most mud volcanoes serve as an efficient, natural source of greenhouse gases and, consequently, play an important role in global climate dynamics [3,5-10].

Among fluid venting structures, mud volcanoes are the most important phenomena related to natural seepage from the earth's surface [11]. Their geometry and size are variable, from one to two meters to several hundred meters in height, and they are formed as a result of the emission of argillaceous material and fluids (water, brine, gas, oil; [3,12-13]). They occur globally in terrestrial and submarine geological settings: most terrestrial mud volcanoes are located in convergent plate margin with thick sedimentary sequences within the Alpine-Himalayan, Carribean and Pacific orogenic belts [14-21]. Mud volcanoes and mud diapirs are responsible for the genesis of many chaotic deposits, such as mélanges, chaotic breccias and various deformed sediments [22-24].

The normal activity of mud volcanoes consists of gradual and progressive outflows of semi-liquid material called mud breccia or diapiric mélange. Explosive and paroxysmal activities are interpreted as responsible for ejecting mud, ash, and decimetric to metric clasts. Mud volcano breccias are composed of a mud matrix, which supports a variable quantity of chaotically distributed angular to rounded rock clasts, ranging in diameter from a few millimeters to several meters [i.e., 14,18]. Clasts are of various lithologies and provenances, derived from the rocks through which the mud passed on its way to the surface or to the sea-floor. Slumps, slides and sedimentary flows can also affect the entire structure of mud volcanoes, even if gradients are very low.

The occurrence of mud volcanoes is controlled by several factors, such as tectonic activity, sedimentary loading due to rapid sedimentation, the existence of thick, fine-grained plastic sediments and continuous hydrocarbon accumulation [13,14,25-28].

A comprehensive study of submarine mud volcanoes is increasing in the last decades due to the wide use of geophysical methods, in particular side scan sonar, and the increased accuracy of the positioning of bottom samplers. Reference [12] presented an up-to-date list of known and inferred submarine mud volcanoes, describing their distribution, the mechanisms by which they form, and associated gas hydrates accumulations (Figure 1). Reference [12] clearly summarized the importance of research on submarine mud volcanoes, which is:

1. they are a source of methane flux from lithosphere to hydrosphere and atmosphere (greenhouse effect and climatic change);
2. they may provide evidence of high petroleum potential in the deep subsurface;
3. useful data about the sedimentary section in mud volcanic areas can be determined by examination of rock fragments incorporated in mud volcanic sediments (breccia);
4. submarine mud volcanic activity may impact drilling operations, ring installations and pipeline routings;
5. gas hydrates associated with deep-water mud volcanoes are a potential energy resource.

In this Chapter, we review the mud volcanism related to gas hydrate system. It is well known that natural gas hydrate occurs worldwide in oceanic sediment of continental and insular slopes and rises of active and passive margins, in deep-water sediment of inland lakes and seas, and in polar sediment on both continents and continental shelves (Figure 2). In marine sediments, where water depths exceed about 300 m and bottom water temperatures approach 0° C, gas hydrate is found at the seafloor to sediment depths of about 1100 m. In polar continental regions, gas hydrate can be present in sediment at depths between about 150 and 2000 m. Thus, natural gas hydrate is restricted to the shallow geosphere where its presence affects the physical and chemical properties of near-surface sediment [29].

Finally, we show an interesting case study in Antarctic Peninsula, where an important gas hydrates reservoir and mud volcanism are associated [30-32].

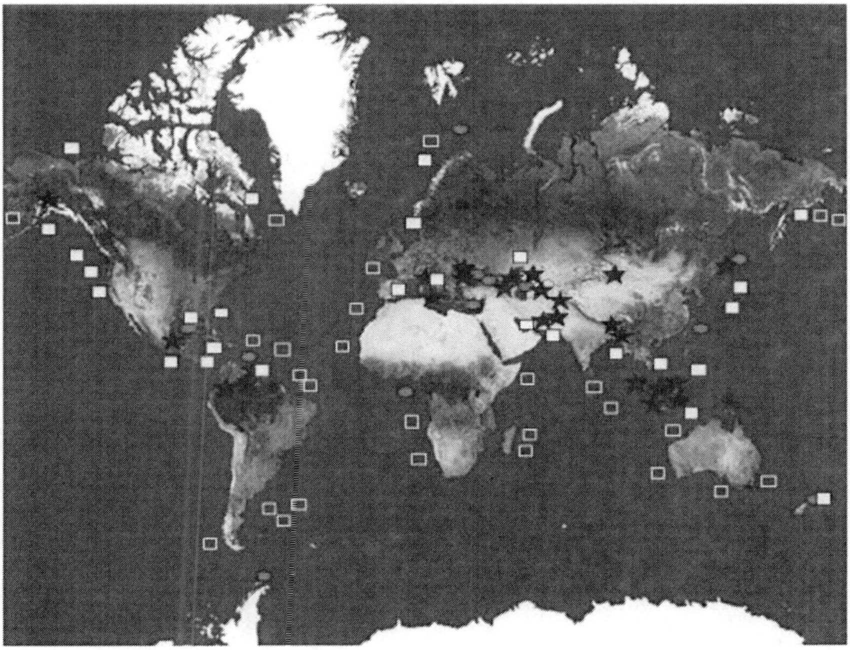

FIGURE 1. Map showing the worldwide locations of onshore (blue stars, after [33], with additions), known (red open oval, without gas hydrates; solid red oval hydrate bearing), and inferred (solid yellow rectangle) submarine mud volcanoes. The "possible sediment diapirs" mapped by [34] are also shown (open yellow rectangle). Modified after [12].

2. THE MUD VOLCANOES

Mud volcanism is not one specific process and mud volcanoes are not uniform feature settings; in fact, driving forces, activity, materials and morphologies may vary significantly [3,12-13,35-37]. About 2000 mud volcanoes have been identified in the worldwide; however, as exploration of the deep seas continues, this number is expected to increase substantially. It is estimated that the total number of submarine mud volcanoes is between 7,000 and 1,000,000 [10,12].

The geographical distribution of mud volcanoes is strongly controlled by geological environments in which they occur, as pointed out by reference [13]. In fact, mud volcanoes are localized within the compressional zones, such as accretionary complexes, thrust and overthrust belts, forelands of Alpine orogenic structures, as well as zones of dipping noncompensating sedimentary basins, which coincide with the active plate boundaries. The few mud volcano areas out of these belts are attached to zones with high rates of recent sedimentation, such as modern fans (including underwater deltas of large rivers) or intensive development of salt diapirism.

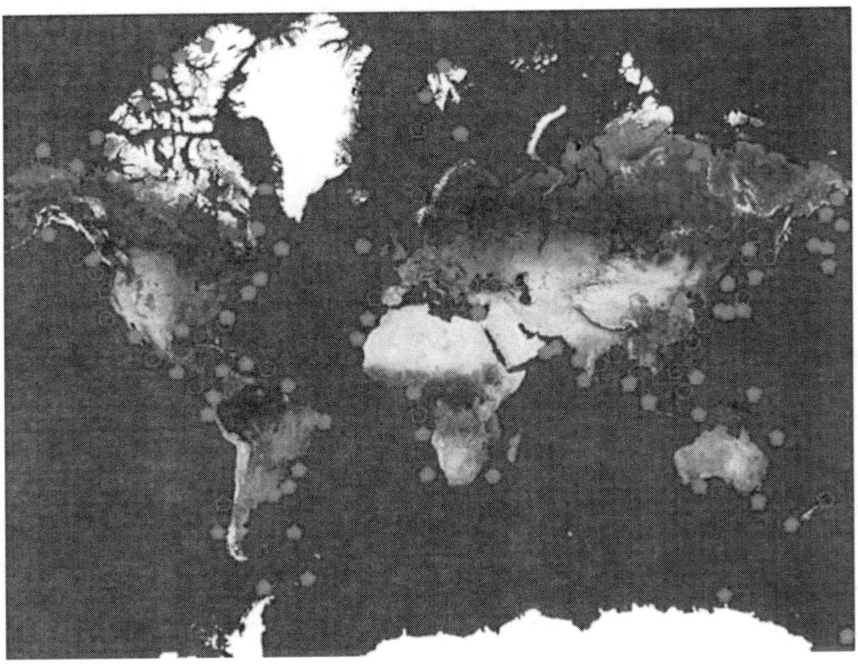

FIGURE 2. Map showing the worldwide locations of gas hydrate obtained by direct (open red pentagons) and indirect (solid red pentagons) measurements. Modified after [29].

Moreover, in all the mud volcano areas, there are suitable "source" layers of muddy sediments in the deeper part of the sedimentary basins or in the vicinity of the decolements in the accretionary complexes. Usually, this source is composed of fine-grained, soft material of low density overlain by at least 1–1.5 km of denser sediments. All collision complexes, where mud volcanoes are abundant, are characterized by thick sediment sequences caused by thickening of accreted sediments. The same is valid for forearc and outer orogenic basins where thickening is caused by thrusting and overthrusting. Enormous thickness is established in the noncompensating sedimentary basins.

Finally, given that the mud volcanoes are situated in areas where hydrocarbons have been, or are, actively being generated, there is a strong connection between the world mud volcano distribution and industrial oil and gas concentrations. In fact, the surrounding facies, below and laterally adjacent to mud volcanoes, may be particularly favorable as both reservoir and source environments for hydrocarbons, very often resulting in multilevel fields. Although this relationship is not valid for many mud volcano areas, for modern accretionary complexes in particular, it is established that present-day or recently active oil and especially gas generation are characteristic features for all of them.

Summarizing these brief comments the occurrence of mud volcanoes is strongly controlled by [12]:

- recent tectonic activity, particularly compressional activity;
- sedimentary or tectonic loading due to rapid sedimentation, accreting or overthrusting;
- continuous active hydrocarbon generation;
- existence of thick, fine-grained, soft, plastic sediments deep in the sedimentary succession.

Here, we summarize the main locations of the mud volcanoes as described in [12-13] and reported in Figure 1. The known or supposed mud volcanoes are irregularly clustered in separated areas forming belts, which almost totally coincide with active areas of the plate boundaries and zones of young orogenic structures. More than half mud volcanoes can be related to the Alpine-Himalaya active belt, where the largest and best cone-shaped mud volcanoes occur, as resumed in [13]. The most active terrestrial mud volcano area with the greatest number of mud volcanoes in the world is the Baku region of the Caspian Coast, Eastern Azerbaijan. Along the Alpine-Himalaya active belt, mud volcanism has been recognized in:

1. Mediterranean Ridge [28] and adjacent land areas (Sicily, Albania and Southern, Central and Northern Italy [38]),
2. the forelands of Eastern Carpathians in Romania, Kerch and Taman Peninsulas [39,40],
3. Great Caucasus [40] and the Black Sea [27],

4. the area of Southern Caspian Sea (Azerbaijan and Turkmenistan [26,41], South Caspian Basin [42], and Gorgon Plain in Iran),
5. the Makran coast of Pakistan [43],
6. Southern Himalayas (India and China), and Burma.

Furthermore, the Alpine–Himalayas mud volcano belt continues to the south in the most NE part of Indian Ocean on and around numerous forearc islands situated along the Indonesia and Banda Arcs [22], Indonesia – Australia accretion and collision complexes [44], as well as within the Banda accretionary complex offshore [22]. The greatest number of mud volcanoes seems to be known on Timor Island at the southeast end of this belt.

The western flank of the Pacific ocean – from the Sakhalin Island/Sea of Ochotsk-area in the north via Japan, Taiwan, the Marianas, Melanesia, Samoa and Australia to New Zealand in the south – holds some 150 onshore individuals [3,7]. The total number of offshore mud volcanoes along this belt is not yet fully determined but can be expected to be even higher, while the eastern flank of the Pacific Ocean is markedly less dense in mud volcanoes. Examples are known from and around the Aleutian Trench, Alaska, British Columbia, California, Costa Rica, Ecuador and inland Peru [3,7].

In the Atlantic Ocean, several hundreds of both onshore and offshore mud volcanoes have been recognized. The vast majority is concentrated along the Caribbean thrust belts and within the Barbados accretionary complex [7], although smaller clusters/individual features have been confirmed in connection to the Amazon and the Niger deltas [7,35], along the Gulf of Cadiz [45], within the southern Canary basin [46] and offshore Portugal and Morocco in the Alboran Basin [47].

Smaller numbers of mud volcanoes have also been described from the Mississippi delta [14], Lake Michigan [3], Greenland [3], the North Sea [48], the Netherlands [49], and areas of salt diapirism, such as in the Gulf of Mexico [50], Buzachi Peninsula (North-Eastern Caspian Sea), where they are related to salt diapirism, and Alboran basin in the Western Mediterranean [47].

The main components, that contribute to mud volcanoes formation and activities, are: mud breccia, water and gas. The relative quantities and the exact qualitative properties of these components vary, depending on local geology and processes at work.

Mud breccia is basically clasts in a clay mineral-rich matrix and is what makes up most mud volcanic features. Whereas the mud typically stems from one specific carrier bed and thus is characterized by a distinct

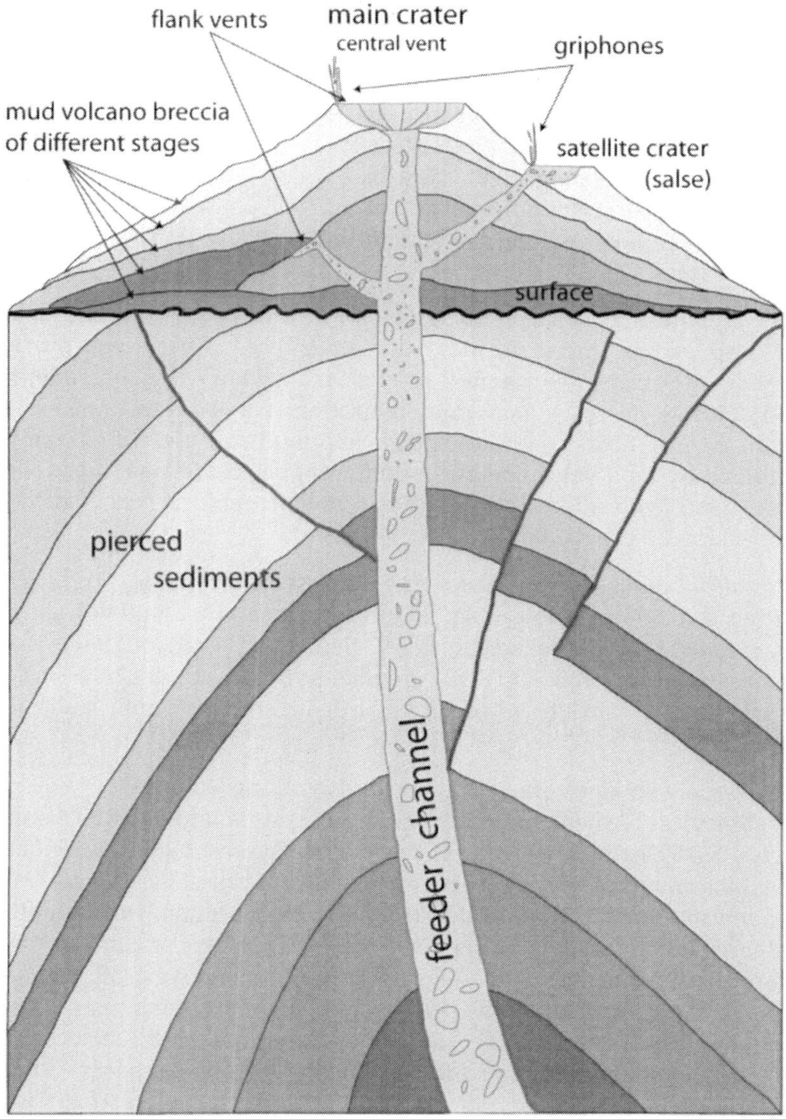

FIGURE 3. Basic structure and main elements of a conical mud volcano. Gryphones are small secondary vents shorter than 3 m, which may form around the craters and in many places on the mud volcano body. These commonly emit gas, mud and water and are characterized by complete absence of solid rock fragments. Modified after [13].

geochemical signature and clast fragments. The first reflects subsurface mud volcanic conditions and processes (clay mineral dehydration/ transformation processes), while the latter is derived from units through which the mud pass on its way to the surface and are consequently of variable lithologies, sizes (up to 5 m) and shapes. Young and forceful mud volcanoes generally extrude mud breccias with a very high clast-matrix ratio (virtually clast-supported deposits), whereas the mud breccia of older mud volcanoes may be virtually clast-free with a mud content of up to 99% [3,13,35]. The latter are often related to the final phase of an eruptive cycle, when loose wall rock along the conduit has already been removed by the ascending mud [3].

The water in mud volcanic extrusions typically stems from both shallow and deep sources and is normally derived through a variety of processes. Consequently, exact geochemical properties may vary virtually indefinitely [4,51]. However, clay mineral-dehydration water often makes up a significant proportion [3]. Mud breccia and mud volcanic water commonly mix whereby mud volcanic flows of different viscosities may form. During fierceful mud volcanic eruptions, up to 5 million cubic metres of such flow-material can be expelled [35].

Methane is almost always the dominated gas (70 - 99%) produced and emitted through mud volcanism. Moreover, since most mud volcanoes are very deeply rooted, thermogenic, 14C-depleted (fossil) methane is more common than biogenic [3,10,52]. Remainders typically include (in falling order) carbon dioxide, nitrogen, hydrogen sulfide, argon and helium [10,13,37].

As clearly explained in [13], a mud volcano comprises two main morphological elements: an internal feeder system and an external edifice (Figure 3). The characteristics of these elements are highly dependent on prevailing mud volcanic processes and in some cases, vice versa. Usually, mud volcano breccia is extruded from one major funnel called central or feeder channel (as summarized in Figure 3). Near the surface, several accompanying smaller flank or lateral pipes may split off the feeder channel. The outcrop of the feeder channel (usually situated on the summit of mud volcano) is called the main vent or central crater being of varied shapes: from planoconvex or flat and bulging plateau circled by a bank to deeply sunk rim depression (a caldera-type crater). Calderas form when a volcano collapses because a large mass of mud volcano breccia has drained through a lower vent, or because of the expulsion of a massive amount of material in an explosive eruption. Such eruptions have been known to destroy the entire structure of a volcano. Craters associated with the lateral vents are called satellite, parasite or secondary craters (Figure 3). Sometimes they collapse and are filled up by water that collects to form small lakes. Such a pool bubbling clay and gas is called salses. Numerous small secondary vents called gryphons may form around the craters and in many places on the mud volcano body. These commonly emit gas, mud and water and are characterized by complete absence of solid rock fragments. Although the numerous visual observations show that the gas seeps are very common on submarine mud volcanoes [53-55], the migrated gas is often captured in the near bottom sediments as gas hydrate [12,56-58] or trapped in shallow reservoirs to erupt when overpressured to form pockmarks on the seafloor [59]. The extruded mud volcano breccia spills in relatively thin sheets from the craters over the landscape in the form of broad fan-shaped or tongue-like flows up to several hundred meters wide and some kilometers long, as explained in reference [41]. This builds up the body of the mud volcano, typically covering some thousands of square meters with each phase of activity, totaling up to few tens of square kilometers. The fluid behavior of the mud volcano breccia is attributed to its high water content, which on land rapidly evaporates to drain the mud over a period of several days. Slumps and slides often form on the entire structure of the mud volcanoes even in very low gradients.

The internal feeder systems of mud volcanoes are not well known. Although studies imply rather large variabilities, typically, they consist of one main, central and deeply (km-scale) rooted feeder channel through which most mud volcanic material is transported. Feeder channels can be everything from cylindrical to irregular shaped to mere slits [3]. Near the surface, feeder channels tend to thin off and split into smaller flanking/lateral pipes [21]. The diameters of volcanic conduits may have a profound impact on mud volcanic activity. Generally, the wider is the conduit, the more voluminous is the expulsions [3].

The external morphology and expression of a mud volcano may vary almost indefinitely. The outcrops (vents/craters) of feeder channels may take on a variety of shapes; from plano-convex or flat and bulging to concave collapse structures of caldera-type [13]. Some mud volcanoes are in fact rather anonymous and quiescent features appearing merely as solitary, mm-scale openings in the ground surface, gently seeping small amounts of high-viscosity mud breccias and/or gas [14,35]. However, some mud volcanoes are really hazardous and expel voluminous amounts of low-viscosity mud-flows through frequent, short but fierce and explosive eruptions. This type of mud volcanoes typically evolve into kilometre-scale, chaotic and complex landscapes that comprise anything from clusters of cone-shaped morphologies rising hundreds of meters above ground to mounds, ravines, pools of bubbling mud and/or water (salses), mud cracks and clastic lobes [14,37,60]. During and following this type of active, hazardous mud volcanism, combustion of emitted gases may produce columns of flames rising up to several hundreds of meters, potentially burning for months or even years [37].

Mud volcanism and mud volcanoes have repeatedly been suggested to be a natural way of degassing the Earth's interior [13,61-63]. Although mud volcanism typically does involve thermogenic formation and expulsion of gas (a natural process which to a certain extent independently would be able to force deeply buried material to the surface), such processes can hardly serve to explain the truly vast extent and scope of worldwide mud volcanism. As resumed by [35], based on the large differences observed in shape, size and eruption styles of mud volcanoes, it is clear that there is no unique model that can explain them all. Ultimately, mud volcanoes form either as clay diapirs that reach and pierce the ground-surface or as fluidized argillaceous sediments, together with water and various amounts of hydrocarbon gases, which are extruded along structural weaknesses (conduits) within subsurface sediments/rocks [12].

Either way, a fundamental requisite for mud volcanism is the existence of a potential source domain; solitary or interconnected argillaceous carrier beds for migrating fluids and gases. Yet, for the actual volcanic processes to commence and continue – for gases to form and/or for the source material to move, rise and eventually extrude from the subsurface – additional forces are needed.

Since a vast majority of the mud volcanoes that are known today exist along active plate boundaries and, more specifically, along the anticlinal crests of accretionary prisms (the major depositional centres), compression through convergent tectonics and associated high sediment accumulation rates are generally considered the major mechanisms of mud volcanic initiation and sustenance. Argillaceous sediments and rocks are typically very weak and therefore, under the influence of compressive forces, prone to various clay mineral alteration and dehydration processes [4] and to brittle deformation through e.g. faulting. Moreover, under these very conditions, thermal and/or biogenic formation of hydrocarbon gases typically increases. Together, this implies formation of potential volcanic conduits, liquefaction, fluidization, gasification, density inversion, pore pressure increase and focused migration of mud volcanic material – i.e. mud volcanism – either through diapirism or along newly created faults/conduits [3,13-14,35].

Finally, the same forces and processes may explain mud volcano formation along passive continental margins. Although tectonic forces are lacking in such settings, compression, fluidization, gasification, overpressuring and mud volcanism may take place due to loading through rapid deposition of large amounts of (argillaceous) sediments [12,35]. A common characteristic for regions of mud volcanism located outside convergent plate boundaries is that mud volcanoes measure greatly in the vertical section (at least 2 km) and that they are a compound of undercompacted sedimentary sequences [7]. Consequently, although local settings may vary, the main mechanism of formation for mud volcanoes and mud volcanism is compression – either through tectonic forces or through high sediment accumulation rates – eventually leading to overpressuring through in situ gas generation, fluidization and liquefaction.

The fact that most mud volcanoes present regular, distinct seasonal changes in activity on a range from weeks to tens of years suggests an influence of more than one external agent, initiating and sustaining some kind of continuous, cyclic, natural pressure-recharging process within the mud volcanoes themselves. Astronomical cycles – e.g. orbital forcing – undoubtedly serve as one explanation. Through altering atmospherical and hydrospherical PT-conditions over a great variety of time-scales, such cycles may also affect and alter PT-conditions in the sediments and thereby mud volcanic processes via e.g. fluid access and bacterial activity (gas formation; [6]). As an example, after studying mud volcanism in the south Caspian Basin, reference [37] concluded that as much as 60% of all eruptions took place during either new or full moon. Moreover, reference [35] suggested a relationship between an 11- year cycle of the sun's activity and the initiation of mud volcano eruptions.

Even though astronomical cycles may explain most of the steady variations in mud volcanic eruption frequencies, they do not explain the rather frequent, more irregular eruptions. These are rather a result of ample, sudden seismic activity. If earthquake hypocenters are located within/in connection to potential carrier beds, shaking of the sediments may induce liquefaction and faulting as well as a significant increase in gas formation and dissociation. Consequently, rather sudden, eruptive mud volcanism may be generated in a normally quiescent or even dormant mud volcanic area [3,13,37,64].

Mud volcanoes can be related also to volcanic basins, which are sedimentary basins with a significant amount of primary volcanic rocks (e.g. sills and dykes). Pierced basins are sedimentary basins with many piercement structures such as mud volcanoes, dewatering pipes and hydrothermal vent complexes. Sills are tabular igneous intrusions that are dominantly layer parallel. They are commonly subhorizontal. Sills may locally have transgressive segments (i.e. segments that cross-cut the stratigraphy). Hydrothermal vent complexes are pipe-like structures formed by fracturing, transport and eruption of hydrothermal fluids. These complexes are dominated by sedimentary rocks with a negligible content of igneous material. Sediment volcanism is surface eruption of mud, sand or sediment breccias through a vent complex [65].

Hydrothermal and phreatomagmatic vent complexes are recognized from several sedimentary basins associated with large igneous provinces, including the Vøring and Møre basins off mid-Norway [66-68], the Faeroe–Shetland Basin [e-g- 69], the Karoo Basin in southern Africa [e.g. 70-75], in the Karoo-equivalent basins of Antarctica [76-78], and the Tunguska Basin in Siberia, Russia [e.g. 79]. Generally, the hydrothermal vent complexes represent conduit zones up to 8 km long rooted in contact aureoles around sill intrusions, where the upper part of the vent complexes comprise eyes, craters or mounds, up to 10 km in diameter [68]. An important consequence of intrusive activity in sedimentary basins is that the magma causes rapid heating of the intruded sediments and their pore fluids, causing expansion and boiling of the pore fluid [75], and metamorphic dehydration reactions. These processes may lead to phreatic volcanic activity by the formation of cylindrical conduits that pierce sedimentary strata all the way to the surface. The hydrothermal vent complexes thus represent pathways for gases produced in contact aureoles to the atmosphere, with the potential to induce global climate changes [80]. Consequently, constraints on processes leading to the formation of hydrothermal vent complexes in sedimentary basins, their abundance and structure may lead to a better understanding of the causes of the abrupt climate changes that are associated with many large igneous provinces [e.g. 81, 82]. References [65, 83]have analyzed the presence of voluminous basaltic intrusive complexes, extrusive lava sequences and hydrothermal vent complexes in the Karoo basin. In this area, the hydrothermal vents pierce the horizontally stratified sediments of the basin. They study have documented that the hydrothermal vent complexes were formed by one or a few phreatic events, leading to the collapse of the surrounding sedimentary strata. They proposed a model in which hydrothermal vent complexes originate in contact metamorphic aureoles around sill intrusions. Heating and expansion of host rock pore fluids resulted in rapid pore pressure build-up and phreatic eruptions. The hydrothermal vent complexes represent conduits for gases and fluids produced in contact metamorphic aureoles, slightly predating the onset of the main phase of flood volcanism.

Reference [84] investigated and understood the mechanisms responsible for the formation of piercement structures in sedimentary basins and the role of strike-slip faulting as a triggering mechanism for fluidization. For this purpose four different approaches were combined: fieldwork, analogue experiments, and mathematical modeling for brittle and ductile rheologies. The results of this study may be applied to several geological settings, including the newly formed Lusi mud volcano in Indonesia [84], which became active the 29th of May 2006. Their integrated study demonstrates that the critical fluid pressure required to induce sediment deformation and fluidization is dramatically reduced when strike-slip faulting is active. The proposed shear-induced fluidization mechanism explains why piercement structures such as mud volcanoes are often located along fault zones. Their results support a scenario where the strike-slip movement of the Watukosek fault triggered the Lusi eruption and synchronous seep activity witnessed at other mud volcanoes along the same fault. The possibility that a drilling, carried out in the same area, contributed to trigger the eruption cannot be excluded. However, so far, no univocal data support the drilling hypothesis, and a blow-out scenario can neither explain the dramatic changes that affected the plumbing system of numerous seep systems on Java after the 27-05-2006 earthquake. Reference [85] have combined satellite images with fieldwork and extensive sampling of water and gas at seeping gryphones, pools and salsa lakes at the Dashgil mud volcano in Azerbaijan in order to investigate the fluid–rock interactions within the mud volcano conduit. The gas geochemistry suggested that the gases migrate to the surface from continuously leaking deep seated reservoirs underneath the mud volcano, with minimal oxidation during migration. However, variations in gas wetness can be ascribed to molecular fractionation during the gas rise. In contrast, the water shows seasonal variations in isotopic composition and surface evaporation is proposed as a mechanism to explain high water salinities in salsa lakes. By contrast, gryphones have geochemical signals suggesting a deep-seated water source. This study has demonstrated that the plumbing system of dormant mud volcanoes is continuously recharged from deeper sedimentary reservoirs and that a branched system of conduits exists in the shallow subsurface. While the gas composition is consistently similar throughout the crater, the large assortment of water present reflects the type of seep (i.e. gryphones versus pools and salsa lakes) and their location within the volcano.

In last decade, many researchers conducted in Karoo have pointed out an interesting link between hydrothermal venting (potential sand/mud volcanoes on the surface) and maar-diatreme volcanism (eg. magma - water interaction driven explosive volcanism). The hydrothermal vent complexes identified in this area have previously been termed diatremes and volcanic necks, and have, since the pioneer work of references [70,71], been interpreted as the result of phreatic or phreatomagmatic activity [e.g. 72,74,86-90]. Some hydrothermal vent complexes are spatially associated with sill intrusions, but a direct relationship between conduit zones and contact aureoles cannot be demonstrated because of lack of exposures and boreholes. A general genetic relationship between sills and vent complexes is, however, supported by interpretations of seismic data from the Vøring and Møre basins of offshore mid-Norway, where it is shown that hydrothermal vent complexes are rooted in aureole segments of sill intrusions [68,75]. The general lack of igneous material in the hydrothermal vent complexes strongly suggests that they are rooted in a zone without major magma disintegration. Reference [65] have recently proposed a model of vent complex formation by heating and boiling of pore fluids in contact aureoles around shallow sills [75]. In this model, boiling of pore fluids may occur at depths as great as c. 1 km, and overpressure and possibly venting occur if the local permeability is low. Thick sills are common in the Stormberg Group sediments, at least in the Molteno Formation, which can be assumed to have caused shallow (1 km) boiling and expansion of pore fluids in contact aureoles. A high-permeability host rock requires a very rapid pressure build-up compared with permeability to initiate hydrofracturing. Following hydrofracturing, the gas phase may expand and lead to a velocity increase during vertical flow through the conduit zone. Thus, the vent formation mechanism bears resemblance to shallow breccia-forming processes in hydrothermal and volcanic systems [e-g- 76,91,92].

In systems dominated by fragmentation of magma (e.g. kimberlite pipes and diatremes), the resulting conduit zone will comprise mixtures of sediments and igneous material, and associated surface deposits dominated by pyroclastic material [e.g. 91-94]. Kimberlite pipes are generally formed from fragmentation of deep dyke complexes [e.g. 92,93], and this mechanism may also explain the formation of the phreatomagmatic complexes in the Karoo Basin [e.g. 95,96]. As very well explained by [97], kimberlitic diatremes are the most important economically, but despite decades of research, numerous open pit and underground mines, and hundreds of kilometers of diamond drilling, they remain poorly understood in volcanological terms, with multiple and strongly conflicting models in place. Reference [97] attempted an evenhanded review of maar diatreme volcanology that extends from mafic to kimberlitic varieties, and from historical maar eruptions to deeply eroded or mined diatreme structures.concentrated their study to convinced that increased understanding of other maar-diatremes will drive advances in kimberlite volcanology, and is best accomplished by integrating information from all parts of all types of maar-diatreme volcanoes, and from both subsurface and surface observations.

Finally, it is important to mention that several study [i.e. 98,99] have highlighted the importance of the emplacement environment of volcanism in causing global environmental climate changes. They suggested that an understanding of the triggering mechanism and consequences of previous climatic changes driven by carbon gas emissions is highly relevant for predicting the consequences of current anthropogenic carbon emissions, as these events are likely of similar magnitude and duration.

3. THE GAS HYDRATES

Natural gas hydrates are a curious kind of chemical compound called a clathrate. Clathrates consist of two dissimilar molecules mechanically intermingled but not truly chemically bonded. Instead one molecule forms a framework that traps the other molecule. Natural gas hydrates can be considered modified ice structures enclosing methane and other hydrocarbons, but they can melt at temperatures well above normal ice [i.e., 100]. At about 3 MPa pressure, methane hydrate begins to be stable at temperatures above 0 °C and at about 10 MPa it is stable at 15 °C [101]. This behavior has two important practical implications. First, it is a nuisance to the gas company. They have to dehydrate natural gas thoroughly to prevent the formation of methane hydrates in high pressure gas lines. Second, methane hydrates will be stable on the sea floor at depths below a few hundred meters and will be solid within sea floor sediments [102]. Masses of methane hydrate "yellow ice" have been photographed on the sea floor. Chunks occasionally break loose and float to the surface, where they are unstable and effervesce as they decompose.

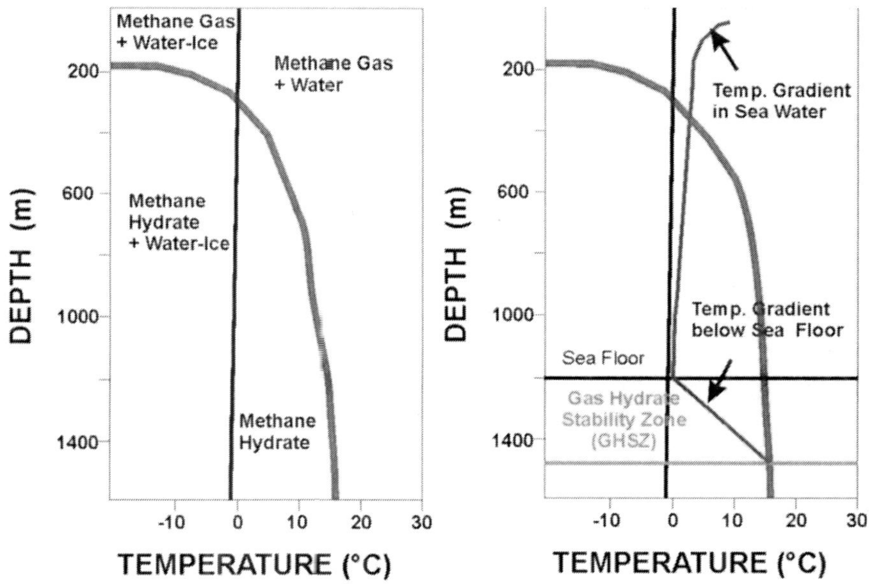

FIGURE 4. Left: Methane hydrate phases. Right: Typical occurrence of the gas hydrate stability zone on continental margins. A water depth of 1,200 m is assumed. The viola line represents the geothermal curve, while the red line is the gas hydrate stability curve. The orange line is the base of gas hydrate stability zone (GHSZ).

Figure 4 shows the combination of temperatures and pressures (the phase boundary) that marks the transition from a system of co-existing free methane gas and water/ice solid methane hydrate, which forms at low temperature and relative high pressure. When conditions move to the left across the boundary, hydrate formation will occur. Moving to the right across the boundary results in the dissociation of the gas hydrate, releasing free water and methane.

The phase diagram reported in Figure 4 shows a typical situation on continental shelves. Assuming a seafloor depth of 1,200 m, temperature steadily decreases with water depth, and a minimum value near 0° C is reached at the ocean bottom. Below the sea floor, temperatures steadily increase, so the top of the gas hydrate stability zone (GHSZ) occurs at roughly 400 m, while the base of the GHSZ is at 1,500 m. From the phase diagram, it appears that hydrates should accumulate anywhere in the ocean-bottom sediments where water depth exceeds about 400 m. Very deep (abyssal) sediments are generally not thought to house hydrates in large quantities. In fact deep oceans lack both the high biologic productivity (necessary to produce the organic matter that is converted to methane) and rapid sedimentation rates (necessary to bury the organic matter) that support hydrate formation on the continental shelves. Note that the conditions for gas hydrate formations are present also in sea water, but gas concentration is always not sufficient for their formations.

The gas hydrate phase is affected by gas mixture and pore-fluids composition (salinity). It is known that the presence of only a small percentage of higher hydrocarbons (such as ethane and propane) shifts the phase boundary to higher temperature (at constant pressure). The effect is that the base of GHSZ is shifted to greater depths [100,103]. Analogous to the effect of salt on the freezing point of water, if pore-fluids composition is brine, the phase boundary is shifted to lower temperatures at a given pressure and thus base of GHSZ will be shallower, as demonstrated by different authors [i.e., 204. Figure 5] shows a comparison among several gas hydrate stability models proposed in literature. The black line indicates the methane hydrate stability in brine water with a salinity of 3.5 % [105]. The red lines are evaluated by using the equations reported in [100] which consider fresh water. The solid red line indicates the methane hydrate stability curve and the dashed red line the hydrate stability curve considering a mixture of methane (90%), ethane (5%) and propane (5%). The blue lines are the hydrate stability in fresh water considering pure methane (solid line) and a mixture of methane (90%) and ethane (10%) by using the empirical expression proposed by [106]. Reference [107] obtained empirical equation for methane hydrate system in function of salinity. The green lines represent the gas hydrate stability considering the following salinity values: 0.0% (i.e. fresh water; solid line), 2 % (dotted line), 3.5 % (dashed line), and 5 % (dashed-dotted line). Finally, the magenta line indicates the stability for system of pure methane and water, following the approach of Reference [108]. Note that the methane hydrate stability curves in fresh water obtained from the proposed models are almost coincident, while the case considering salt water shows differences between the considered approaches.

The stability of methane hydrates on the sea floor has several implications [i.e., 109,110]. First, they may constitute a huge energy resource [111]. Second, natural and man-made disturbances may cause their destabilization causing the release of huge amounts of fluids (gas and water) and affecting slope stability. Finally, methane is an effective greenhouse gas (26 times more powerful than carbone dioxide), and large methane releases may explain sudden episodes of climatic warming in the geologic past. Some authors suggested that gas hydrate dissociation influenced significantly climate changes in the late Quaternary period [112-115]. The Clathrate Gun Hypothesis [116] suggests that past increases in water temperatures near the seafloor may have induced such a large-scale dissociation, with the methane spike and isotopic anomalies reflected in polar ice cores and in benthic foraminifera [115]. Reference [117] suggested that methane would oxidize fairly quickly in the atmosphere, but could cause enough warming that other mechanisms (for example, release of carbon dioxide from carbonate rocks and decaying biomass) could keep the temperatures elevated.

Gas hydrates in marine environments have been mostly detected from analysis of seismic reflection profiles, where they produce remarkable bottom-simulating reflectors (BSRs; [100,118]). Generally, the BSR is a very high-amplitude reflector that is associated with a phase reversal that approximately parallels the seafloor [119]. This phase reversal, which results from a strong acoustic impedance contrast between the layers, may indicate that sediments above the BSR are extensively filled with gas hydrates and sediments below it are filled with free gas in the pore space [i.e., 120-122]. Because the BSR follows a thermobaric surface rather than a structural or stratigraphic interface, it is normally observed to crosscut other reflectors [123].

Several studies [i.e. 30] revealed a seismic reflector below the BSR that can be associated with the base of the free gas zone, called base of the free gas reflector (BGR). The scientific community have been devoted much effort in studying marine sediments containing gas hydrates to characterize the hydrate reservoir and to quantify the gas trapped within sediments from seismic data analysis [i.e., 31,122,124,125]. To reach this goal advanced techniques have been developed. In fact, the BSR, detected from seismic data, is an easily recognizable indicator of the presence of hydrate, but it does not provide information directly on the concentration of hydrate and free gas or their distribution. One approach to estimate hydrate and free gas concentration is from seismic velocity (primarily P-wave velocity, Vp), obtained through advanced seismic analysis and/or modeling of data from a multi-channel seismic streamer, using techniques such as common-image gathers analysis [i.e., 32], one dimensional waveform inversion [i.e., 126], and amplitude versus offset analysis [i.e., 127]. The obtained velocity can be translated in terms of concentration by using theoretical models [i.e., 127,128]. Figure 6 reports an example of compressional and the shear (Vs) velocity versus gas hydrate and free gas saturation in pore space by using two models: the Biot theory [129] and the approximation for seismic frequency [128]. Note that in presence of high hydrate concentration, the velocity increases significantly, while, if we suppose uniform distribution of free gas in the pore space, it is sufficient a small content of free gas to reduce drastically the velocity.

Recently, the international community has considered CO_2 sequestration as a possible means of offsetting the emission of greenhouse gases into the atmosphere [130]. Some studies have considered confining CO_2 hydrate directly to shallow sediments on the deep sea floor, but this approach would not be permissible under the above international conventions. In the case of hydrates, several studies have investigated the use of injected CO_2 to liberate methane gas from hydrate in sediments, and in the process lock up CO_2 in CO_2 hydrate [i.e., 131]. The CO_2 storage program is a further reason to assess the feasibility of mapping and monitoring the reservoir by means of an efficient seismic analysis [132,133] and to obtain information about hydrate and free gas concentrations in a time-effective way.

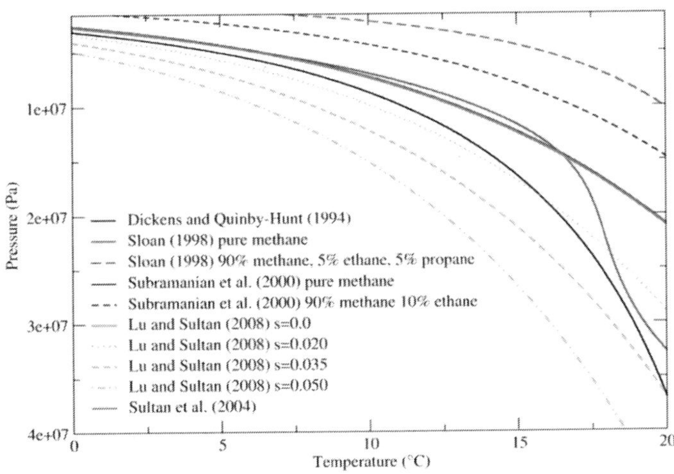

FIGURE 5. Comparison among different gas hydrate stability models. See text for details.

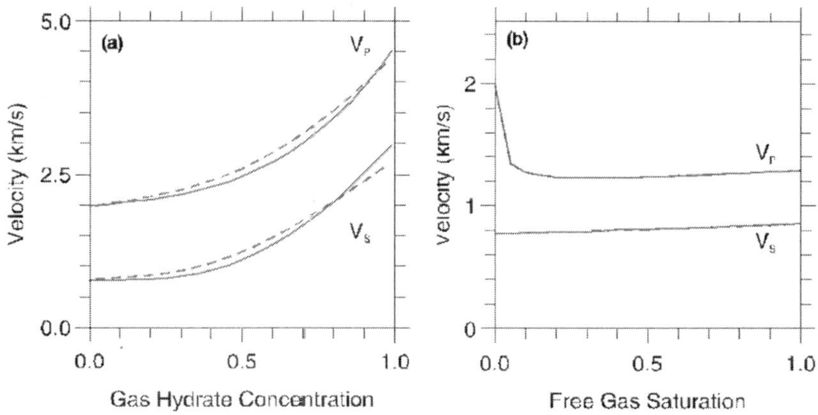

FIGURE 6. Comparison between two models: the Biot theory (solid red lines; [129]) and the approximation for seismic frequency (green dashed lines; [128]). (a) Compressional (Vp) and shear (Vs) velocity versus gas hydrate concentration in pore space. (b) Vp and Vs versus free gas saturation in pore space, supposing uniform distribution.

Finally, it is worth to mention that gas hydrates cannot survive for long geological times either because buried sediments are submitted to increasing geothermal conditions or are tectonically uplifted. In fossil sediments, the study of the interaction between seep-carbonates, hydrate destabilization and sediment instability is particularly difficult, owing to the lack of a direct recognition of fluid seepages and to the absence of precise quantifications of paleoenvironmental factors (pressure, temperature, paleodepth) conditioning hydrate stability conditions [134]. Following present-day analogues, the only means to infer a possible role of gas hydrates in fossil seep-carbonates are geochemical (oxygen isotope signature) and textural (presence of distinctive sedimentary features such as breccias, pervasive non-systematic fractures, soft sediment deformation) described in clathrites. Additional evidences can derive from the close association between seep-carbonates and sedimentary instability, and the large dimensions of seep-carbonate masses bearing brecciated structures. Recently, [134] studied cold seep-carbonates and associated lithologies in the northern Apennines and highlighted the seepage activity and the possible relationships with gas hydrate destabilization. In this geological context, many seep-carbonates are characterized by negative ∂13C and positive ∂18O values, by various types of brecciated structures and fluid-flow conduits, and are associated with intense sediment instability such as slumps, intraformational breccias and olistostromes.

Many authors have focused their attention on the possible modes of gas hydrate formations. Here, we report the study proposed by [135], which clearly summarized the possible mode of gas hydrate formation and produced a cartoon of gas hydrate system (Figure 7). They envisioned three possible modes of hydrate formation.

First, dipping permeable layers may focus gas flow and drive large amounts of free gas into the regional gas hydrate stability zone (Figure 7, number 1, inset). This is illustrated with a dipping stratigraphic layer in Figure 7; however, the permeability conduit could also be a fault or fracture. Beneath the GHSZ, the permeable layer draws gas from the surrounding material over an extensive source region, because of its high permeability and resultant low capillary entry pressure [136]. In this environment, gas rapidly enters the GHSZ and salinity rises as hydrate forms. The increased salinity inhibits further hydrate formation, which allows free gas to coexist with hydrate within the GHSZ. This process is repeated and the gas chimney rapidly propagates to the seafloor. Within the chimney, hydrate concentration increases upward toward the seafloor, where the system is furthest from equilibrium (Figure 7, number 1, inset). At the base of the gas chimney, the BSR will be diminished because gas is continuously present across the base of the GHSZ. These types of gas chimneys may be present, for example, at South Hydrate Ridge [109,137,138], at Blake Ridge [139-142] and along the Norwegian margin [143,144].

A second form of focused gas flow is illustrated in Figure 7 (number 2, inset). In this case, gas concentrates beneath the topographic crest of the seafloor structure. On the flanks of the structure, gas is trapped beneath the low-permeability base of the hydrate stability zone. Buoyancy drives the gas laterally toward the shallowest zone beneath the regional hydrate stability zone. The gas pressure is at a maximum at this location and ultimately the gas will drive its way through the GHSZ creating a gas chimney. As illustrated, the gas vent does not penetrate to the seafloor.

At the flanks of the topographic structures, a low permeability hydrate seal rapidly develops at the base of the GHSZ as illustrated in case 3 (Figure 7, number 3, inset). Hydrates are formed when water flows up through the GHSZ. Even in this low-flux example, the gas supply is large enough to create a separate gas phase that migrates upward by buoyancy. The changes in salinity during hydrate formation are too small for a three-phase zone to develop, and hence all free gas is crystallized as hydrate at the base of the GHSZ (Figure 7, number 3, inset). As hydrate forms, the permeability drops and a capillary seal to gas is formed. In these circumstances, either the gas pressure will build until it fractures the overlying column, or if there is another pathway present, the gas will flow upward but underneath the low-permeability cap of the base of the GHSZ (Figure 7). Finally, far from where any methane gas flow is focused but where there is upward flow of water, low concentrations of hydrate may be deposited within the RHSZ but not at its base (Figure 7, number 4, inset).

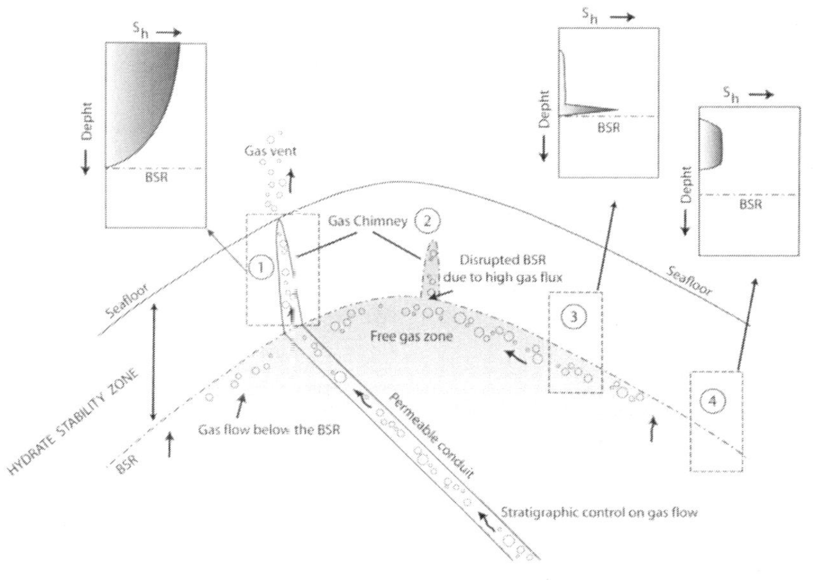

FIGURE 7. Cartoon of the gas hydrate reservoir system. Four characteristic forms of hydrate deposit are shown. (1) Gas chimney sourced by permeable conduits. Gas is focused along permeable conduits beneath the hydrate stability zone. Focused flow penetrates the GHSZ and self-generates a three-phase pathway to vent gas to the seafloor. (2) Gas chimney sourced by gas trapped beneath the GHSZ. Gas is focused along the base of the GHSZ and trapped beneath the crest of the structure. Gas builds up until it begins to form a chimney through the GHSZ. (3) Capillary sealing and lateral migration. On the flanks of the structure, hydrate formation rapidly forms capillary seals to gas and the gas is driven laterally to the highest structural point. (4) Aqueous flow and hydrate formation. Far from the crest, water with dissolved methane migrates upward and deposits hydrate within the GHSZ. Modified after [109].

The simulations of [109] provide insight into how gas chimneys form and sustain themselves within the GHSZ. The penetration of gas into the GHSZ is controlled by a competition between the basal supply of gas and the lateral diffusion of salt. The gas flow is driven primarily by buoyancy: as a result, the natural tendency for gas is to flow vertically even when permeability is reduced by hydrate formation. In addition, salt diffusivity is extremely low; thus high salinity zones within chimneys can be maintained for long times, particularly if there is continued supply of gas to form more hydrate and maintain salinity. Finally, lateral salt diffusion concentrates hydrate at the margins of the chimney; this further lowers the salt diffusivity and further limits salt loss.

Reference [135] showed that at South Hydrate Ridge, gas is supplied at a rate 10 times greater than is depleted by hydrate formation due to salt diffusion. Salt loss by diffusion, and hence the amount of methane needed to form hydrate to replace the salt, is independent of the vent half-width. In general, if the flux of methane supplied is greater than the loss due to diffusion, a chimney will be created and maintained at three-phase equilibrium (Figure 7, number 1, inset). However if the gas flux supplied is less than the loss due to diffusion, the chimney will only penetrate a short distance within GHSZ and free gas will not reach the seafloor (Figure 7, number 2, inset).

4. THE GAS HYDRATE IN SUBMARINE MUD VOLCANOES

The jointly occurrence of submarine mud volcanoes and gas hydrate has been reported by many authors in world-wide [i.e. 26,42,145-148]. For example, on the upper continental slope of the Gulf of Mexico, active gas migrations along faults or at mud volcanoes have been identified and their sources attributed to accumulated gas hydrates [53,149]. Reference [12] estimated that methane accumulated in gas hydrate associated with mud volcanoes is about $10^{10} \sim 10^{12}$ m^3 at normal temperature and pressure. Reference [150] estimated that up to 40% of total United Kingdom methane emission was from the continental shelf around UK. It is therefore important to investigate marine gassy sediments and submarine mud volcanoes to better understand the dynamics of shallow-water methane transport, fluid migration and the relationship of these phenomena to gas hydrate.

Firstly, we recall an important review of submarine mud volcanoes reported in Reference [12], reporting the main points. Evidence for submarine mud volcanoes exists in many regions showing the following features:

1. subcircular structures up to several kilometers in diameter elevated above the surrounding seafloor and visible on bathymetric maps and/or sonar images;
2. seafloor-piercing shale diapirs visible on seismic profiles;

3. fluid expulsion above elevated seafloor structures revealed by acoustic profiles and through visual observations from a submersible, remote operated vehicle (ROV), or by underwater video-camera;
4. transient mud islands in shallow waters;
5. gas bubbles at the surface of water that may be related to mud volcanoes.

Submarine mud volcanoes occur world-wide on continental shelves, slopes and in the abyssal parts of inland seas. Some studies (i.e. from the Barbados accretionary complex [54,151]) have linked the morphology of submarine mud volcanoes to different development stages and processes of mud liquefaction. Conical-shaped mud volcanoes ('mud-mounds' or gryphons), which do not have any central summit 'mud lakes' (or salses), are formed by the expulsion of plastic mud breccia in concentric radial flows. In contrast, shearing with the feeder conduit liquefies the mud leading to the formation of flat-top mud volcanoes (mud-pies) with central 'mud lakes' and elongated, radial mud-flow tongues. In both types, the mud is found to have a plastic behavior in which its yield strength decreases with increasing porosity. Thixotropy is associated with high porosity (e.g. more than 70%), which is often related to the dissociation of gas hydrate [151]. Often, mud volcanoes are associated with methane fluxes, either as free gas or, depending on ambient temperature and pressure conditions, as gas hydrate [54,152]. On this basis, Reference [14] argue that the global flux of methane to the atmosphere from the world's terrestrial and submarine mud volcanoes is highly significant. The relative difficulty in studying submarine mud volcanoes, compared with their terrestrial counterparts, leaves substantial gaps in our knowledge about their modes of formation, the duration and frequency of eruptions and the fluxes of mud and volatile phases from the subsurface. For this reason, several efforts are spend to simulate numerically the formation of submarine mud volcanoes [i.e., 153].

At the present time, evidence for submarine mud volcanoes has been found in all oceans (Figure 1). For instance, in shallow water areas (shelves) where mud islands are recorded, submarine mud volcanoes are likely to be present. However, there are regions where the existence of submarine mud volcanoes is unexpected, such as in the Baltic Sea, where sediments are only 10 m thick, but miniature mud diapirs/volcanoes (1.5 m in diameter, 30 cm high above the surrounding seafloor) are reported [154,155]. Figure 1 includes sediment diapirs reported in [33]; however, it is unclear whether all these sediment diapirs are mud volcanoes. Submarine mud volcanoes are more extensive than their sub-aerial analogs.

All the regions where submarine mud volcanoes or evidence for them have been observed are confined to shelves, continental and insular slopes, and abyssal parts of inland seas (e.g. Black and Caspian). The examination of geologic and tectonic features of these areas is crucial for the understanding of the mechanisms of mud volcanic activity. In the abyssal parts of inland seas, mud volcanoes have been found in the Caspian Sea and in the Black Sea where the sedimentary cover is typically thick (10–20 km) for these regions. Sediments are mostly terrigenous and were deposited during Tertiary and recent times under high subsidence and accumulation rates. Shale diapirs and faults deform many sedimentary sequences.

On the continental slopes of passive margins, submarine mud volcanoes have been found in the Norwegian Sea, offshore Nigeria and in the Gulf of Mexico. In the first two cases, mud volcanic activity is confined to submarine fans (the Bear Island Trough mouth fan and the Niger Delta are notable examples), composed of Tertiary terrigenous sediments deposited at a high accumulation rate. In the Niger Delta there are many diapirs and faults [156]. The continental slope of the Gulf of Mexico is an extremely complex deep-water region characterized by a combination of rapid sediment influx, faulting, and diapiric (salt and shale) tectonism [157].

At active margins, submarine mud volcanoes have been reliably identified in the Mediterranean Sea and offshore Barbados, both characterized by different geologic and tectonic settings. In the Mediterranean Sea, mud volcanoes have been found only within the limits of the accretionary prism, characterized by a complex fabric composed of many thrusts [28]. Offshore Barbados mud volcanoes are located within the limits of the accretionary prism as well as in front of the prism where the thickness of sedimentary cover is only 2.3 km [158]. The majority of reported evidence of submarine mud volcanoes presents at active margins.

The association of gas hydrates with submarine mud volcanoes was first noted by Reference [145] and has since observed in the Caspian [42], Black [27,146-148], Mediterranean [57,148], Norwegian seas [48,58], offshore Barbados [159], offshore Nigeria [160] and in the Gulf of Mexico [161,162].

There are many common features of gas hydrates associated with mud volcanoes. Gas hydrate content in sediments varies from 1–2% to 35% by volume and changes through a mud volcano area as well as through depth. Methane is the major gas component of gas hydrates and can be thermogenic, biogenic or mixed in origin. Reference [12] proposed a model for the formation of gas hydrates within a mud volcano: (a) hydrothermal process dominates around the central part of the mud volcano; (b) metasomatic process dominates at the peripherical part of the mud volcano (Figure 8). This generalized model is based mainly on data from the Haakon

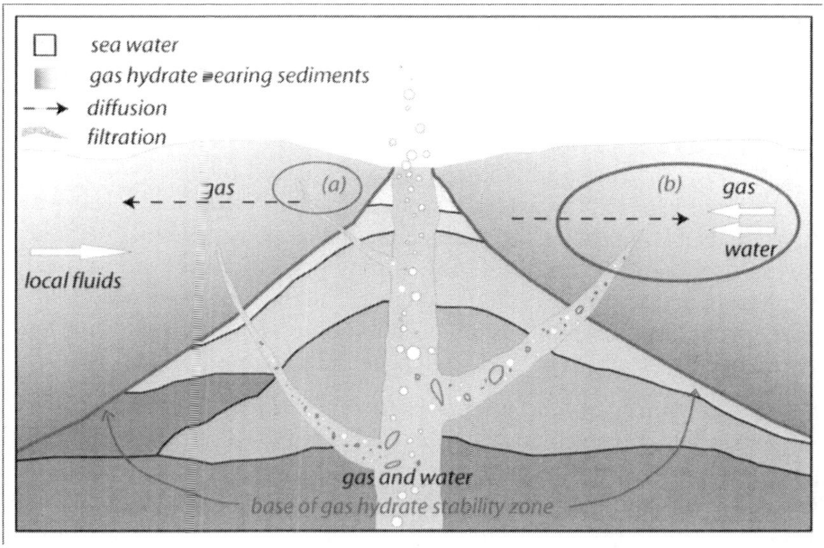

FIGURE 8. Cartoon showing the proposed model of the formation of gas hydrates within a mud volcano: (a) hydrothermal process dominates around the central part of the mud volcano; (b) metasomatic process dominates at the peripherical part of the mud volcano. Modified after [12].

Mosby mud volcano in the Norwegian Sea, which is the most famous mud volcano characterized by a concentric-zonal distribution of gas hydrates [58,163]. Gas hydrate accumulation is controlled by the ascending flow of warm fluids. The water from the mud volcanic fluid as well as from the surrounding recent sediments is involved in the formation of gas hydrates. So, gas hydrates can occur within the edifice of a mud volcano (crater and hummocky periphery) as well as outside in the host marine sediments. However, the processes that are responsible for the formation of gas hydrates differ from point to point. Note that in Figure 8 there are no gas hydrates in the central part of the model mud volcano (usually in the central part of a crater where mud and fluid flow out) because of the high temperature. Around the central part of the mud volcano, gas hydrates form from the fluids that have risen from the deep subsurface. This fluid is warmer than the surrounding sediments (by up to 15–20° C at the sub-bottom depth of 1 m) and contains gas in solution and perhaps as a free phase. Gas hydrates crystallize from this warm fluid when it becomes cold and the solubility of gas decreases [164]. Both water and gas participating in the formation of gas hydrates have come from the deep, external fluid that filters through mud volcanic sediments. This process is analogous to the conventional low-temperature hydrothermal process of mineral formation [165].

At the peripheral part of the mud volcano, gas hydrates form from the gas that emanates from the central part of the mud volcano, and is transported in solution by diffusion. On the other hand, the water participating in the formation of gas hydrates is contained in the host sediments (local water). In addition, some local biochemical gas may be captured in gas hydrates. Thus, in this case the local water is partly replaced by gas hydrates due to the supply of gas from an external source (mud volcanic fluid). This process of gas hydrate formation is analogous to the conventional metasomatic process of mineral formation [165]. At any point between the central and peripheral parts of the mud volcano, mixing of hydrothermal and metasomatic processes is possible. The source of water (mud volcanic or local) determines which of these two processes is dominant.

The close proximity of mud volcanoes to zones where BSRs crop out on the seafloor deserves particular attention. Seismic records strongly suggest that much of the gas in mud volcanoes originates from levels deeper than that of the gas hydrates and faulting could be responsible for this unique situation. It has been argued that, in the case that a concentric zonal distribution of hydrate is present, the gas hydrates have probably been formed by gas emanating from the central part of the mud volcano, and transported into solution by diffusion [12]. For example, a strong BSR and the presence of mud volcanoes have recently been detected by seismic data along the southwest African margin, which is a passive margin [166]. This region, located in the distal part of the Orange River delta, is also characterized by overpressure which results in active fluid expulsion, as shown by the existence of mud volcanoes, pockmarks, and possibly cold-water corals thriving on methane gas seeps [167].

5. AN EXAMPLE: THE ANTARCTIC PENINSULA

The global climate change is particularly amplified in transition zones, such as the peri-Antarctic regions. For this reason, the gas hydrate reservoir present offshore Antarctic Peninsula was studied in the last 20 years acquiring a quite extensive geophysical dataset. The presence of a diffused and discontinuous BSR was discovered during the Italian Antarctic cruises of 1989–1990 [i.e., 30] and 1996–1997 [122], onboard the R/V OGS Explora. Seismic data showed the existence of a potential gas hydrate reservoir [32,168] along the South Shetland margin. Along this margin, the extent of the BSR was mapped based on about 1,000 km of seismic lines [e.g.,168 and references therein]. Ocean bottom seismometers (OBSs) deployed during the 1996–1997 cruise provided energy arrivals from the BSR and the refraction and the converted waves from the base of the free-gas zone, the so-called base of the free gas reflector or BGR [122]. During the austral summer 2003-2004, additional data were acquired in the same area: multibeam bathymetry, seismic profiles, chirp, and sediment gravity cores [31]. Figure 9summarizes the position of the three acquisition legs.

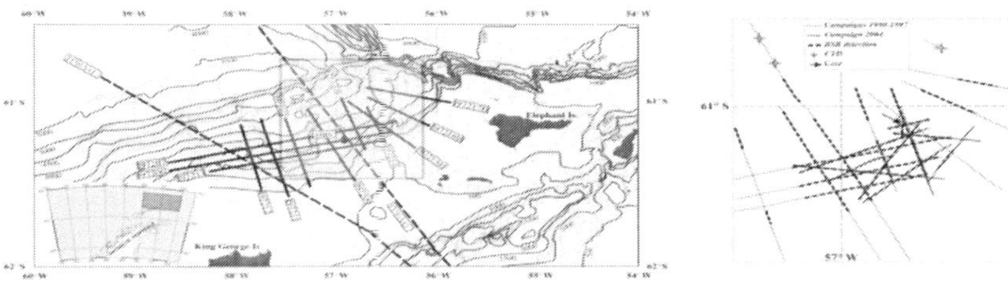

FIGURE 9. Left: Multibeam bathymetry map after [169], showing the locations of the seismic lines acquired in 1990 (dashed lines) and 1997 (continuous lines). The dot indicates the location of the OBS data acquired in 1997. The yellow box presents the 2004 study area. Right: 2004 study area. Continuous blue lines: locations of the airgun seismic lines. Dotted black lines: location of seismic lines of 1990 and 1997 legs. Thick dashed segments indicate the presence of particularly well developed BSRs. Also annotated are the CTD measurements (red stars), and the two coring sites (grey and black arrows cores GC01 and GC02 respectively). Modified after [31].

As already explained, seismic velocity obtained from advanced seismic analysis can be translated in terms of concentrations of gas hydrate and free gas. This procedure has been applied in the Antarctic Peninsula, where seismic velocities obtained from advanced analysis of multichannel seismic data were analyzed to determine gas hydrate and free-gas distributions and to estimate the methane volumetric fraction trapped in the sediments [170]. The elastic properties of the layers across the BSR were modeled applying the approximation of the Biot equations for seismic frequency in order to quantify the concentrations of gas hydrate and free gas in the pore space [128]. This theory considers two solid phases - grains and hydrates - and two fluid phases - water and free gas - including an explicit dependence on differential pressure and depth, and the effects of cementation by hydration on the shear modulus of the sediment matrix. So, the seismic velocities of the 2D seismic lines were translated in terms of concentrations of gas hydrate and free gas in the pore space, obtaining 2D models. The jointly interpolation of the 2D models allowed obtaining a 3D model of gas hydrate concentration from the seafloor to the BSR. The total volume of hydrate, estimated in the area (600 km^2) where the interpolation is reliable, is 16 × 10^9 m^3. The gas hydrate concentration is affected by errors that could be equal to about ±25%, as deduced from sensitivity tests [31,171] and from error analysis related to the interpolation procedure. The estimated amount of gas hydrate can vary in a range of 12 × 10^9 - 20 × $10^9 m^3$. Moreover, considering that 1 m^3 of gas hydrate corresponds to 140 m^3 of free gas in standard conditions, the total free gas trapped in this reservoir ranges between 1.68 × 10^{12} and 2.8 × 10^{12} m^3. This estimation does not take into account the free gas contained within pore space below the hydrate layer, so this values could be underestimated.

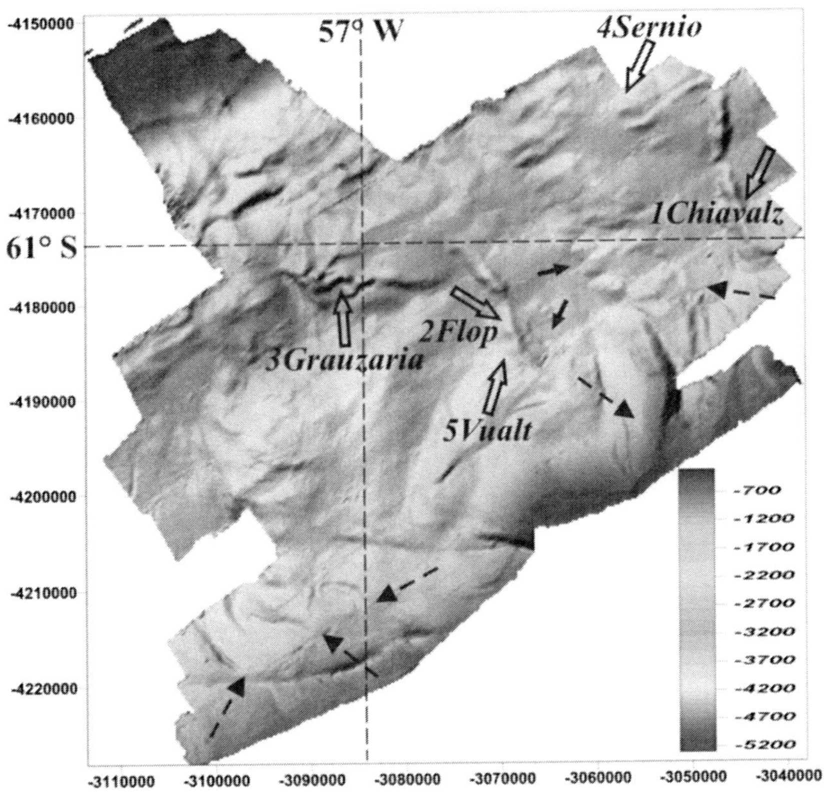

FIGURE 10. Multibeam bathymetry map of the study area, showing evidence of mud volcanoes (open arrows), collapse troughs (closed arrows) and slides (dashed arrows). The numbers indicate the four mud volcano ridges described in the text and in Table 1. Modified after [31].

Reference [31] reported the main results obtained by the analysis of bathymetric data, CHIRP data and gravity core analysis. In particular, the bathymetric map provided the evidences of mud volcanoes, collapse troughs and slides (Figure 10). It is well known [i.e., 172] that these features are generally associated to the presence of gas hydrate, as already explained. Reference [31] have recognized five main mud volcano ridges, named Chiavalz, Flop, Grauzaria, Sernio and Vualt (see location map in Figure 10). Table 1 reports the main characteristics of the mud volcanoes (Figure s 11-15). The Vualt mud volcano is the highest detected in our study area; its top is at 2,216 m below sea level, with an elevation of about 255 m above the seafloor and an extension of 9.4 km^2 (Figure 10, label 5; Figure 11). On the flank, a gravity core (GC02) was recovered. The Flop mud volcano has its top at 2,363 m below sea level and a relief of about 115 m. Its extension is 7.5 km^2 (Figure 10, label 2; Figure 12). The Grauzaria ridge, oriented W–E and located at around 61 S–57 W (Figure 10; label 3; Figure 13), can be considered an alignment with several culminations and a total extension of 45.9 km^2. The highest culmination is at 2,594 m below sea level and exhibits a relief of about 185 m above the seafloor. The Sernio mud volcano ridge is located in the proximity of core GC01 and is oriented SW–NE (Figure 10, label 4; Figure 14). Its top is at 1,990 m below sea level with an elevation of about 185 m. It presents several culminations for a total extension of 23.9 km^2. Finally, the Chiavalz mud volcano ridge is located in the northeast of our survey area (Figure 10; label 1; Figure 15) and is oriented S–NE. It has its top at 1,615 m below sea level, a maximum elevation of about 210 m, and an extension of 14.5 km^2.

TABLE 1. Details of the mud volcanoes offshore Antarctic Peninsula: latitude and longitude of the midpoint, water depth at the top, elevation with respect to the bathymetry and extension. Grauzaria is a group of several mud volcanoes.

Mud Volcano	Lat. (WGS84)	Lon. (WGS84)	Meter below seafloor of the top	Elevation (m)	Extension (km^2)
1 Chiavalz	60 52 29.31 S	56 18 46.88 W	1615	210	14.5
2 Flop	61 01 40.52 S	56 45 11.88 W	2363	115	7.5
3 Grauzaria	61 01 31.44 S	56 56 36.64 W	2594	185	45.9
4 Sernio	60 51 54.73 S	56 28 19.10 W	1990	185	23.9
5 Vualt	61 04 30.63 S	56 43 02.71 W	2216	255	9.4

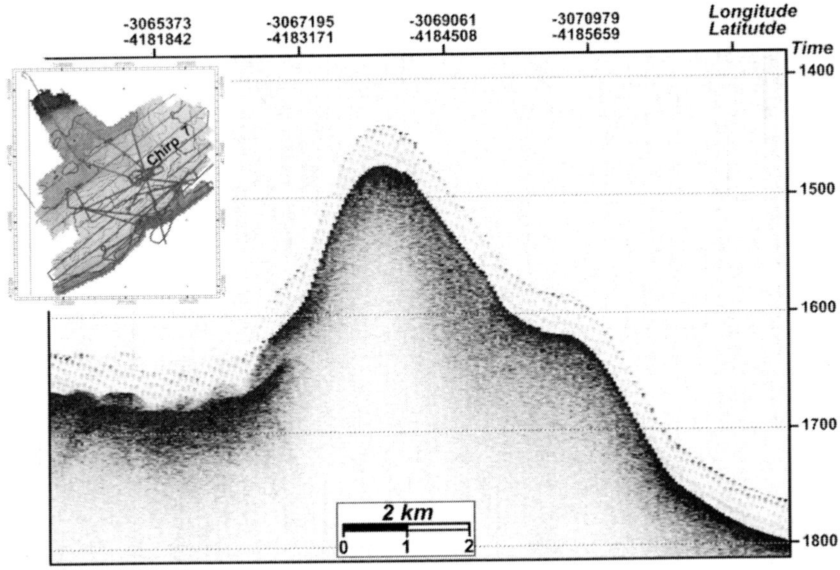

FIGURE 11. Comparison between chirp (a) and airgun (b) data across the Vualt mud volcano, in which the BSR is evident. The arrow indicates the location of core GC02. CHIRP image of the Vualt. Insert: Bathymetric map after BSR project. The red line indicates the CHIRP location. After [31].

Fluid analyses performed on the two gravity cores [31] revealed the presence of several hydrocarbon gases, i.e. methane, ethane, propane, butane, pentane and hexane, and traces of aromatic hydrocarbons of > C12 carbon chain length, suggesting a thermogenic origin of the gas. The major difference in gas contents between the two cores is that methane and propane are totally absent in core GC02. On the contrary, pentane is present at all analyzed depths in both cores, with quite similar contents. Below the upper 1 m of sediment in core GC02, the interstitial gases are essentially composed of pentane. The average total gas content amounts to 150.54 and 49.30 μg/kg for the two cores, respectively. The gas content measured in core GC01 is therefore about three times higher than that measured in core GC02. Downcore profiles for specific gases showed that core GC01 has a quite uniform gas type and content along the whole core; on the contrary, core GC02 has variable gas content. Even if both cores are located in the proximity of mud volcanoes, Reference [31] suggested that the sediment permeability of core GC01 is lower than that of core GC02, in which the fluids can easily escape and produce a collapse trough (see closed arrows in Figure 10). Moreover, the sediment stiffness in core GC01 is higher than that of core GC02, as suggested also by the different core length (1.07 and

2.98 m respectively); this is in agreement with the hypothesis of different permeability values between the two cores.

In conclusion, interpretation of the data acquired on the South Shetland margin confirmed the crucial role of tectonics controlling the extent of the hydrate reservoir, and active venting of fluids and mud through faults bordering and crossing the gas hydrate field. Mud volcanoes and fluid expulsion events are likely located in close association with faults, through which they are connected to the reservoir located beneath the BSR. Their activity is probably episodic [31]. Moreover, the different sediment stiffness at the two coring sites can be related to the temporal frequency of expulsion events, where the hardness of the mud volcano flanks is directly proportional to the interval between expulsion events, as suggested by [173]. Finally, the hydrocarbons trapped in our sediment cores possibly indicate the existence of deeper reserves.

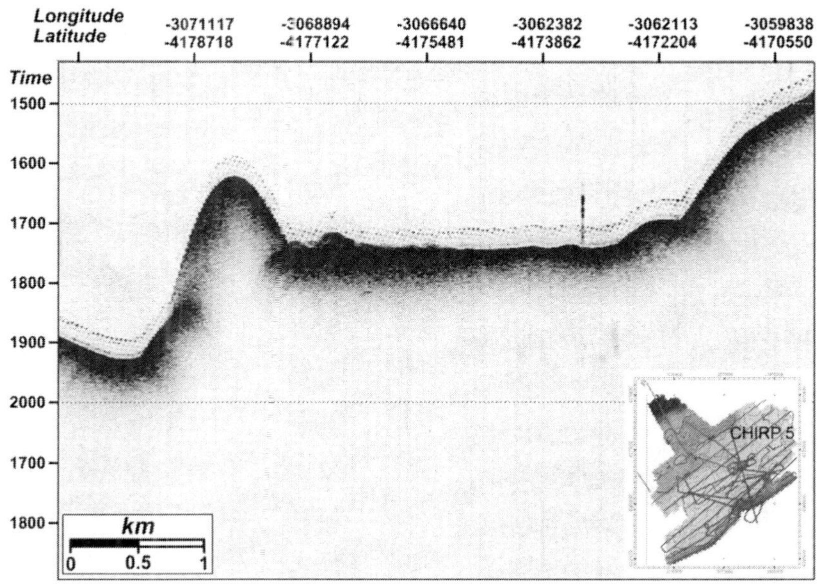

FIGURE 12. CHIRP image of the Flop. Insert: Bathymetric map after BSR project. The red line indicates the CHIRP location.

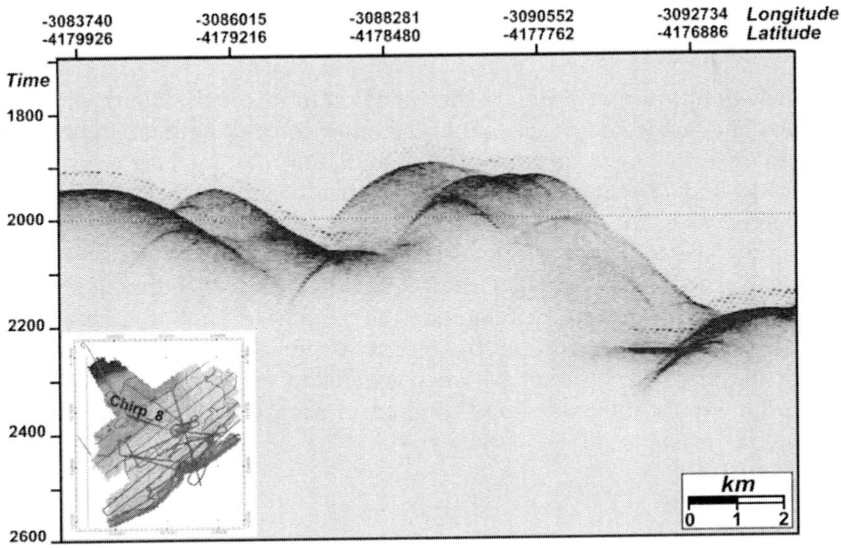

FIGURE 13. CHIRP images of the Grauzaria group. Insert: Bathymetric map after BSR project. The red line indicates the CHIRP location.

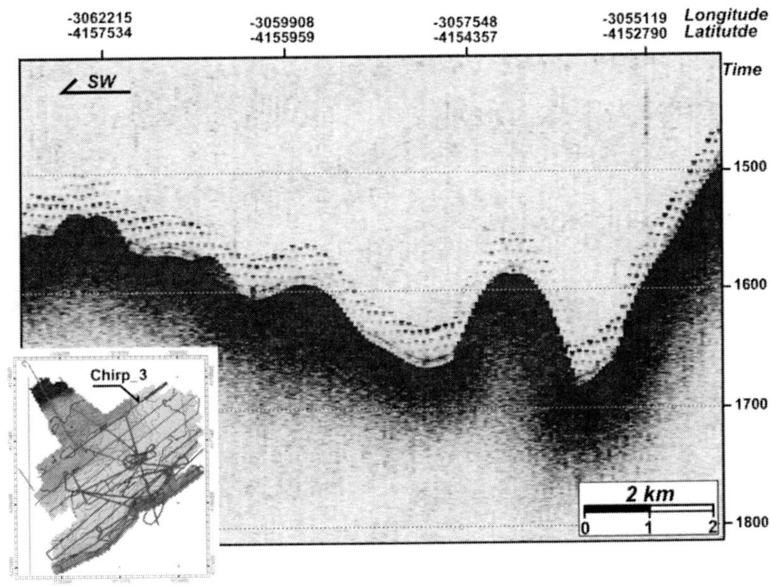

FIGURE 14. CHIRP images of the Sernio mud volcano. Insert: Bathymetric map after BSR project. The red line indicates the CHIRP location.

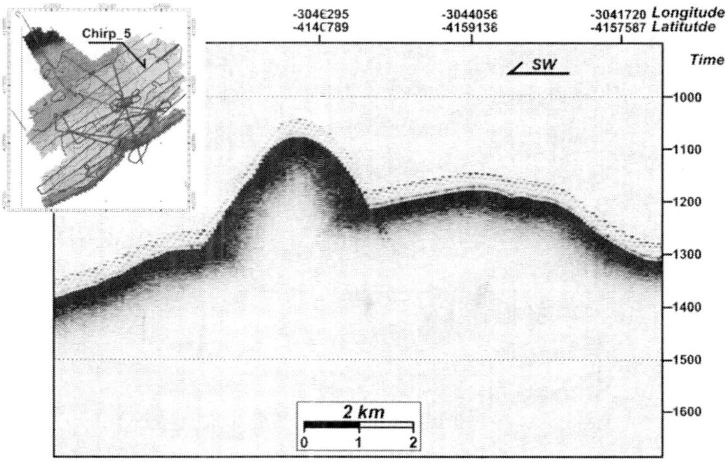

FIGURE 15. CHIRP images of the Chiavalz mud volcano. Insert: Bathymetric map after BSR project. The red line indicates the CHIRP location.

6. CONCLUSIONS

Knowledge of natural occurring gas hydrate is increasing rapidly in the last years; however commercialization of gas hydrate remains unproven. Great uncertainty of the global gas hydrate resource and imitated estimates of hydrate system retard economic analysis of hydrate recovery [174]. In this context, the gas hydrate associated to mud volcanoes is a very interesting topic because this system contains high gas hydrate concentration in a very small area. In addition, gas hydrate accumulations related to fluid discharges sites (including mud volcanoes) occur at very shallow depths or on the seafloor and show the maximum hydrate content in their upper parts. These features may consider as natural reactors, in which part of the migrating gas from the surrounding areas is stabilized in gas hydrates. Gas resources in such accumulations are therefore renewable and could become important gas hydrate formations to be exploited [11].

Moreover, the wide and extensive literature about hydrate, mud volcanism and their interaction suggests that this topic is timely because gas hydrates may play an important role in the global carbon cycle and global climate dynamics through emissions of methane and in affecting stability of geological features, including mud volcanoes. The role of gas hydrates in above-mentioned processes cannot be assessed accurately without a better understanding of the hydrate reservoir and their interactions with geological features and meaningful estimates of the amount of methane it contains. In conclusion, lack of knowledge hampers the evaluation of the resource potential of gas hydrates and the hazards related to gas hydrates, requiring efforts to improved knowledge about gas hydrate and their interaction with mud volcanoes.

7. ACKNOWLEDGEMENTS

We wish to thank Manuela Sedmach for graphic support. This work is partially supported by Programma Nazionale di Ricerche in Antartide, project CLISM.

REFERENCES

1. E. Canon-Tapia, A. Szakacs, 2010 What is a volcano? Boulder, Colo.: Geological Society of America 140
2. A. J. van Loon, 2010 Sedimentary volcanoes: Overview and implications for the definition of a volcano on Earth. Canon-Tapia E, Szakacs A, Editors. What is a volcano?. Boulder, Colo.: Geological Society of America 31 41
3. A. J. Kopf, 2011 Significance of mud volcanism. Reviews of Geophysics 40 2
4. C. Hensen, M. Nuzzo, E. Hornibrook, L.M. Pinheiro, B. Bock, V. H. Magalhães, W. Brückmann, 2007 Sources of mud volcano fluids in the Gulf of Cadiz- indications for hydrothermal imprint. Geochimica et Cosmochimica Acta 71 1232 1248
5. G. Etiope, R. W. Klusman, 2002 Geologic emissions of methane to the atmosphere Chemosphere 49 777 789
6. A. G. Judd, M. Hovland, L. I. Dimitrov, Gil. S. García, V. Jukes, 2002 The geological methane budget at Continental Margins and its influence on climate change Geofluids 2 109 126
7. L. I. Dimitrov, 2003 Mud volcanoes-a significant source of atmospheric methane. Geo-Mar Lett 23 155 161
8. A. V. Milkov, R. Sassen, T. V. Apanasovich, F. G. Dadashev, 2003 Global gas flux from mud volcanoes: a significant source of fossil methane in the atmosphere and the ocean. Geophysical Research Letters 30
9. G. Etiope, A. V. Milkov, 2004 A new estimate of global methane flux from onshore and shallow submarine mud volcanoes to the atmosphere. Environmental Geology 46 997 1002
10. A. Judd, 2005 Gas emissions from mud volcanoes. Significance to Global Climate Change Martinelli G., Panahi B., (ed.) Mud Volcanoes, Geodynamics and Seismicity 51 chapter 4 147 157
11. L. L. Mazurenko, V. A. Soloviev, 2003 Worldwide distribution of deep-water fluid venting and potential occurrences of gas hydrate accumulations. Geo-Marine Letters 23 162 176

12. A. V. Milkov, 2000 Worldwide distribution of submarine mud volcanoes and associated gas hydrates. Marine Geology 167 29 42
13. L. I. Dimitrov, 2002 Mud volcanoes-the most important pathway for degassing deeply buried sediments. Earth-Science Reviews 59 49 76
14. M. Hovland, A. Hill, D. Stokes, 1997 The structure and geomorphology of the Dashgil mud volcano, Azerbaijan. Geomorphology 21 1 15
15. A. Kopf, D. Klaeschen, J. Mascle, 2001 Extreme efficiency of mud volcanism in dewatering accretionary prisms. Earth and Planetary Science Letters 189 295 313
16. G. Delisle, Rad. U. von, H Andruleit, Daniels. C. H. von, A. R. Tabrez, A. Inam, 2002 Active mud volcanoes on- and offshore eastern Makran, Pakistan. International Journal of Earth Sciences 91 93 110
17. G. Etiope, A. Caracausi, R. Favara, F. Italiano, C. Baciu, 2002 Methane emissions from the mud volcanoes of Sicily (Italy). Geophysical Research Letters 29 8 56-1-56-4
18. E. Deville, A. Battani, R. Criboulard, S. Guerlais, J. P. Herbin, J. P. Houzay, C. Muller, A. Prinzhofer, 2003 The origin and processes of mud volcanism: new insights from Trinidad. Van Rensbergen P, Hillis RR, Maltman AJ, Morley CK. (ed.) Subsurface Sediment Mobilization. Geol. Soc. London, Spec. Publs. 216 475 490
19. N. Yassir, 2003 The role of shear stress in mobilizing deep-seated mud volcanoes: geological and geomechanical evidence from Trinidad and Taiwan. Van Rensbergen P, Hillis RR, Maltman AJ, Morley CK. (ed.) Subsurface Sediment Mobilization. Geol. Soc. London, Spec. Publs. 216 461 474
20. R. Shakirov, A. Obzhirov, E. Suess, A. Salyuk, N. Biebow, 2004 Mud volcanoes and gas vents in the Okhotsk Sea area. Geo-Marine Letters 24 140 149
21. S. A. Stewart, R. J. Davies, 2006 Structure and emplacement of mud volcano systems in the South Caspian Basin. AAPG Bulletin 90 5 771 786
22. A. J. Barber, S. Tjokrosapoetro, T. R. Charlton, 1986 Mud volcanoes, shale diapirs, wrench faults and melanges in accretionary complexes, eastern Indonesia. AAPG Bull 70 1729 1741
23. D. L. Orange, 1990 Criteria helpful in recognizing shear-zone and diapiric melenges, examples from the Hoh accretionary complex, Olympic Peninsula, Washington. Geol. Soc. Am. Bull. 102 935 951
24. K. M. Brown, D. L. Orange, 1993 Structural aspects of diapiric melange emplacement: The Duck Creek diaper. Journal of Structural Geology 13 831 847
25. B. Treves, 1985 Mud volcanoes and shale diapirs, their implications in accretionary processes. A review. Acta Naturalia de l'Ateneo Parmense 21 31 37
26. I. S. Guliyiev, A. A. Feizullayev, 1997 All About Mud Volcanoes. Azerbaijan, Baku: Publ. House, Nafta Press
27. M. K. Ivanov, A. F. Limonov, J. M. Woodside, 1998 Extensive deep fluid flux through the sea floor on the Crimean continental margin (Black Sea). Henriet JP, Mienert J. (ed.) Gas hydrates: relevance to world margin stability and climate change. Geol. Soc. London, Spec. Publs. 137 195 213
28. A. F. Limonov, J. M. Woodside, M. B. Cita, M. K. Ivanov, 1998 The Mediterranean Ridge and related mud diapirism: a background. Mar. Geol. 132 7 19
29. K. A. Kvenvolden, T. D. Lorenson, 2000 The global occurrence of natural gas hydrate. Paull C.K., Dillon W. P. (ed.) Natural Gas Hydrates: Occurrence, Distribution, and Dynamics, AGU Monograph 55
30. U. Tinivella, E. Lodolo, A. Camerlenghi, G. Boehm, 1998 Seismic tomography study of a bottom simulating reflector off the South Shetland Islands (Antarctica). Henriet J-P, Mienert J, (ed.) Gas hydrate: relevance to world margin stability and climate change. Geol. Soc. London, Spec. Publs. 147 141 151
31. U. Tinivella, F. Accaino, B. Della Vedova, 2007 Gas hydrates and active mud volcanism on the South Shetland continental margin, Antarctic Peninsula. Geo-Mar Lett doi s00367-007-0093-z.
32. U. Tinivella, M. F. Loreto, F. Accaino, 2009 Regional versus detailed velocity analysis to quantify hydrate and free gas in marine sediments: the South Shetland margin target study. Long D., Lovell M.A., Ress J.G., Rochelle CA. (ed.) Sediment-Hosted Gas

Hydrates: New Insights on Natural and Synthetic Systems. Geol. Soc. London, Spec. Publs. 319 103 119

33. R. R. Rakhmanov, 1987 Mud volcanoes and their importance in forecasting of subsurface petroleum potential. Nedra (in Russian).

34. Y. Lancelot, R. W. Embley, 1977 Piercement structures in deep oceans. Bull. Amer. Assoc. Pet. Geol. 61 1991 2000

35. K. Graue, 2000 Mud volcanoes in deep water Nigeria. Marine and Petroleum Geology 17 959 974

36. C. Huguen, J. Mascle, E. Chaumillon, A. Kopf, J. Woodside, T. Zitter, 2004 Structural setting and tectonic control of mud volcanoes from the Central Mediterranean Ridge (Eastern Mediterranean). Marine Geology 209 245 263

37. D. A. Huseynov, I. S. Guliyev, 2004 Mud volcanic natural phenomena in the South Caspian Basin: geology, fluid dynamics and environmental impact. Environmental Geology 46 1012 1023

38. G. Martinelli, 1998 Mudvolcanoes of Italy: proceedings of V Int. Conference on "Gas in Marine Sediments" 40 42

39. A. Arhangelski, 1932 Some words about genesis of mud volcanoes on the Apsheron peninsula and Kerch-Taman area. Bull. MOIP, Ser. Geol. 3: 269-285 (in Russian).

40. I. Gubkin, S. Feodorov, 1932 Mud volcanoes of the USSR in connection with oil and gas prospects Proceedings of 27th Int. Geol. Congr. Moscow 4 33-67(in Russian)

41. A. A. Jakubov, A. A. Ali-Zade, M.M. Zeinalov, 1971 Mud volcanoes of the Azerbaijan, SSR-Atlas Baku: Publishing house of the Academy Sciences of the Azerbaijan SSR245

42. G. D. Ginsburg, V. A. Soloviev, 1994 Mud volcano gas hydrates in the Caspian Sea Bull. Geol. Soc. Denm 41 95 100

43. G. E. Higgins, J. B. Saunders, 1973 Mud volcanoes-their nature and origin: contribution to the geology and paleobiology of the Carribbean and adjacent areas Verh. Naturforsch. Geschel. Basel 84 101 152

44. P. Williams, C. Pigram, D. Dow, 1984 Melange production and the importance of shale diapirism in accretionary terrains Nature 309 145 146

45. L. L. Mazurenko, V. A. Soloviev, I. Belenkaya, M. K. Ivanov, L. M. Pinheiro, 2002 Mud volcano gas hydrates in the Gulf of Cadiz. Terra Nova 14 321 329

46. C. Müller, F. Theilen, B. Milkereit, 2001 Large gasprospective areas indicated by bright spots World Oil 222 1

47. F. Perez-Belzuz, B. Alonso, G. Ercilla, 1997 History of mud diapirism and triggering mechanisms in the Western Alboran Sea Tectonophysics 282 399 423

48. P. R. Vogt, G. Cherkashev, G. Ginsburg, G. Ivanov, A. Milkov, K. Crane, A. Lein, E. Sunvor, N. Pimenov, A. Egorov, 1997 Haakon Mosby mud volcano provides unusual example of venting EOS Trans Am Geophys Union 78 48 556 557

49. W. R. Paine, 1968 Recent peat diapirs in the Netherlands: A comparison with Gulf Coast salt structures. In: Braunstein G.,. O'Brien G.D. (ed.) Diapirism and Diapirs., AAPG Mem. 8 271 274

50. T. W. Neurauter, H. H. Roberts, 1994 Three generations of mudvolcanoes on the Louisiana continental slope Geo-Mar. Lett. 14 120 125

51. S. Planke, H. Svensen, M. Hovland, D. A. Banks, B. Jamtveit, 2003 Mud and fluid migration in active mud volcanoes in Azerbaijan Geo-Marine Letters 23 258 268

52. G. Etiope, 2005 Methane emission from mud volcanoes. Martinelli G, Panahi B (ed.) Mud Volcanoes, Geodynamics and Seismicity. Springer

53. R. Sassen, S. L. Losh, L. Cathles, H. H. Roberts, J. K. Whelan, A. V. Milkov, S. T. Sweet, D. A. De Freitas, 2001 Massive vein filling gas hydrate: relation to ongoing migration from the deep subsurface in the Gulf of Mexico. Mar. Pet. Geol. 18 5 551 560

54. A. F. Limonov, T. C. E. van Weering, N. H. Kenyon, M. K. Ivanov, L. B. Meisner, 1997 Seabed morphology and gas venting in the Black Sea mudvolcano area: observations with the MAK-1 deep-tow sidescan sonar and bottom profiler Mar. Geol. 137 121 136

55. J. P. Foucher, G. De Lange, 1999 Submersible observations of cold seeps on eastern Mediterranean mud volcanoes Proceedings of J. Conf. Abstr. EUG 10, Strasbourg, France B13 41
56. D. L. Reed, E. A. Silver, E. Tagudin, H. Shipley, P. Volijk, 1990 Relations between mud volcanoes, thrust deformation, slope sedimentation, and gas hydrate, offshore north Panama Mar. Pet. Geol. 7 44 54
57. J. M. Woodside, M. K. Ivanov, A. F. Limonov, 1998 Shipboard Scientists of the Anaxiprobe Expeditions. Shallow gas and gas hydrates in the Anaximander Mountains regions, eastern Mediterranean Sea. Henriet JP, Mienert J. (ed.) Gas Hydrates: Relevance to World Margin Stability and Climate Change. Geol. Soc. London, Spec. Publ. 137 177 193
58. G. D. Ginsburg, A. V. Milkov, V. A. Soloviev, A. V. Egorov, G. A. Cherkashev, K. Crane, T. D. Lorenson, Khutorskoy, 1999 Gas hydrate accumulation at the Haakon Mosby mud volcano. Geo-Mar. Lett. 19 57 67
59. L. I. Dimitrov, J. Woodside, 2003 Deep sea pockmark environments in the eastern Mediterranean Marine Geology 195 263 276
60. R. J. Evans, R. J. Davies, S. A. Stewart, 2006 Internal structure and eruptive history of a kilometrescale mud volcano system, South Caspian Sea Basin Research 19 153 163
61. H. D. Hedberg, 1974 Relation of methane generation to undercompacted shales, shale diapirs and mud volcanoes Bull. Am. Assoc. Pet. Geol 58 661 673
62. A. A. Ali-Zade, E. F. Shnyukov, B. V. Grigoryants, A. A. Aliyev, R. R. Rakhmanov, 1984 Geotectonic conditions of mud volcano manifestation in the world and their role in prediction of gas and oil content in the earth's interior: proceedings of 27th International Geological Congress 13 377 393
63. I. S. Guliyiev, 1992 A review of mud volcanism Translation of the report by: Azerbaijan Academy of Sciences Institute of Geology 65
64. M. Manga, E. Brodsky, 2006 Seismic triggering of Eruptions in the Far Field: Volcanoes and Geysers Annual Review of Earth and Planetary Science 34 263 291
65. H. Svensen, B. Jamtveit, S. Planke, L. Chevallier, 2006 Structure and evolution of hydrothermal vent complexes in the Karoo Basin,South Africa Journal of the Geological Society 163 671 682
66. J. Skogseid, T. Pedersen, O. Eldholm, B. T. Larsen, 2010 Tectonism and magmatism during NE Atlantic continental break-up: the Vøring Margin. Storey B C, Alabaster T, Pankhurst, R J, Editors. Magmatism and the Causes of Continental Break-up. London, Geological Society 68 305 320
67. H. Svensen, S. Planke, B. Jamtveit, T. Pedersen, 2003 Seep carbonate formation controlled by hydrothermal vent complexes: a case study from the Vøring Basin, the Norwegian Sea Geo-Marine Letters 23 351 358
68. S. Planke, T. Rassmussen, S. S. Rey, R. Myklebust, 2005 Seismic characteristics and distribution of volcanic intrusions and hydrothermal vent complexes in the Vøring and Møre basins. Dore A, Vining B, Editors. Petroleum Geology: North-West Europe and Global Perspectives. Proceedings of the 6th Geology Conference. London 833 844
69. B. Bell, H. Butcher, 2002 On the emplacement of sill complexes: evidence from the Faroe-Shetland Basin Jolley D W, Bell B R, Editors. The North Atlantic Igneous Province: Stratigraphy, Tectonic, Volcanic and Magmatic Processes. London Geological Society, Special Publications 197 307 329
70. Toit. A. L. Du, 2002 Geological Survey of Elliot and Xalanga, Tembuland Annual Report of the Geological Commission Cape of Good Hope for 1903 8 169 205
71. Toit. A. L. Du, 1912 Geological Survey of Part of the Stormbergen Annual Report of the Geological Commission Cape of Good Hope for 1911 16 112 136
72. T. W. Gevers, 1928 The volcanic vents of the Western Stormberg Transactions of the Geological Society of South Africa 31 43 62
73. G. M. Stockley, 1947 Report on the Geology of Basutoland. Maseru, Authority of the Basutoland Government.

74. R. V. Dingle, W. G. Siesser, A. R. Newton, 1983 Mesozoic and Tertiary Geology of Southern Africa. Rotterdam, Balkema.

75. B. Jamtveit, H. Svensen, Y. Y. Podladchikov, S. Planke, 2004 Hydrothermal vent complexes associated with sill intrusions in sedimentary basins. Breitkreuz C, Petford N, Editors. Physical Geology of High-Level Magmatic Systems. London, Geological Society of London, Special Publications 234 233 241

76. R. H. Grapes, D. L. Reid, J. G. Mc Pherson, 1973 Shallow dolerite intrusions and phreatic eruption in the Allan Hills region, Antarctica. New Zealand Journal of Geology and Geophysics 17 563 577

77. R. E. Hanson, D. H. Elliot, 1996 Rift-related Jurassic basaltic phreatomagmatic volcanism in the central Transantarctic Mountains: precursory stage to floodbasalt effusion Bulletin of Volcanology 58 327 347

78. J. D. L. White, M. K. Mc Clintock, 2001 Immense vent complex marks floodbasalt eruption in a wet, failed rift: Coombs Hills, Antarctica Geology 29 935 938

79. V. V. Zolotukhin, A. I. Al'mukhamedov, 1988 Traps of the Siberian Platform. In: Macdougall JD, Editors Continental Flood Basalts. Kluwer, Dordrecht 273 310

80. H. Svensen, S. Planke, A. Malthe-Sørenssen, B. Jamtveit, R. Myklebust, T. Eidem, S. S. Rey, 2004 Release of methane from a volcanic basin as a mechanism for initial Eocene global warming Nature 429 542 545

81. P. B. Wignall, 2001 Large igneous provinces and mass extinctions Earth-Science Reviews 53 1 33

82. V. E. Courtillot, P. R. Renne, 2003 On the ages of flood basalt events. Comptes Rendus de l'Acade´mie des Sciences Geoscience 335 113 140

83. H. Svensen, G. Bebout, A. Kronz, L. Li, S. Planke, L. Chevallier, B. Jamtveit, 2008 Nitrogen geochemistry as a tracer of fluid flow in a hydrothermal vent complex in the Karoo Basin, South Africa Geochimica et Cosmochimica Acta 72 4929 4947

84. A. Mazzini, A. Nermoen, M. Krotkiewski, Y. Podladchikov, S. Planke, H. Svensen, 2009 Strike-slip faulting as a trigger mechanism for overpressure release through piercement structures Implications for the Lusi mud volcano, Indonesia. Marine and Petroleum Geology 26 1751 1765

85. A. Mazzini, H. Svensen, G. G. Akhmanov, G. Aloisi, S. Planke, A. Malthe-Sorenssen, B. Istadi, 2007 Triggering and dynamic evolution of the LUSI mud volcano, Indonesia Earth and Planetary Science Letters 261 3-4 375 388

86. B. V. J. Botha, J. C. Theron, 1966 New evidence for the early commencement of Stormberg volcanism Tydskrif vir Natuurwetenskappe 7 469 473

87. C. B. Coetzee, 1966 An ancient volcanic vent on Boschplaat 369 in the Bloemfontein district, Orange Free State Transactions of the Geological Society of South Africa 69 127 137

88. N. C. Taylor, The volcanic vents of the Stormberg series BSc(Hons) project, Rhodes University, Grahamstown, South Africa 1970

89. U. T. Seme, 1997 Diatreme deposits near Rossouw, north Eastern Cape:sedimentology-volcanology and mode of origin BSc(Hons) project, Rhodes University, Grahamstown, South Africa

90. A. C. Woodford, J. F. Botha, L. Chevallier, et al. 2001 Hydrogeology of the main Karoo Basin: Current Knowledge and Research Needs Water Research Commission Pretoria Report, 860

91. L. A. Navikov, R. M. Slobodskoy, 1979 Mechanism of formation of diatremes International Geology Review 21 1131 1139

92. V. Lorenz, 1985 Maars and diatremes of phreatomagmatic origin: a review Transactions of the Geological Society of South Africa 88 459 470

93. C. R. Clement, A. M. Reid, The origin of kimberlite pipes: an interpretation based on a synthesis of geological features displayed

94. K. J. Webb, B. H. Smith, J. L. Paul, C. M. Hetman, 2004 Geology of the Victor Kimberlite, Attawapiskat, Northern Ontario, Canada: cross-cutting and nested craters Lithos 76 29 50

95. Surtee southern African occurrences. 1989 Ross, J. et al. Editors. Kimberlites and Related Rocks. Geological Society of Australia, Special Publications 14 632 646
96. M. K. Mc Clintock, B. F. Houghton, I. P. Skilling, J. D. L. White, 2002 The volcaniclastic opening phase of Karoo flood basalt volcanism, Drakensberg Formation, South Africa EOS Transactions of the American Geophysical Union 83 47
97. J. D. L. White, P. S. Ross, 2011 Maar-diatreme volcanoes: A review J Volcanol Geotherm Res 2011-4 1 29
98. H. Svensen, S. Planke, L. Chevallier, A. Malthe-Sorenssen, F. Corfu, B. Jamtveit, 2007 Hydrothermal venting of greenhouse gases triggering Early Jurassic global warming Earth Planet. Sci. Lett. 256 3-4 554 566
99. H. Svensen, S. Planke, A. G. Polozov, N. Schmidbauer, F. Corfu, Y. Y. Podladchikov, B. Jamtveit, 2009 Siberian gas venting and the end-Permian environmental crisis Earth Planet. Sci. Lett. 277 3-4 490 500
100. E. D. Sloan, C. A. Koh, 2008 Clathrate Hydrates of Natural Gases, third ed. CRC Press, New York, Taylor and Francis Group, Publishers
101. E. A. Smelik H. E. King, 1997 Crystal-growth studies of natural gas clathrate hydrates using a pressurized optical cell American Mineralogist 82 88 98
102. C. K. Paull, W. Dillon, 2001 Natural gas hydrates: occurrence, distribution, and detection. AGU Monogr 124 1 315
103. G. D. Holder, R. D. Malone, W. F. Lawson, 1987 Effects of gas composition and geothermal properties on the thickness and depth of natural-gas-hydrate zone Journal of Petroleum Technology 1147 1152
104. J. F. Wright, S. R. Dallimore, F. M. Nixon, C. Duchesne, 2005 In situ stability of gas hydrate in reservoir sediments of the JAPEX/JNOC/GSC et al. Mallik 5L-38 gas hydrate production research well. Paper presented at Scientific Results from the Mallik 2002 Gas Hydrate Production Research Well Program, Mackenzie Delta, Geol. Surv. Canada, Northwest Territories, Canada
105. G. R. Dickens, M. S. Quinby-Hunt, 1994 Methane hydrate stability in seawater Geophys. Res. Lett 21 2115 2118
106. S. Subramanian, R.A. Kini, S.F. Dec, E. D. Sloan, 2000 Evidence of structure II hydrate formation from methane + ethane mixtures Chemical Engineering Science 55 1981 1999
107. Z. Lu, N. Sultan, 2008 Empirical expressions for gas hydrate stability law, its volume fraction and mass-density at temperature 273.15 K to 290.15 K Geochemical Journal 42 163 175
108. N. Sultan, P. Cochonat J. P. Foucher, J. Mienert, 2004 Effect of gas hydrates melting on seafloor slope instability Marine Geology 213 379 401
109. X. Liu, P. B Flemings, 2007 Dynamic multiphase flow model of hydrate formation in marine sediments Journal of Geophysical Researcher 112 B03101
110. J. P. Henriet, J. Mienert, 1998 Gas hydrates: relevance to world margin stability and climate change Geol Soc Spec Publ 137
111. G. D. Holder, P. R. Bishnoi, 2000 Gas Hydrates: Challenges for the Future. Ann. N.Y. Acad. Sci.912
112. J. P. Kennett, K. G. Cannariato, I. L. Hendy, R. J. Behl, 2007 Carbon isotopic evidence for methane hydrate stability during Quaternary Interstadials Science 288 128 133
113. J. M. Brooks, H. B. Cox, W. R. Bryant, M. C. Kennicutt, R. G. Mann, T. J. Mc Donald, 1986 Association of gas hydrates and oil seepage in the Gulf of Mexico Organic Geochemistry 10 221 234
114. J. P. Severinghaus, T Sowers, E. J. Brook, R. B. Alley, M. L. Bender, 1998 Timing of abrupt climate change at the end of the Younger Dryas interval from thermally fractionated gases in polar ice Nature 391141
115. M. T. Reagan, G. J. Moridis, 2007 Oceanic gas hydrate instability and dissociation under climate change scenarios. Geophysical Research Letters 34L22709
116. J. P. Kennett, K. G. Cannariato, L. L. Hendy, R. J. Behl, 2002 Methane Hydrates in Quaternary Climate Change: The Clathrate Gun Hypothesis Spec. Publ., 54, AGU, Washington, D. C.

117. B. U. Haq, 1998 Natural gas hydrates: searching for the long-term climatic and slope-stability records. Geological Society, London, Special Publications 137 303 318
118. T. H. Shipley, M. H. Houston, R. T. Buffler, et al. 1979 Seismic reflection evidence for widespread occurrence of possible gas-hydrate horizons on continental slopes and rises AAPG Bull 63 2204 2213
119. T. H. Shipley, B. M. Didyk, 1982 Occurrence of methane hydrates offshore southern Mexico. Watkins JS, Moore JC, et al., Editors. Init. Repts. DSDP, 66. Washington, U.S. Govt. Printing Office 547 556
120. T. A. Minshull, S. C. Singh, G. K. Westbrook, et al. 1994 Seismic velocitystructure at a gas hydrate reflector,offshore western Colombia, from full waveform inversion J. Geophys. Res. 99 4715 4733
121. K. Sain, T. A. Minshull, S. C. Singh, R. W. Hobbs, 2000 Evidence for a thick free gas layer beneath the bottom simulating reflector in the Makran accretionary prism Marine Geology 164 37 51
122. U. Tinivella, F. Accaino, 2000 Compressional velocity structure and Poisson's ratio in marine sediments with gas hydrate and free gas by inversion of reflected and refracted seismic data (South Shetland Islands, Antarctica) Mar Geol 164 13 27
123. J,. Dai, H,. Xu, F,. Snyder, N. Dutta, 2004 Detection and estimation of gas hydrates using rock physics and seismic inversion: Examples from the northern deepwater Gulf of Mexico The Leading Edge.
124. S. Chand, T. A. Minshull, 2003 Seismic constraints on the effects of gas hydrate on sediment physical properties and fluid flow: a review Geofluids 3 1 15
125. M. Zillmer, 2006 A method for determining gas-hydrate or free-gas saturation of porous media from seismic measurements Geophysics 71 21 32
126. S. C. Singh, T. A. Minshull, G. D. Spence, 1993 Velocity structure of a gas hydrate reflector Science 260 204 207
127. U. Tinivella, 2002 The seismic response to overpressure versus gas hydrate and free gas concentration J Seismic Explor 11 283 305
128. U. Tinivella, 1999 A method to estimate gas hydrate and free gas concentrations in marine sediments Bollettino di Geofisica Teorica ed Applicata 40 19 30
129. U. Tinivella, M. Carcione, 2001 Estimation of gas-hydrate concentration and free-gas saturation from log and seismic data. The Leading Edge 20
130. T. S. Ledley, E. T. Sundquist, S. E. Schwartz, D. K. Hall, J. D. Fellows, T. L. Killeen, 1999 Climate Change and Greenhouse Gases. EOS, Transactions of the American Geophysical Union 80 453 458
131. S. Nakano, M. Moritoki, K. Ohgaki, 1998 High-Pressure Phase Equilibrium and Raman Microprobe Spectroscopic Studies on CO2 Hydrate System. J. Chem. Eng. Data; 43 807 810
132. K. S. Lackner, 2003 A Guide to CO2 Sequestration Science 300 1677 1678
133. R. A. Chadwick, R. Arts, O. Eiken, 4D seismic quantification of a growing CO2 plume at Sleipner, North Sea. Geological Society, London, Petroleum Geology Conference series 6 1385 1399
134. S. Conti, D. Fontana, 2011 Possible Relationships between Seep Carbonates and Gas Hydrates in the Miocene of the Northern Apennines Journal of Geological Research 20727
135. X. Liu, P. B. Flemings, 2007 Dynamic multiphase flow model of hydrate formation in marine sediments Journal of Geophysical Research 112: B03101
136. T. T. Schowalter, 1979 Mechanics of secondary hydrocarbon migration and entrapment American Association of Petroleum Geologists Bulletin 63 723 760
137. A. V. Milkov, G. E. Claypool, Y. Lee-J, R. Sassen, 2005 Gas hydrate systems at Hydrate Ridge offshore Oregon inferred from molecular and isotopic properties of hydrate-bound and void gases, Geochim. Cosmochim. Acta 69 1007 1026
138. A. M. Trehu, P. B. Flemings, N. L. Bangs, J. Chevallier, E. Gra`cia, J. E. Johnson, C. Liu-S, X. Liu, M. Riedel, Torres, 2004 Feeding methane vents and gas hydrate deposits at south Hydrate Ridge Geophys. Res. Lett. 31: L23310

139. P. B. Flemings, X. Liu, W. Winter, 2003 Critical pressure and multiphase flow in Blake Ridge gas hydrates Geology 31 1057 1060
140. A. R. Gorman, W. S. Holbrook, M. J. Hornbach, K. L. Hackwith, 2002 Migration of methane gas through the hydrate stability zone in a low-flux hydrate province Geology 30 327 330
141. M. Hornbach, D. M. Saffer, W. S. Holbrook, 2003 Critically pressured free-gas reservoirs below gas-hydrate provinces Nature 427 142 144
142. J. Nimblett, C. Ruppel, 2003 Permeability evolution during the formation of gas hydrates in marine sediments J. Geophys. Res. 108B92420
143. S. Bouriak, M. Vanneste, A. Saoutkine, 2000 Inferred gas hydrates and clay diapirs near the Storegga slide on the southern edge of the Voring Plateau, offshore Norway Mar Geol 163 125 148
144. S. Buenz, J. Mienert, 2004 Acoustic imaging of gas hydrate and free gas at the Storegga Slide J. Geophys. Res 109B04102
145. G. D. Ginsburg, V. L. Ivanov, V. A. Soloviev, 1984 Natural gas hydrates of the World's Oceans. In: Oil and gas content of the World's Oceans PGO Sevmorgeologia 141 158in Russian).
146. G. D. Ginsburg, A. N. Kremlev, M. N. Grigor'ev, G. V. Larkin, A. D. Pavlenkin, N. A. Saltykova, 1990 Filtrogenic gas hydrates in the Black Sea (twenty-first voyage of the research vessel Evpatoria). Sov. Geol. Geophys 31 8 16
147. A. F. Limonov, J. M. Woodside, M. K. Ivanov, 1994 Mud volcanism in the Mediterranean and Black Seas and shallow structure of the Eratosthenes Seamount Initial results of the geological and geophysical investigations during the Third UNESCO-ESF "Training Through Research" Cruise of RV Gelendzhik, June-July 1993. UNESCO Rep. Mar. Sci. 64
148. J. M. Woodside, M. K. Ivanov, A. F. Limonov, 1997 Neotectonics and fluid flow through seafloor sediments in the Eastern Mediterranean and Black Seas- Parts I and II IOC Tech. Ser. 48
149. Donald. R. I. Mac, W. W. Sager, M. B. Peccini, 2003 Gas hydrate and chemosynthetic biota in mounded bathymetry at mid-slope hydrocarbon seepsNorthern Gulf of Mexico Marine Geology 198 133 158
150. A. Judd, G. Davies, J. Wilson, R. Holmes, G. Baron, I. Bryde, 1997 Contribution to atmospheric methane by natural seepages on the UK continental shelf Mar. Geol 137 165 189
151. S. Lance, P. Henry, X. Le Pichon, S. Lallemant, H. Chamley, F. Rostek, J. Faugeres-C, 1998 Submersible study of mud volcanoes seaward of the Barbados accretionary wedge: sedimentology, structure and rheology Mar. Geol 145 255 292
152. B. T. Cronin, M. K. Ivanov, A. F. Limonov, A. Egorov, G. G. Akhmanov, A. M. Akhmetjanov, E. Kozlova, 1997 Shipboard Scientific Party TTR-5, New discoveries of mud volcanoes on the Eastern Mediterranean Ridge J. Geol. Soc. 154 173 182
153. B. J. Murton, J. Biggs, 2003 Numerical modelling of mud volcanoes and their flows using constraints from the Gulf of Cadiz Marine Geology 195 223 236
154. P. Soderberg, T. Floden, 1991 Pockmark developments along a deep crustal structure in the northern Stockholm Archipelago, Baltic Sea Beitr. Meereskd 62 79 102
155. P. Soderberg, T. Floden, 1992 Gas seepages, gas eruptions and degassing structures in the seafloor along the Stromma tectonic lineament in the crystalline Stockholm Archipelago, east Sweden Cont. Shelf Res. 12 1157 1171
156. H. A. Cohen, K. Mc Clay, 1996 Sedimentation and shale tectonics of the northwestern Niger Delta front Mar. Pet. Geol. 13 313 328
157. Bouma A.H., Roberts H.H. 1990 Northern Gulf of Mexico continental slope Geo-Mar. Lett. 10 177 181
158. M. G. Langseth, G. K. Westbrook, A. Hobart, 1988 Geophysical survey of mud volcano seaward of the Barbados Ridge Complex J. Geophys. Res. 93 1049 1061

159. J. B. Martin, M. Kastner, P. Henry, X. Le Pichon, S. Lallemant, 1996 Chemical and isotopic evidence for sources of fluid in a mud volcano field seaward of the Barbados accretionary wedge J. Geophys. Res. 101 20325 20345

160. R. Heggland, E. Nygaard, 1998 Shale intrusions and associated surface expressions-examples from Nigerian and Norwegian deepwater areas: proceedings Offshore Technology Conference. Houston, TX 1 111 124

161. J. E. Corthay, 1998 Delineation of a massive seafloor hydrocarbon seep, overpressured aquifer sands, and shallow gas reservoir, Louisiana continental slope Proceedings Of Offshore Technology Conference.Houston, Texas 37 56

162. T. W. Neurauter, W. R. Bryant, 1998 Gas hydrates and their association with mud diapirs/mud volcanoes on the Louisiana continental slope Proceedings of Offshore Technology Conference, Houston, Texas 1 599 607

163. A. V. Milkov, Y. Lee-J, W. S. Borowski, Xu. W. Torres, H. Tomaru, A. M. Tre´hu, P. Schultheiss, G. R. Dickens, G. E. Claypool, 2004 Co-existence of gas hydrate, free gas, and brine within the regional gas hydrate stability zone at Hydrate Ridge (Oregon margin): Evidence from prolonged degassing of a pressurized core Earth Planet. Sci. Lett. 222 829 843

164. O. Y. Zatsepina, B. A. Buffett, 1997 Phase equilibrium of gas hydrate: Implications for the formation of hydrate in the deep sea floor Geophys. Res. Lett. 24 1567 1570

165. S. I. Tomkeieff, 1983 Dictionary of Petrology WileyNew York

166. Z. Ben-Avraham, G. Smith, M. Reshef, E. Jungslager, 2002 Gas hydrate and mud volcanoes on the southwest African continental margin off South Africa Geology (Boulder) 30 927 930

167. E. H. A. Jungslager, 1999 Petroleum habitats of the South Atlantic margin. Cameron NR, Bate RH, Clure VS. (ed.) The Oil and Gas Habitats of the South Atlantic, Spec. Publ. geol. Soc. Lond 153 168

168. U. Tinivella, F. Accaino, A. Camerlenghi, 2002 Gas hydrate and free gas distribution from inversion of seismic data on the South Shetland margin (Antarctica) Mar Geophys Res 23 109 123

169. K. Klepeis, L. A. Lowver, 1996 Tectonics of the Antarctic-Scotia plate boundary near Elephant and Clarence Islands, West Antarctica J. Geophys Res 101 20211 20231

170. M. F. Loreto, U. Tinivella, F. Accaino, M. Giustiniani, 2011 Offshore Antarctic Peninsula Gas Hydrate Reservoir Characterization by Geophysical Data Analysis Energies 4 11 39 56

171. I. Vargas-Cordero, U. Tinivella, F. Accaino, M. F. Loreto, F. Fanucci, 2010 Thermal state and concentration of gas hydrate and free gas of Coyhaique, Chilean Margin (44°30′ S) Marine and Petroleum Geology 27 1148 1156

172. H. H. Roberts, 2001 Fluid expulsion on the Northern Gulf of Mexico continental slope: mud-prone to mineral-prone responses. Paull CK, Dillon WP. (ed.) Natural gas hydrates: occurrence, distribution, and detection. AGU Monogr 124 145 161

173. H. H. Roberts, Shedd. W. W. Hardage, J. Hunt, 2006 Seafloor reflectivity-an important seismic property for interpreting fluid/ gas expulsion geology and the presence of gas hydrate Leading Edge 25 620 628

174. A. V. Milkov, R. Sassen, 2002 Economic geology of offshore gas hydrate accumulations and provinces Marine and Petroleum Geology 19 1 11

CHAPTER 3

Mud volcanoes in the South Caspian basin: Nature and estimated depth of its products

Akper A. Feyzullayev

Geology Institute of the Azerbaijan National Academy of Sciences, Baku, Azerbaijan;

ABSTRACT

In paper based on the results of previous studies and latest investigations of mud volcanoes in the South Caspian basin the depth occurrence of fluid, mud and rocks are calculated and modeled. Major factors causing formation of diapirism/mud volcanism in the South Caspian basin are Pliocene-Quaternary high sedimentation rates (up to 3 km/my), super thick sedimentary cover (up to 25 - 30 km), predominance of clayey rocks (reaching 80%) in the section, low temperatures (with 15°C - 18°C/km gradient), overpressures reaching lithostatic, the onset of petroleum generation lowered to considerable depths. The majority of the mud volcanoes are associated with the petroleum bearing structures. Depths where the liquid, gaseous and solid products of mud volcanoes are sourced appear to be different. The gases have the deepest roots (7 - 15 km) which are the main force in formation and activity of MVs. Source of the fluidized clayey mass does not lie below 3 - 4 km. Oils emitted by mud volcanoes are the product of destruction of petroleum accumulations occurring beneath them.

Keywords: mud volcano; fluid; geochemistry; temperature; depth; South Caspian basin

1. INTRODUCTION

Many years of study of mud volcanoes (MVs) in Azerbaijan has led to revealing main regularities in their distribution, morphological patterns, activity characteristics, as well as compositional properties of their products (breccia, rock fragments, gas, oil, water) and also character of their reflection in geophysical fields were discovered [1-5, etc.].

Remarkable achievements in studies of Azerbaijan mud volcanoes using modern geophysical, geochemical and other methods were obtained during the last two decades [6-16, etc].

This will promote better understanding of the conditions and mechanism of their formation.

The petroleum bearing South Caspian basin (SCB) is located within the Alpine-Himalayan mobile tectonic belt where earthquakes and intense modern earth's crust movements have been observed. SCB is a relic of the Tethys Ocean originated in the post orogenic stage of the regional development. It is typically intermountain basin surrounded by mountain systems: Great and Lesser Caucasus, Talysh, Elburs, Kopetdag and Balkhan (**Figure 1**).

The present-day structure of SCB is still controlled by the ongoing collision of the Arabian and Russian plates [17-19]. The orogeny during the Oligocene, continuing up to now and causing a large area, is dominated by right lateral strike slip deformation [20] and reactivation of many of the earlier extensional faults, which resulted in weakening of the overburden sediments.

The main geological peculiarities of the SCB, some of which are unique are the following:

1) Very rapid Pliocene-Quaternary subsidence and sedimentation rate (in average about 2 km/Ma). Over 80% of these sediments consist of clayey rocks;
2) The thickness of the sedimentary cover (MesozoicCenozoic) is 25 - 30 km and has no analogues in the world (**Figure 2**);

3) Characteristic the split-level structure of the sedimentary section in SCB, including the upper Paleogene-Quarternary complex presented mainly by plastic (undercompacted and decompacted) rocks and the lower Mesozoic, consisting of the dense rocks;

4) Anomalous low temperature regime: heat flow— (25 - 51) mW/m.sq., temperature gradient—15°C - 18°C/ km;

5) Displacement of the onset of oil generation on considerable depths (oil to 4 - 5 km, gas to 7 km) (see **Figure 2**);

6) High recent geodynamic activity: horizontal and vertical movements of Earth crust and shallow seismicity. Within the SCB tens of earthquakes with different magnitude are annually recorded. It is necessary to mention that most earthquakes have shallow focus which do not exceed 10 - 15 km (**Figure 3**).

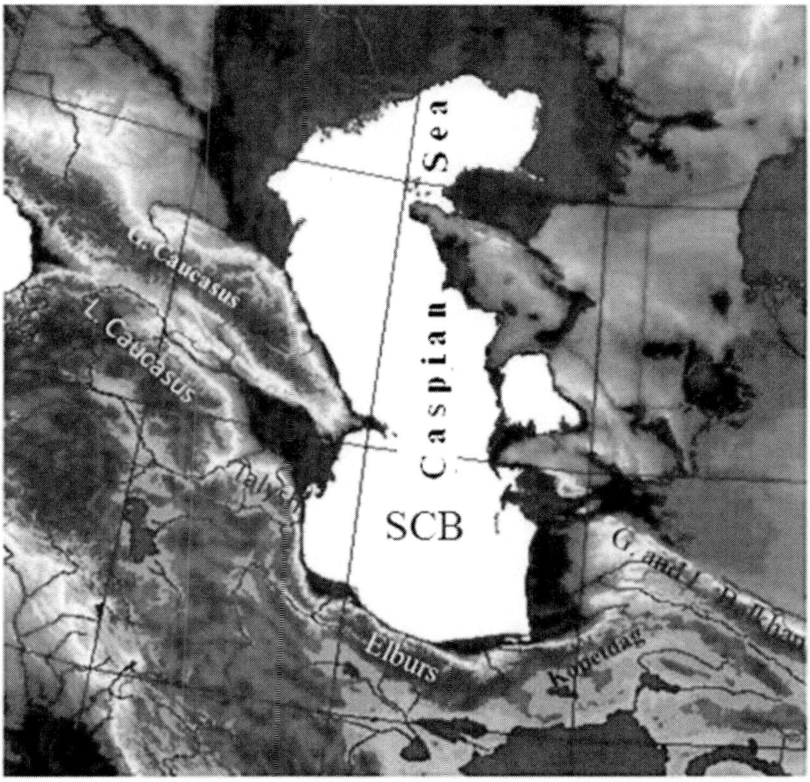

Figure 1. Morphology and position of the SCB.

All above-mentioned geological factors contribute to formation and activity of diapirs/mud volcanoes.

In spite of achieved successes, scientists continue to argue about the origin of this amazing phenomenon.

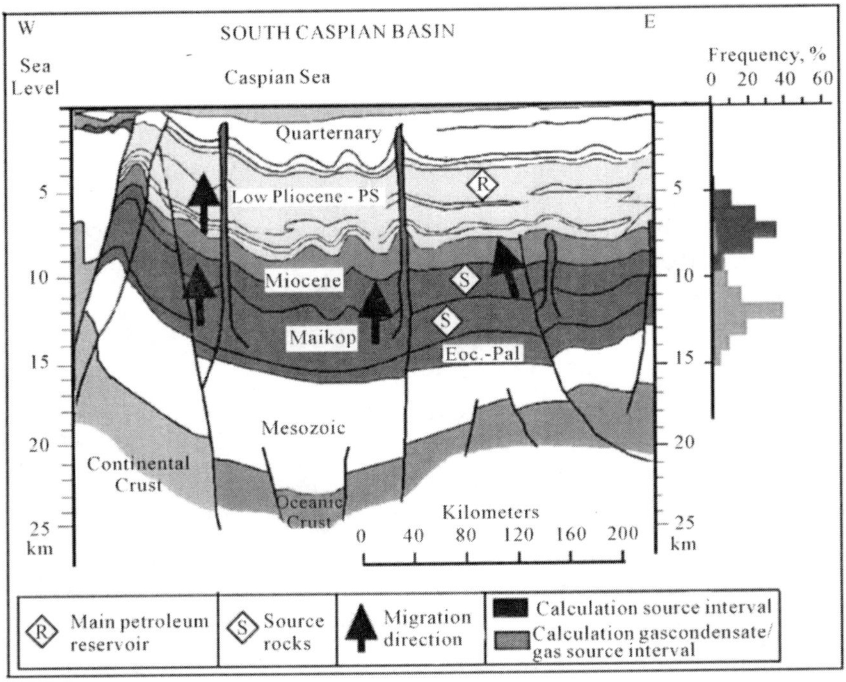

Figure 2. Regional geological profile along Caspian Sea and location in the sedimentary.

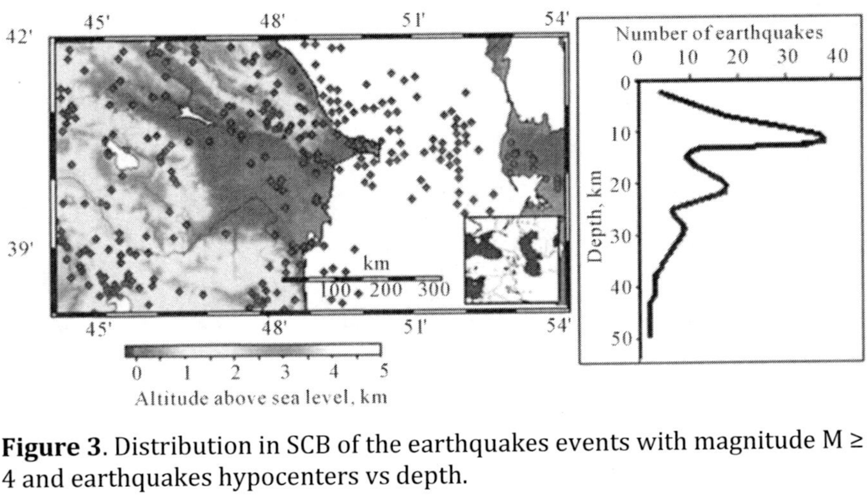

Figure 3. Distribution in SCB of the earthquakes events with magnitude M ≥ 4 and earthquakes hypocenters vs depth.

Some suggest that the process takes place wholly within the sedimentary succession [1,14,21], whereas others envisage a connection with the deeper processes operating within the mantle [3]. Therefore results of estimation of the depth occurrence of the MVs products in SCB are presented in this paper.

2. INVESTIGATION APPROACH

A number of isotopic-geochemical parameters are applied for direct assessments of oil and hydrocarbon gas maturity [22,23].

2.1. The Gas Maturity

The gas maturity estimates were made using experimentally established interrelation between isotopic composition of carbon in ethane and vitrinite reflectance (R_o, %) [22]. Methane carbon isotopic signatures are less reliable in view of possible bacterial methane input. Using this relation in combination with the R_o versus depth change in the SCB depths of HC gas source are evaluated.

2.2. The Oil Maturity

Sterane isomerisation degree {αααC29(20S/S + R)}, monoaromatic sterane aromatization degree {C28triaromatic sterane/C28triarom. + C29monoaromatic sterane} and methyl-phenantrene index {formulae MPI – 1 = (1.5 (2 – MP + 3 – MP))/(P + 1 – MP + 9 – MP + 4 – MP) and are most often applied biomarker ratios for oil maturity assessments. Based on empirically derived relationships values of these parameters are converted to equivalent vitrinite reflectance values (R_{eq}). In this study for oil maturity assessment MPI – 1 parameter was used [23].

For estimation of the depth occurrences of mud, water and rock-ejects the complex data (recent and paleo-temperatures, seismic records, geological section, etc.) were applied.

3. RESULTS AND DISCUSSION

3.1. Estimation of Depth of the Liquid and Solid Products of MVs

MVs emit the fluids and mud during both dormant and explosive activity. The nature of its products is great scientific interest. Based on the isotopic composition of helium [2] argon and carbon in methane [6,9] the Azerbaijan mud volcanoes have a sedimentary origin and they are products of different stages of a single process of hydrocarbon generation within the sedimentary succession. The chemical and isotopic signatures of MV gases are predominantly thermogenic with some admixture of microbial methane [14].

Based on the isotopic composition of the C_1 to C_4, that the hydrocarbon gases on the Regnano mud volcano (northern Italy) have a thermogenic origin and they were formed by thermal cracking (>150°C) [24]. They are later products of oil and/or kerogen alterations in overmature environments [25].

The results of determination of the hypsometric and stratigraphic source of liquid and solid products of MVs in the SCB are given below.

3.1.1. Gas

Calculated gas maturity values (at the vitrinite reflectance equivalent -Req.) vary between 1.2% - 2.0%, corresponding to the depth interval of 10 - 16 km (peak - to 11 - 12 km). On MVs are almost absent low mature gases (less than 1.4% Req.) in comparison with reservoir gases (**Figure 4**). Except one mud volcano (Zaakhtarma), gases of all studied MVs refer to gas-condensate—wet gas zone.

It is established dependence between isotope composition of helium and calculated maturity of gases of the Azerbaijan mud volcanoes (**Table 1**).

There is a direct relationship between ICC of CH_4 and $^{40}Ar/^{36}Ar$ values too, reflecting consistent increase of gas maturity with increasing of age of rocks (**Figure 5**).

ICC of CO_2 in MVs changes from −49.6‰ to +23.1‰, showing at least two source of its origin (**Figure 6**). It is especially interesting nature of isotopically super heavy CO_2 (>+5‰), which have been controversial for a long time. However, the results of recent studies of ICC in CO_2 from petroleum reservoirs revealed their occurrence on the depths not exceeding 2000 m and related with the oil biodegradation/fermentation at shallow depths [16] (see **Figure 6**). In [26] were also coming to conclusion, that "oil biodegradation with CO_2 reduction likely occurs at a shallower depth along the seepage system of the mud volcano".

Table 1. Dependence between $^3He/^4He$ values and calculated maturity of gases in the Azerbaijan mud volcanoes.

Volcano	$^3He/^4He$, 10^{-8}		Maturity, Req at ICC ethane, %	
	Limits	Average	Limits	Average
Group I (main)	2.8 - 9.7	6.0	1.47 - 1.71	1.59
Group II	15 - 23	19.0	1.72 - 1.77	1.74
Group III	26 - 30	27.7	2.02	2.02

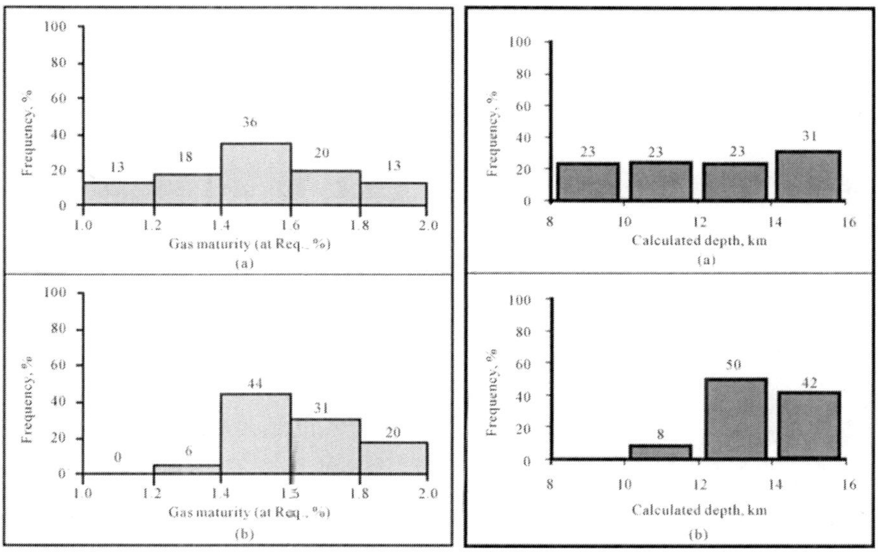

Figure 4. Histograms of distribution of calculated maturity values and source depths of the reservoir (a) and mud volcanic (b) gases in SCB.

Existence of several stratigraphic and hypsometric sources of CH_4 and CO_2 is confirmed by change within wide limits of its ICC from different vents at the same mud volcano (**Table 2**).

3.1.2. Gas-Hydrate

Gas-hydrate accumulation, associated with active submarine mud volcanoes was discovered first time in the Southern Caspian Sea [5]. According to isotope composition of carbon in hydrocarbon gases the gas of gashydrate located on Buzdag MV is thermogenic, coming along volcanic channel from deep-sited reservoir towards seabed (**Figure 7**).

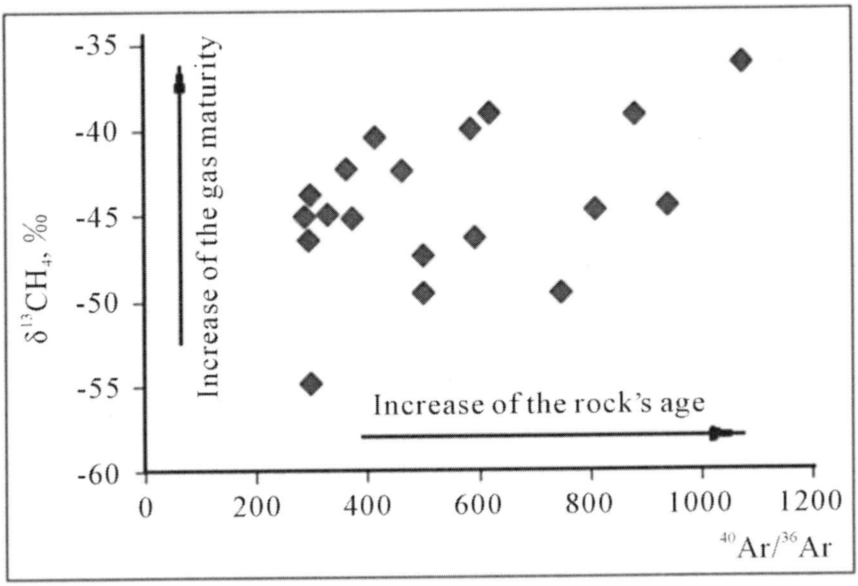

Figure 5. ICC of methane vs $^{40}Ar/^{36}Ar$ of the Azerbaijan mud volcanoes.

3.1.3. Oil

The evaluated maturity of oils on mud volcanoes is lower in comparison with gas. The present day maturity of the MV oils is 0.65% - 0.80% in Req. and well agreed with maturity of the reservoir oils (**Figure 8**). This allows concluding that the MV oils are a product of the destruction of oil accumulation in structures, pierced by MV. Therefore the maximal depth of the MV oils does not exceed depth of corresponding oil reservoir, which according to studied reservoir oils reaches about 4.9 km in offshore (8 March field, well 580) and 5.3 km in onshore (Kalamaddyn field, well 3).

The stratigraphic location of the HC sources changes towards regional subsidence of strata (from southern slope of the Great Caucasus to the central deep buried part of SCB) from Mesozoic to Paleogene-Miocene—for gas and from Oligocene to Miocene-for oil [9,27].

3.1.4. Water

According to hydrochemical and isotope analyses of the volcanic and formation waters and their correlation they may represent complex mixtures of deep waters (slightly salty, mainly alkaline and fresh condensate waters) and relatively shallower brines (predominantly chlorine) or meteoric waters. This corresponds to the results of other investigations [26,28-30].

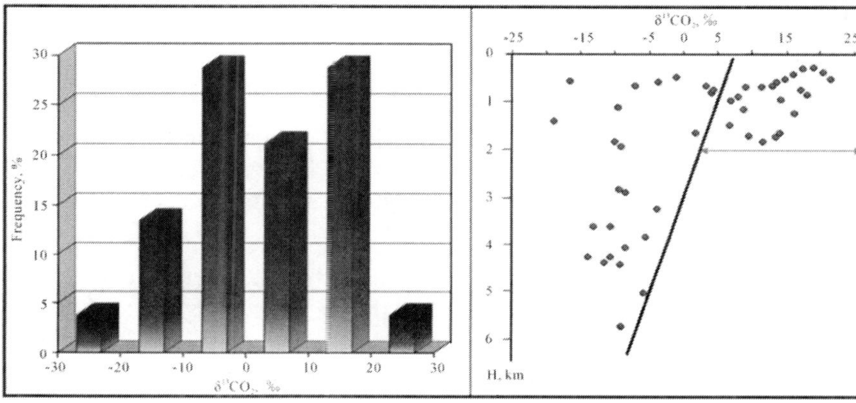

Figure 6. Histogram of distribution of the ICC of CO_2 from mud volcanoes and ICC of CO_2 vs depth for the reservoir gases.

Table 2. Isotope-chemical composition of gases from different vents of the same volcano.

MV	CH_4, %	C_2H_6+, %	N_2, %	CO_2, %	$^{13}C_{CH_4}$, ‰	$^{13}C_{CO_2}$, ‰
Shorbulag:						
a)	75.4	nd	23.6	1	−41.5	0.5
b)	96.6	nd	1.1	2.3	−50.6	−26.8
Melikchobanly:						
a)	84.1	0.9	2.8	12.1	−45.2	20.9
b)	72.3	nd	20.4	7.3	−44.7	−14
Cheildag:						
a)	80.3	nd	7.8	11.9	−43.3	2.4
b)	91.3	3.5	2	3.2	−37.3	23.1
Ayrantekan:						
a)	94.4	1.4	1.3	2.3	−49.8	−8.1
b)	85.2	nd	13.2	1.6	−47.9	−0.1

3.1.5. Mud

There are complex monitoring indices proving shallow origin of the mud, erupted by MVs. They are the following:

3.1.5.1. Rock compaction criterion Depending on a geological history of development, normal and non-equilibrium compaction of sediments occurs in various basins of the world. Therefore the intensity of the worldwide sediments compaction changes over a wide range (especially within 0 - 2 km). Below the depth of about 3 km rate of compaction is minimized and stabilized. This allows concluding that the sediments and, especially clays, undergo the greatest compaction (up to 60% - 70% of the initial volume) in the interval of 0 - 4 km depths during the subsidence of any basin. It is mean that clays reaching the depth of 3 - 4 km lose the large part of sediment waters. The loss of waters by the clay leads to decreasing of its plasticity and mobility and therefore the focus of the clayey diapir formation, most probably, occurs at the depth not greater than 4 km. Bulk of clayey breccia erupted by the mud volcano is predominantly connected with this interval of depths. It is suggested a shallow occurrence of mud chamber on the Lokbatan mud volcano [31].

Results of investigations of MVs in northern Italy [24] well correspond to this statement. They came to conclusion, that mud is younger and shallower (about 1 km) than methane released by MVs from deep Miocene deposits and Mesozoic carbonate source rocks. In [32] under the maturity data depth of a mud mobilization about 1.7 km was estimated.

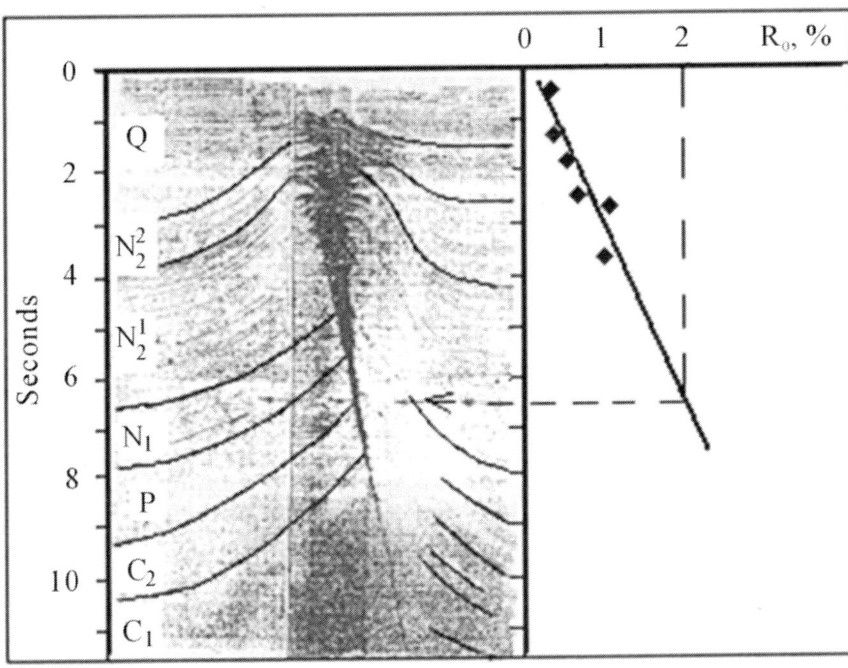

Figure 7. Fragment of the seismic time section across Buzdag submarine mud volcano in the Southern Caspian Sea with calculated depth of location of the gas source of gas hydrate.

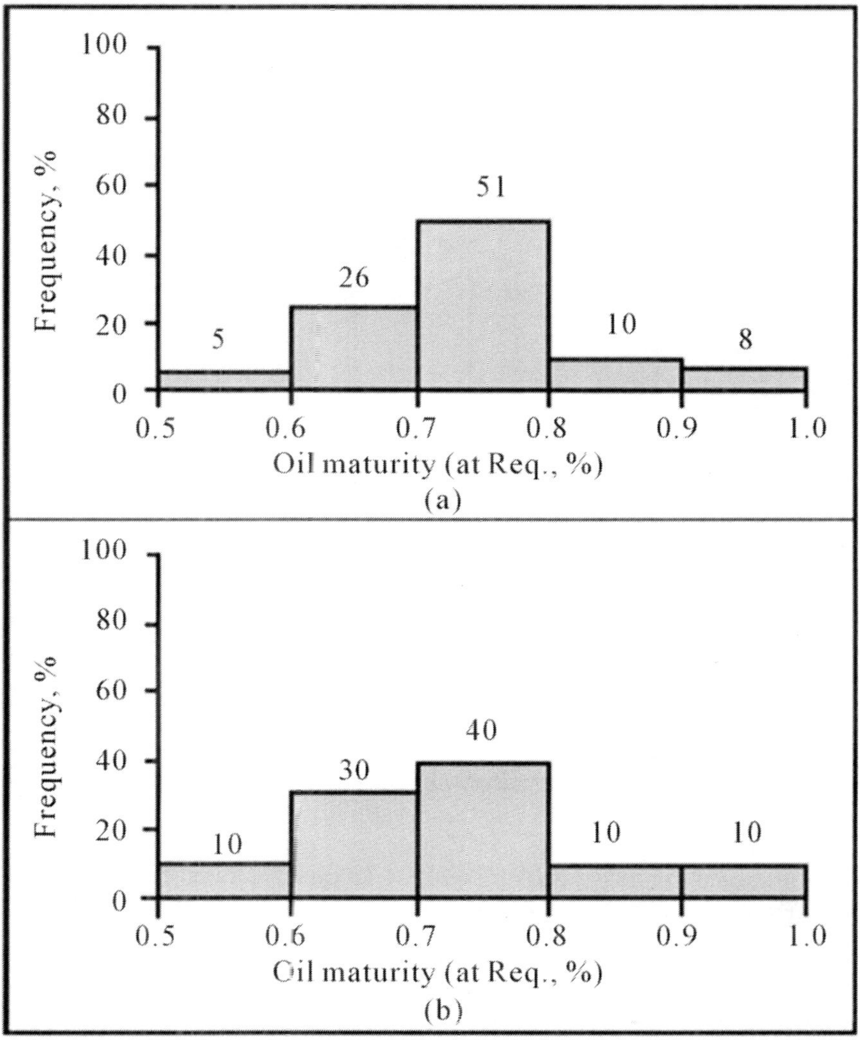

Figure 8. Histograms of distribution of maturity values of the reservoir (a) and mud volcanic (b) oils (at Req., calculated on MPI-1).

3.1.5.2. Temperature criterion Geothermal measurements on Lokbatan MV had started three days after the eruption in October 2001 and continued for about 7 month [7,8]. The temperature variations at depths of 0.5, 1.0, and 1.5 m following the Lokbatan mud volcano eruption are shown on **Figure 9**. The temperature of erupted mud, on the first day attained about

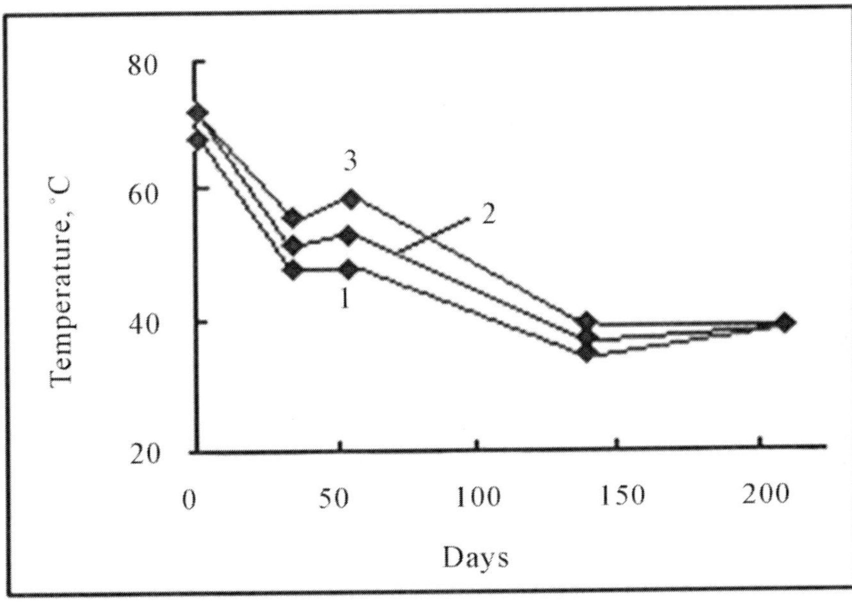

Figure 9. The change of mud temperature after eruption of Lokbatan mud volcano in 2000: at depths of (1) 0.5; (2) 1.0; and (3) 1.5 m.

70°C at all measured levels. The observations revealed a continuously decrease temperature and its stabilization about 150 days after eruption (due to cooling of the mud). The temperature of the liquid mud in the crater of the Touragay volcano was 43°C after six days of eruption. According to the assumed temperature gradient 20°C/km calculated maximal depth occurrence of the mud source is about 3.5 km.

3.1.5.3. Seismic records of the mud volcano eruptions and the eyewitness evidences Eruptions of mud volcanoes, like magmatic volcanoes, are followed by local ground oscillations. Unfortunately these oscillations are documented only by regional seismic network; there is no special direct (or close) local seismic monitoring on mud volcano.

At least, the existence of two phases of mud volcano eruption was established on the basis of seismic records. The hypocenter of the relatively weak first phase of earth tremors corresponds to the depths of approximately 2 - 3 km and is related with the outflow to the surface of the fluidized clayey mass. As a result the effusion of mud disrupted the established dynamic equilibrium in the volcano channel, which provokes ejection of gas and fragments of rocks from the depth of approximately 12 km to the surface. The eruption of gas is accompanied by the second energetically more powerful phase of earth tremors.

This two-phase process of eruption agrees well with the observations of eyewitnesses indicating that initial rise of the breccia mass covering volcano was followed by eruption of fresh mud. The second phase of eruption began with the strong subsurface roar, which was precursor to a powerful eruption of gas.

3.1.6. Rock-Ejections

Maturity of the Paleogene-Neogene rock-ejection varies in range of 0.71% - 1.49% Ro (**Table 3**), which corresponds to the depth interval of 5.5 - 14.5 km (**Figure 10**).

Table 3. Vitrinite reflectance data of the mud volcanoes rockejections, SCB.

Volcano	Age of rock ejecta	Ro, %
Bozdag-Gyuzdek	Paleogene	1.27
Bozdag-Gyuzdek	Miocene	0.86
Jagirli	Paleogene	1.35
Gushchi	Miocene	0.73
Gushchi	Miocene	0.71
Gushchi	Paleogene	1.03
Gushchi	Paleogene	0.85
Suleyman-Akhtarma	Neogene	0.99
Maraza	Miocene	0.72
Tourogay	Miocene	0.88
Tourogay	Paleogene	1.49

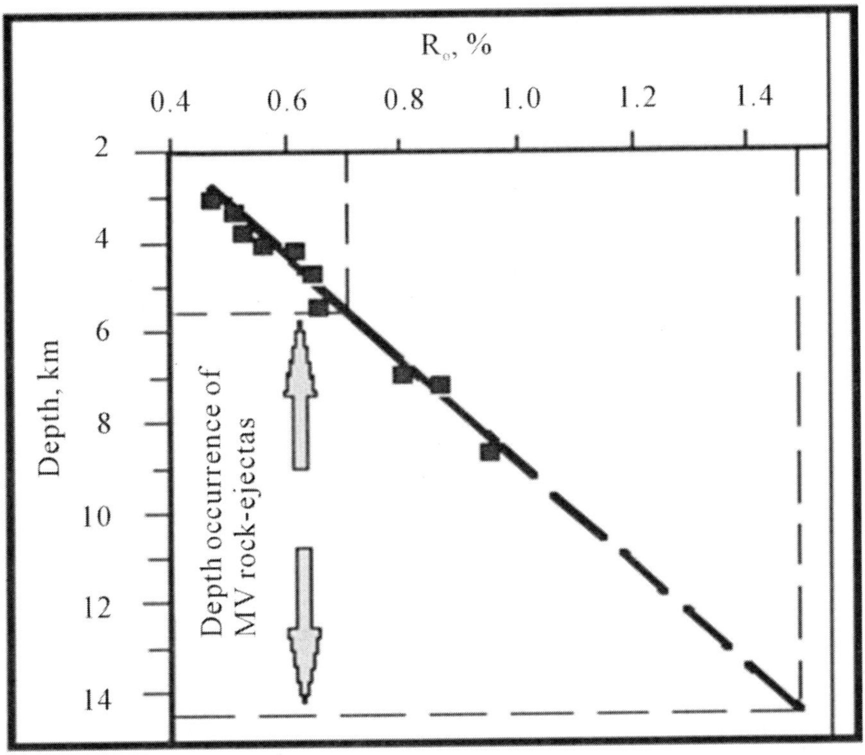

Figure 10. Estimation of the depth occurrence of the mud volcanic rock-ejectas, based on its maturity degree and paleotemperature model of the SCB.

The estimated depth occurrence of the mud volcanoes rock-ejections is commensurable with the depth occurrence of gas. As a rule rock fragments are blown out to the earth surface simultaneously with erupted gas, both of which originated from the same feeding source.

3.2. The Mud Volcano Modeling

Despite the numerous studies, the mechanism controlling mud volcano formation and activity has not been studied adequately and it is still debatable. But attempts to create a model of a mud volcano formation have already been made by some researchers [10,29,30,33 etc.].

Gravitational modeling on the Lokbatan, AkhtarmaPuta and Qushkhana mud volcanoes in Azerbaijan indicates the presence of decompacted mud in 2 - 3 km depth interval [10] (**Figure 11**).

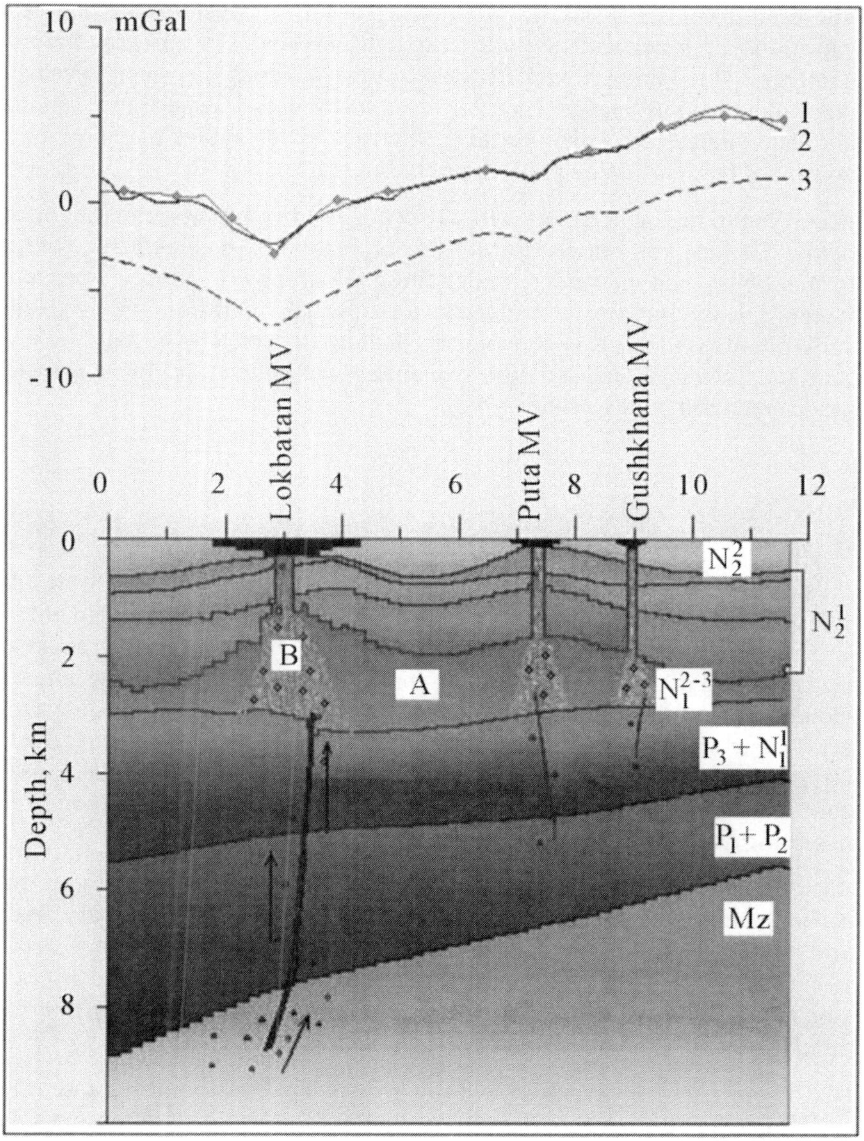

Figure 11. Model of the mud volcano formation, based on results of gravity survey over Lokbatan (erupted 22 times)-Puta (erupted 1 time)-Gushkhana (not erupted) mud volcanoes: A. Plastic clayey bed; B. Fluidized, undercompacted (low dense) and metastable clayey mass; 1. Observed gravity anomaly; 2. Computed gravity anomaly from initial model; 3. Selected configuration and density of rocks. Arrows are showed upward migration of deep overpressured gas-vapor mixture.

These mud volcano localities are characterized by considerable non-tidal variations of gravitational field. The relationship between gravitational anomaly values and activity of these mud volcanoes has been revealed: volcanoes having higher eruption frequency have larger gravitational anomaly values (e.g. Lokbatan MV), whereas lower values of the inactive volcanoes (e.g. Qushkhana MV) are noted.

According to mud volcano model [29] in deep sedimentary strata (at depths below 10 km) the supercritical state of water plays significant role in formation of mud volcanoes. Under these conditions dramatic decrease of water density initiates a vapor-driven upward migration process with subsequent condensation of water at shallow depths. The authors believe that such a mechanism give an explanation of subvertical mass transfer associated with mud volcanoes.

4. CONCLUSIONS

MVs are characteristic to young non-equilibrium basins with rapidly subsiding, high sedimentation rate and great thickness of sedimentary cover. Formation of diapirs/mud volcanoes happen as a result of accumulation of critical energy in overpressured mud volcanoes system and violation of equilibrium in this system provoked by seismicity, tide waves, etc. The primary mechanism of the upward flow of plastic mud (mud diapir formation) is a reaction to buoyancy forces caused by the bulk density contrast between overpressured decompacted mud at depth and the overlying deposits of greater density.

According to studies of the isotopic composition of helium, argon and carbon in methane the Azerbaijan mud volcanoes have a sedimentary origin and they are products of different stages of a single process of hydrocarbon generation within the sedimentary succession. The chemical and isotopic signatures of MV gases are predominantly thermogenic with some admixture of microbial methane.

On the ground of the complex isotope-geochemical, temperature and other criterions the difference of the source depth of the liquid, gaseous and solid products of MVs has been established. The gases and rock fragments (rock-ejections) have the deepest roots (within depth interval of 5 - 15 km). Overpressured gas is the main force in formation and activity of MVs. But origin of MV gases is remaining debatable, because the methane can be derived from both organic-reach mud (primary accumulations in sand interlayer within source rocks) and from larger reservoirs just above or just below such shales.

MV oils are product of destruction of oil accumulations in reservoirs pierced by volcanoes. Therefore, source of MV oils is thought to be limited by the depth of associated oil reservoirs, which in the studied region does not exceed 5.5 km.

The source of the fluidized clayey mass does not exceed 3 - 4 km. It is confirmed by qravitational modeling on some mud volcanoes in Azerbaijan, which indicates the presence of decompacted mud in 2 - 3 km depth interval.

According to seismic records of volcanic earthquakes and the eyewitnesses two-phase mechanism of MV eruptions occurs. The first phase is related to outflow of mud from depth of 2 - 3 km and the next more powerful phase — with blowout of the overpressured gas (predominantly methane with admixture of its gaseous and liquid homologues, produced in deeply buried, organic-rich sediments). As a rule rock fragments are blown out to the earth surface simultaneously with erupted gas, both of which originated from the same feeding source.

5. ACKNOWLEDGEMENTS

Akper A. Feyzullayev The results presented in this paper are based on many field studies on mud volcanoes fulfilled in cooperation with my colleagues from Azerbaijan and other countries. All of them are greatly thanked.

REFERENCES

1. Yakubov, A., Ali-Zade A. and Zeinalov, M. (1971) Mud volcanoes of the Azerbaijan Republic. Nafta Press, Baku (in Russian).
2. Aliyev, A.A. and Kabulova, A.Y. (1980) Isotope of helium in gases of mud volcanoes in Azerbaijan. Reports of the Azerbaijan National Academy of Sciences, 36, 52-56 (in Russian).
3. Valyaev, B.M., Grinchenko, Y.I., Erokhin, V.E., Prokhorov, V.S. and Titkov, G.A. (1985) Isotopic composition of gases from mud volcanoes. Lithology and Minerals, 20, 62-75 (in Russian).
4. Rakhmanov, R.R. (1987) Mud volcanoes and their implications for oil and gas prediction. Nedra, Moscow (in Russian).
5. Ginsburg, G.D., Guseynov, R.A., Dadashev, A.A., Ivanova, G.A., Kazantsev, S.A., Solovyev, V.A., Telepnev, E.V., Askeri-Nasirov, R.Ye., Esikov, A.D., Maltseva, V.I., Mashirov, Yu.G. and Shabayeva, I.Yu. (1992) Gas hydrates of the Southern Caspian. International Geology Review, 34, 765-782.
6. Guliyev, I.S. and Feyzullayev, A.A. (1996) Geochemistry of hydrocarbon seepages in Azerbaijan. In: Shumacher, D. and Abrams M., Eds., Hydrocarbon Migration and Its Near-Surface Expression, AAPG Memoir, 66, 63-70.

7. Mukhtarov, A., Kadirov, F., Guliyev, I., Lerche, I. and Feyzullayev, A. (2003) Temperature variations in the Lokbatan mud volcano crater (Azerbaijan) after the eruption in 25 October 2001. Energy Exploration and Exploitation, 21, 187-207.
8. Kadirov, F.A. and Mukhtarov, A.Sh. (2004) Geophysical fields, deep structure and dynamics of the Lokbatan mud volcano. Fizika Zemli, 40, 327-333 (in Russian).
9. Guliyev, I.S., Feyzullayev, A.A. and Husseynov, D.A. (2004) Geochemical features and source of fluids of mud volcanoes in South Caspian basin in lights of new data at isotopes C, H, and O. Geochimiya, 7, 792-800 (in Russian).
10. Feyzullayev, A.A., Kadirov, F.A. and Aliyev, Ch.S. (2005) Mud volcano model resulting from geophysical and geochemical research. In: Martinelli G. and Panahi B., Eds., Mud Volcanoes, Geodynamics and Seismicity, Springer, Berlin, 251-261.
11. Guliyev, I.S., Feyzullayev, A.A., Aliyev, A.A. and Movsumova, U.A. (2005) Composition of gases and organic matter of rocks ejecta of mud volcanoes in Azerbaijan. Geologiya nefti i Gaza, 3, 27-30 (in Russian).
12. Isaksen, G., Aliyev, A., Guliyev, I., Barboza, S. and Puls, D. (2007) Regional evaluation of source rock quality in Azerbaijan from the geochemistry of organic-rich rocks in mud-volcano ejecta. AAPG Special Volumes, 51-64.
13. Mellors, R., Kilb, D., Aliyev, A., Gasanov, A. and Yetirmishli, G. (2007) Correlations between earthquakes and large mud volcano eruptions. Journal of Geophysical Research, 112, B04304.
14. Etiope, G., Feyzullayev, A.A., Milkov, A.V., Waseda, A., Mizobe, K. and Sun, C.H. (2009) Evidence of subsurface anaerobic biodegradation of hydrocarbons and potential secondary methanogenesis in terrestrial mud volcanoes. Marine and Petroleum Geology, 26, 1692-1703.
15. Kopf, A., Stegmanna, S., Delisle, G., Panahic, B., Aliyev, C. and Guliyev, I. (2009) In situ cone penetration tests at the active Dashgil mud volcano, Azerbaijan: Evidence for excess fluid pressure, updoming, and possible future violent eruption. Marine and Petroleum Geology, 26, 1716- 1723.
16. Feyzullayev, A.A. and Movsumova, U.A. (2010) The nature of the isotopically heavy carbon composition of carbon dioxide and bicarbonates in the waters of mud volcanoes in Azerbaijan. Geochemistry International, 48, 517- 522.
17. Philip, H., Cisternas, A., Gvishiani, A. and Gorshkov, A. (1989) The Caucasus: An actual example of the initial stages of continental collision. Tectonophysics, 161, 1-21.
18. Axen, G.J., Lam, P.S., Grove, M. and Stockli, D.F. (2001) Exhumation of the west-central Alborz mountains, Iran, Caspian subsidence, and collision-related tectonics. Geology, 29, 559-562.
19. Jackson, J.K., Priestley, K., Allen, M.B. and Berberian, M. (2002) Active tectonics of the South Caspian Basin. Geophysical Journal International, 148, 214-245.
20. Guliyev, I.S., Kadirov, F.A., Reilinger, R.E., Gasanov, R.I. and Mamedov, A.R. (2002) Active tectonics in Azerbaijan based on geodetic, gravimetric, and seismic data. Earth Science Section, 383, 174-177.
21. Feyzullayev, A.A. (2010) Physik-chemical interaction in the rock-fluid system in connection with ontogenesis of oil and gas (an example of the South Caspian basin). News of ANAS, Earth Sciences Series, 4, 28-45 (in Russian).
22. Faber, E.Z. (1987) Isotopengeochemie gasformiger Kohlenwasserstoffe. Erdole, Erdgas and Kohle, 103, 210-218 (in German).
23. Peters, K. and Moldovan, J. (1993) The biomarker guide: Interpreting molecular fossils in petroleum and ancient sediments. Prentice Hall, Englewood Cliffs.
24. Capozzi, R. and Picotti, V. (2002) Fluid migration and origin of a mud volcano in the Northern Apennines (Italy): The role of deeply rooted normal faults. Terra Nova, 14, 363-370.
25. Mattavelli, L., Ricchiuto, T., Grignani, D. and Schoell, M. (1983) Geochemistry and habitat of natural gases in Po basin, Northern Italy. AAPG Bulletin, 67, 2239-2254.

26. Nakadaa, R., Takahashia, Y., Tsunogaib, U., Zhengc, G., Shimizua, H. and Hattorid, K. (2011) A geochemical study on mud volcanoes in the Junggar Basin, China. Applied Geochemistry, 26, 1065-1076.

27. Feyzullayev, A.A. and Aliyeva, Es.A. (2003) Estimation of the various source rocks contribution in oil pools formation. Proceedings of EAGE 65 Conference and Exhibition, Stavanger, 2-5 June 2003.

28. Planke, S., Svensen, H., Hovland, M., Banks, D.A. and Jamtveit, B. (2003) Mud and fluid migration in active mud volcanoes in Azerbaijan. Geo-Marine Letters, 23, 258-268.

29. Hovland, M., Ruesletten, H., Lifhseth, H., Fichler, C. and Johnsen, H. (2006) Mud volcanoes—A result of supercritical water formation at depth? Proceedings of AAPG/ GSTT Hedberg Conference Mobile Shale Basins—Genesis, Evolution and Hydrocarbon Systems, Trinidad and Tobago, 4-7 June 2006.

30. Mazzini, A. (2009) Mud volcanism: Processes and implications. Marine and Petroleum Geology, 26, 1677-1680.

31. Cooper, C. (2001) Mud volcanoes of Azerbaijan visualized using 3D seismic depth cubes: The importance of overpressured fluid and gas instead of non extant diapirs. Proceedings of Subsurface Sediment Mobilization Conference, Ghent, 10-13 September 2001. [Citation Time(s):1]

32. Kopf, A., Robertson, A.H. and Volkmann, N. (2000) Origin of mud breccia from the Mediterranean ridge accretionary complex based on evidence of maturity of organic matter and related petrographic and regional tectonic evidence. Marine Geology, 166, 65-82.

33. Arafin, S. (2005) Three-dimensional gravity modeling of a Trinidad mud volcano. West Indies Exploration Geophysics, 36, 329-333.

CHAPTER 4

Damage from Lava Flows: Insights from the 2014–2015 Eruption of Fogo, Cape Verde

S. F. Jenkins[1,2], S. J. Day[3], B. V. E. Faria[4] and J. F. B. D. Fonseca[5]

[1] *School of Earth Sciences, University of Bristol, Bristol, UK.*

[2] *Earth Observatory of Singapore, Nanyang Technological University, Singapore, Singapore.*

[3] *Institute for Risk and Disaster Reduction, University College London, London, UK.*

[4] *National Institute of Meteorology and Geophysics, São Vicente, Cape Verde.*

[5] *CERENA, Instituto Superior Tecnico, University of Lisbon, Lisbon, Portugal.*

ABSTRACT

Fast-moving lava flows during the 2014–2015 eruption of Fogo volcano in Cape Verde engulfed 75% ($n = 260$) of buildings within three villages in the Chã das Caldeiras area, as well as 25% of cultivable agricultural land, water storage facilities and the only road into the area. The eruption had a catastrophic impact for the close-knit communities of Chã, destroying much of their property, land and livelihoods. Volcanic risk assessment typically assumes that any object - be it a building, infrastructure or agriculture - in the path of a lava flow will be completely destroyed. Vulnerability or fragility functions for areas impacted by lava flows are thus binary: no damage in the absence of lava and complete destruction in the presence of lava. A pre-eruption field assessment of the vulnerability of buildings, infrastructure and agriculture on Fogo to the range of volcanic hazards was carried out in 2010. Many of the areas assessed were subsequently impacted by the 2014–2015 eruption and, shortly after the eruption ended, we carried out a post-eruption field assessment of the damage caused by the lava flows. In this paper, we present our findings from the damage assessment in the context of building and infrastructural vulnerability to lava flows. We found that a binary vulnerability function for lava flow impact was appropriate for most combinations of lava flow hazard and asset characteristics but that building and infrastructure type, and the flow thickness, affected the level of impact. Drawing on these observations, we have considered potential strategies for reducing physical vulnerability to lava flow impact, with a focus on buildings housing critical infrastructure. Damage assessments for lava flows are rare, and the findings and analysis presented here are important for understanding future hazard and reconstruction on Fogo and elsewhere.

Keywords: Fogo Cape Verde Lava flows Building, infrastructural and agricultural damage eruption impact assessment hazard and risk assessment mitigation

INTRODUCTION

Lava flows are Earth's most common volcanic feature and one of the most easily-recognised products of a volcanic eruption (Kilburn 2015). They can be categorised into three main types, according to their surface features: pāhoehoe, ʻaʻā and blocky and all three may occur within the same eruption. Blocky flows are associated with more viscous lava and can be tens of metres thick, while pāhoehoe and ʻaʻā can be produced by chemically identical mafic lavas moving at different velocities (faster in the case of ʻaʻā, at least initially). These types of flow are more commonly 2–10 m thick, although pāhoehoe flows can be thinner, particularly on initial emplacement (Kilburn 2015). Their subsequent thickening, or inflation, due to continued movement of magma into the interior of the flow is common, but varies from case to case according to the history of magma input to the flow (Calvari and Pinkerton 1998; Walker 2009). During relatively high effusion rate eruptions, sheet flows can also form where fluid lava ponds in low-lying areas or where individual lobes of lava coalesce, or at lava flow breakouts where local effusion rates are transiently elevated (Kilburn 2015). The area impacted by a lava flow depends upon a number of factors including the vent location and local topography, and the effusion rate and duration.

Descriptions of past lava flow impacts are limited and relatively few studies (e.g. Behncke et al. 2005; Felpeto et al. 2001; Rhodes et al. 2013) have directly assessed the future threat of lava flows to buildings, infrastructure or agriculture. For an overview of the historical impacts of lava flows see

Blong (1984) and the updated review of Harris (2015). Over the past decades there have been a number of lava flow impacts in populated areas (Table 1). The largest lava flow impact of recent times took place during the 2002 eruption of Nyiragongo, Democratic Republic of Congo, in which lava flows destroyed approximately 15% of the city of Goma and the homes of more than 120,000 people (Baxter et al. 2002). Nyiragongo had previously erupted in 1977 destroying the homes of around 800 people (Blong 1984). The number of buildings damaged or destroyed by lava flows in the Chã das Caldeiras area during the 2014–2015 eruption of Fogo volcano in Cape Verde was much smaller ($n = 260$) than at Goma. However, a greater proportion of buildings in Chã das Caldeiras were affected, and the community's isolation relative to the rest of the island's inhabitants and their reliance upon now-buried agricultural land means that the short- and long-term consequences of the lava flows for the community are likely to be significant.

Table 1. Notable building damage (>20 houses destroyed) by lava flows in the period 1965–2015. Data sourced from Blong (1984) and Harris (2015) and supplemented with references noted in the table

Volcano	Date	Building damage description	Source
Vestmannaeyjar, Iceland	1973–1974	Approximately 300 houses in Heimaey town destroyed by lava or fire.	Williams and Moore (1983)
Karangetang (Api Siau), Indonesia	1976	24 houses destroyed (and a further 44 in the path of the flow dismantled)	Global Volcanism Program (1976)
Nyiragongo, Democratic Republic of Congo	1977	Approximately 400 houses in two villages destroyed, and 12 km^2 of agricultural land buried.	Global Volcanism Program (1977a)
Piton de la Fournaisse, Réunion	1977	33 houses and a church in Piton Sainte Rose were destroyed. The road was also buried.	Global Volcanism Program (1977b)
Kilauea, USA	1986, and 1990–1991	181 buildings – the majority of Kalapana town –destroyed, mostly during the 1990–1991 lava flows. More than 10 km of public highway was buried.	Global Volcanism Program (1992)
Nyiragongo, Democratic Republic of Congo	2002	4500 buildings destroyed.	Baxter et al. (2002)
Fogo, Cape Verde	2014–2015	170 buildings destroyed, 90 damaged. The only road was buried, along with ~2 km^2 of agricultural land (25% of all cultivable land)	This study and PDNA, 2016

In volcanic risk assessment, anything impacted by a lava flow is expected to be completely destroyed so that fragility functions, which relate the hazard intensity to a probability of damage or disruption, are considered binary: lava results in complete destruction; no lava results in no damage (Jenkins et al. 2014). This is in contrast to most volcanic hazards, which show a gradational fragility (or vulnerability) function or curve. As a result, little attention has been paid to possible relationships between morphological, kinematic and mechanical features of lava flows and the intensities of the resulting hazards to different assets, such as land, buildings and infrastructure. In 2010, an assessment of the physical vulnerability of Fogo's building stock and agricultural infrastructure to volcanic hazards was carried out as part of the European FP7 MIA-VITA (MItigate and Assess risk from Volcanic Impact on Terrain and human Activities) project (Jenkins et al. 2014). For lava flows, the likely loss was assumed to be complete should an area be inundated by lava. However, some of the assessed areas have since been impacted by lava during the 2014–2015 eruption, allowing us to test this assumption. A good knowledge and understanding of the pre-eruption infrastructure, context and setting helped us in carrying out a field study to assess the impact of the 2014–2015 eruption on buildings, infrastructure and agriculture. This paper describes and discusses our findings, with the text split into four main components:

- Section 2: We introduce Fogo volcano and its pre-2014 eruption history, describe the main characteristics of buildings, infrastructure and agriculture as determined through the pre-eruption vulnerability assessment, and provide a detailed overview and timeline for the 2014–2015 eruption;
- Section 3: We outline the methods employed in our remote and field impact assessments, describing how satellite images, media reports and field data were used to quantify the impacts of the 2014–2015 Fogo lava flows;
- Section 4: We present our findings from the post-eruption impact assessment undertaken in early 2015. Building damage is described according to the different levels of damage sustained by buildings in Chã, while damage to infrastructure and agriculture is described for each sector, e.g. roads, telecommunications, in turn;
- Section 5: We discuss our findings and their implications, in particular the assumption of a binary fragility function in lava flow risk assessment, and the lava flow parameters and properties that control the fragility functions. In this section we also propose some building design strategies that could reduce the physical vulnerability of typical buildings and buildings housing critical infrastructure.

BACKGROUND

Fogo volcano

The approximately 25 km wide island of Fogo is formed almost entirely from a single large and steep-sided shield volcano (Fig. 1). It is the most active volcano in the Cape Verde hotspot archipelago and lies around

800 km west of Senegal in Africa. The steep summit cone of Pico do Fogo rises to ~2830 m above sea level and sits within the more than 80,000 years old Monte Amarelo lateral collapse structure that is 9 km wide, open towards the east and associated with debris avalanche deposits on the ocean floor (Day et al. 1999; Fonseca et al. 2003; Masson et al. 2008). The upper part of the collapse scar, the Bordeira cliff, is still up to 1 km high in places and effectively protects most of the island from lavas erupted within the scar (Fig. 1). The east side of the island, which is within the collapse scar, and areas in the northeast and southeast, which are adjacent to sections of the collapse scar cliff that have been completely buried by post-collapse lavas, are not protected in this way and are susceptible to lava flows from Pico and within the collapse scar. Between Pico and the Bordeira cliffs, there is a flat area called Chã das Caldeiras ("plain of craters") formed by ponding of lavas between Pico and the cliffs (Fig. 1). Chã is one of the most productive agricultural areas in semi-arid Cape Verde, owing to relatively reliable orographic rainfall in the summer and autumn months. It has been inhabited since the mid-19th century, with the main villages of Portela and Bangaeira assuming their present organisation in the early 20th Century (Fernandes and Faria 2015).

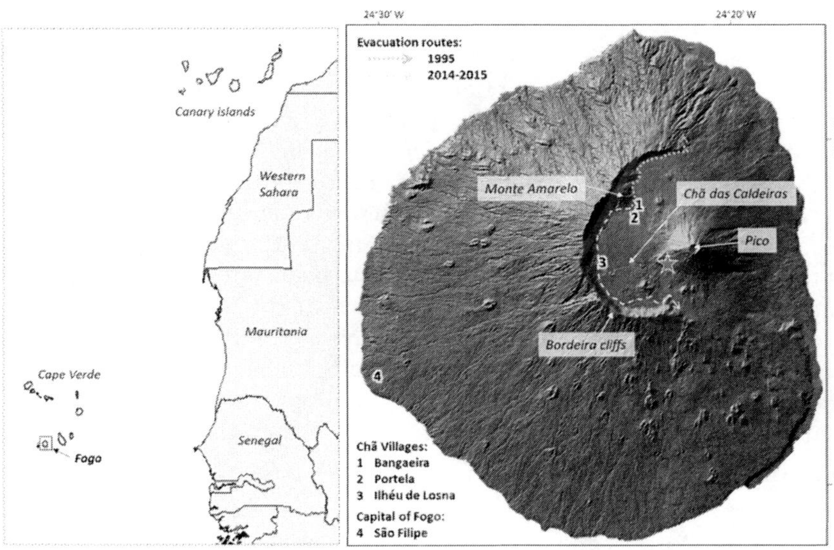

Figure 1. *Left*: The location of Fogo island within the Cape Verde archipelago off the western coast of Africa; *Right*: Digital elevation model for the island of Fogo showing the location of the main villages in Chã, the capital São Filipe, Monte Amarelo and the 2014–2015 vents (*red star*). The evacuation routes used during the 1995 and 2014–2015 eruptions are also shown

Major natural hazards on the island as a whole are from floods and rockfalls, although in Chã and along the eastern flank, effusive eruptions and lava flows are also important. However, by far the most important natural hazard in Fogo, as in Cape Verde as a whole, is drought. Although the effects of droughts since 1950 have been effectively mitigated by food imports, in previous decades large scale mortality occurred in Cape Verde as a result of multi-year droughts; most notably in 1810 AD, 1830 AD, and from 1941 through to the late 1940s (WorldAtlas 2011). In 2014, Cape Verde received 65% less rain than 2013 (FAO 2015) and the region was experiencing severe drought at the time of the 2014–2015 eruption. This history of droughts goes some way towards explaining the strong attachment of the inhabitants of Chã das Caldeiras to their land, and their reluctance to be resettled on the more arid south side of Fogo in the aftermath of recent eruptions as discussed below.

Previous eruptions

Historical accounts from the early stages of Portuguese settlement in the 15th Century through to about 1725 AD indicate that Fogo volcano experienced frequent eruptions from the summit and from flank fissures on all sides of the Pico (Day et al. 2000; Fonseca et al. 2003; Ribeiro 1960). In 1680 AD, a major summit eruption of Pico do Fogo produced large ash falls over the whole island rendering agricultural lands temporarily unusable and triggering mass emigration from the island, to Brava in particular (Ribeiro 1960). Most eruptions from 1725 AD up to and including that in 1857 AD were from fissure vents on the northern and southern flanks of the Pico and Fogo (Faria and Fonseca 2014). In the 150 years or so since first settlement of Chã das Caldeiras there have been three effusive eruptions that occurred from subsidiary flank vents near the base of Pico: in 1951, 1995 and 2014–2015 (Fig. 2). All three of these eruptions were from fissure vent arrays located farther west than any other post-1725 AD vents and, in the case of the 1995 and 2014–2015 eruptions, from mainly northeast-southwest trending fissures within Chã (Faria and Fonseca 2014).

The 1951 eruption, from vents both north-west and south of the Pico do Fogo, had relatively little net impact upon Chã das Caldeiras. Some fields to the east of Portela and Bangaeira were covered by lavas, which also destroyed a few houses in Bangaeira (Pers. Comm. to Simon Day from Sr. Antonio Teixeira, 2002). More damage due to lava flows occurred outside rather than inside Chã das Caldeiras, as lava flows from the southern vents covered fields and houses in the south east of the island (Ribeiro 1960; Pers. Comm. to Simon Day from Sr. Antonio Teixeira, 2002). On the other hand, lapilli falls around the north-western vents covered some older lava flows, providing areas that proved suitable for small-scale commercial agriculture (producing apples, quince and grapes) in later years.

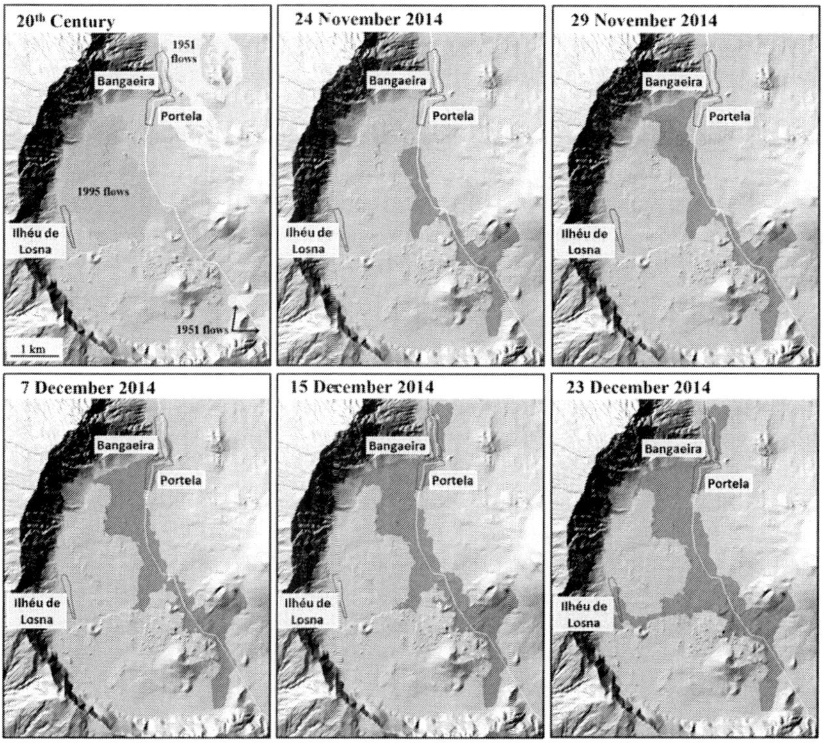

Figure . 2. Eruption evolution shown by lava flow maps sourced from satellite images on the dates shown, with outlines created from Orthophoto images sourced in the MIA-VITA project for the 20th century outlines, and provided by the Copernicus Emergency Management Service (http:// emergency.copernicus.eu EMSR111) for the 2014–2015 outlines. We have revised 2014–2015 lava flow outlines using field observations where possible. The three main villages of Chã (Bangaeira, Portela and Ilhéu de Losna) and the only road into the area are shown on each map

At the time of the 1995 eruption, Chã was home to approximately 1300 people (Bulletin of the Global Volcanism Network 1995) in three villages: the larger and densely packed villages of Portela and Bangaeira and the smaller and less densely populated Ilhéu de Losna (Figs. 1 and 2) as well as scattered houses around Chã. The 1995 lava flows cut the main road into Chã early in the eruption, leading to a difficult evacuation of the villages via footpaths to the north coast of the island (Fig. 1). The lavas subsequently destroyed buildings and a water reservoir and, most importantly, over 3 km^2 of fertile agricultural land. The agricultural land covered included most of an area west of Portela village that formed the best land in Chã das Caldeiras for growing a variety of food crops. The sale of most of these food crops (such as peas and beans) formed an important source of cash income to families not involved in tourism or in fruit and vine cultivation. After the 1995 eruption, communities and associated services such as health and education were permanently relocated outside of Chã to planned villages on the arid south side of Fogo with the aim of preventing future impacts. These planned villages lacked agricultural land, although they did allow easier access to public and commercial services in the main town of Fogo, São Filipe. Within 2 years Chã was being repopulated, mostly within Portela and Bangaeira villages, beginning with adults returning to work the fields. Over the following two decades, a burgeoning wine, agricultural and tourism industry developed.

By 2010, a census recorded 697 Chã das Caldeiras inhabitants, but estimates of resident numbers prior to the 2014–2015 eruption were as high as nearly 1500 (Global Volcanism Program 2014). The exposure and vulnerability to future lava flows was therefore, in broad terms, as high as it had been before the 1995 eruption. Furthermore, another 11,000 people in a number of villages on the steep eastern flanks of Fogo were exposed to over-spilling flows such as those that reached the east coast of the island in all eruptions between 1785 AD and 1857 AD. Indeed, because of the potential for lava flows to descend the steep eastern slope of Fogo at speed, coupled with the dependence of these villages upon a single road for normal communications and emergency evacuation, managing lava flow risk in the east flank communities presented as much of a problem as it did for the communities in Chã das Caldeiras.

Buildings, infrastructure and agriculture

The pre-eruption vulnerability assessment in Chã das Caldeiras (Jenkins et al. 2014) found building types to be dominated by unreinforced masonry buildings, with approximately 40% of all buildings constructed from large locally-sourced lava blocks 40–60 cm thick. These include distinctive cylindrical one-storey rubble-stone traditional buildings, which are often used for shelter while tending agricultural lands or for storage or annex homes. Cooled and solidified lava blocks make very strong construction material for roads and buildings, and are valuable construction assets across Fogo. The remaining buildings were constructed using squared masonry or breeze blocks, around half with and half without a reinforced concrete frame. The construction quality of the frames and mortar, where present, was variably poor with the cement used as mortar or within the frame often showing large air gaps that would significantly weaken any steel reinforcement bars and the building's resistance to lateral pressures. Almost all of the buildings had flat reinforced concrete slab roofs that were used to collect rainwater, the predominant source for water in Chã, which was then stored in concrete, plastic or metal tanks. A small number of buildings had tiled or metal sheet roofs (Jenkins et al. 2014). The pre-eruption vulnerability assessments forecast complete loss in areas affected by lava flow; however, the thick walls of rubble stone construction were recognised to be particularly resistant to lateral loading from a flow (Jenkins et al. 2014).

Very little infrastructure existed in Chã prior to 2014. The only road into the area constructed of cobble-sized lava blocks sourced from previous lava flows and raised up to 2 m above the surrounding flat topography in places, cutting levees and other elevated parts of old lava flows in other places. Communication cables carried to the villages on timber utility poles followed the route of the road. Generators provided power for residents at night and cooking was fuelled by bottled gas transported by road from São Filipe. In addition to rainwater, small springs at the foot of the Bordeira cliffs between Portela and Boca de Fonte, which are tens of metres above the level of Chã, and a 370 m deep borehole a few hundred metres south west of Portela completed in 2013 provided water for the community.

Residents were reliant upon income generated from tourism, remittances from relatives who had joined the Cape Verdean diaspora, and the local to national level export of vegetables and fruit as well as produce made at the agricultural co-operative within Chã, such as crab apple and quince jams as well as wine (the most important by value). Fruits and vegetables were enclosed within tephra 'wells' that protected seedlings from the winds and provided them with access to better soils and moisture. Grapes are grown on the ground and harvested for wine, an increasingly successful export activity prior to the 2014–2015 eruption.

2014–2015 eruption

Increased unrest at Fogo in late 2014 lead to progressive raising of the alert level within the established scheme (Faria 2010; Faria and Fonseca 2014) (Table 2). Following the initial unrest and alert level raise, the official meteorological and volcanological hazard monitoring agency for Cape Verde (National Institute of Meteorology and Geophysics: INMG) issued a formal alert of impending eruption to the Civil Protection on 17 October. On 22 November, the permanent monitoring network recorded numerous shallow volcano-tectonic earthquakes (Fernandes and Faria 2015). That evening, residents of Chã das Caldeiras felt earthquakes, and at approximately 10 am (~11:00 UTC) the next day, 23 November 2014, a new eruption began. It was fed by a fissure array orientated northeast-southwest at the western base of Pico, sub-parallel to and approximately 100 m to the southeast of the 1995 vents. The following day, the Copernicus Emergency Management Service satellite image acquisition was activated (http://emergency.copernicus.eu; Activation ID: EMSR-111), which mapped lava flow outlines throughout the eruption (Fig. 2) to facilitate crisis management. A number of social and professional images of the eruption and its impact were also available online throughout. The flows initially advanced at approximately 35–40 m/h (Worsley 2015) and had travelled ~1.3 km to the southwest by the second day of the eruption (24 November: Fig. 2), where they reached the eastern edge of the 1995 lava flow and split into two flows. The flow to the south-southeast travelled a further 1.6 km at a much-reduced rate. The flow to the north-northwest reduced to ~20 m/h and then 2–3 m/h (Worsley 2015) as it advanced along the eastern edge of the 1995 ‘a‘ã flow and the route of the only road towards the main settlements of Portela and Bangaeira (Fig. 2). These lava flows appear to have been fed from separate vents, with flows to the south sourced from a breach on the southernmost rim of the upper vent and flows to the north-northwest sourced from the lower vents, with the flows and flow sources clearly visible in Google Earth Digital Globe images acquired on 25 and 26 November.

Table 2. The timing of alert level activations during the 2014–2015 eruption, and the alert level scheme (Faria 2010; Faria and Fonseca 2014) used by the National Institute of Meteorology and Geophysics (INMG), the official meteorological and volcanological hazard monitoring agency for Cape Verde. Times are local (UTC - 1 h)

Level	Criteria	Interpretation	Depth	Time window for possible eruption	Activation during 2014–2015 eruption	
					Escalation	Decline
1	• Usual records	Normal state	-	-	Background	25 February
2	• Long term ground deformation (GPS, InSAR) • Seismic noise modification	An eruption is possible soon	5 to 13 km	10 days to 5 months	Early October	-
3	• Peak of seismic activity • Variation of the tilt	Probable eruption	4 to 5 km	4 to 40 days	21 November	20 February
4	• Peak of the tilt • Long-period events with greater magnitude and number	Very probable eruption	2 to 4 km	To be determined as the dike progresses	20:00, 22 November	AM, 8 February
5	• Seismic activity is maintained • Continuous tremor	Imminent eruption	0 to 2 km		08:30, 23 November	

A number of buildings along the road to the northwest of the vent were destroyed in the first 2 days, including the newly completed (April 2014) Fogo National Park Headquarters. All of the Chã residents were evacuated during this time by the National Civil Protection Service (SNPC) and Cape

Verdean Military using a track along the foot of the Bordeira cliff (Fig. 1). The Red Cross of Cape Verde and the United Nations Office Cabo Verde reported that in total 1076 people were evacuated, with 838 relocated to temporary accommodation centres and houses built in the aftermath of the 1995 eruption (IFRC 2015; UN 2014).

Lava entered the southwest portion of Portela on 2 December 2014. Over the following few days, the lava flow fronts slowed down to less than 1 m/h (Worsley 2015), raising hopes that further destruction of the villages could be avoided; however, between 6 and 8 December 2014 the flows advanced and destroyed most (>90%) of the remainder of Portela. A sheet flow and pāhoehoe lava breakout in the east of Portela, which evolved into a thin rubbly ʻaʻā flow on the slope between Portela and Bangaeira, led to most of Bangaeira also being destroyed (Fig. 2) (Global Volcanism Program 2015) by fast-moving flows with flow front velocities of up to 180 m/h (Worsley, 2015). At this time there was a concern that the lava flows would continue towards the northeast, exiting the collapse scar and threatening villages, and the circum-island road, along the steep eastern flanks of the island. Rapid calculations of flow length (following Kilburn 1996) using effusion rates estimated from satellite infrared images (Ferrucci et al. 2015) suggested that this would not be the case, and in line with these estimates, the flow stalled a few hundred metres beyond the northernmost house in Bangaeira and was inactive after 18 December (Ferrucci et al. 2015; Pers. Comm. to Simon Day from Sr. Jose Antonio of Portela Village). Slow growth of a lava flow front to the west of Portela along the south side of Monte Amarelo, also in the form of a pāhoehoe breakout, continued through to mid-December but had ended by 14 December (Fig. 2). This flow covered the very last area of flat ground in Chã das Caldeiras that had fertile soil derived from the 1680 AD phreatomagmatic ash, although similar soil is still present on the slopes of Monte Amarelo (that derives its name from the characteristic yellowish-brown colour of the phreatomagmatic ash).

In mid-December, new lava flow breakouts occurred to the west of Pico, from breaches on the west side of the main flow channel about 2 km from the vent, with small initial breakouts from 9 to 12 December and a large breakout by 14 December (Copernicus 2014; Ferrucci et al. 2015). These breakouts produced a west-directed flow that destroyed many agricultural fields. On 21 December, buildings within the small settlement of Ilhéu de Losna in the west of Chã were destroyed as the flow reached the talus cones at the foot of the Bordeira cliff and turned north (Fig. 2). The eruption continued at a reduced pace through late December, January and early February 2015 with periodic explosions and tephra plumes, and pulses of increased effusion rate that caused slow, intermittent lava encroachment, including a discrete flow that was emplaced near the vent in mid-January (Ferrucci et al. 2015) and mainly covered lavas from earlier in the eruption.

The formation of this last flow implies that the earlier lava flows, including any internal lava tubes, were inactive by that time and no longer capable of transporting lava to the main flow fronts. The eruption ended on the 7 February at about 20:00 (local time) with the cessation of volcanic tremor and explosions. The alert level was gradually reduced from level 4 on the 8 February to level 3 on the 20 February, and to level 1 (normal state) on the 25 February (Table 2).

The lava flows were, in broad terms, alkaline basaltic in composition (basanites), as in most previous eruptions, and were mostly characterised by ʻaʻā or pāhoehoe surfaces. ʻAʻā flows were 0.5–2.5 m thick as they exited the vents and flowed down a relatively steep slope (~30°), thickening to between 3 to more than 8 m thick as they flowed across the flatter topography between Pico and the collapse scar cliffs. Pāhoehoe flows were typically 2 m thick or less. Lava channels and tubes were formed throughout the flow. A lava flow volume of approximately 45×10^6 $m^3$was derived from comparison of pre- and post- eruption DEMs (Bagnardi et al. 2016; Richter et al. 2016). This value includes the volume of cones formed during the eruption (>55 m and 33 m height) and the comparison found that lavas ponded in low-lying areas ~1 km west of Portela, producing thicknesses of up to 26 m. This thickening is likely to have occurred by internal inflation of the flows with lava that travelled through lava tubes. The flow area south and west of Portela was still cooling and feeding local fumaroles at the time of our field visit in early 2015; limited fumarolic activity continued as late as early 2016.

METHODS

The areas in Chã assessed during our 2010 vulnerability survey have since been inundated by lava during the 2014–2015 eruption. The eruption was declared over on 7 February 2015 and, with the approval and collaboration of INMG, we carried out our field impact assessment from 24 February to 4 March 2015. The main aims of the mission were to assess the extent and nature of the damage (from both lava and ash) for buildings, infrastructure and agriculture and to sample and characterise the properties of the 2014–2015 lava flows. The damage assessment included georeferencing and cataloguing the nature of the damage and making measurements where appropriate, e.g. length and width of wall cracks, distance from flow to heat-affected item. Building or infrastructure characteristics such as construction material, wall span and size and number of openings were always recorded, and, if appropriate, tied back to the pre-eruption building survey that had recorded similar information in 2014 for a subset of buildings in Chã. The properties of the lava flow and the setting of the building in the area were also recorded to give some context, for example the thickness and type of lava flow, and if the building walls were oblique or perpendicular to the flow direction and in-between or within the main flow channels. Importantly, a large number of georeferenced photos from multiple angles were taken so that post-mission analysis could easily refer back to the field observations and compare our records with media images. In addition to damage descriptions and insights, which are described over the following sections, field observations were used to ground-truth our remotely derived assessments of damage, particularly for those buildings or objects near the peripheries of the flow outline for which projection or resolution issues with the satellite images may have led to some mismatch. Below we describe our use of satellite imagery, GIS, media reports and the field data in interpreting damage observations.

Pre-eruption building locations were plotted from a Google Earth Digital Globe satellite image acquired 9 November 2014. Lava flow outlines obtained by the Copernicus Emergency Management Service (based on satellite imagery of resolution 0.5 m, 1 m or 3 m) were then overlain on the building locations in GIS to establish the numbers of buildings affected. Plotted buildings included traditional circular scoria buildings often used as shelter by livestock or workers when tending crops. In the more densely populated villages of Portela and Bangaeira, one building footprint may have included more than one family home, where buildings are joined. Following the remote and field assessment, we classified buildings into three impact categories:

1. Destroyed buildings were completely buried, destroyed or transported from their original location by the lava flows;
2. Damaged buildings were impacted by the lava flows but still visible in situ in post-eruption satellite images (acquired 3 February 2016) and during the field mission;
3. Unaffected buildings were not in areas impacted by the lava flows and were therefore not damaged.

Syn- and post-eruption satellite images acquired 23, 25, 29 November 2014, 4 December 2014 and 3 February 2016 were used to verify the estimates of number of buildings in each category by independently mapping the condition of the building in post-eruption images. Professional media reports and photographs from professional and social media were used along with the syn-eruption satellite images to correlate reports of lava flow advancement and the reported destruction of certain buildings and/or infrastructure during the eruption. Post-eruption satellite images could also be combined with media reports to identify further resettlement and rebuilding that occurred after our field mission.

Lava flow damage

No deaths or injuries associated with the eruption were reported but most of the buildings and important areas of agricultural land, communication poles and lines and much of the only surfaced road into Chã were covered by lava. Intermittent explosions dispersed and deposited ash in the main city of São Filipe and across agricultural crops on the island, as well as affecting the airport and tourism industry. During the more intense periods of the eruption, tourism viewing of the lava flows continued to provide income for some residents of Chã; however, the time and use of vehicles involved in this prevented some residents from removing all contents and fixtures from their homes before the lava flows reached them. Other residents, however, stripped everything of value from their homes that could be moved, including plumbing fixtures, doors and windows. Equipment and stored produce from the co-operative buildings concerned with wine-making and fruit-processing were salvaged by residents and the military, and stored along with the contents and fixtures of residential homes on the hillside of Monte Amarelo adjacent to Portela and Bangaeira villages. Some equipment and energy systems were also removed from the Park headquarters before it was destroyed. Despite these valuable actions, the loss of buildings, livelihoods and (above all) important areas of agricultural land in Chã has resulted in significant economic losses and an uncertain future for the local population. The total economic impact has been estimated as 2832 million CVE (~US$28 million), with the contribution by sector as follows: Agriculture and livestock (42%); Housing (27%); Tourism (7%); the remaining is attributed to social and cross-cutting sectors such as governance, environment and health (PDNA, 2016). In an effort to better understand the impacts of the lava flows on buildings and infrastructure, and the future hazards for displaced populations, we carried out a field impact assessment less than 3 weeks after the end of the eruption and approximately 2.5 months after the main villages were destroyed by lava. Our remote and field observations are described below and discussed in Discussion section.

Buildings

Of the approximately 350 buildings in Chã das Caldeiras, around 260 were in areas covered by lava flows during the 2014–2015 eruption (Fig. 3). Approximately 85% ($n = 210$) of the affected buildings were concentrated within the main villages of Portela and Bangaeira to the northwest of the vents, with a further 5% (10–15 buildings) in the small village of Ilhéu de Losna, due west of the vents. Fig. 3 shows the distribution of destroyed, damaged and unaffected buildings in Chã. The majority ($n = 170$) of the 260 affected buildings were completely destroyed and/or buried by the advancing flow. Destroyed buildings were particularly concentrated in the southwestern parts of Portela (Fig. 3), which was covered by the main 'a'ā flow that had a flow front up to several metres high.

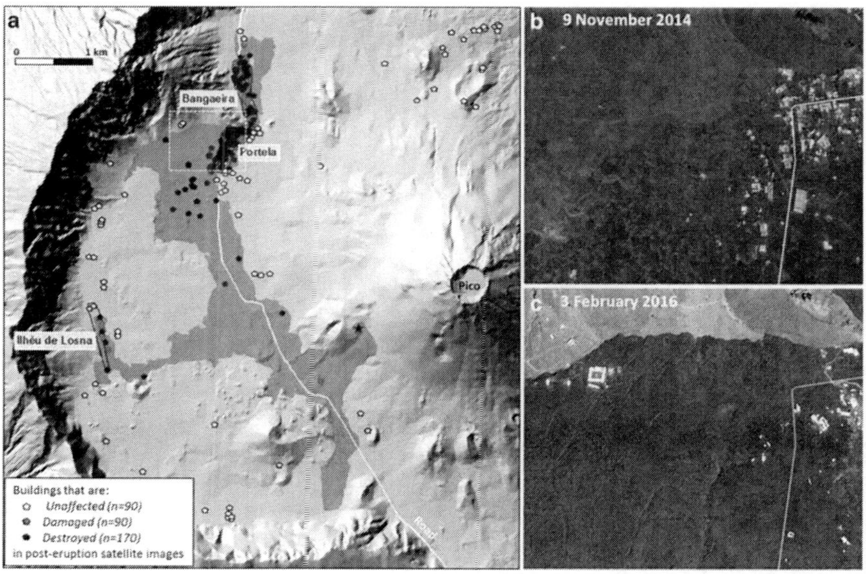

Figure 3. a Buildings in Chã that were unaffected (*yellow symbols*: $n = 90$), damaged (*blue symbols*: $n = 90$) and destroyed (*black symbols*: $n = 170$) by the 2014–2015 lava flows; *Yellow dashed box* shows inset area of **b** Pre-eruption Google Earth Digital Globe satellite image (9 November 2014) showing the western portion of Portela village and **c** Post-eruption Google Earth Digital Globe satellite image (3 February 2016) showing the destruction of Portela and subsequent reconstruction of a small number of buildings on the 2014–2015 lava flows to the west of Portela

Of the destroyed buildings, a small number also had parts of the building transported within the flows. Such components were typically slab reinforced concrete. For example, a number of reinforced concrete roofs were sheared from building walls by the flow and transported within the flow, some for many tens of metres (Fig. 4a). In at least fifteen cases, the roof sheared cleanly from the walls along the connection between the two, and in a further seven observed instances the wall was broken along the mortar-block interface. Household concrete water tanks in particular were observed to remain relatively intact when transported within the flow. The most impressive example was that of a water storage tank from within the Park Headquarters, which was transported more than 100 m from its original location by the lava flows (Fig. 4b). While the building suffered damage to the roofing, walls and support beams, the overall structural envelope remained relatively intact. The remainder of the Park Headquarters was destroyed and the debris buried, although media images and reports and satellite images from the 25 November show that the flow was at first redirected around the oblique walls, before the south wall failed and flows inundated the building, entering and exiting through large existing openings.

Figure . 4. a A building roof completely sheared from the walls and transported ~100 m, which showed surprisingly little damage to the concrete slab roof or the concrete water tank; **b** The water storage tank from within the newly constructed Fogo National Park Headquarters and cultural venue. The tank was transported more than 100 m northeast from its original location. Photographs: S.F. Jenkins, February 2015

Damaged buildings ($n = 90$: Fig. 3) were not limited to the periphery of flows and showed varying degrees of damage when visited on the ground. We observed four main modes of damage that were difficult to distinguish from each other in the remote assessment of satellite images, but that could be seen in the field, as described below.

Partial burial and/or inundation

In the more eastern portions of Portela and in Bangaeira, where the pāhoehoe and rubbly ʻaʻā flows had fronts of only 2 to 3 m high, many buildings were simply inundated by lava with the structure remaining relatively intact and in situ (Fig. 3). For buildings with strong, or very few openings, such as the traditional, cylindrical scoria buildings characteristic of Fogo, partial burial was common (Fig. 5a) and, where visible, the structure did not seem to be obviously damaged. 'A'ā flows entering through smaller openings and into buildings was limited and consisted of scoria fall from a flow outside the entrance, rather than flow ingress; pāhoehoe flows by contrast were able to flow through relatively thin openings like doors and windows. Multiple and weaker openings often permitted ingress of both types of lava flow into, or through, a building. Despite the lack of structural damage in some cases, these buildings were left uninhabitable and of limited value to the owners. Where lava flows had entered a building through an opening, diagonal fractures from each opening corner radiated outwards along the mortar-block interface. In some instances, where lava flows reached a great enough thickness, ingress caused doming of reinforced concrete slab roofs (Fig. 5b).

Figure 5. **a** Burial of a traditionally-constructed community centre in Portela, which comprised semi-circular scoria walls up to 40 cm thick, with few openings; **b** Doming of a reinforced concrete slab roof within Portela: the building envelope remained relatively intact but lava flows inundated all of the building footprint to approximately roof height. Photographs: S.F. Jenkins, February 2015

Minor structural damage

Where lava flows reached a building but did not destroy, bury or enter it, there was evidence of structural damage in the form of wall cracks. This was only observed at the peripheries of the flow in Bangaeira and Ilhéu de Losna, and is likely a result of the flow slowing down and 'resting' upon the structure as it reached the end of its path. Wall cracks appeared to form most prominently at the junction between the top of the wall and the reinforced concrete roof slab. Cracks within reinforced concrete slab roofs were diagonally orientated, likely related to the direction of pressure of the lava flow.

Fire and explosions

Lava flows from the 2014–15 eruption did not cause many fires, reflecting the minimal use of flammable materials in building construction, a general lack of vegetation, and the efficient removal of flammable building contents and fittings by the residents of most buildings before the lava flows arrived. An exception was the co-operative facility, where lava had surrounded and partially entered the building, and which suffered a small fire in one part of the building. At the time of our visit, there was no evidence of what flammable material had been ignited by the lava flow.

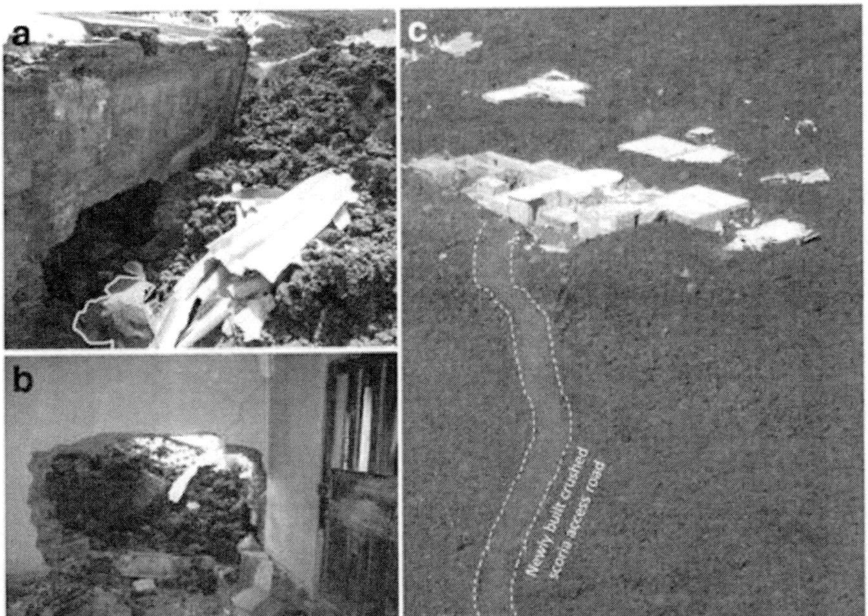

Figure 6. a The explosion of a gas canister used for cooking (oxidised remains highlighted in *orange* in *lower middle* of photo), presumably as a result of proximity with the lava flow, **b** punctured a large hole in the side of this squared masonry building (photo taken inside building); The building remained otherwise intact; **c** A building in Bangaeira that was not inundated or destroyed by the surrounding lava flow but was left isolated within the flow. By the time of our visit, approximately 11 weeks after impact, a road had been prepared to allow the family to return and people and donkeys to transport provisions to the building. Photographs: S.F. Jenkins, February 2015

The 2014–2015 eruption occurred near the start of the dry season (November–July) and water storage tanks in Chã are normally near full at this time of year; however, Cape Verde was experiencing severe drought in 2014 and many storage tanks were low or empty. Had they been full, then contact with lava flows could have feasibly initiated steam explosions, but we saw no evidence for that. Some concrete storage tanks showed evidence of cracks from the pressure of the lava flows, and these may have allowed the water to seep out gradually. Only one example of the explosion of a gas canister was observed (Fig. 6a and b). The explosion punctured a large hole in the wall of an otherwise intact building isolated within the lava flow.

Isolation

In Bangaeira, two parallel rubbly ‘a‘ā flows, only about 2–3 m thick and fed by breakouts from the thicker ‘a‘ā flow that had covered most of Portela, swept through the village. Some buildings that were on slightly raised ground resisted destruction, burial or inundation by the flows but were left isolated within the flow. In one example in Bangaeira (Fig. 6c), the surrounding agricultural land, livestock pens and water storage tanks had been destroyed leaving the residents without their livelihoods. In spite of this, the residents moved back into the building within weeks of the lava flow passing because it provided more comfortable housing than the temporary shelters. By the time of our visit (~11 weeks after impact), a new access route across the lava had been built by crushing the scoria surface, and people and donkeys were able to reach the property with ease (Fig. 6c). A small hotel in Bangaeira was also preparing to open at the time of our visit; lava had surrounded and isolated the building, but not entered into the courtyard or adjoining rooms.

Some buildings experienced more than one mode of damage, e.g. isolation with minor structural damage to one wall, or partial burial with an explosion caused by inundation of the lava. From our field visit, we deduced that the vast majority of buildings (~80 to 90%) are best described by the most severe damage class: partial burial and/or inundation.

Infrastructure and agriculture

Chã had relatively little community infrastructure, with power and water generally sourced on a building-by-building basis prior to the 2014–2015 eruption. Our field observations of the impacts upon infrastructure and agriculture have been synthesised and are described over the following subsections.

Road access

The 2014–2015 lava flows crossed the road within the first day of eruption onset, with flows then travelling south-southeast and north-northwest. The latter, and largest flow lobe extended initially along the face of the obstruction provided by the eastern edge of the 1995 lava flow but later expanded to bury the road as well as parts of the 1995 flow. Approximately 5.7 km (PDNA, 2016) of the road is now covered by lava flows up to 6 m thick. Once the road had been cut by the lava, access to Portela and Bangaeira was gained by a dirt road between older lava flows and the Bordeira cliffs.

Telecommunications

Prior to the 2014–2015 eruption, telecommunication lines were carried into buildings in Chã from timber utility poles that ran alongside the road. These were destroyed within the first few days of the eruption as lava flows ignited, felled and buried the poles and lines west and northwest of the vents. Replacement mobile communications were in place by the time of our visit to provide temporary mobile phone coverage. No issues with breakdown of telecommunications outside of Chã as a result of ash clouds were recorded.

Water storage and sources

Prior to the eruption, the majority of water in Chã was sourced through harvesting of rain water on concrete slab roofs. A borehole to the southwest of Portela supplemented this source, but was covered by the 2014–2015 lava flows. An open concrete rainfall trap reservoir between Bangaeira and Portela was also completely buried underneath the flows. At least three examples of melted or partially melted plastic water storage containers, and the pipes connecting the tanks to the building interior, were observed even though the containers did not come into direct contact with the flow. Thermal damage of plastic tanks typically had a rapid attenuation, with only the portion nearest the flow melted. In the case of a plastic water tank on top of the co-operative facility, the melted tank edge was more than 5 m above the flow that surrounded the building below. Concrete water storage tanks remained relatively intact upon inundation by a flow, with many transported whole on top of the flow. The large reinforced concrete rainwater storage tank in the Fogo Park Headquarters was the only structure from the building that could still be seen following the eruption: it was transported within the flow more than 150 m away from the headquarters. Metal structures showed strong corrosion from contact or proximity with lava flows, which would render the tanks unusable.

Other infrastructure

As discussed in Fire and explosions section, power was supplied through diesel generators and gas canisters; however, residents were able to recover most of these prior to the lava flows reaching them, so explosion of a gas canister was only observed in one building.

Agriculture

Fortunately, the eruption did not happen while grapes were due for harvest and there seemed to be no adverse impacts on observed crops from the basaltic tephra falls. The slow advance of the lava flows offered some residents the opportunity to dig up plants ahead of the flow, but 2.08 km^2 (~25% of the total: PDNA, 2016) of important agricultural land was buried underneath the 2014–2015 lava flows. In particular, the flat area of fertile soil derived from 1680 AD phreatomagmatic ash, west of Portela, where the inhabitants used to grow vegetables has now been completely covered by lavas. In contrast, areas to the east where fruit trees and vines grown on more recent lapilli deposits were not affected by the lavas although smaller areas of vine growing south west of Portela were covered. A United Nations Post-Disaster Needs Assessment (PDNA, 2016) found that no agricultural equipment was damage or destroyed as they were salvaged before the lava flows reached them. The agricultural sector suffered the greatest economic impact from the 2014–2015 eruption as a result of damage to land and facilities and the disruption to future production. The vast majority of the losses (98%; 1171 million CVE; US$11.6 million) were sustained by private individuals and businesses rather than the public sector (PDNA, 2016).

DISCUSSION

This discussion qualitatively interprets our field and remotely derived observations (Qualitative trends in observed damage section). We then examine the implications for developing non-binary fragility functions (Some pointers towards quantitative fragility functions for lava flow hazards section) before putting forward potential strategies for reducing physical vulnerability to lava flow impact, with a focus on buildings typical of Chã and those housing critical infrastructure (Building design concepts for resistance to destruction by lava flows section).

Qualitative trends in observed damage

As suggested in the pre-eruption vulnerability assessment (Jenkins et al. 2014), the physical vulnerability of a structure – as in its construction age, quality and type, the material and its orientation relative to flow – did affect the level, and perhaps method, of damage or destruction. For example, the shearing of roof slabs from a building and the development of cracks at the roof slab/wall interface, as noted in Building section, suggests weak horizontal connections between the wall and roof. Observed failure along the mortar-block interface is suggestive of poor cohesion between mortar and block, as has been suggested for wall failure in pyroclastic density currents (Jenkins et al. 2013). A building wall perpendicular to flow and facing up-flow will experience the largest lateral pressures (either static or dynamic) and is most likely to collapse through wall failure (Blong 1984). In contrast, buildings where the flow can be diverted either side of the corner of a building were more likely to remain standing but be buried, especially for less viscous flows with low yield strength, i.e. the pāhoehoe flows. This was especially true for the traditional cylindrical buildings, potentially because the weight of a lava flow against a convex wall will place it in compression, making the same materials as used for a planar or concave wall (that would be placed into tension by the weight of a flow) significantly stronger. For example, the two-storey squared block masonry school in Portela was completely destroyed while a single-storey community centre, ~ 180 m to the northeast, and of traditional construction with ~40 cm thick cylindrical scoria walls, was partially buried to roof level but not transported or destroyed. Water storage tanks also exhibited low vulnerability to lateral pressures because they had been designed to withstand large lateral loads from water storage, with metal reinforcement bars and relatively small span (1–3 m) walls. We observed many water storage tanks and reinforced concrete roof slabs, because of their high structural integrity relative to masonry walls and because of their position at the top of buildings, intact but transported tens of metres at the top of the flow. We suggest that higher flow density, either in the case of pāhoehoe flows as compared to ʻaʻā, or in the case of inflating flows, is likely to increase the vulnerability of structures to failure as a result of their buoyancy, causing them to shear off their foundations or even to float, as in the cases of the transported concrete water tanks.

Buildings in Chã are quite widely spaced and in the most densely populated areas of Portela and Bangaeira there was no evidence for buildings channelling the flows before being destroyed or partially buried. For the traditional cylindrical buildings to remain standing once inundated with lava, flows must have been diverted around the walls, although we could not find any syn-eruption images of this happening. Google Earth Digital Globe satellite images dated 25 November 2014 show that the oblique angled strong squared scoria western walls of the National Park Headquarters initially resisted the flow, redirecting flows around the building before it was destroyed and buried. From the satellite and media images available it is not possible to see if the flow inflated as it redirected around the walls. In the eastern portions of Portela and in Bangaeira, where flows were two to three metres thick, some slightly elevated buildings that avoided lava flow impact became isolated within the flows.

Evidence of thermal damage was limited because of the dominance of stone and concrete structures, the very arid climate and lack of vegetation, and the general removal of flammable liquids and contents from buildings before lava flow impact. Buildings in Chã relied upon diesel generators for electricity and gas canisters for cooking. As a result, the potential for explosions from the storage of fuel coming into contact with lava flows was expected to be relatively high. However, the distance of the villages from the vent (~3 km straight line; nearly 5 km flow path) and the rate of lava flow advancement gave enough warning time for most residents to remove all of the flammable building contents and fittings, including expensive commodities such as gas canisters and diesel generators. The few instances in which fires were ignited and the single instance of a gas explosion emphasise the significance of this hazard in areas with more vegetation or flammable construction types, or in cases where fuels are not removed. Where metal structures were exposed to the lava, including reinforcing bars exposed in collapsed structures as well as abandoned metal objects, corrosion damage as well as direct thermal damage was evident. Thermal and corrosion damage to metal reinforcement of concrete structures may therefore be significant in weakening of these if they are exposed to hot lava and fumarolic gases for long periods, as may decarbonation and dehydration of cement in extreme heat.

The number of buildings impacted during the Fogo eruption ($n = 260$) is comparable to most recent impacts (in the hundreds), however, there are relatively few recent examples of lava flows that have destroyed such a large proportion of a community's buildings and infrastructure (Table 1). By the time of our visit to Fogo, approximately 2 months after the villages had been impacted by lava, a number of shelters had been constructed on the slopes of Monte Amarelo and residents were returning daily to tend to the remaining crops. By February 2016, just over 1 year after the eruption of Fogo, Google Earth Digital Globe images showed that more substantial routes had been built across the lava flows to access isolated buildings and agricultural land, as well as new buildings constructed on top of the 2014–

2015 lava flows (Fig. 3c). Thus, at least some residents are already rebuilding in Chã das Caldeiras, demonstrating the importance of attachment to their community and land for many residents of Chã das Caldeiras. The return of residents to areas destroyed by lava flows is not novel: residents have returned repeatedly to the Italian village of San Sebastiano, which was destroyed in 1855 AD, 1872 AD and most recently in 1944 by lava flows from Vesuvius (Kilburn 2015).

In Chã, the loss of land and the co-operative buildings will severely affect future community and commercial activities, such as tourism and the export of wine and produce. The balance of possible future agricultural production has been permanently changed by the combined effects of the 1995 and 2014–2015 lava flows, with much of the land devoted to food crops for local consumption or sale in Sao Filipe having been covered by lava flows whilst most of the land devoted to vines and fruit was unaffected. This pattern has exacerbated dependence on these cash crops and so increased the economic impact of the loss of the wine and jam production facilities in Chã during the 2014–2015 eruption. Agricultural land is assigned within Chã on a family-by-family basis so that some families have lost large swathes of agricultural land to lava flows while others have been less affected. However, land rights are spread through extended families in Cape Verde, and so many of those who lost their primary land will have some claim upon other plots owned by their relatives. It is not clear how or if redistribution of resources will be instigated, or to what extent complete resettlement of the villages will be possible. Richter et al. (2016) found that the former sites of Portela and Bangaeira villages are still characterised by high future lava hazard. Should relocation be appropriate, identifying the most suitable site will require consideration of the range of possible locations of future vent opening (as in Richter et al. 2016) and the likely lava flow hazard relative to other natural hazards, and in the socio-economic and cultural context of strong livelihood ties to the agricultural land. In preparing for an effusive eruption, these aspects will be more important than assessing the fragility of the building stock to lava flow impact.

Tourism was becoming an increasingly important source of income for Chã residents prior to the eruption and, with the loss of agricultural land, it is conceivable that there may be a greater focus on tourism in the future. However, significant reconstruction and rehabilitation will be required to return the villages to their pre-2014–2015 eruption status/conditions. Lava covering the road during the 2014–2015 eruption meant that transport in and out of the villages and adjacent agricultural lands passed close to the Bordeira cliffs through an area prone to rockfall. Access to the villages via this track was slow, imposed additional wear on vehicles and was only easy in lightly loaded four-wheel drive vehicles leading to restrictions in the numbers of available vehicles that could use it and to additional economic costs. As in 1995, reconstructing a road into Chã will require construction over, or excavations through, the recently emplaced lava flows. In early 2016, plans were underway to develop a track and footpaths that run from Bangaeira to the northeastern corner of Chã and onto the northern flank of the island (used during the 1995 evacuations: Fig. 1) into a road. This will provide a second access (and evacuation) route from Chã and potentially support future redevelopment and rehabilitation of communities.

Given the observed damage from the Fogo eruption, we believe that a binary fragility function is still considered appropriate under most combinations of lava flow hazard and building characteristics. However, in the following subsection we consider what aspects of the flow and building characteristics are important for the development of non-binary fragility functions, i.e. for critical or costly assets. Building design concepts for resistance to destruction by lava flows section then uses these considerations to suggest how physical vulnerability may be reduced in lava flows.

Some pointers towards quantitative fragility functions for lava flow hazards

The general assumption to date of binary fragility functions for lava flow hazards means that little consideration has been given to the form of vulnerability variations and fragility functions for buildings and other exposed assets impacted by lava. These impacts may be divided into 1) Gravitational-mechanical or static load forces; 2) Dynamic-mechanical forces; 3) Permanent inundation by lava; and 4) Thermal and thermo-chemical effects. Buildings impacted by the thicker (~4 m or more) but slow moving (~1 m/h flow front velocity) 'a'ā flows were completely destroyed, while some buildings impacted by thinner (~2 m) but faster moving (up to ~180 m/h flow front velocity) 'a'ā flows remained intact. This implies that for 'a'ā flows at least the gravitational forces, linked to flow thickness, dominate over dynamic forces linked to flow velocity and rheology. Therefore, fragility functions consistent with this pattern of destruction should relate to flow thickness and density more than to flow velocity and rheology. However, we recognise that the latter are controls upon flow thickness and so an indirect dependence will exist.

In contrast to the analogy of water inundation, where water drains away and buildings can be dried out, inundation with lava is likely to completely destroy the value of a building (or at least its inundated floors). This is especially true in the case of pāhoehoe lava, which requires much effort to break up and may be impossible to remove without damaging the building structure. Furthermore, our observations indicate that ingress of 'a'ā into a building through small openings is much less rapid than ingress of pāhoehoe through openings of similar size. Pāhoehoe lava and inflating lava flows of all types therefore present a much greater inundation risk than do ʻaʻā flows of similar thickness.

The survival of some masonry and concrete buildings impacted by lavas in Chã das Caldeiras, in contrast to the complete destruction of wooden buildings by fire in communities inundated by lava in Hawai'i supports the obvious point that heat resistance in construction materials is essential for avoidance of complete vulnerability of buildings to lava flows. More subtly, our observations indicate that flammable and potentially explosive fuels and other liquids must be removed or else carefully protected from heat (for example in underground tanks) prior to lava flow impact; and that thermochemical impacts such as corrosion of metal objects by hot reactive gases may be an important hazard to otherwise protected buildings and contents.

Building design concepts for resistance to destruction by lava flows

The previous section indicates some clear trends that could form the basis for building design concepts. For typical structures built by Chã residents, such as domestic housing, agricultural buildings and small commercial buildings, a key consideration is capital cost of materials. The first and simplest measure, following existing local practice for levelling ground before construction of houses on gentle slopes, would be to build upon a rubble pedestal raised some 1–2 m above the surrounding ground. Resistance to lava flows of existing buildings could be increased by reverting to the roundhouse style of construction, or thickening rectilinear walls to the 40 cm typical of roundhouses. Closely spaced internal load-bearing walls would significantly buttress such thick external rectilinear walls, by analogy with buttress dam construction (Schnitter 1994). Additionally, outward-facing openings at the ground floor level should be on the downslope side and could be strengthened or recessed and blocked with rubble stone in the event of an eruption. Roof level (or upper storey) access should be possible so that the building can be accessed post-eruption, even if the entrance has been blocked by lava.

Critical structures and infrastructure built by the government or other external agencies may require high levels of protection either because of political and economic costs associated with their provision leading to a low tolerance of risk of loss, or because of the wider impacts of their loss. Greater expenditures to reduce those risks might therefore be acceptable. For those infrastructure elements that can be co-located but not easily placed on high ground, such as deep wells and heavy generators, a single large multi-use structure would be preferable. Given the advantages of outward curved walls, convex toward the pressure as in arch dams, this would likely resemble a circular tower with thick and high quality (pre-stressed reinforced concrete or masonry) external walls. The building would also benefit from outward-sloping walls as in slab buttress dams (Schnitter 1994) and strong ground anchoring to withstand buoyancy forces, and an outward-sloping pedestal to raise any openings and promote escape of hot gases in the lava. Service pipes (for example, water pipes leading from a well head) should be buried below the surface or else cut and sealed in the event of an eruption. If only limited access to the structure was required, the entrance could be at roof level and accessed by an external ladder or spiral stair, a jib crane or hose (for liquids such as generator fuel).

CONCLUSIONS

The Fogo eruption was one of the most devastating lava flow impacts in recent times. The majority (~75%) of buildings and 25% of important agricultural land (~2 km^2 out of a total of 8.5 km^2 of cultivable land: PDNA, 2016) in the isolated Chã das Caldeiras community was damaged or destroyed. Post-eruption field assessments confirmed that a binary vulnerability function for lava flows is generally appropriate, especially for agricultural land, but that the mode of damage differs. Fire and explosions from interaction of lava flows with vegetation, buildings or contents such as gas canisters were limited at Fogo, because of the arid environment and the very effective removal of building contents and fittings by residents and the military prior to lava flow inundation. The principal observed impact of lava flows was the burial of physical structures and land. Depending upon the nature of the structure impacted by the lava, the object was buried, inundated or transported by the flow. However, the traditional cylindrical buildings constructed with thick scoria walls were found to be particularly resistant to lava flows because of their very strong resistance to lateral pressures, and potentially their ability to divert lava around their walls, with those that could still be seen partially buried but not destroyed by the flow. Even where buildings remained intact, lava flows still rendered them uninhabitable and unusable through burial or flow ingress through openings such as doors or windows; the loss of original function was therefore still total. A small number of buildings in Bangaeira were not damaged by the lava flows but were isolated between flow channels, making access difficult and destroying land associated with the property.

The major impact for infrastructure and agriculture was their burial by lava. Land used for agriculture or settlements has been rendered unusable for many decades and retrieval of infrastructure is mostly impossible. Fortunately, the timing of the eruption did not interfere with grape harvesting, the most valuable crop in Chã.

The greatest damage occurred in areas affected by thick (several metres) 'a'ā flows, which produced higher forces upon the structures that they encountered, than the thinner (few metres) but faster-moving pāhoehoe flows, where damage was limited to partial burial and/or inundation by lava. The pre-eruption vulnerability and post-eruption damage assessments were used in combination to identify the key lava and building characteristics that influence impact. For the basaltic pāhoehoe and 'a'ā lava flows that dominate at Fogo, the curvature, thickness and short span of the walls of traditional cylindrical buildings, as well as the lack of multiple openings such as doors or windows, seemed to prevent destruction or the ingress of lava into the building interior. Our observations enabled us to propose building design strategies that could be employed in Chã and similar locations on other dominantly effusive volcanoes to increase the resistance of both typical structures and special purpose structures housing critical infrastructure. The rapid rebuilding of the villages after the 1995 eruption, largely on their old sites, shows a high level of acceptance of lava flow risk by Chã residents. In contrast, the slow restoration of services after 1995 and the initial opposition to resettlement of Chã das Caldeiras by the government in 2015 shows that there is, in general, a lower level of lava flow risk acceptance for public buildings and infrastructure. The proposed mitigation strategies are tentative but challenge the notion of purely binary fragility functions. We emphasise that in the absence of strong imperatives for resettlement of high lava flow hazard zones, the best method for reducing lava flow risk to buildings and infrastructure remains the re/location of structures in areas of relatively low lava flow hazard.

ACKNOWLEDGEMENTS

We thank Sr. Jose Antonio Fonseca of Portela village and Brendon Rolfe-Betts for assistance in fieldwork. We are also grateful to Chris Kilburn (UCL), Laura Connor, Jacob Richardson, Chuck Connor and Sylvain Charbonnier (USF) who supported emergency management efforts by assisting in lava flow modelling and flow path and runout estimation during the crisis. Giuseppe Cornaglia, from the Portuguese Civil Protection Agency, was instrumental in the activation of the Copernicus Emergency Management Service at the early stage of the eruption. We are grateful to Earth Observatory of Singapore for covering the open access fees. Finally, we sincerely thank Jim Kauahikaua and an anonymous reviewer, as well as the guest editor Natalia Deligne, for their insightful coments and feedback that greatly improved the manuscript.

Funding

SFJ is grateful for funding from the Earth Observatory of Singapore and NERC's Impact Accelerator Fund for hazard and risk assessment at high-risk volcanoes. JFBDF acknowledges funding from Fundação para a Ciência e Tecnologia, Portugal, as emergency support immediately after the onset of the eruption, and through project FIRE (contract PTDC/GEOGEO/1123/2014).

Authors' contributions

SFJ carried out 2010 pre-eruption vulnerability assessments, and SFJ and SD carried out post-eruption fieldwork to investigate the 2014–2015 lava flows and their impact. BF and JFBD aided both field visits, associated studies and interpretations and have supported a long-standing collaboration on Fogo. All authors have read and approved the final manuscript.

Competing interests

The authors declare that they have no competing interests.

REFERENCES

1. Bagnardi M, González PJ, Hooper A. High-resolution digital elevation model from tri-stereo Pleiades-1 satellite imagery for lava flow volume estimates at Fogo Volcano. Geophys Res Lett. 2016;43(12):6267–75.
2. Baxter P, Allard P, Halbwachs M, Komorowski JC, Woods A, Ancia A. Human health and vulnerability in the Nyiragongo volcano eruption and humanitarian crisis at Goma, Democratic Republic of Congo. Acta Vulcanol. 2002;14–15(1–2):109–14.
3. Behncke B, Neri M, Nagay A. Lava flow hazard at Mount Etna (Italy): New data from a GIS-based study. In: Manga M, Ventura G, editors. Kinematics and Dynamics of Lava Flows. Colorado: Geological Society of America: Special Paper 396 Boulder; 2005.
4. Blong RJ. Volcanic hazards: a sourcebook on the effects of eruptions. Academic Press Australia, North Ryce, New South Wales; 1984. p. 424.
5. Bulletin of the Global Volcanism Network (BGVN 20:03). New eruption on 2 April generates lava flows within the caldera. Washington DC: Smithsonian Institution (editor). BGVN 20:03; 1995.

6. Calvari S, Pinkerton H. Formation of lava tubes and extensive flow field during the 1991–1993 eruption of Mount Etna. J Geophys Res Solid Earth. 1998;103 (B11):27291–301.
7. Copernicus. Emergency Management Service, EMSR111: Volcanic eruption at Fogo Island (Cape Verde). 2014. http://emergency.copernicus.eu/mapping/list-of-components/EMSR111.
8. Day SJ, Heleno da Silva SIN, Fonseca JFBD. A past giant lateral collapse and present-day flank instability of Fogo, Cape Verde Islands. J Volcanol Geotherm Res. 1999;94(1–4):191–218.
9. Day SJ, Carracedo JC, Guillou H, Pais Pais FJ, Badiola ER, Fonseca JFBD, Heleno SIN. Comparison and cross-checking of historical, archaeological and geological evidence for the location and type of historical and sub-historical eruptions of multiple-vent oceanic island volcanoes. Geol Soc Lond, Spec Publ. 2000;171(1):281–306.
10. FAO. FAO provides urgent assistance to drought-stricken Cape Verde. In: Food and Agriculture Organization of the United Nations. 2015. http://www.fao.org/news/story/en/item/278021. Accessed 8 Mar 2016.
11. Faria B. Monitorização Geofísica do Vulcão do Fogo e Níveis de Alerta. Ph.D thesis. Universidade Técnica de Lisboa. In, Portuguese. 2010.
12. Faria B, Fonseca J. Investigating volcanic hazard in Cape Verde Islands through geophysical monitoring: network description and first results. Nat Hazards Earth Syst Sci. 2014;1:1–36.
13. Felpeto A, Araña V, Ortiz R, Astiz M, García A. Assessment and modelling of lava flow hazard on Lanzarote (canary islands). Nat Hazards. 2001;23(2):247–57.
14. Fernandes R, Faria B. FOGO-2014: monitoring the Fogo 2014 eruption, Cape Verde. In: European Geoscience Union General Assembly. Vienna: Copernicus; 2015.
15. Ferrucci F, Day S, Hirn B, Faria B, Zoffoli S. Multi-payload Multi-platform Tactical Monitoring and Evaluation of the 2014 Eruption of Fogo, Cabo Verde. In: European Geoscience Union General Assembly. Geophysical Research Abstracts, Vienna; 2015.
16. Fonseca JFBD, Faria BVE, Lima NP, Heleno SIN, Lazaro C, d'Oreye NF, Ferreira AMG, Barros IJM, Santos P, Bandomo Z, Day SF, Osorio JP, Baio M, Matos JLG. Multiparameter monitoring of Fogo Island, Cape Verde, for volcanic risk mitigation. J Volcanol Geotherm Res. 2003;2605:1–18.
17. Global Volcanism Program. Report on Fogo (Cape Verde). In: Sennert SK, editor. Weekly Volcanic Activity Reports, 19 November 2014–24 February 2015. Smithsonian Institution and US Geological Survey, Washington DC; 2015.
18. Global Volcanism Program. Weekly Reports: 3 to 9 December 2014. In, Weekly Reports of the Global Volcanism Network: Smithsonian Institution, Washington DC; 2014.
19. Global Volcanism Program (1976) Report on Karangetang (Indonesia). In: Squires D (ed) Natural Science Event Bulletin, 1:13. Smithsonian Institution. http://dx.doi.org/10.5479/si.GVP.NSEB197610-267020.
20. Global Volcanism Program (1992) Report on Kilauea (United States). In: McClelland L (ed) Bulletin of the Global Volcanism Network, 17:12. Smithsonian Institution. http://dx.doi.org/10.5479/si.GVP.BGVN199212-332010.
21. Global Volcanism Program (1977a) Report on Nyiragongo (DR Congo). In: Squires, D (ed.), Natural Science Event Bulletin, 2:2-3. Smithsonian Institution. http://dx.doi.org/10.5479/si.GVP.NSEB197703-223030.
22. Global Volcanism Program (1977b) Report on Piton de la Fournaise (France). In: Squires, D (ed.), Natural Science Event Bulletin, 2:4. Smithsonian Institution. http://dx.doi.org/10.5479/si.GVP.NSEB197704-233020.
23. Harris AJL. Basaltic lava flow hazard. In: Papale P, Eichelberger JC, Loughlin SC, Nakada S, Yepes H, editors. Volcanic hazards, risks and disasters. Elsevier; Amsterdam, 2015. p. 17–46.
24. IFRC. Emergency plan of action: Preliminary final report; Cape Verde: Volcano (Fogo). International Federation of Red Cross and Red Crescent Societies. In: DREF

no. MDRCV002; 2015. http://reliefweb.int/sites/reliefweb. int/files/ resources /MDRCV002pfr.pdf. Accessed 16 Mar 2017.

25. Jenkins S, Komorowski JC, Baxter P, Spence R, Picquout A, Lavigne F, Surono. The Merapi 2010 eruption: An interdisciplinary impact assessment methodology for studying pyroclastic density current dynamics. J Volcanol Geotherm Res. 2013;261:316–29.
26. Jenkins SF, Spence RJS, Fonseca JFBD, Solidum RU, Wilson TM. Volcanic risk assessment: Quantifying physical vulnerability in the built environment. J Volcanol Geotherm Res. 2014;276 105–20.
27. Kilburn CRJ. Patterns and predictability in the emplacement of subaerial lava flows and flow fields. In: Scarpa R, Tilling RI, editors. Monitoring and mitigation of volcano hazards. Berlin: Springer; 1996. p. 491–537.
28. Kilburn CRJ. Lava flow hazards and modeling. In: Sigurdsson H, Houghton BF, McNutt SR, Rymer H, Stix J, editors. Encyclopedia of volcanoes. San Diego: Academic; 2015. p. 957–69.
29. Masson DG, Le Bas TP, Grevemeyer I, Weinrebe W. Flank collapse and large-scale landsliding in the Cape Verde Islands, off West Africa. Geochem Geophys Geosyst. 2008;9(7):n/a-n/a.
30. PDNA (2016) Post-Disaster Needs Accessment Report, Gabinete de Reconstrução do Fogo, Paia, Cape Verde. https://www.gfdrr.org/acp-eu/cabo-verde-post-disaster-needs-assessment-2015
31. Rhodes JM, Lockwood JP, Trusdell FA. Lava flow hazards and risk assessment on Mauna Loa volcano, Hawaii. In: Rhodes JM, Lockwood JP, editors. Mauna Loa revealed: structure, composition, history, and hazards. Washington, D. C: American Geophysical Union; 2013.
32. Ribeiro O. A ilha do Fogo e as suas erupções. Junta de Investigaçôes do Ultramar. 1960.
33. Richter N, Favalli M, de Zeeuw-van Dalfsen E, Fornaciai A, Fernandes R, Perez Rodriguez N, Levy J, Victoria SS, Walter TR. Lava flow hazard at Fogo volcano, Cape Verde, before and after the 2014–2015 eruption. Natural hazards and earth system sciences discussions. 2016.
34. Schnitter N. A history of dams: the useful pyramids. Balkema, Amsterdam; 1994.
35. UN. Initial Assessment Report. Volcano eruption, Republic of Cabo Verde. In: Prepared by UNDAC and the United Nations Office Cabo Verde. 2014. http://reliefweb.int/report/cabo-verde/initial-assessment-report-cabo-verde-volcano-eruption. Accessed 16 Mar 2017.
36. Walker G. The endogenous growth of pahoehoe lava lobes and morphology of lava-rise edges. Studies in volcanology—the legacy of George Walker (Special Publications of IAVCEI No. 2). The Geol Soc. 2009;17–32.
37. Williams, RS, and Moore, JG, 1983, Man Against Volcano: The Eruption on Heimaey. Vestmannaeyjar, Iceland. United States Geological Survey Open File Report. https://pubs.usgs.gov/gip/heimaey/heimaey.pdf.
38. WorldAtlas. In, Cape Verde History Timeline. 2011. http://www.worldatlas. com/webimage/countrys/africa/capeverde/cvtimeln.htm. Accessed 16 Mar 2017.
39. Worsley (2015) Physical geology of the Fogo volcano (Cape Verde Islands) and its 2014–2015 eruption. Geology Today 31(4).

CHAPTER 5

Global Recording Rates for Large Eruptions

Jonathan Rougier[1]*, Stephen R. Sparks[2] and Katharine V. Cashman[2]

[1]School of Mathematics, University of Bristol, University Walk, BS8 1TW Bristol, England.

[2]School of Earth Sciences, University of Bristol, Bristol, England.

ABSTRACT

A non-parametric statistical approach is used to assess the global recording rate for large (M4+) stratovolcano eruptions in a modern database, LaMEVE (v3.1). This approach imposes minimal structure on the shape of the recording rate through time. We find that recording rates have declined rapidly, going backwards in time. Prior to the year 1600 they are below 50 %, and prior to 1100 they are below 20 %. Even in the recent past, e.g. the 1800s, they are likely to be less than 100 %. The assessment for very large (M5+) eruptions is more uncertain, due to the scarcity of events.

Keywords: Under-recording Non-parametric Confidence set

BACKGROUND

Under-recording is an unavoidable issue in empirical studies of large volcanic eruptions, as has been discussed by, e.g., Simkin (1993), Siebert et al. (2010, pp. 31–34) and Brown et al. (2014). As Simkin (1993, p. 436) puts it, the rise in the number of recorded eruptions follows from "more observers, in a wider geographical distribution, with better communication, and broader publication." Statisticians refer to this type of absence as 'missing not at random', and it must be accounted for when making inferences from observations, in order to remove, as far as possible, systematic biases. But this under-recording poses a major challenge, because we have few quantitative beliefs about the way in which the timing, location, and magnitude of an eruption combine together to affect its missingness from the modern record.

For an initial analysis, we can make one drastic simplification at the outset, and consider global volcanism above a specified large magnitude threshold. This reduces three dimensions of missingness to just one—timing. It also simplifies the statistical modelling: first, because we avoid complicated issues regarding the modelling and accuracy of magnitude measurements; second, because the process of global volcanism is likely to be more stable than its component parts. These issues are discussed further below (Section "Our statistical non-parametric approach").

This paper provides an empirical analysis of the global recording rate of large eruptions of stratovolcanoes; the definition of the global recording rate is discussed and clarified in Section "Definition of the recording rate". 'Large' is taken to be at least 100 million tonnes of ejected matter, or M4 according to the magnitude scale of Mason et al. (2004), for which $M :=$ log10m−7, where m is the mass erupted in kilograms. M thus provides a continuous measure of eruption magnitude based on estimates of erupted mass; more common is the use of the Volcanic Explosivity Index (VEI), which is discrete and assigned on the basis of either erupted volume or eruption intensity (average mass eruption rate).

Our key innovation is to use a non-parametric approach (Section "Our statistical non-parametric approach"), and to present our results visually in terms of a 95 % confidence set (Section "Results and discussion"). We are able to quantify recording rates through time in terms of upper bounds of 95 % confidence intervals, and show, for example, that this upper bound is below 50 % prior to the year 1600.

METHODS

Definition of the recording rate

The notion of a 'global recording rate' is imprecise. We provide a definition which highlights the difference between the global recording rate and the mean recording rate across volcanoes.

Imagine a process, extensive in space and time, which indicates with a 1 or a 0 whether there is an information link from a large eruption at a given location and date, through to today and an appearance in the database. Index the active stratovolcanoes with $i=1,\ldots,n$, and write $c_i(t)$ for this linkage function at the location of volcano i and time t. If we observed this function for a given i going backwards through time, we would see it at 'on' (i.e. 1) for today, but then at some point in the past it would start to flicker between 'on' and 'off', indicating that a link to today is contingent on there being a suitable person present at the eruption, or traces (e.g. a near-field tephra deposit) that have survived and been discovered and interpreted. As we go back further, the function spends longer and longer amounts of time 'off', until finally it is permanently 'off'. This sequence, from currently 'on', through flickering, to 'off', varies from volcano to volcano. A global recording function reflects the amalgamation of such functions for the global collection of volcanoes.

With the linkage function specified, define the global recording rate at time t as

$$\pi(t) := \frac{c_1(t) \cdot \lambda_1 + \cdots + c_n(t) \cdot \lambda_n}{\lambda} \tag{1}$$

where λ_i is the large-eruption rate of volcano i, and $\lambda := \lambda_1 + \cdots + \lambda_n$, the global large-eruption rate; these rates are taken to be constant in time—see the next paragraph for the generalisation. The numerator of (1) is the database's global large-eruption rate for time t, which can be written as $\pi(t)\cdot\lambda$, where λ is today's global large-eruption rate. Note that $\pi(t)\to 1$ as t tends to today, because each $c_i(t)\to 1$. In (1), we are not assuming that large eruptions follow a Poisson process for each volcano, although we are assuming that we can combine rates in the same way that we would combine rates were the volcanoes to be following independent Poisson processes. In Section "Our statistical non-parametric approach", we will treat the global large-eruption process as a time-homogeneous Poisson process, but we will not need to assert that the individual processes are either Poisson or stationary. This issue has been discussed by De la Cruz-Reyna (1991).

In the case where the large-eruption rates fluctuate in time, this fluctuation can be absorbed into the $c_i(\cdot)$'s. Let λ_i be today's large-eruption rate for volcano i, and then write

$$c_i(t) \cdot \lambda_i(t) = c_i(t) \cdot \lambda_i(t) \cdot \frac{\lambda_i}{\lambda_i} = c_i'(t) \cdot \lambda_i$$

where $c_i'(t) := c_i(t) \cdot \lambda_i(t)/\lambda_i$. We expect that $c_i(t) \approx c_i'(t)$, because this is exactly true when $c_i'(t)$, or t tends to today, and approximately true elsewhere, if the fluctuations in $\lambda_i(t)$ are not too large, in proportionate terms.

Now consider period j of width Δ centered at time t_j. Starting with (1), take the time-average over the period to find the global large-eruption rate for period j:

$$\pi_j := \frac{\pi_{1j} \cdot \lambda_1 + \cdots + \pi_{nj} \cdot \lambda_n}{\lambda} \tag{2a}$$

where

$$\pi_{ij} := \Delta^{-1} \int_{t_j-\Delta/2}^{t_j+\Delta/2} c_i(t)\, dt. \tag{2b}$$

The value π_{ij} is the recording rate of volcano i in period j. It is not a probability, but it can be interpreted as one, under additional conditions. If a large eruption of volcano i is randomly and uniformly distributed in period j, then π_{ij} is the probability of that eruption being recorded.

Consider (2a) in the special case where all of the large-eruption rates are approximately the same. In this case

$$\pi_j \approx \frac{\pi_{1j} + \cdots + \pi_{nj}}{n}$$

and the global recording rate for period j is approximately the mean recording rate across all volcanoes. In general, though, (2a) shows that the global recording rate is biased away from the mean recording rate towards the recording rates of the most active volcanoes. So, for example, if recordings of a very active volcano start at the end of period j, then π_j to π_{j+1} will show a large increase, larger than the increase that would be expected for the addition of one extra volcano. Just monitoring π_j alone, it would not be possible to distinguish between, say, three newly-recorded volcanoes with modest eruption rates, and one newly-recorded volcano with a high eruption rate.

This distinction between the global recording rate and the mean recording rate is an important part of the definition and interpretation of the global recording rate, but whether it really matters is moot. Large eruptions are scarce. We do not expect to constrain our uncertainty about π_j to within the scale of the distinction, and from this point of view our uncertainty about π_j is more-or-less the same as our uncertainty about the mean recording rate, which is how we will interpret our results below.

Review of previous approaches

Naïve estimators of the global recording rates are easy to derive. Consider a time period centred on time t_j, of width Δ_j, which has x_j recorded large eruptions. Equating the expectation with the observation suggests that $\Delta_j \cdot \pi_j \cdot \lambda \approx x_j$, where λ is the global large-eruption rate, defined after (1), and taken to be effectively constant over the time-frame of the analysis; π_j is defined in (2a). Let the most recent period, period k say, have a recording rate of $\pi_k = 1$. Then $\Delta_k \cdot \lambda \approx x_k$, and substituting for λ gives

$$\pi_j \approx \frac{x_j}{\Delta_j} \Big/ \frac{x_k}{\Delta_k} \qquad j = 1, \ldots, k. \tag{3}$$

Naïve though this is, it will turn out to be the Maximum Likelihood (ML) estimator in the model proposed in Section "Our statistical non-parametric approach". It is, of course, a woeful estimator, being very over-fitted: it has one parameter per observation, and no constraints.

There are several ways to proceed. One approach is to smooth the x_j's, or equivalently the π_j's. This approach was implemented in Guttorp and Thompson (1991), who discussed the critical issue of choosing a smoother and a bandwidth (their p. 579). There was no assessment of uncertainty in their estimated π_j's, although this was not their main concern. We present a similar estimate below, at the start of Section "Results and discussion".

A second approach is to propose a specific functional form for $\pi(\cdot)$, which is controlled by a small number of unknown parameters; this imposes smoothness on $\pi(\cdot)$ and reduces over-fitting. This approach was originally suggested in a more general context by Solow (1993), who proposed a monotonic four-parameter functional form based on the distribution function of the Beta distribution. The parameters of the proposed function are fitted to the x_j's, which typically requires a statistical model for the number of recorded large-eruptions, if uncertainty is to be assessed. This is the parametric approach that has been used in the volcanological literature to date.

One functional form for $\pi(\cdot)$, used by Coles and Sparks (2006) and Deligne et al. (2010), is

$$\pi(t; t_0, b) = \left(\frac{t - t_0}{t_k - t_0} \right)^b \tag{4}$$

where t_0 and b are the parameters, $t_0 \leq t_1$, and $b > 0$. This functional form encodes two basic beliefs: that the recording rate is currently 1, and that it decreases smoothly backwards in time. A very similar approach is used by Kiyosugi et al. (2015), who propose an exponential curve.

Another, used by Furlan (2010) and Mead and Magill (2014), is

$$\pi(t; \pi_0, t_s) = \begin{cases} \pi_0 & t \leq t_s \\ 1 & t > t_s \end{cases} \tag{5}$$

where π_0 and t_s are the parameters, with $0 \leq \pi_0 \leq 1$ and $t_1 \leq t_s \leq t_k$. Equation (5) defines a step function that moves abruptly from π_0 to 1 at time t_s. This functional form encodes the same type of belief, except replacing 'smoothly backwards in time' with 'dramatically at some time previous to today'. Furlan (2010) uses a single global π_0 and t_s, while Mead and Magill (2014) allow π_0 and t_s to vary by region.

Mead and Magill's (2014) model is consistent with smooth functions such as (4) at the global level, as our analysis in Section "Definition of the recording rate" shows. In our terms, their model asserts that πij=π(i)0πij=π0(i) for tj≤t(i)stj≤ts(i), and 1 thereafter, the superscript *i*'s indicating that these parameter values differ by volcano or region. If different regions have different t(i)sts(i) values, then the total effect on the global large-eruption rate may look like a smooth curve, or a smooth curve with kinks, where a volcano or region with a high λ_i starts to be fully recorded. But a smooth curve would also arise in the more general case where, instead of jumping directly from π(i)0π0(i) to 1, the π_{ij}'s increased more gradually.

The existence of these two different functions for $\pi(\cdot)$ indicates a difficulty with this parametric approach: there is little guidance about the shape of the functional form. In fact there is more disagreement than we have presented here, because both functions also allowed for the effect of magnitude, but in different ways.

A third approach, adopted in this paper and new to volcanology, is to use a non-parametric representation of $\pi(\cdot)$, subject to some simple conditions. Because the conditions do not prescribe a tight relationship between the π_j's, this option is sometimes known as a 'soft' constraint, compared to the second (parametric) approach, which would be a 'hard' constraint. A hard constraint is more effective if the parametric form of $\pi(\cdot)$ is known, but a soft constraint is more widely applicable, and more suitable for dealing with observations affected by numerous complex factors related to recording of historical and geological events. As we show in Section "Our statistical non-parametric approach", a soft constraint can generalize both of the models given above. And as shown in Section "Results and discussion", although the uncertainty is large, the soft constraint still entails useful statements about temporal changes in the global recording rate for large eruptions.

Our statistical non-parametric approach

The notation is the same as the previous sections. Let there be k non-overlapping periods, with known lengths $\Delta_1,\ldots,\Delta_k$, uncertain recording rates $\boldsymbol{\pi} := (\pi_1,\ldots,\pi_k)$, and for which the number of recorded large eruptions is $\boldsymbol{x} := (x_1,\ldots,x_k)$. A statistical assessment of $\boldsymbol{\pi}$, including a quantification of uncertainty, requires a probabilistic model for $\boldsymbol{x}$.

We are considering global sums of large eruptions, and therefore we treat $\boldsymbol{x}$ as arising from a homogeneous Poisson process with unknown rate denoted λ (units of /yr). This simple model is suggested by De la Cruz-Reyna (1991), who also noted that it would be less applicable for individual volcanoes. It is not possible to test the time-homogeneity of this model if the recording rate varies by period. Instead, homogeneity becomes the condition under which the recording rates are identified (Solow 1993, makes a similar point).

Most of the papers discussed in Section "Review of previous approaches" (Coles and Sparks 2006; Deligne et al. 2010; Furlan 2010; Mead and Magill 2014) used a 2D Poisson process in time and magnitude, which is homogeneous in time. Because this implies a 1D Poisson process which is homogeneous in time, our model is more general. In lumping all eruptions of at least M4 together, we have ignored information in the database which these other papers have used. This is a deliberate choice on our part, due to the difficulty of incorporating the magnitude information, which we consider more than offsets the benefits of the extra information.

First, a 2D Poisson process over time and magnitude has many more possible implementations than a 1D process over time alone (which has only one). The common choice of all four papers cited above is to parameterise this process according to the peaks over threshold (POT) model for extremes, based on asymptotic theory (Pickands 1971). This requires three parameters rather than just one, and there is also the delicate issue of whether the asymptotic conditions hold, or whether some other justification might be found for the POT model.

Second, to implement this model, the papers cited above use a likelihood function defined with observed times and magnitudes, whereas we only use times. (Kiyosugi et al. 2015, use a separate recording function for each for each magnitude class using the Volcanic Explosivity Index). Thus all papers implicitly assume that magnitudes have been accurately recorded, and that, *a fortiori*, this accuracy has not changed through time. But magnitude is always imperfectly inferred. Mead and Magill (2014) and Kiyosugi et al. (2015), for example, implicitly assume that VEIs have been accurately categorized for several thousands of years, despite the very different ways in which VEIs have been assessed for eruptions occurring at different times over the Holocene and the Quaternary, respectively. Evidence of misrecording of magnitudes over the last millenium is given in Fig. 1, which shows the magnitudes 'piling up' on the integer values 4 and 5. Furlan (2010, sec. 5.4) shows that inferences are sensitive to misrecording of magnitudes.

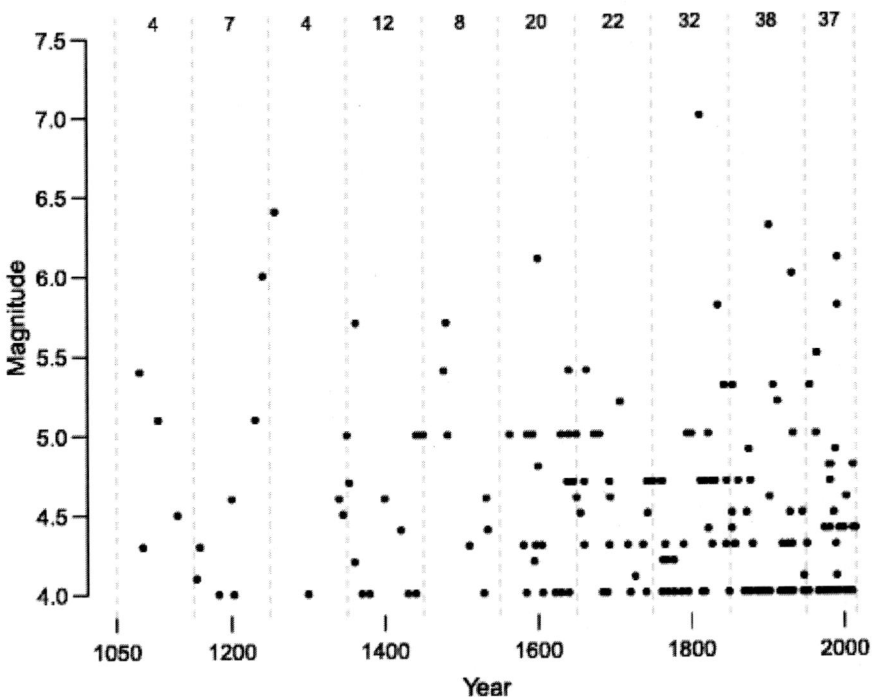

Figure 1. The dataset, of large (M4+) recorded stratovolcano eruptions, 1050–2014. The number of eruptions in century-wide intervals (excepting the most recent, which is 1950–2014) are shown at the *top* of the Figure

So, in suppressing the magnitude information, our results are robust to the specification of the 2D Poisson process over time and magnitude, and robust to magnitude measurement errors. This is not the case for the results of the other papers. The smooth-in-time model used by Coles and Sparks (2006) and Deligne et al. (2010) is also sensitive to timing errors, while the abrupt-in-time model used by Furlan (2010) and Mead and Magill (2014) is insensitive to timing errors, as is our model (because we bin the times into intervals).

Under our model, the likelihood function for ($\boldsymbol{\pi}$,λ) is

$$\begin{aligned} p(x \mid \pi, \lambda) &= \prod_{j=1}^{k} e^{-\Delta_j \pi_j \lambda} \frac{(\Delta_j \pi_j \lambda)^{x_j}}{x_j!} \\ &= e^{-\lambda \sum_j \Delta_j \pi_j} \cdot \lambda^s \cdot \prod_{j=1}^{k} \frac{(\Delta_j \pi_j)^{x_j}}{x_j!} \end{aligned} \tag{6}$$

where s $:= \sum_j x_j$. The rate λ is a nuisance parameter which can be eliminated, in order to focus on $\boldsymbol{\pi}$. Let λ have a Gamma distribution with shape a and rate b (units of yr), where both these values are specified. Then integrating out λ gives the likelihood function for $\boldsymbol{\pi}$:

$$\begin{aligned} p(x \mid \pi; a, b) &= \int p(x \mid \pi, \lambda) p(\lambda; a, b) \, \mathrm{d}\lambda \\ &= \frac{b^a}{\Gamma(a)} \prod_j \frac{(\Delta_j \pi_j)^{x_j}}{x_j!} \cdot \left(\frac{\left(\sum_j \Delta_j \pi_j + b\right)^{s+a}}{\Gamma(s+a)} \right)^{-1} \\ &\propto \prod_j (\pi_j)^{x_j} \cdot \left(\sum_j \Delta_j \pi_j + b \right)^{-(s+a)} \end{aligned} \tag{7}$$

where Γ is the Gamma function, and the final line discards all multiplicative constants not involving $\boldsymbol{\pi}$ see, e.g., Lunn et al. (2013, ch. 3). A similar approach was used by Solow (2001).

We need to specify a and b in the Gamma distribution for λ. The functional form of (7) suggests that a and b have a natural interpretation, in terms of dataset augmentation. The marginal distribution for λ carries the same weight as an additional period of length $b$yr which has a large eruptions. Informed by previous studies (e.g. Deligne et al. 2010), we assess the global rate of large eruptions at about one every two years, although acknowledge that this value is 'contaminated' by our exposure to the dataset $\boldsymbol{x}$. So we set a=1 and b=2yr, with this small value for $b$indicating that, putting aside the information in the dataset, we have only vague beliefs about λ. Very similar results would follow in passing to the (improper) limit of a=0 and b=0yr, which is completely vague. However, it is not possible to simulate λ from this limit, and simulation is required for code verification and for finding confidence sets.

This likelihood function in (7) has a closed-form maximum in $\boldsymbol{\pi}$, namely

$$\hat{\pi}_j = \frac{b}{a}\frac{x_j}{\Delta_j} \qquad j = 1, \ldots, k, \tag{8}$$

which can be verified by substituting into the first-order conditions. This expression is intuitive, because a/b is the expectation of λ, and so Δj·π^j·E(λ)=xjΔj·π^j·E(λ)=xj, similar to the starting-point in Section "Review of previous approaches". Replacing $E(\lambda)$ with the estimate x_k/Δ_k then leads to (3), although, as already discussed, this is a very poor estimator. As a Reviewer has noted, the ML estimator in (8) is insensitive to the strength of the information in the marginal distribution for λ, because it depends only on a/b, rather than on the value of b directly.

We adopt a more sophisticated approach and constrain the possible values that $\boldsymbol{\pi}$ can take, what we termed a 'soft constraint' in Section "Review of previous approaches". Our basic beliefs about $\boldsymbol{\pi}$ are

1. *1.* The global large-eruption recording rate since 1980 is 1.
2. *2.* For sufficiently long periods, the global large- eruption recording rate of an earlier period is never higher than that of a later period.

The choice of 1980 is determined empirically and we use periods of 100 years, see Section "Results and discussion". These two beliefs define the parameter space as

$$\Omega := \left\{\pi : 0 \leq \pi_1 \leq \cdots \pi_{k-1} \leq \pi_k = 1\right\}, \tag{9}$$

taking the final period to be 1980–2014. This parameter space generalizes the parametric models given in (4) and (5), since it is possible for the π_j's to decline smoothly, or abruptly, or both. The constraints in the parameter space allow observations from later periods to influence the estimates of π_j in earlier periods; in this way these early π_j's can 'borrow strength' from the later periods, to make up for the paucity of observations.

The ML estimator for $\boldsymbol{\pi}$ is now

$$\hat{\pi}(x) := \max_{\pi \in \Omega} p(x \mid \pi). \tag{10}$$

Although Ω is hard to explore in a sequential fashion, not having a simple interior, it is easy to enumerate a large set of elements of Ω, from which π^π^ can be approximated using the maximum value from the set. This algorithm is described in the Appendix, along with a method for computing a 95 % confidence set for $\boldsymbol{\pi} \in \Omega$, which implies a 95 %confidence interval for each π_j.

As the Appendix demonstrates, it is quite hard to assess uncertainty about $\boldsymbol{\pi}$. A Bayesian approach is more straightforward, but requires a prior distribution on Ω. However, it is not possible to specify such a prior distribution without introducing additional beliefs to the two basic beliefs given above. 'Obvious' priors such as uniform on Ω are actually highly constricting, since they impose a linear relationship on the prior expectations of the π_j's, namely $E(\pi_j)=j/k$. Thus, for a large k, the prior distribution for π_1 is forced towards zero. We do not want to incorporate any additional beliefs about $\boldsymbol{\pi}$, and therefore we have eschewed a Bayesian approach.

RESULTS AND DISCUSSION

The dataset is drawn from the LaMEVE database, see Crosweller et al. (2012) and Brown et al. (2014), version 3.1, downloaded Oct 2015. Our code, including the dataset, is available from the first author, implemented in the statistical computing environment R (R Core Team 2013).

The dataset is visualised in Fig. 1, with simple summaries in 100-year intervals. Figure 1 shows that most of the under-recording has happened in the distant past and at the lower end of the magnitudes, consistent with the narrative evidence on under-recording (e.g., Siebert et al. 2010). Suppressing the magnitude information gives the cumulative eruptions sequence, shown in Fig. 2: the convex shape indicates an increasing recording rate. There is a distinct change in the recording rates just prior to 1600 but, otherwise, the shape of the curve looks quite smooth.

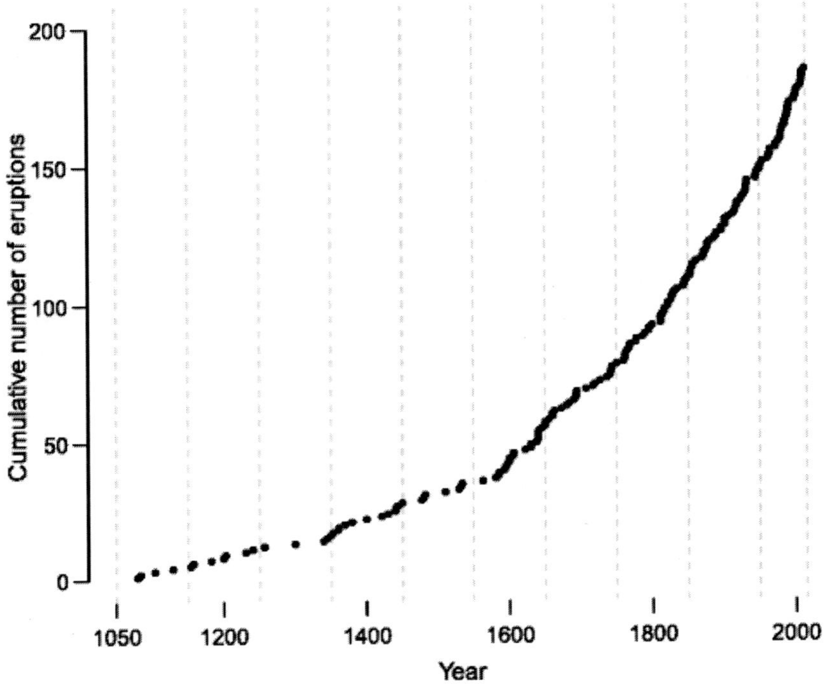

Figure 2. Cumulative eruption sequence, taken from Fig. 1. 'Elbows' between *straight lines* would indicate step changes between constant recording rates, taking the underlying eruption rate to be constant

Figure 3 shows a crude point estimate based on the relative slope of a smooth line fitted through the cumulative curve shown in Fig. 2. After inspection of the recent part of the cumulative curve we decided that the era of no under-recording could begin at 1980, and hence the average gradient of the curve for the period 1980–2014 was used to estimate the global large-eruption rate. Figure 3 includes both the raw estimate, and the estimate after monotonic smoothing, as suggested by Solow (1993). We regard this point estimate of the recording rate as very unreliable, being sensitive to choices such as the smoother and its bandwidth, and the way in which numerical gradients are computed. Nevertheless, its low value of about 0.06 in the year 1100 is striking.

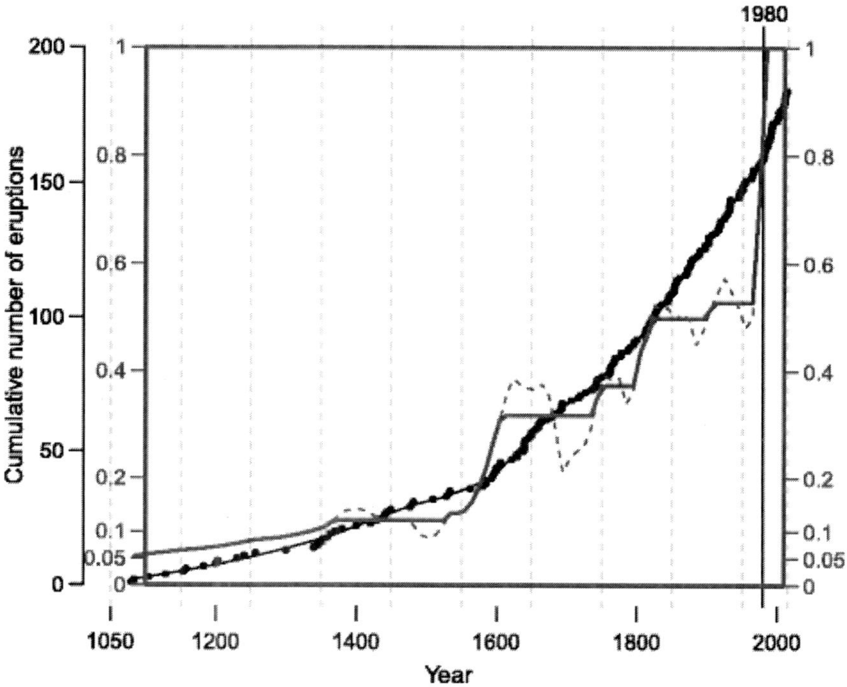

Figure 3. A crude point estimate of the global recording rate for large eruptions, superimposed on Fig. 2 (*dark blue* when reproduced in colour). After inspection of Fig. 2, the era of no under-recording is started at 1980. The point estimate is the ratio of the gradient of the smoothed curve to the average gradient of the smoothed curve post-1980. The raw values are shown as a *dashed line*, and the monotonically smoothed values as a *solid line*

Figure 4 visualises the 95 % confidence set for $\boldsymbol{\pi}$ for large ($M \geq 4$) eruptions. Technically this is a Parallel Coordinates Plot (see, e.g., Venables and Ripley 2010, sec. 11.1); each grey line represents a point in the 95 %confidence set. The initial value of 1 and non-positive slope (going back in time) are imposed by Ω, but the broadly exponential shape reflects the dataset, as does the low recording rate c1100. These features are consistent with the Coles and Sparks (2006) model, given in (4) above, with a value of $b>1$. As anticipated, there seems to be a discontinuity around 1550, which is not a feature of the Coles and Sparks (2006) model. This discontinuity was picked up in the parametric step-change model fitted by Furlan (2010), given in (5) above, which put the step-change just prior to 1600. Furlan's (2010) dataset is similar but not identical to ours, as she uses all large eruptions, not just those for stratovolcanoes, and we have extended the dataset to include recent eruptions. But her claim that "the under-recording process largely disappears in the most recent 400 years" (Abstract, p. 113) is refuted in our analysis.

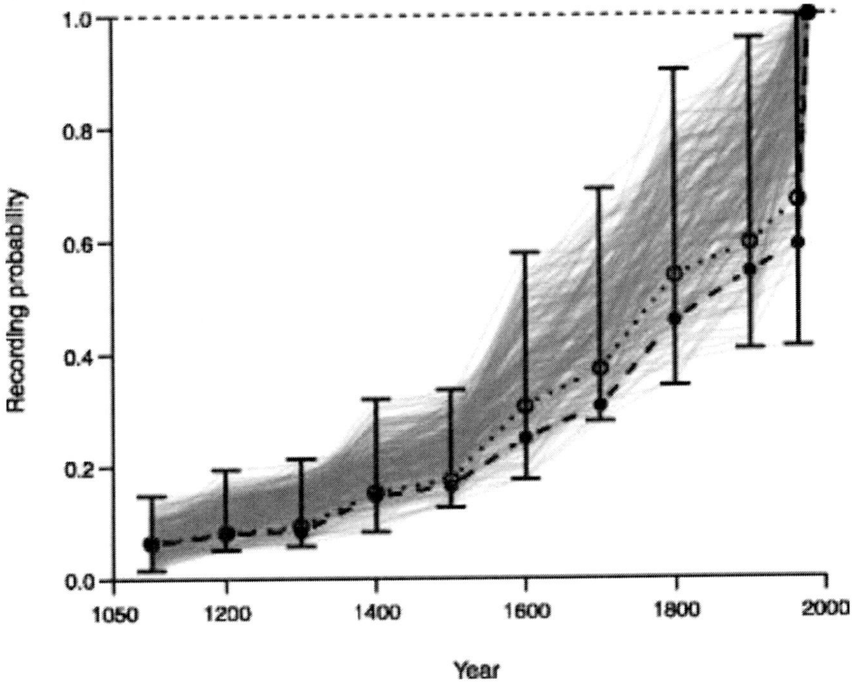

Figure 4. Visualisation of the 95 % confidence set for $\pi \in \Omega$ for large ($M \geq 4$) stratovolcano eruptions. Each *grey line* represents a point in the confidence set; the error bars are superimposed for convenience. The dots joined by a *dashed line* show the ML estimate. As an alternative point estimate, the open dots joined by a *dotted line* show the centroid of the 50 %confidence set

The confidence intervals for individual π_j's are large, but the main message from Fig. 4 is clear: the global recording rates for large eruptions decline rapidly going back in time. Prior to 1600 they are below 50 %, and prior to 1100 they are below 20 %. Even in the recent past, e.g. the 1800s, they are likely to be significantly less than 100 %. As a technical point, the error bars in Fig. 4 are margins of a 95 % confidence set, and as such their marginal coverage is *at least* 95 %. So the actual 95 % confidence intervals for each π_j are likely to be narrower than the error bars, and these statements are made with at least 95 % confidence.

Close inspection of Fig. 1 shows that the recorded magnitudes 'pile up' at M=4 (and also at M=5). So possibly some of the eruptions recorded as M=4 eruptions are smaller than this, and should be excluded. To check the sensitivity of our results to this source of mis-recording, we repeated the analysis for eruptions with recorded M≥4.1; the results are shown in Fig. 5. The uncertainty is larger, compared to Fig. 4, because 57 observations out of 184 have been dropped. But the qualitative conclusions are unchanged, and we are satisfied that our simple quantitative assessment of recording rates is robust to this source of mis-recording.

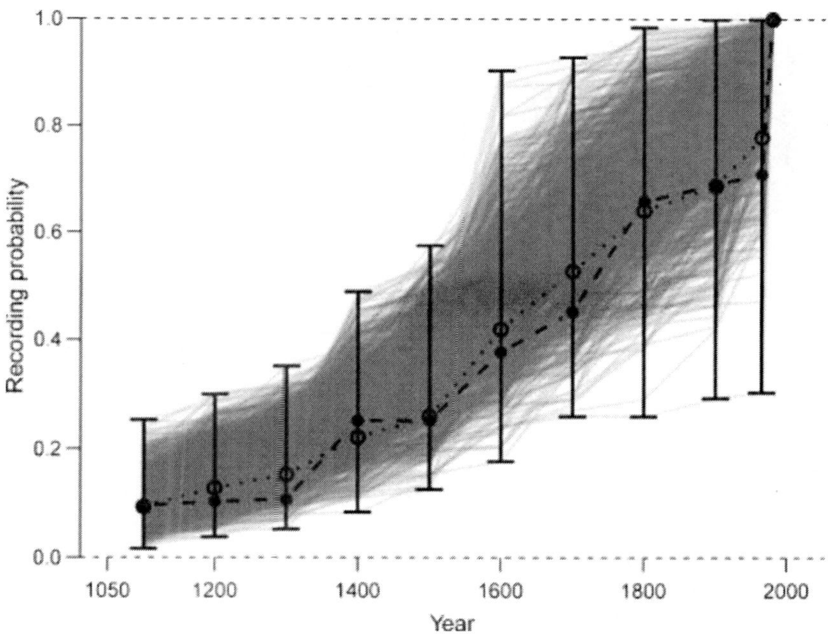

Figure . 5. Same as Fig. 4, but for magnitudes M≥4.1, to investigate sensitivity to 'piling up' at M=4 shown in Fig. 1

It is highly likely that most of the under-recording occurs for eruptions at the lower end of the large magnitude range. To test this assertion we restricted the dataset to very large magnitude eruptions (M≥5.0), and redid the calculation. Figure 6 shows the result: the number of M5+ eruptions in the database is low (42), and our uncertainty about the recording rates is large. We also found that using eruptions M≥5.1 (24 eruptions) gave a similar confidence set, although the point estimates were different. According to Fig. 6, the recording rate is likely to be higher for M5+ than for M4+, but we do not think the uncertainty assessment in Fig. 6 is reliable.

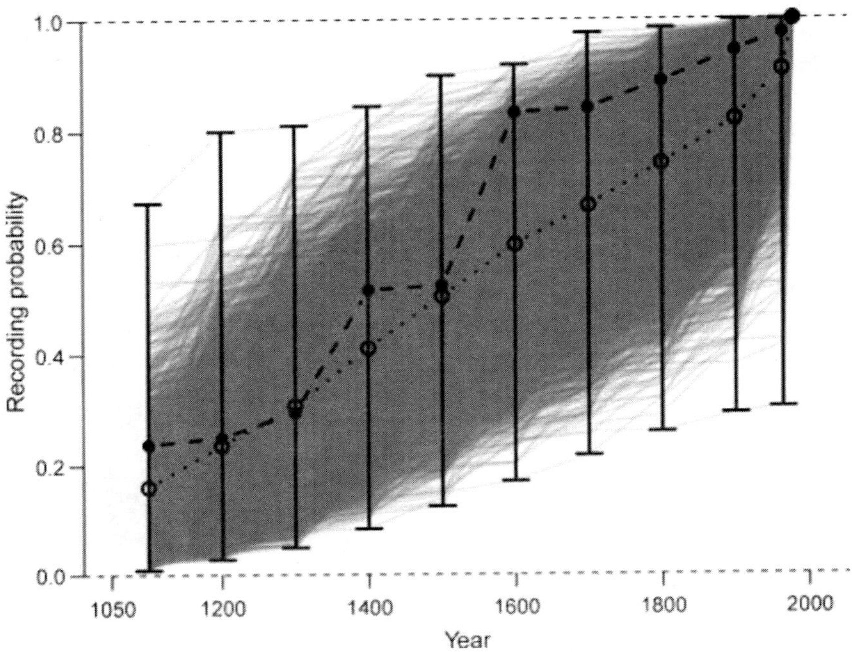

Figure . 6. Same as Fig. 4, but for magnitudes $M \geq 5.0$

The confidence set Figure s include an additional curve, the centroid of the 50 % confidence set, as an alternative point estimate to the ML estimate. This centroid will tend to be smoother than the ML estimate, which can get closer to the edges of Ω. But the difference between the two point estimates is mostly less than 10 percentage points for M4+, which is small in relation to the width of the confidence set. Overall, we hesitate to provide a point estimate for the recording rates from this dataset, given the volume of the 95 % confidence set.

CONCLUSIONS

There is empirical and narrative evidence that the recording rate for volcanic eruptions was lower in the past that it is today. We would like to estimate historical under-recording rates, to better interpret the historical record, and, by extension, to better predict the future. Yet we have few beliefs about the quantitative features of the under- recording process. Hence in this paper we have adopted a non-parametric approach to estimate the recording rate as a function of time, in which the global large-eruption rate is treated as constant in time, and the only two constraints on the recording rates are that the global recording rate is now one (since 1980), and that the global recording rates are non-increasing, going backwards in time. Compared to the parametric approach used previously in the volcanological literature (Sections "Review of previous approaches" and "Our statistical non-parametric approach"), our approach has the advantage of being far more general, but may not reduce our uncertainty much below its current levels. This possibility is especially acute given our focus on large (M4+) stratovolcano eruptions, which occur at a global rate of only about one every two years.

Fortunately, the dataset from the LaMEVE database is sufficiently rich that we can derive a clear message about historical under-recording (Section "Results and discussion"), including that the global recording rate drops below 100 % even in the recent past (e.g. in the 1800 s), is below 50 % before 1600, and is below 20 % before 1100 (all with 95 % confidence). These statements, which are formally about the global recording rate, apply, approximately, to the average recording rate across volcanoes (Section "Definition of the recording rate"). We also find that the sequence of global recording rates appears to combine aspects of both of the proposed parametric models, being generally exponential in shape, but also having more abrupt changes, notably around 1550. We extended our approach to very large (M5+) eruptions, but we judged our results to be unreliable.

Our results are consistent with the narrative evidence in Siebert et al. (2010, pp. 31–34) and Brown et al. (2014). Siebert et al. (2010) attribute the step-change around 1550 to the Spanish and Portuguese explorations at the end of the 15th century, which "opened Latin America and much of the western Pacific to European record-keeping" (p. 31), and also to the development of printing.

Finally the methods we have used here may also be useful in assessing under-recording of other hazards, e.g. earthquakes and floods. Like volcanoes, these infrequent hazards suffer from a sparse record of larger events, that is further thinned by low recording rates in the past. Our methods and computer code should be easily adaptable to other applications, and we would be happy to discuss this further.

Appendix: Computing a *95 %* confidence set

Putting aside π_k, which is equal to 1, the order statistics of $k-1$ independent uniform random quantities are uniformly distributed on Ω, see Cox and Hinkley (1974, Appendix 2). This result provides a (stochastic) method to generate a point uniformly in Ω: generate $k-1$ independent uniform random quantities, order them from smallest to largest, and then append 1.

There is a simple improvement on this method, if the intention is to generate a sequence of such points. Instead of using a sequence of random points, the points are generated according to a deterministic space-filling design on $[0,1]^{k-1}$, in this case a Sobol sequence. This sequence, which does not suffer from random variability, will have better coverage of any large connected subset of Ω, because it decreases the clumpiness of the points, relative to a random sample. Figure 7 illustrates this property in 2D.

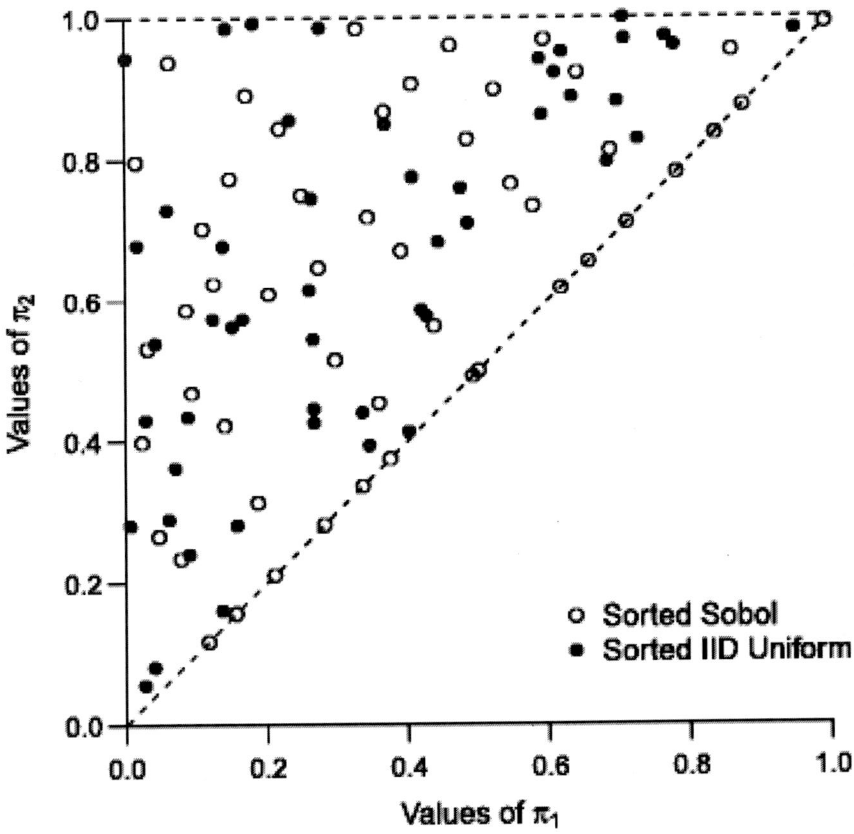

Figure . 7.2D illustration of the superior space-filling properties of a sorted Sobol sequence, compared to a sorted IID uniform sequence

The following algorithm enumerates a set of points occupying a particular level set of the log-likelihood function, using the space-filling approach described in the previous paragraph.

1. Generate an m-point Sobol sequence in $k-1$ dimensions. Order the values in each point from smallest to largest, and then append 1. This is now a deterministic m-spoint space-filling design on Ω.

2. Compute the log-likelihood for each point using (7), denoted ℓ_i for point $\boldsymbol{\pi}^{(i)}$. Let $\hat{i} = \arg\max_i \ell_i$.

3. For some value c, keep all points $\boldsymbol{\pi}^{(i)}$ for which $\ell_i \geq \ell_{\hat{i}} - c$. That is, all points for which the log-likelihood is within c of the maximum log-likelihood. These points form a space-filling design in some subset of Ω, denoted CC.

A Sobol sequence can be generated using the sobol function from the randtoolbox package in the statistical computing evironment R (R Core Team 2013).

Let χ^{-2}_d be the quantile function of the chi-squared distribution with d degrees of freedom. According to the standard asymptotics, setting

$$c \leftarrow \chi^{-2}_{k-1}(0.95)/2 \qquad (11)$$

will imply that CC is a 95 % confidence set for $\boldsymbol{\pi}$; see Cox (2006, ch. 6) or van der Vaart (1998, ch. 16). But the standard asymptotics do not hold here, because the sample size is small, and the space Ω is irregular. In a nutshell, the 95 % level set of the likelihood function is not elliptical, because it is clipped by the edges of Ω. Therefore c must be adjusted to correct for 'level error', which is where the actual coverage is not equal to the nominal coverage.

The adjusted value, say c^*, is found empirically, in a simulation experiment using synthetic datasets; see DiCiccio and Efron (1996, notably sec. 7) for bootstrap methods for approximating confidence sets. Set $\hat{\pi} := \pi^{(\hat{i})}$, and generate n independent samples $\boldsymbol{x}^{(1)}, \ldots, \boldsymbol{x}^{(n)}$ from the model with $\pi \leftarrow \hat{\pi}$. This involves, for each sample:

1. Simulate $\lambda \sim \text{Gamma}(a, b)$,

2. For $i=1,\ldots,k$, simulate $x_i \sim \text{Poisson}(\Delta_i \cdot \hat{\pi}_i \cdot \lambda)$.

For each sample, evaluate the log-likelihood for all points $\boldsymbol{\pi}^{(1)}, \ldots, \boldsymbol{\pi}^{(m)}$, denote this $\ell_{ij} := \log p(\boldsymbol{x}^{(j)} \mid \boldsymbol{\pi}^{(i)})$. Also identify the maximum log-likelihood value, $\hat{\ell}_j := \max_i \ell_{ij}$. The value c^* is chosen so that 95 % of the sample contain the 'true' value $\hat{\pi}$. In other words, c^* solves

$$n^{-1}\sum_{j=1}^{n}\mathbb{1}\{\ell_{ij} \geq \hat{\ell}_j - c^*\} = 0.95$$

where $\mathbb{1}\{\cdot\}$ is the indicator function. A simple rearrangement shows that c^* is the 95th percentile of the histogram of {ℓ^j−ℓi^j}mj=1{ℓ^j−ℓi^j}j=1m.

Denote the resulting level set based on c^* as C∗C∗. This empirical approach ensures that the coverage of C∗C∗ is very close to 95 % at π=π^π=π^, and, one hopes, approximately 95 % in the region of Ω around π^π^. Thus the resulting C∗C∗ is approximately an exact 95 % confidence set.

In our calculation we used $m \leftarrow 10^4$ and $n \leftarrow 1001$, and the 95 % confidence set cut-off at π^π^ was found to be $c^*=4.61$. By way of contrast, the asymptotic cut-off from (11) is $c=9.15$. In order to check the coverage of this cut-off at plausible values other than π^π^, we chose ten other points at random from the 95 % confidence set, and evaluated the coverage of the confidence set at these $\boldsymbol{\pi}$'s using the above values of m, n, and c^*; the coverages were 94,92,94,92,94,93,95,93,95,92 %, which is reassuring.

ACKNOWLEDGEMENTS

This work was supported by the Natural Environment Research Council (NERC) funded Consortium on Risk in the Environment: Diagnostics, Integration, Benchmarking, Learning and Elicitation (CREDIBLE); grant number NE/J017450/1. Sparks acknowledges funding from the European Research Council in the VOLDIES project which led to the development of the LaMEVE database. Cashman acknowledges support from the AXA Research Fund. We would like to thank two anomymous Reviewers who helped improve the description of our approach and how it differs from past parametric approaches.

AUTHORS' CONTRIBUTIONS

JR carried out the statistical analysis, SS and KC provided volcanological expertise. All authors read and approved the final manuscript.

COMPETING INTERESTS

The authors declare they have no competing interests.

REFERENCES

1. Brown, S, et al. Characterisation of the Quaternary eruption record: Analysis of the Large Magnitude Explosive Volcanic Eruptions (LaMEVE) database. J Appl Volcanol. 2014; 3(5). http://www.appliedvolc.com/content/3/1/5.
2. Coles, S, Sparks R. In: Mader H, Coles S, Connor C, Connor L, (eds).Extreme value methods for modelling historical series of large volcanic magnitudes. Geological Society of London: Special Publication of IAVCEI; 2006, pp. 47–56.
3. Cox, D. Principles of Statistical Inference. Oxford University Press. Oxford: UK; 2006.
4. Cox, D, Hinkley D. Statistics Theoretical. Chapman and Hall. London: UK; 1974.
5. Crosweller, H, et al. Global database on Large Magnitude Explosive Volcanic Eruptions (LaMEVE). J Appl Volcanol. 2012; 1(4). http://www. Applied volc.com/content/1/1/4.
6. De la Cruz-Reyna, S. Poisson-distributed patterns of exposive eruptive activity. Bull Volcanol. 1991; 54:57–67.
7. Deligne, N, Coles S, Sparks R. Recurrence rates of large explosive volcanic eruptions. J Geophys Res. 2010; 115(B06):203.
8. DiCiccio, T, Efron B. Bootstrap confidence intervals. Stat Sci. 1996; 11(3):189–212. with discussion and rejcinder, 212–228.
9. Furlan, C. Extreme value methods for modelling historical series of large volcanic magnitudes. Stat Modell. 2010; 10(2):113–32.
10. Guttorp, P, Thompson M. Estimating second-order properties of volcanicity from historical data. J Am Stat Assoc. 1991; 86:578–83.
11. Kiyosugi, K, Connor C, Sparks R, Crosweller H, Brown S, Siebert L, Wang T, Takarada S. How many explosive eruptions are missing from the geologic record? Analysis of the quaternary record of large magnitude explosive eruptions in Japan. J Appl Volcanol. 2015; 4(1).
12. Lunn, D, Jackson C, Best N, Thomas A, Spiegelhalter D. The BUGS Book: A Practical introduction to Bayesian Analysis. Boca Raton FL, USA: CRC Press; 2013.
13. Mason, B, Pyle D, Oppenheimer C. The size and frequency of the largest explosive eruptions on earth. Bull Volcanol. 2004; 66(8):735–48.
14. Mead, S, Magill C. Determining change points in data completeness for the Holocene eruption record. Bull Volcanol. 2014; 76(14).
15. Pickands, J. The two-dimensional Poisson process and extremal processes. J Appl Probab. 1971; 8:745–53.
16. R Core Team. R: A Language and Environment for Statistical Computing. R Foundation for Statistical Computing. Vienna, Austria, ; 2013. http://www.R-project.org/.
17. Siebert, L, Simkin T, Kimberly P. Volcanoes of the World. Berkeley and Los Angeles CA, USA: University of California Press; 2010.
18. Simkin, T. Terrestrial volcanism in space and time. Ann Rev Earth Planetary Sci. 1993; 21:427–52.
19. Solow, A. Estimating record inclusion probability. Am Stat. 1993; 47(3):206–8.

20. Solow, A. An Empirical Bayes analysis of volcanic eruptions. Math Geology. 2001; 33(1):95–102.
21. van der Vaart, A. Asymptotic Statistics. Cambridge, UK : Cambridge University Press; 1998.
22. Venables, W, Ripley B. Modern Applied Statistics with S, 4th edn. USA: Springer, New York NY; 2010.

CHAPTER 6

Forecasting Volcanic Ash Deposition using HYSPLIT

Tony Hurst[1]* and Cory Davis[2]

[1] *GNS Science, Lower Hutt, New Zealand.*

[2] *MetService, Wellington, New Zealand.*

ABSTRACT

A major source of error in forecasting where airborne volcanic ash will travel and land is the wind pattern above and around the volcano. GNS Science, in conjunction with MetService, is seeking to move its routine ash forecasts from using the ASHFALL program, which cannot allow for horizontal variations in the wind pattern, to HYSPLIT, which uses a full 4-D atmospheric model. This has required some extensions to the standard version of the HYSPLIT program, both to get appropriate source terms and to handle the fall velocities of ash particles larger than 100 microns.

Application of the modified HYSPLIT to ash from the Te Maari eruption of 6 August 2012 from Tongariro volcano gives results similar to the observed ash distribution. However, it was also apparent that the high precision of these results could be misleading in actual forecasting situations, and there needs to be ways in which the likely errors in atmospheric model winds can be incorporated into ash models, to show all the areas in which there is a significant likelihood of deposited ash with each particular volcanic eruption model.

Keywords: Volcanic hazard Volcanic ash Ashfall Atmospheric models Mt Tongariro Te Maari

INTRODUCTION

One aspect of GeoNet, the geological hazard monitoring system for New Zealand run by GNS Science, is forecasting the expected thickness distribution of volcanic ash if any volcano was to erupt. As well as the four volcanoes that have erupted in the last 50 years (Ruapehu, Ngauruhoe, Tongariro and White Island), possible ash distributions are calculated daily for six other volcanoes seen as significant risks, as shown in Table 1. There are six cases calculated for each volcano, with two column heights and three eruption volumes. If any volcano is erupting or has erupted recently, the forecasts for that volcano are distributed to relevant organisations and the public, whereas during normal times the information is kept within GNS Science to avoid causing confusion.

Table 1. The standard eruption model input parameters used in New Zealand by GNS Science, with three volumes and two column heights for each volcano

Volcano	column height (high) km ASL	column height (low) km ASL	tephra volume large km^3	tephra volume medium km^3	tephra volume small km^3	ash size distribution
Auckland	5	3	0.1	0.01	0.001	Basaltic
Mayor Island	10	3	0.5	0.05	0.005	Rhyolitic
White Island	5	3	0.01	0.001	0.0001	Andesitic
Haroharo	16	4	1	0.1	0.01	Rhyolitic
Tarawera	16	4	1	0.1	0.01	Rhyolitic
Taupo	15	3	1	0.1	0.01	Rhyolitic
Tongariro	11.5	4.5	0.1	0.01	0.001	Andesitic
Ngauruhoe	12	6	0.1	0.01	0.001	Andesitic
Ruapehu	12.5	6.5	0.1	0.01	0.001	Andesitic
Taranaki	12	6	0.5	0.05	0.005	Andesitic

These forecasts are currently produced by the program ASHFALL, originally described in Hurst (1994), which is now somewhat out-dated. ASHFALL follows the ash particles from an initial vertical eruption column as they are moved horizontally by the wind, and at the same time fall vertically with a velocity which depends on their size and density. It was designed to have a minimal processing requirement, based on the processing power of mid-1990s Personal Computers. To obtain reduced processing times, ASHFALL ignored vertical diffusion of the ash, although its effect was partly accounted for by using a large value for the horizontal diffusion coefficient. This horizontal diffusion coefficient is effectively a "fudge factor", which spreads out the expected ash distribution to incorporate uncertainties in the wind pattern and other parameters. ASHFALL uses MetService (New Zealand Meteorological Service) forecasts of the wind at various levels above each volcano. However, it does not allow for any horizontal variations in the wind pattern, i.e. ASHFALL can only use 1-D spatial wind fields, although it does allow for wind changes with time.

Hurst and Turner (1999) found that the forecast wind field was the major source of error in the volcanic ash distribution from Ruapehu eruptions. During investigations as to how the wind information used by ASHFALL might be improved, we looked at the atmospheric dispersion model HYSPLIT (Stein et al., 2015), already in use by MetService for the movement of fine particles travelling in the atmosphere, such as volcanic ash, and found it has the capability for estimating the distribution of ash particles on the ground. As well as capturing the influence of realistic atmospheric motion in ash transport, HYSPLIT can also model in-cloud (rainout) and below-cloud (washout) wet deposition processes.

HYSPLIT is a hybrid Lagrangian dispersion model, developed by NOAA/ARL which is used by MetService, in its role as a Volcanic Ash Advisory Centre (VAAC), to model airborne volcanic ash, with meteorological data provided by external and in-house NWP (Numerical Weather Prediction) models, which operate with 3 spatial dimensions and time. The spatial resolution of these models is 4 km, and the temporal resolution one hour. HYSPLIT is operationally used by several of the 9 Volcanic Ash Advisory Centres for aviation forecasting. A by-product of the HYSPLIT volcanic ash dispersion simulations is the ash deposition at the ground surface. The ground surface elevation is modelled at the same resolution as the atmospheric model. Aside from the dispersion of volcanic ash, HYSPLIT is used in several other atmospheric transport applications, including the dispersion of hazardous materials (e.g. nuclear material after the Fukushima reactor accident), air quality modelling (e.g. ozone, visibility, haze, and dioxin), dust storms, smoke, and the transport of biological material (e.g. pollen and mould spores) (Stein et al., 2015). Rather than using a horizontal dispersion coefficient, the dispersion was calculated from the friction velocity, height and boundary layer height following Kantha and Clayson (2000), where there are different equations for stable/neutral and unstable layers for both the surface and boundary layer. Above the boundary layer a mixing coefficient obtained from mixing length theory is used to calculate the velocity variances (Draxler and Hess, 1997).

We initially compared the ash deposition results from HYSPLIT with ASHFALL for similar eruptions and wind patterns. This showed that alterations to the standard fall velocity model of HYSPLIT were required to deal with ash particles larger than about 100 microns, which make up the bulk of ash deposits near a volcano. This is not a consideration in the aviation applications of HYSPLIT, as these particles are the ones that rapidly fall out of ash clouds.

Ash sizes and fall velocities

HYSPLIT was originally designed to use atmospheric models to calculate the tracks of pollutants in the atmosphere (Stein et al., 2015). In regard to volcanic ash, it is used for tracking the small ash particles with long residence times that are hazardous to aircraft generally at considerable distances from the volcano. Accordingly by default it only considers ash particles less than 100 microns in diameter (e.g. in VAFTAD, Heffter and Stunder (1993)). The terminal velocity for these small particles can be calculated by Stokes Law, which applies in the laminar flow region, with Reynolds Number less than 0.4.

The ash size distributions used for calculating ash deposit thicknesses in the work presented here were based on historic and well-preserved ash deposits, after making allowance for the fraction of very fine particles that do not fall out within the area of identifiable ash deposits. In New Zealand, we use three size distributions, based on the rock type of the volcano (Fig. 1 and Table 1). For rhyolitic volcanoes the distribution is based on the Hatepe Ash (Walker, 1981), for andesitic volcanoes it is based on the AD 79 eruption of Vesuvius (Macedonio et al., 1988), while for the basaltic volcanoes it is based on the 1973 eruption of Heimaey (Self et al., 1974). These ash size distributions are similar to those used for hazard analysis in Hurst and Smith (2010). Size-dependent densities are used to convert size to mass for each of the three ash types. The distribution shown as VAAC in Fig. 1 is derived from Heffter and Stunder (1993). The normal practice at MetService, who operate the Wellington VAAC, when using HYSPLIT for Aviation Hazard purposes, is to assume that 5% of any total eruption volume is comprised of fine ash, with this VAAC size distribution, and to ignore the 95% of coarser ash as it is falling out quickly.

Aggregation of ash particles can have significant effects on ash distribution (Carey and Sigurdsson, 1982, Mastin et al., 2016), however this is not considered in either HYSPLIT or ASHFALL, and the focus of this work is the incorporation of realistic wind fields.

The larger (>100 micron) ash particles fall at higher terminal velocities, which produce turbulent air flow, with Reynolds Numbers greater than 500, and their actual terminal velocities are much less than that given by Stokes Law (Rose, 1993). Volcanologists typically use empirical methods or a range of equations for particles of different sizes, based on the Reynolds Number (Bonadonna et al., 1998). Figure 1 also shows how the three equations used for different ash sizes by Bonadonna et al. (1998) can all be matched by an equation derived by Ganser (1993) covering all particle sizes. This Ganser equation has recently been incorporated in HYSPLIT (Dare, 2015) and we have used this in our ash modelling.

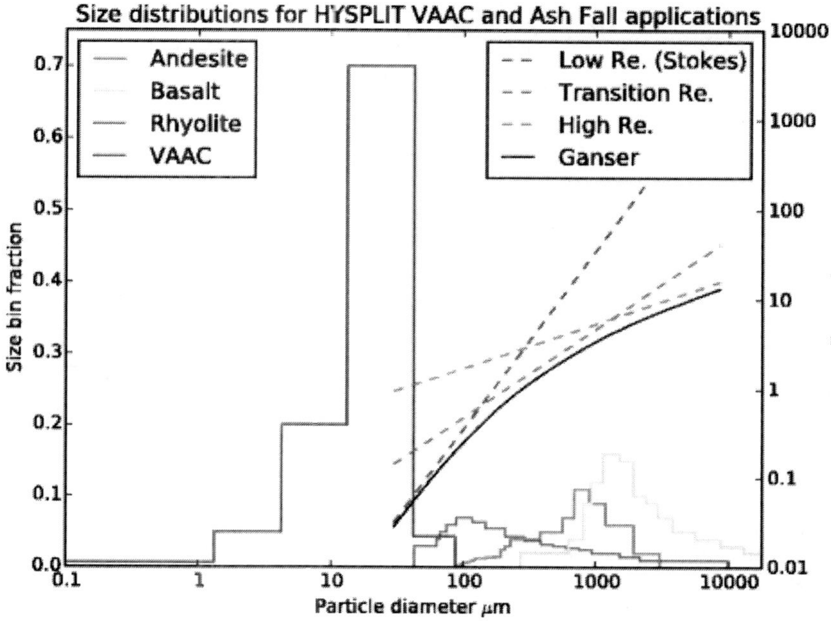

Figure 1. The three ash size distributions used by ASHFALL, for Andesitic, basaltic and Rhyolitic volcanoes, compared to the ash size distribution used for VAAC modelling of ash as an aviation hazard. On the same particle diameter scale are also shown particle fall speeds for the three equations of Bonadonna et al. (1998) and the Ganser equation

It was also necessary to provide a more accurate initial ash concentration as a function of height, with a concentration based on a typical umbrella-shaped eruption column, rather than uniform across all levels. The distribution used was that suggested by Suzuki (1983), in which the ash concentration at height h is proportional to $(1\text{-}h/H)\ e^{-A(1\text{-}h/H)}$, where H is the maximum column height, and A is the Suzuki constant. We used a constant of 4, commonly used for sub-plinian and plinian eruption columns (Pfeiffer et al, 2005), for which the distribution is plotted in Fig. 2.

Comparisons of HYSPLIT and ASHFALL showed broadly similar results on days when there were uniform wind patterns, but had large differences on days with more complex flows, as in the example of Fig. 3, for 5 August 2016. In this case, with a complex low pressure system over the central North Island, although there was a consistent north-westerly wind above Ruapehu, once ash emitted from Ruapehu had gone some distance to the east it would have encountered a completely different wind pattern, with much of it blown back towards the west.

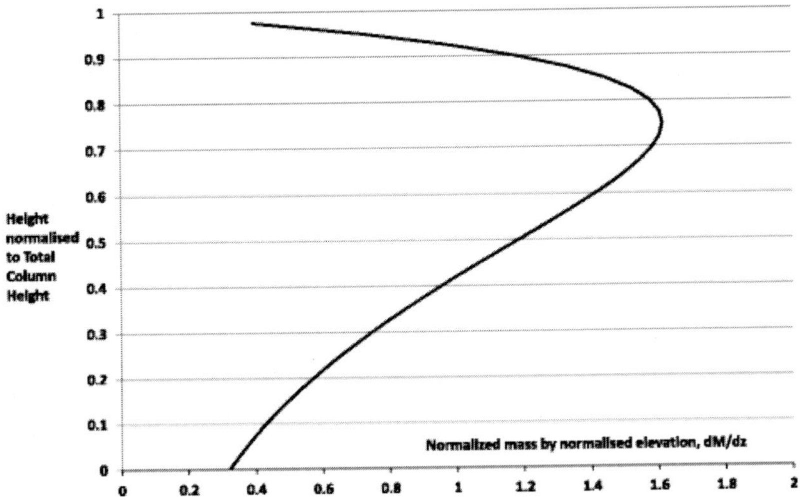

Figure 2. Suzuki distribution with constant of 4, representing how ash is concentrated near the top of an eruption plume as it slows down due to its decreasing temperature and a lower density atmosphere

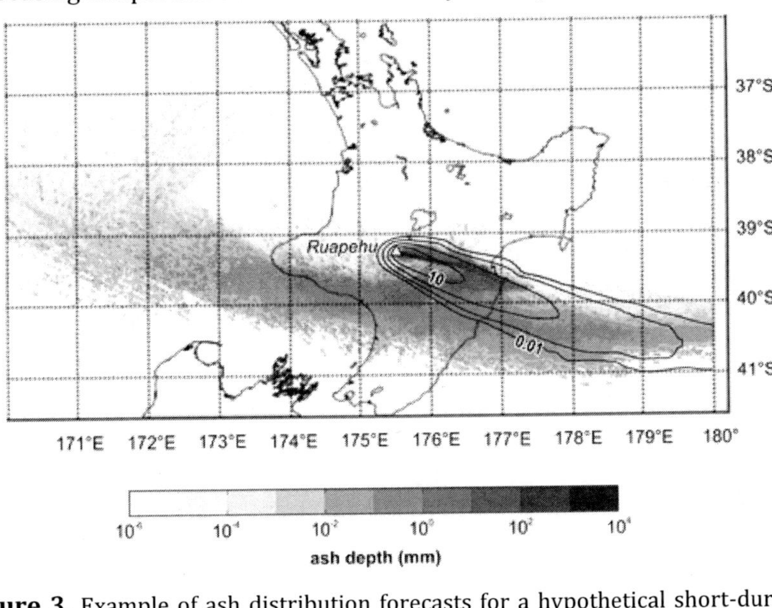

Figure 3. Example of ash distribution forecasts for a hypothetical short-duration Ruapehu eruption at 0500UT on 8 August 2016, with a 12.5 km column height and a total volume of 0.1 km^3. Colours and grey 0.01 mm contour are from HYSPLIT, black 0.01, 0.1, 1 and 10 mm contours from ASHFALL. The ASHFALL distribution is entirely based on the expected wind at Ruapehu, whereas HYSPLIT forecasts that a significant amount of ash will be deposited to the south and west under the influence of a complex wind pattern

Accordingly, we have started the implementation of routine ash forecasts from HYSPLIT, using the same ten volcanoes, three sizes of eruptions, and two eruption column heights as we use for ASHFALL.

Ash distribution comparisons for Te Maari eruption of 6 August 2012

We did not compare a HYSPLIT model with the actual volcanic ash thicknesses from the 1995 and 1996 eruptions of Ruapehu, used to evaluate the performance of ASHFALL predictions by Hurst and Turner (1999), because of the difficulty in obtaining an accurate 4 dimensional wind data for that time. However, there was a small ash eruption from the Upper Te Maari crater of Mt Tongariro on 6 August 2012, and NWP data for this date was archived by MetService, making it comparatively easy to calculate a hindcast for this event with HYSPLIT and ASHFALL. This NWP data was from the operational 8 km resolution WRF-ARW model, using European Centre for Medium Range Weather Forecasting (ECMWF) Integrated Forecasting System (IFS) data for initial and boundary conditions.

This eruption, which followed a period of low-level seismicity (Hurst et al., 2014), was the first eruption of Mt Tongariro for over 75 years (Scott and Potter, 2014). It occurred at night, commencing at 11:52 pm NZST (11:52 UTC), so there were no visual observations, but weather radar observations showed an approximate ash column height of 7800 m above sea level (the altitude of Upper Te Maari is 1485 m), and a duration of less than 15 min (Crouch et al., 2014). Ash samples were collected along roads in the vicinity of the volcano, and downwind to the east, by researchers from Massey University (Pardo et al., 2014). The ash cloud was observed in both visible and infra-red satellite images taken during the early hours of 7 August (NZST), as shown in Fig. 4.

Because relevant atmospheric model data had been archived, it was comparatively easy to run HYSPLIT with the same information for the Te Maari eruption as would be available for an eruption today. The results have been compared with the main plume that travelled eastwards. Some ash travelled north from Te Maari, but it was clear (Lube et al., 2014; Pardo et al., 2014; Turner et al., 2014) that these ash deposits were produced by low-angle directed blasts.

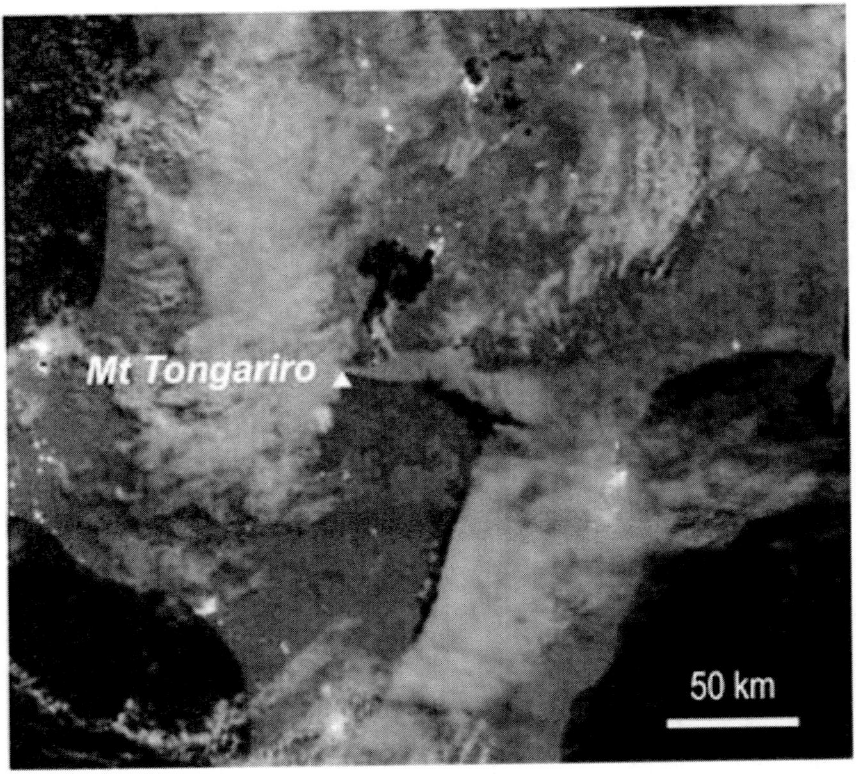

Figure 4. A moonlight image of part of the North Island, New Zealand from the NASA-NOAA Suomi satellite using the VIIRS (Visual Infrared Imaging Radiometer Suite) showing the ash cloud travelling east from the Te Maari eruption (white triangle) taken at 0055 NZST on 7th August 2012. NOAA photo

The initial result from the HYSPLIT model with a point ash source had an extremely narrow ash plume heading east. With such a narrow plume we need to take account of the horizontal extent of the umbrella portion of the initial eruption column, as the width of this will significantly affect the near-source ash distribution. This has not been necessary with ASHFALL, because of the large horizontal diffusion coefficient it used. The GNS webcam south of Ruapehu, with an infra-red sensitive camera which observed the top portion of the Te Maari eruption column (Crouch et al., 2014), suggested a diameter of the initial eruption column of approximately 1.5 km, so we modelled the initial ash distribution as a vertical cylinder with this diameter, with uniform ash concentration horizontally, and a Suzuki distribution vertically. This technique was used by Crawford et al. (2016), but on a much larger scale, when modelling the

2008 eruption of Kasatochi. It was also apparent from the radar images in Crouch et al., (2014) that the eruption column was not vertical, but directed northwards, with its top region between one and two kilometres north of the source vent. We did not try to model a sloping column, but since most of the ash is in the top region of the column, we simply shifted the source one kilometre north of the Upper Te Maari crater.

Figure 5 shows a comparison of the actual ash distribution from (Pardo et al., 2014) along the approximately north-south State Highway 1 and nearby roads with the output from both ASHFALL and HYSPLIT for a north-south profile at 175.81° E (about 12 km east of Te Maari), using a point source for ASHFALL and a 1.5 km diameter cylindrical source for HYSPLIT. Both models used a source centred 1 km north of the Upper Te Maari vent, and had their maxima close to the actual maximum ash thickness. The width of the actual ash distribution was intermediate between the two models, but generally closer to the very narrow distribution predicted by HYSPLIT. This is confirmed by the narrow initial plume seen heading east from Te Maari in Fig. 4, which can also be seen in Fig. 6.

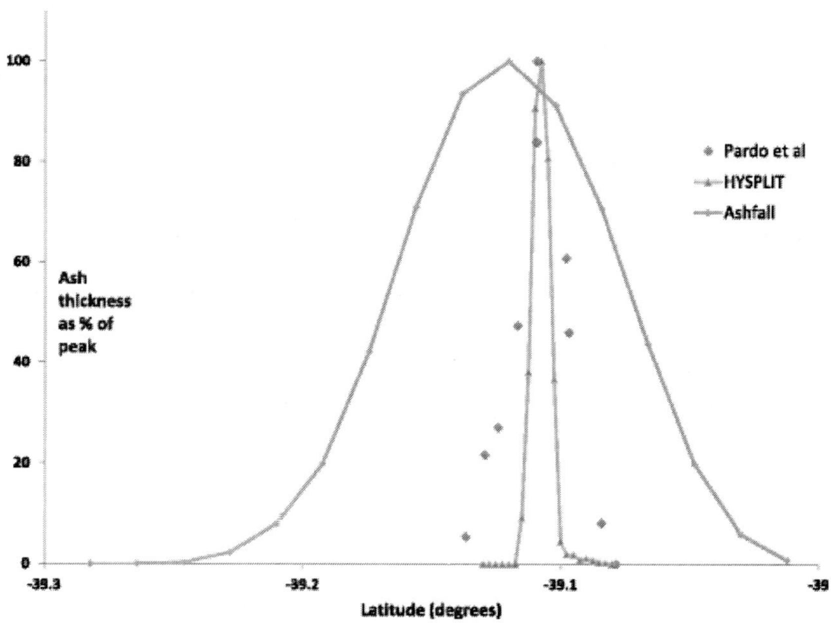

Figure 5 North-south ash distribution profile from the Te Maari eruption of Tongariro at approximately 12 km downwind at 175.81° E, showing modelled thicknesses from ASHFALL and HYSPLIT, and actual ash thicknesses from Pardo et al. (2014), with maximum thicknesses normalised to 100

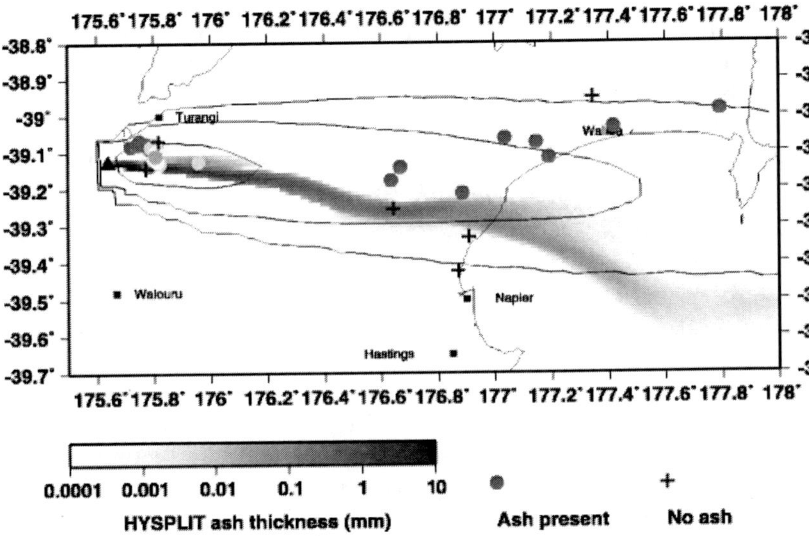

Figure 6. Ash thickness observations from Te Maari eruption (black triangle), compared to forecasts from ASHFALL and HYSPLIT. ASHFALL contours are 10 μm, 100 μm and 1 mm. Ash observations from Pardo et al. (2014), with blue for trace amounts, and other colours approximately equivalent to thicknesses of 0.25 mm (dark green), 0.5 mm (light green), 1 mm (yellow) and 2 mm (orange)

The ash distribution predicted by ASHFALL for the Te Maari eruption has already been discussed by (Turner et al., 2014), which also compared both the ground and satellite measurements of where ash particles travelled with the results of the UK Met Office aerial dispersion model NAME-III (Jones, 2004), which used the NWP model NZLAM-12. The NAME and HYSPLIT models are very similar; both being Lagrangian dispersion models employed for a range of atmospheric pollution problems. NAME-III and HYSPLIT gave fairly similar ash distributions from this eruption, and in particular both had narrow ash plumes compared to ASHFALL. We will not discuss the differences between the ash distributions from these modelling programs further, other than to note that they probably are as much due to using different atmospheric models as to differences between the programs.

The region further downwind from this eruption was mountainous and virtually unpopulated until at least 80 km from Te Maari, and by then only trace amounts of volcanic ash were present, so all that could be said was that it was present or not present. Figure 6 shows which of the small settlements from 80 to 200 km reported either the presence or absence of ash. We have compared these reports to the outer limits of the ASHFALL and HYSPLIT predictions, effectively for an ash thickness of a few microns.

In Fig. 6 it can be seen that the narrow range of ash directions predicted by HYSPLIT does not cover all the places where ash was observed, suggesting that the wind was directed slightly more to the north than predicted by the atmospheric model. The available ground truth data in the far downwind area is too limited to let us distinguish between the narrow ash distribution of the HYSPLIT model and the much wider distribution from ASHFALL. However, the narrowness of the ash distribution in the near-source region does suggest that we have a narrow plume, but directed slightly north of where the atmospheric model used for HYSPLIT predicted it to be.

DISCUSSION

The ash predictions produced by HYSPLIT from a 4-D atmospheric model use much more information about the atmospheric wind pattern, and so are potentially much more accurate than the ASHFALL model that is currently used which assumes a horizontally uniform wind profile. For the 6 August 2012 eruption from the Upper Te Maari crater of Tongariro, the HYSPLIT forecast using an archived atmospheric model agreed well with the actual ash distribution recorded about 12 km downwind. There was not such good agreement between HYSPLIT and the reports of where small amounts of ash were present in areas over 80 km downwind. As found by Hurst and Turner (1999) for the 1995 and 1996 eruptions of Ruapehu, the main reason for the difference between forecast and actual ash distributions was the difference between the forecast and actual wind pattern.

HYSPLIT should provide a much better match to the ash distribution of an actual eruption than ASHFALL, because of the more detailed atmospheric model, and because it more completely accounts for ash removal processes (e.g. wet deposition). However, when using these programs for forecasting purposes, there are two factors that need to be considered. Firstly, HYSPLIT outputs have more detail, compared to the simple ellipses produced by ASHFALL, so they give a higher impression of accuracy. Secondly, without the non-physically large horizontal diffusion coefficient used in ASHFALL, HYSPLIT sometimes forecasts narrow ash distributions, as was seen for the Te Maari eruption, for which a very small change in wind direction will displace the forecast ashfall area to the extent that it no longer overlaps with the actual ashfall area some distance from the vent. This difficulty in demonstrating value in more detailed forecasts is something already encountered in NWP for normal weather forecasts, where for instance high resolution forecasts of severe convection are seen subjectively as being more accurate, but traditional verification metrics strongly penalise displacement errors, which high resolution models by nature are more prone to (e.g. Mass et al 2002). Human forecasters (e.g. Roebber et. al. 2004) and/or the use of model ensembles (e.g. Peralta et. al., 2012) or other post-processing techniques (e.g. Theis et. al., 2005)) are needed to account for uncertainty in this high resolution model output, and appropriately convey risk to end users.

To provide more useful automated quantitative ashfall forecasts from a HYSPLIT/WRF system that account for uncertainty in the NWP winds and eruption parameters, we could follow the example of high resolution NWP and use an ensemble of model runs based on multiple NWP forecasts and a range of eruption parameters.

For airborne ash the Wellington VAAC does this to some extent by running a small HYSPLIT ensemble, with a range of eruption parameters, and with meteorological data from 3 NWP models. Dare et al., (2016) examined the impact of atmospheric uncertainty with a 24 member NWP model ensemble, again for airborne ash. Large ensemble systems are not yet used operationally by VAACs (WMO 2012, 2015). We believe similar ensemble based approaches that also include eruption source uncertainty are needed for ash fall forecasts.

An immediately achievable way of including uncertainty into HYSPLIT predictions would be to follow the Wellington VAAC approach and make the predictions based on multiple NWP models for each hypothetical eruption. Further investigations would be required to determine how best to present this output; options include the most conservative approach of outputting the union of ensemble ash fall areas, the contour of a specific percentile of ash depth (e.g. Draxler et al 2003), or the probability of exceedance for a specific ash depth.

By moving to the HYSPLIT/WRF we have eliminated the gross forecast errors that ASHFALL will produce in a complex wind field. The approach that the existing ASHFALL model takes to address meteorological uncertainty (an artificially large diffusion coefficient) apportions too much of this uncertainty close to the vent, and even upwind, whereas NWP wind errors should grow downwind from the vent.

It should also be noted that as well as the uncertainties of the wind pattern for ash forecasting, there will be large uncertainties in the eruption parameters. We cannot predict what eruption (i.e. magnitude, plume height) will next take place from any particular volcano, and when an eruption starts we often have very limited information.

CONCLUSIONS

We have successfully set up the use of HYSPLIT to give forecasts of volcanic ash deposition, were a New Zealand volcano to erupt. These forecasts use a much more detailed atmospheric model than the forecasts produced by ASHFALL, in particular the wind pattern varies in 3 spatial dimensions, rather than one. This gives potentially more accurate forecasts, particularly in situations with complex wind patterns.

Comparison of HYSPLIT hindcasts of the ash deposition from the 8 August 2012 Te Maari eruption from Mt Tongariro showed that HYSPLIT would have accurately forecast the narrow ash plume observed about 12 km downwind of the vent, but there was apparently a discrepancy between the forecast and actual wind direction further downwind, that significantly altered where ash was deposited. This emphasizes that we need to give an indication of what areas are at risk of volcanic ash, rather than just giving a single best estimate of the ash distribution.

FUNDING

The work leading to this paper was funded from GNS Science Core Funding, and MetService Internal Funding.

AUTHORS' CONTRIBUTIONS

TH provided all ASHFALL simulation output and performed the comparison of both models with deposition observations for the Te Maari case. CD provided all HYSPLIT output. Both authors drafted the manuscript. Both authors read and approved the final manuscript.

COMPETING INTERESTS

The authors declare that they have no competing interests.

REFERENCES

1. Bonadonna C, Ernst GGJ, Sparks RSJ. Thickness variations and volume estimates of tephra fall deposits: the importance of particle Reynolds number. J Volcanol Geotherm Res. 1998;81:173–87.
2. Carey S, Sigurdsson H. Influence of particle aggregation on deposition of distal tephra from the May 18, 1980, eruption of Mount St. Helens volcano. J Geophys Res. 1982;87(B8):7061–72.
3. Crawford A, Stunder B, Ngan F, Pavolonis MJ. Initializing HYSPLIT with satellite observations of volcanic ash: A case study of the 2008 Kasatochi eruption. J Geophys Res Atmos. 2016;121:10786–803.
4. Crouch JF, Pardo N, Miller CA. Dual polarisation C-band weather radar imagery of the 6 August 2012 Te Maari eruption, Mount Tongariro, New Zealand. J Volcanol Geotherm Res. 2014;286:415–36.
5. Dare RA. Sedimentation of volcanic ash in the HYSPLIT dispersion model. CAWCR Technical Report No. 079. Melbourne: Centre for Australian Weather and Climate Research; 2015.
6. Dare RA, Smith DH, Naughton MJ (2016) Ensemble Prediction of the Dispersion of Volcanic Ash from the 13 February 2014 Eruption of Kelut, Indonesia. J Appl Meteor Climatol
7. Draxler RR. Evaluation of an ensemble dispersion calculation. J Appl Meteor. 2003;42:308–17.
8. Draxler RR, Hess GD. DESCRIPTION OF THE HYSPLIT_4 MODELING SYSTEM, NOAA Technical Memorandum ERL ARL-224. 2010 revision. Silver Spring: NOAA Air Resources Laboratory; 1997.
9. Ganser GH. A rational approach to drag prediction of spherical and nonspherical particles. Powder Technol. 1993;77:143–52.
10. Heffter JL, Stunder BJB. Volcanic Ash Forecast Transport And Dispersion (VAFTAD) model. Wea Forecasting. 1993;8:533–41.
11. Hurst AW. ASHFALL - a computer program for estimating volcanic ash fallout : report and users guide. Wellington: Institute of Geological & Nuclear Sciences. Institute of Geological & Nuclear Sciences science report 94/23; 1994. p. 22.
12. Hurst AW, Smith WD. Volcanic ashfall in New Zealand : probabilistic hazard modelling for multiple sources. N Z J Geol Geophys. 2010;53:1–14.
13. Hurst AW, Turner R. Performance of the program ASHFALL for forecasting ashfall during the 1995 and 1996 eruptions of Ruapehu volcano. N Z J Geol Geophys. 1999;42:615–22.
14. Hurst AW, Jolly AD, Sherburn S. Precursory characteristics of the seismicity before the 6 August 2012 eruption of Tongariro volcano, North Island, New Zealand. J Volcanol Geotherm Res. 2014;286:294–302.
15. Jones A. Atmospheric dispersion modelling at the Met Office. Weather. 2004;59:311–6.
16. Kantha LH, Clayson CA. Small Scale Processes in Geophysical Fluid Flows, vol. 67. San Diego: International Geophysics Series, Academic Press; 2000. p. 883.
17. Lube G, Breard ECP, Cronin SJ, Procter JN, Brenna M, Moebis A, Pardo N, Stewart RB, Jolly A, Fournier N. Dynamics of surges generated by hydrothermal blasts during the 6 August 2012 Te Maari eruption, Mt. Tongariro, New Zealand. J Volcanol Geotherm Res. 2014;286:348–66.

18. Macedonio G, Pareschi MT, Santacroce R. A numerical simulation of the plinian fall phase of 79 A.D. eruption of Vesuvius. J Geophys Res. 1988;93(B12):14817–27.
19. Mass C, Ovens D, Westrick K, Colle B. Does Increasing Horizontal Resolution Produce More Skillful Forecasts? Bull Amer Meteor Soc. 2002;83:407–30.
20. Mastin LG, Van Eaton AR, Durant AJ. Adjusting particle-size distributions to account for aggregation in tephra-deposit model forecasts. Atmos Chem Phys. 2016;16:9399–420.
21. Pardo N, Cronin SJ, Nemeth K, Brenna M, Schipper CI, Breard E, White JDL, Procter J, Stewart B, Agustin-Flores J, Moebis A, Zernack A, Kereszturi G, Lube G, Auer A, Neall V, Wallace C. Perils in distinguishing phreatic from phreatomagmatic ash : insights into the eruption mechanisms of the 6 August 2012 Mt. Tongariro eruption, New Zealand. J Volcanol Geotherm Res. 2014;286:397–414.
22. Peralta C, Ben Bouallègue Z, Theis SE, Gebhardt C, Buchhold M. Accounting for initial condition uncertainties in COSMO-DE-EPS. J Geophys Res. 2012;117:D07108.
23. Pfeiffer T, Costa A, Macedonio G. A model for the numericalsimulation of tephra fall deposits. J Volcanol Geotherm Res. 2005;140:273–94.
24. Roebber P, Schultz D, Colle B, Stensrud D. Toward Improved Prediction: High-Resolution and Ensemble Modeling Systems in Operations. Wea Forecasting. 2004;19:936–49.
25. Rose WI. Comment on `Another look at the calculation of fallout tephra volumes' by Judy Fierstein and Manuel Nathenson. Bull Volcanol. 1993;55:372–4.
26. Scott BJ, Potter SH. Aspects of historical eruptive activity and volcanic unrest at Mt. Tongariro, New Zealand : 1846–2013. J Volcanol Geotherm Res. 2014;286:263–76.
27. Self S, Sparks RSJ, Booth B, Walker GPL. The 1973 Heimaey Strombolian Scoria deposit, Iceland. Geol Mag. 1974;111:539–48.
28. Stein AF, Draxler RR, Rolph GD, Stunder BJB, Cohen MD, Ngan F. NOAA's HYSPLIT atmospheric transport and dispersion modeling system. Bull Amer Meteor Soc. 2015;96:2059–77.
29. Suzuki T. A theoretical model for distribution of tephra. In: Shimozuru D, Yokohama I, editors. Arc volcanism: physics and tectonics. Tokyo: Terra Scientific Publishing; 1983. p. 95–113.
30. Theis SE, Hense A, Damrath U. Probabilistic precipitation forecasts from a deterministic model: a pragmatic approach. Met Apps. 2005;12:257–68.
31. Turner R, Moore S, Pardo N, Kereszturi G, Uddstrom M, Hurst AW, Cronin S. The use of Numerical Weather Prediction and a Lagrangian transport (NAME-III) and dispersion (ASHFALL) models to explain patterns of observed ash deposition and dispersion following the August 2012 Te Maari, New Zealand eruption. J Volcanol Geotherm Res. 2014;286:437–51.
32. Walker GPL. The Waimiha and Hatepe plinian deposits from the rhyolitic Taupo Volcanic Centre. N Z J Geol Geophys. 1981;24:305–24.
33. World Meteorological Organization. VAAC "Inputs and Outputs" (Ins and Outs) Dispersion Modelling Workshop: Final Report. Washington DC; 2012.
34. World Meteorological Organization. 7th International Workshop on Volcanic Ash, Workshop report. Alaska; 2015.

CHAPTER 7

Probabilistic Hazard Modelling of Rain-Triggered Lahars

Stuart R. Mead[1,2,3]* and Christina R. Magill[1]

[1] *Risk Frontiers, Department of Environmental Science, Macquarie University, Sydney, Australia.*

[2] *Commonwealth Scientific and Industrial Research Organisation, Melbourne, Australia.*

[3] *Volcanic Risk Solutions, Institute of Agriculture and Environment, Massey University, Palmerston North, New Zealand.*

ABSTRACT

Probabilistic quantification of lahar hazard is an important component of lahar risk assessment and mitigation. Here we propose a new approach to probabilistic lahar hazard assessment through coupling a lahar susceptibility model with a shallow-layer lahar flow model. Initial lahar volumes and their probabilities are quantified using the lahar susceptibility model which establishes a relationship between the volume of mobilised sediment and exceedance probabilities from rainfall intensity-frequency-duration curves. Rainfall-triggered lahar hazard zones can then be delineated probabilistically by using the mobilised volumes as an input into lahar flow models. While the applicability of this model is limited to rain-triggered lahars, this approach is able to reduce the reliance on historic and empirical estimates of lahar hazard and creates an opportunity for the generation of purely quantitative probabilistic lahar hazard maps. The new approach is demonstrated through the generation of probabilistic hazard maps for lahars originating from the Mangatoetoenui Glacier, Ruapehu volcano, New Zealand.

Keywords: Probabilistic lahar hazard Susceptibility Rain-triggered lahar Numerical modelling

Lahars are among the most hazardous volcanic processes and are responsible for a large proportion of volcanic fatalities (Auker et al. 2013). Reliable information on the likelihood of lahar occurrence and the resulting inundation area is critical for the mitigation of risks posed by lahars (Pierson et al. 2014). Lahar hazard assessments typically provide this information in the form of hazard maps quantifying the probability and extent of potential lahars to varying degrees (Calder et al. 2015). In these assessments, models that approximate lahar behaviour and/or run-out are frequently used (e.g. Aguilera et al. 2004; Carranza and Castro 2006; Darnell et al. 2013; Thouret et al. 2013; Pistolesi et al. 2014; Córdoba et al. 2015). Commonly used models include Laharz (Iverson et al. 1998; Schilling 2014), an empirical model relating lahar volume to cross-sectional and planimetric inundation area; the two-phase shallow layer model of Pitman and Le (2005) implemented in the Titan2D toolkit (Pitman et al. 2003; Patra et al. 2005); and single phase rheology approaches such as Flo-2D and Delft3D, used in Caballero et al. (2006) and Carrivick et al. (2009, 2010). Regardless of methodology, the accuracy and output of these methods is dependent on the value and accuracy of model inputs. The initial lahar volume or volumetric flux is one such input that determines initial lahar size and gravitational potential energy. Since lahar size and energy transfer are important factors controlling lahar behaviour and runout (Lube et al. 2012), an accurate estimate of initial volume is therefore crucial for accurate and reliable hazard footprint estimates. Commonly, initial lahar volumes are estimated from volumes of previous lahar events or rely on expert judgement. However, historic data is often incomplete or can be irrelevant under different environmental conditions which limits probabilistic lahar hazard estimates.

Recently, Mead et al. (2016) presented a physically based model for determining initial volumes of rain-triggered lahars. Lahar susceptibility, defined as the probability of an initial lahar volume at a specific location, was determined through assigning annual probabilities to lahar volumes using rainfall intensity-frequency-duration (IFD) curves. The potential of coupling lahar susceptibility model outputs to lahar flow model inputs in order to quantify lahar hazard probabilistically was discussed, but not demonstrated. Here, the work of Mead et al. (2016) is extended by coupling lahar susceptibility outputs to the two-phase shallow layer model of Pitman and Le (2005). Outputs of the combined model in the form of probabilistic hazard maps are demonstrated for lahar scenarios originating from

Mangatoetoenui Glacier, Mt. Ruapehu, New Zealand. This work is intended to demonstrate an approach for probabilistic estimates of lahar hazard and highlight areas of research needed to enhance the reliability of these hazard estimates.

METHODS

The probabilistic lahar hazard approach proposed here relies on the combination of two modelling methods. Lahars are simulated as a two-phase flow using Titan2D with initial volumes determined using the lahar susceptibility model of Mead et al. (2016). Details of each method are summarised below; for full details readers are referred to Patra et al. (2005) and Pitman and Le (2005) for the Titan2D toolkit and Mead et al. (2016) for the rain-triggered lahar susceptibility model.

Rain-triggered lahar susceptibility

In the rain-triggered lahar susceptibility approach described in Mead et al. (2016), lahar volumes are estimated from deposit properties, rainfall intensity and rainfall duration using a combined shallow landslide and overland erosion model. The total initial volume of the lahar is calculated in the combined model as the sum of the material mobilised through overland erosion and shallow landsliding for a given rainfall intensity and duration. In the overland erosion model, the height and motion of infiltration excess rainfall (i.e. net rainfall after infiltration into the deposit) is simulated using a depth averaged shallow water (SW) approximation. The entrainment of volcanic sediment is calculated from the height and velocity of the overland flow using the Meyer-Peter and Müller (MPM) bedload transport model (see Castro Díaz et al. (2009)), and the total volume of mobilised volcanic sediment is calculated at the end of the rainfall duration. The volume of volcanic material mobilised through shallow landsliding is calculated using the approach developed by Iverson (2000) where shallow slope failures are assumed to occur when gravitational forces are greater or equal to the resisting Coulomb stresses. In the Iverson (2000) model, resisting stresses are reduced in proportion to the rainfall infiltration rate (i.e. deposit permeability) and hydraulic diffusivity, which controls the transmission of pore pressure through the deposit. The depth, and consequently volume, of the shallow failure is calculated at the end of the rainfall duration. The probability of specific lahar volumes can be estimated using this method through the use of rainfall intensity-frequency-duration (IFD) tables. These tables express the probability of rainfall intensities occurring over a given duration and can be used to estimate the occurrence probability of specific lahar volumes being triggered by rainfall. These lahar volumes, when used as an input in Titan2D, can then be used to express lahar inundation in a probabilistic manner. Here, the mobilised volume estimates and recurrence intervals calculated in Mead et al. (2016) are used as the probabilistic input for the lahar flow model.

Lahar flow modelling

The choice of lahar flow model can have a large effect on the quality and reliability of probabilistic hazard estimates and therefore needs to be considered in conjunction with the needs of the hazard assessment. The prediction of lahar flow is difficult due to complex physical processes such as entrainment and deposition of sediment, changing solid concentrations and flow transformations (e.g. Doyle et al. 2009). While empirical models such as Laharz can be used to estimate lahar inundation areas (Schilling 2014), numerical models are able to provide more information (e.g. velocities and pressures) useful for estimating the intensity of the hazard (McDougall 2016). A range of numerical models have been used or proposed (e.g. Pudasaini 2012) for lahar flow modelling; however, the selection of numerical models and their parameters is complicated by the lack of universal constitutive laws governing lahar flow behaviour (McDougall 2016).

While input parameters for simple models that rely on bulk measures of the lahar (e.g. internal and basal friction, solid concentration) can be calibrated to previous lahar flows, in practice, the complexity of lahar events mean the calibration can only be optimised on one characteristic of the lahar to the detriment of others. For example, a reduction of friction values to match long runout lahars will result in excessively mobile lahars with high momentum in proximal (near source) zones. Recent, more advanced models of two-phase flow (e.g. Pudasaini 2012; Iverson and George 2014; Iverson and George 2016; Mergili et al. 2017) may improve predictions of lahar flow. However, these models rely on a large array of input parameters that are often not known or measurable a priori and, as yet, do not demonstrate the ability to model transformations in flow behaviour.

Probabilistic lahar hazard estimates are obtained through the use of volume and location outputs from the lahar susceptibility method as the input for lahar flow models. Consequently, the flow modelling approach can be chosen according to an assessment of modelling capabilities needed to accurately simulate the lahar hazard. To demonstrate this approach, we chose to use the two-phase material model of Pitman and Le (2005) to model lahar runout and dynamics. This model, implemented in the Titan2D toolkit (Pitman et al. 2003; Patra et al. 2005), simulates the flow of mixtures of soil, rock and interstitial fluid (water) over natural terrains. The main inputs required for this method is a digital elevation model (DEM), initial height and area of the mixture (pile height) and basal and internal friction values. The Pitman and Le (2005) model has previously been applied to simulate lahars (e.g. Williams et al. 2008; Procter et al. 2010; Córdoba et al. 2015). Crucially, this model does not consider the effects of erosion, deposition or transformation of the lahar and therefore will not reproduce all features of a lahar. Consequently, simulations using this method will not exactly match the outcomes of actual lahars, but will provide a probabilistic guide to the lahar hazard. In order to couple the susceptibility model material volumes to the inputs in the Titan2D toolkit, the Titan2D code was modified to add support for user defined pile heights in the form of a geographical information system (GIS) raster file. The modified source code is provided in the Zenodo repository https://doi.org/ 10.5281 /zenodo.153993.

STUDY AREA AND SIMULATION INPUTS

The process outlined in the previous section is demonstrated through the generation of probabilistic hazard maps for lahars originating from the Mangatoetoenui Glacier, Ruapehu volcano, New Zealand. The Mangatoetoenui Glacier and Stream are located on the north-eastern flank of Ruapehu volcano, draining eastwards into Tongariro River (Fig. 1). A hydro-electric dam, trout fisheries and several towns located along Tongariro River are at risk from lahars originating on the north-eastern flanks of the volcano (Cronin et al. 1997). In this study, deposit characteristics are chosen to create a scenario with similar conditions to those in Mangatoetoenui Stream prior to the October 28, 1995 lahar described in Hodgson and Manville (1999) and used in lahar susceptibility estimations by Mead et al (2016). The initial conditions in this study only consider a limited range of initial volumes and solid volume fractions, with all other inputs fixed, which means the generated hazard maps only represent a subset of the possible parameter space. A complete hazard assessment would consider the range of other inputs (such as basal and internal friction), their uncertainty and model suitability (Calder et al. 2015).

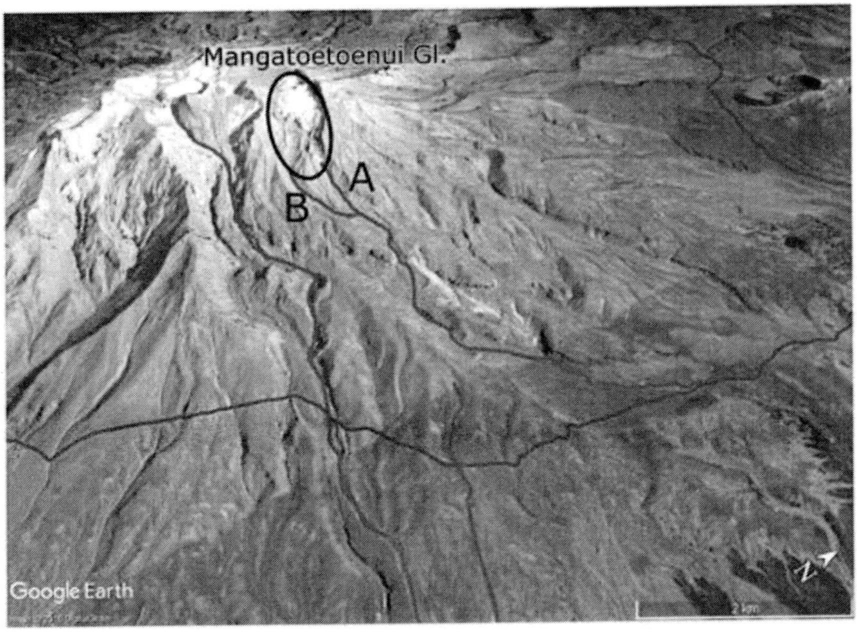

Figure 1. Study area overview showing Ruapehu volcano (left), round the mountain track (purple), Mangatoetoenui Glacier (circled) and the North and South branches of Mangatoetoenui Stream ('**A**' and '**B**' respectively)

Lahar simulations for were run for 30 min of simulation time using a 25 m resolution digital elevation model (DEM) sourced from Landcare Research NZ (2010). The DEM spanned the region from Mangatoetoenui Glacier (see Fig. 1) in the west to the confluence of Mangatoetoenui Stream and Tongariro River in the east, and was chosen because it best represented hydrological features of the terrain, despite the lower resolution compared to the contour-derived Land Information New Zealand DEM (Stevens et al. 2003).

Mobilised material volumes in the source study area were calculated for rainfall durations of 30 min, 2 h, 6 h, 12 h and 24 h at annual exceedance probabilities (AEP) of 0.5 (2 years annual recurrence interval (ARI)), 0.1 (10 years ARI), 0.02 (50 years ARI) and 0.01 (100 years ARI), using the susceptibility simulations first presented in Mead et al. (2016). We assumed most material is mobilised in a single event to match the observations of Hodgson and Manville (1999). GIS raster files of mobilised material depths from the susceptibility simulations were used as the pile height input in Titan2D. The internal friction angle of the material was given as 32°, which is between the minimum and maximum material angles of repose (Procter et al. 2010), basal friction was set to 20° to represent a reasonably fluid granular material.

In our simulations, the solids volume fraction of each modelled lahar was computed from the volumes of mobilised material and rainfall. The solid volume fraction was estimated as the volume fraction of solids in the total volume of mobilised material and water. Volume fractions, shown for each scenario in Table 1, decrease as ARI increases since higher rainfall intensities result in proportionally greater volumes of rainfall relative to the amount of mobilised material.

Table 1. Solid volume fractions of mobilised material for each scenario at Mangatoetoenui Stream

ARI (y)	AEP	Duration				
		30 min	2 h	6 h	12 h	24 h
2	0.5	0.67	0.67	0.70	0.64	0.53
10	0.1	0.61	0.60	0.62	0.56	0.46
50	0.02	0.55	0.52	0.54	0.49	0.39
100	0.01	0.60	0.50	0.51	0.46	0.36

Generation of probabilistic lahar hazard maps

Lahar hazard maps derived from overall annual exceedance probabilities are shown in Fig. 2 for lahar height thresholds of 0.5 and 0.1 m. While lahar properties such as velocity and sediment concentration are also important to consider when evaluating overall lahar hazard, height is used here to express the hazard due to its ease in interpretation and use in delineation of hazard zones. Overall exceedance probabilities were calculated from simulated maximum lahar height at each grid cell using the complementary function

$$1 - p_{(h \ngtr x)} \tag{1}$$

where $p_{(h \ngtr x)}$ is the probability of observing no lahar heights (h) greater than the threshold height x, defined using the formula of Tonini et al. (2016) as

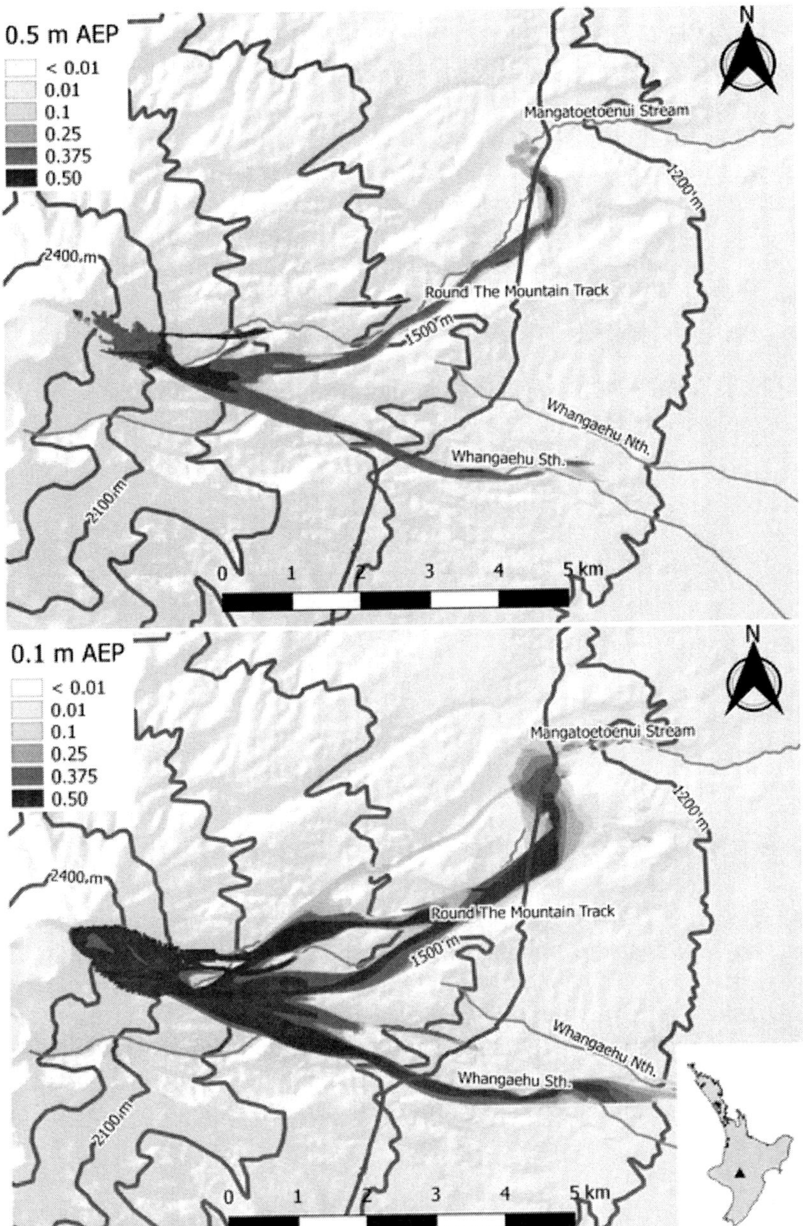

Figure 2. Annual exceedance probabilities (AEP) for lahars exceeding 0.5 m and 0.1 m (bottom) in height

$$p_{(h>x)} = \prod_i \left(1 - p^i(h > x)\right) \tag{2}$$

where $p^i(h > x)$ is the probability of lahar heights exceeding the threshold for each AEP value (in this instance i = 0.5, 0.1, 0.02, 0.01)

$$p^i(h > x) = i \cdot p(h > x) \tag{3}$$

and $p(h > x)$ is determined from the cumulative distribution of simulated lahar heights for each AEP.

DISCUSSION AND LIMITATIONS

The deposit characteristics for this demonstration were chosen to create a scenario similar to the tephra deposit in the Mangatoetoenui Catchment prior to the October 28, 1995 lahars. A rainfall event (23 mm total in 24 h, 9 mm in one 6-h interval) triggered lahars in multiple catchments of Ruapehu (Manville et al. 2000), including the Mangatoetoenui catchment. The Mangatoetoenui lahar, described in Hodgson and Manville (1999), travelled downstream as a debris flow for the first 5 km (the proximal zone) and then progressively transformed into a hyperconcentrated flow between 5 and 9 km from the source due to the entrainment of streamflow and deposition of sediment. The lahar continued to dilute downstream, eventually reaching the Tongariro River. The lahar was confined to the Mangatoetoenui catchment, although a small deposit was observed in a distributary valley of the Whangaehu catchment (Hodgson and Manville 1999).

The similarities in initial conditions between the October 28 lahar and simulations mean that the hazard maps in Fig. 2 should show a reasonable degree of similarity between the hazard outlines and observations of the lahar. However, these hazard maps indicate that a large proportion of the lahar source material enters the Whangaehu catchment (shown in Fig. 2). Lahar heights also appear to fall below 0.1 m downstream of the walking track encircling Ruapehu (~8 km from source, purple line in Fig. 2). This differs from the extent of lahar deposits identified in Hodgson and Manville (1999), but is close to the distance (9 km) where flow fully transformed from a debris flow to hyperconcentrated flow. Differences in the DEM representation and initial deposit characteristics contribute to these observed errors, but the main source of error is the lack of representation of dilution, entrainment and flow transformation processes in current numerical models. The friction parameters used here provide a runout prediction near to the transition to hyperconcentrated flow, but also represent a highly mobile initial mass of material that causes a large proportion of the lahar volume to enter the Whangaehu catchment. This highlights the current state of lahar models as a key limitation of the proposed methodology and shows that operational hazard maps generated using this approach would still require expert guidance.

While differences in AEPs are visible in the vicinity of the walking track, there is little difference in lahar extent between AEPs in upstream portions of the Mangatoetoenui catchment (Fig. 2). This lack of difference can be attributed to the deeply incised channel walls present in upper reaches of Mangatoetoenui stream, but can also be caused by similar initial volumes of material being mobilised for all rainfall ARIs (see Mead et al. (2016) for explanation). However, the level of detail is limited in this demonstration as only 5 rainfall duration scenarios are simulated for each rainfall ARI, affecting the validity and resolution of the cumulative distribution used to determine event AEP. A complete hazard mapping exercise would need to sample the entire spectrum of rainfall durations in order to increase the resolution of the cumulative distribution.

CONCLUSION

The probabilistic lahar hazard maps generated here by combining numerical modelling with initial volumes determined through the susceptibility approach of Mead et al. (2016) have demonstrated a potential methodology for probabilistic hazard mapping of lahars. However, through this demonstration, some key limitations and simplifications have been identified that could affect the feasibility of this technique. Importantly, lahar numerical modelling approaches require further research to develop advanced methods capable of representing entrainment, deposition and flow transformations typical of lahars. Recent and ongoing research in this area (e.g. Pudasaini 2012; Pudasaini and Krautblatter 2014; Iverson and George 2016) may, in the future, provide alternative modelling approaches capable of accurately predicting additional features of lahar flows. Another limitation seen in this demonstration is the reduced parameter space used to generate the hazard maps. A complete lahar hazard assessment would need to quantify the range and uncertainty of all inputs into the model (e.g. basal and internal frictions) in addition to simulating a larger number of rainfall durations. However, simulations spanning the entire range of input parameters increases the computational requirements of this approach, which could impact on the feasibility of the method for lahar hazard assessment.

ACKNOWLEDGEMENTS

Stuart Mead is supported by an Australian Postgraduate Award (APA). Computational resources for this study were provided by the Commonwealth Scientific and Industrial Research Organisation (CSIRO).

Digital elevation data is reproduced with the permission of Landcare Research New Zealand Limited.

AVAILABILITY OF DATA AND MATERIALS

The modified Titan2D source code used in this article is available in the Zenodo repository https://doi.org/10.5281/zenodo.153993. Any future versions and updates will be released at https://github.com /stuartmead /titan2d.

AUTHORS' CONTRIBUTION

SM modified Titan2D code, analysed simulation results and wrote the manuscript. CM assisted in development of probabilistic hazard maps, reviewed and edited the manuscript. All authors read and approved the final manuscript.

COMPETING INTERESTS

The authors declare that they have no competing interests.

REFERENCES

1. Aguilera E, Pareschi MT, Rosi M, Zanchetta G. Risk from Lahars in the Northern Valleys of Cotopaxi Volcano (Ecuador). Nat Hazards. 2004;33(2):161–89.

2. Auker M, Sparks R, Siebert L, Crosweller H, Ewert J. A statistical analysis of the global historical volcanic fatalities record. J Appl Volcanol. 2013;2(1):1–24.
3. Caballero L, Macías JL, García-Palomo A, Saucedo GR, Borselli L, Sarocchi D, et al. The September 8–9, 1998 Rain-Triggered Flood Events at Motozintla, Chiapas, Mexico. Nat Hazards. 2006;39(1):103–26.
4. Calder E, Wagner K, Ogburn SE. Volcanic hazard maps. In: Loughlin SC, Sparks RSJ, Brown SK, Jenkins SF, Vye-Brown C, editors. Global Volcanic Hazards and Risk: Cambridge University Press; 2015.
5. Carranza E, Castro O. Predicting Lahar-Inundation Zones: Case Study in West Mount Pinatubo, Philippines. Nat Hazards. 2006;37(3):331–72.
6. Carrivick JL, Manville V, Cronin SJ. A fluid dynamics approach to modelling the 18th March 2007 lahar at Mt. Ruapehu, New Zealand. Bull Volcanol. 2009;71(2):153–69.
7. Carrivick JL, Manville V, Graettinger A, Cronin SJ. Coupled fluid dynamics-sediment transport modelling of a Crater Lake break-out lahar: Mt. Ruapehu, New Zealand. J Hydrol. 2010;388(3–4):399–413.
8. Castro Díaz MJ, Fernández-Nieto ED, Ferreiro AM, Parés C. Two-dimensional sediment transport models in shallow water equations. A second order finite volume approach on unstructured meshes. Comput Methods Appl Mech Eng. 2009;198(33–36):2520–38.
9. Córdoba G, Villarosa G, Sheridan MF, Viramonte JG, Beigt D, Salmuni G. Secondary lahar hazard assessment for Villa la Angostura, Argentina, using Two-Phase-Titan modelling code during 2011 Cordón Caulle eruption. Nat Hazards Earth Syst Sci. 2015;15(4):757–66.
10. Cronin SJ, Neall VE, Palmer AS. Lahar history and hazard of the Tongariro River, northeastern Tongariro Volcanic Centre, New Zealand. N Z J Geol Geophys. 1997;40(3):383–93.
11. Darnell AR, Phillips JC, Barclay J, Herd RA, Lovett AA, Cole PD. Developing a simplified geographical information system approach to dilute lahar modelling for rapid hazard assessment. Bull Volcanol. 2013;75(4):1–16.
12. Doyle E, Cronin S, Cole S, Thouret J The challenges of incorporating temporal and spatial changes into numerical models of lahars. In: Proceedings of the 18th world international congress on modelling and simulation, modelling and simulation society of Australia and New Zealand, Cairns, 2009. pp 13–17.
13. Hodgson KA, Manville VR. Sedimentology and flow behavior of a rain-triggered lahar, Mangatoetoenui Stream, Ruapehu volcano, New Zealand. Geol Soc Am Bull. 1999;111(5):743–54.
14. Iverson RM. Landslide triggering by rain infiltration. Water Resour Res. 2000;36(7):1897–910.
15. Iverson RM, George DL (2014) A depth-averaged debris-flow model that includes the effects of evolving dilatancy. I. Physical basis, vol 470. vol 2170.
16. Iverson RM, George DL. Modelling landslide liquefaction, mobility bifurcation and the dynamics of the 2014 Oso disaster. Géotechnique. 2016;66(3):175–87.
17. Iverson RM, Schilling SP, Vallance JW. Objective delineation of lahar-inundation hazard zones. Geol Soc Am Bull. 1998;110(8):972–84.
18. Landcare Research NZL. NZDEM North Island 25 metre: Landcare Research NZ Ltd.; 2010.
19. Lube G, Cronin SJ, Manville V, Procter JN, Cole SE, Freundt A. Energy growth in laharic mass flows. Geology. 2012;40(5):475–8.
20. Manville V, Hodgson KA, Houghton BF, Keys JR, White JDL. Tephra, snow and water: complex sedimentary responses at an active snow-capped stratovolcano, Ruapehu, New Zealand. Bull Volcanol. 2000;62(4–5):278–93.
21. McDougall S. Landslide Runout Analysis - Current Practice and Challenges. Can Geotech J. 2016;
22. Mead S, Magill C, Hilton J. Rain-triggered lahar susceptibility using a shallow landslide and surface erosion model. Geomorphology. 2016;273:168–77.

23. Mergili M, Fischer JT, Krenn J, Pudasaini SP. r.avaflow v1, an advanced open-source computational framework for the propagation and interaction of two-phase mass flows. Geosci Model Dev. 2017;10(2):553–69.
24. Patra AK, Bauer AC, Nichita CC, Pitman EB, Sheridan MF, Bursik M, et al. Parallel adaptive numerical simulation of dry avalanches over natural terrain. Journal of Volcanology and Geothermal Research. 2005;139(1–2):1–21.
25. Pierson TC, Wood N, Driedger C. Reducing risk from lahar hazards: concepts, case studies, and roles for scientists. J Appl Volcanol. 2014;3(1):16.
26. Pistolesi M, Cioni R, Rosi M, Aguilera E (2014) Lahar hazard assessment in the southern drainage system of Cotopaxi volcano, Ecuador: Results from multiscale lahar simulations. Geomorphology 207 (0):51-63.
27. Pitman EB, Le L (2005) A two-fluid model for avalanche and debris flows, vol 363. vol 1832.
28. Pitman EB, Nichita CC, Patra A, Bauer A, Sheridan M, Bursik M. Computing granular avalanches and landslides. Phys Fluids. 2003;15(12):3638–46.
29. Procter JN, Cronin SJ, Fuller IC, Sheridan M, Neall VE, Keys H (2010) Lahar hazard assessment using Titan2D for an alluvial fan with rapidly changing geomorphology: Whangaehu River, Mt Ruapehu Geomorphology 116 (1–2):162-174.
30. Pudasaini SP (2012) A general two-phase debris flow model. Journal of Geophysical Research: Earth Surface 117 (F3):n/a-n/a.
31. Pudasaini SP, Krautblatter M. A two-phase mechanical model for rock-ice avalanches. Journal of Geophysical Research: Earth Surface. 2014;119(10):2272–90.
32. Schilling SP (2014) Laharz_py: GIS tools for automated mapping of lahar inundation hazard zones. US Geological Survey,
33. Stevens NF, Manville V, Heron DW. The sensitivity of a volcanic flow model to digital elevation model accuracy: experiments with digitised map contours and interferometric SAR at Ruapehu and Taranaki volcanoes, New Zealand. J Volcanol Geotherm Res. 2003;119(1–4):89–105.
34. Thouret J-C, Enjolras G, Martelli K, Santoni O, Luque J, Nagata M, et al. Combining criteria for delineating lahar-and flash-flood-prone hazard and risk zones for the city of Arequipa, Peru. Nat Hazards Earth Syst Sci. 2013;13:339–60.
35. Tonini R, Sandri L, Rouwet D, Caudron C, Marzocchi W, Suparjan (2016) A new Bayesian Event Tree tool to track and quantify volcanic unrest and its application to Kawah Ijen volcano. Geochem Geophys Geosyst 17 (7):2539-2555
36. Williams R, Stinton AJ, Sheridan MF. Evaluation of the Titan2D two-phase flow model using an actual event: Case study of the 2005 Vazcún Valley Lahar. J Volcanol Geotherm Res. 2008;177(4):760–6.

CHAPTER 8

Volcanic Natural Resources and Volcanic Landscape Protection: An Overview

Jiaqi Liu[1], Xiaoyu Chen[1] and Wenfeng Guo[1]

[1] Institute of Geology and Geophysics, Chinese Academy of Sciences, Beijing, China

1. INTRODUCTION

Many land resources are formed by volcanoes. In the vast oceans, there are numerous sporadic islands, big and small. Take Kosrae and Azores Archipelago for example. Kosrae (5°9′N, 163°00′E), Federated States of Micronesia, is a small (112 km^2) volcanic island in the west-central Pacific Ocean [1]. Compared with Kosrae, Azores Archipelago (36°55′~39°43′N, 25°01′~31°07′W) is a much larger volcanic island. It consists of nine volcanic islands and covers 2,247 km^2 [2].

These volcanic islands not only offer space for humans to live on, they have also become courier stations for shipping and communication. They were particularly more important in ancient times when seamanship was not well developed. Some typical lands which have been created by volcanic activities will be introduced below.

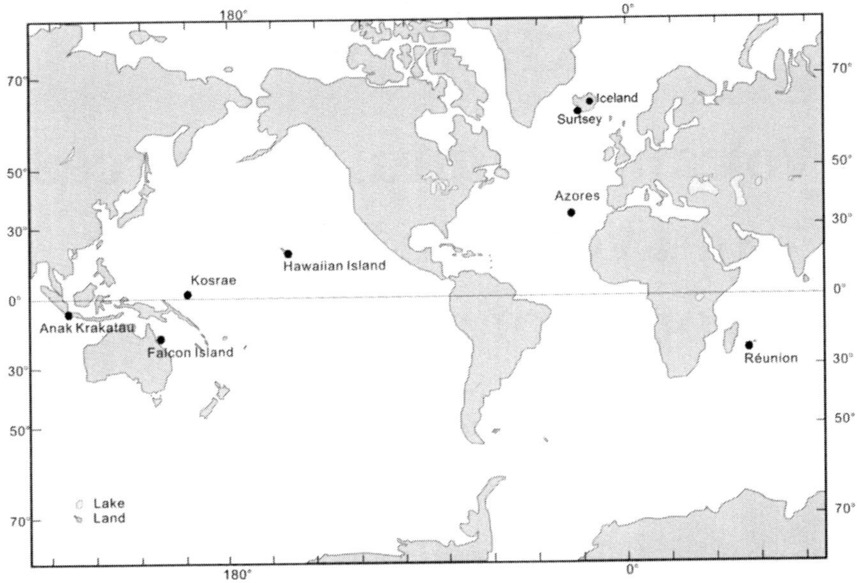

FIGURE 1. Locations of the volcanic islands referred to in the article

The Hawaiian Islands in the Pacific Ocean are a typical example of islands constructed by volcanoes. With continuous oceanic volcano eruptions and magma pouring into the ocean constructing islands, the young Hawaiian Islands keep growing. This situation also happens in Iceland and Reunion.

Indonesia is the biggest archipelagic state in the world and also a "volcano country", most of the islands there have been constructed by volcanoes, for example, Anak Krakatau is called Krakatoa's son. It was formed both by volcanic activity and wave-cut erosion. Eventually the growth speed of the volcano exceeded the wave-cut erosion and emerged in 1930, piled up from an ocean floor of over 100 m depth, forming islands of cinder and lava covering more than 2 km^2 in area [3]. This new island, with an elevation of 9 m in 1930, grew relatively quickly in the first decade of its existence, to 67 m in 1933, and 132 m by 1941 and subsequently to 170 m by 1966. Changes in height were accompanied by enlargement of the island's area. By 1981 the diameter of Anak Krakatau was reported to be about 2,000 m, with the highest elevation at about 200 m above sea level [4].

Falcon Island (Australia) is a small, uninhabited, rocky island, bounded by a fringing reef, at the south of the Palm Islands, formed under the joint action of submarine volcanic eruptions and waves. It is approximately 35 km offshore, 40 km from the nearest city and river estuary, and 8 km from an aboriginal settlement on Great Palm Island [5]. It is in the Pacific Ocean about 1600 km from the eastern coast of Australia. In 1915, Falcon Island disappeared suddenly due to the dominant role of seawater erosion [6]. Eleven years later, the accumulative effect of submarine volcanic eruptions made it re-emerge out of the seawater.

Surtsey (63.3°N, 20.6°W) is also a new island which was formed by volcanic eruptions in 1963-1967. It is a volcanic island off the southern coast of Iceland. It was formed in a volcanic eruption which began 130 metres below sea level and reached the surface on 15 November 1963. The eruption lasted until 5 June 1967, when the island reached its maximum size of 2.7 km^2. Since then, wind and wave erosion have caused the island to steadily diminish in size: as of 2002, its surface area was 1.4 km^2 [7]. The heavy seas around the island have been eroding it ever since the island appeared and since the end of the eruption almost half its original area has been lost. The island currently loses about 1.0 hectare of its surface area each year [7]. This island is unlikely to disappear entirely in the near future. The eroded area consisted mostly of loose tephra, easily washed away by wind and waves. Most of the remaining area is capped by hard lava flows, which are much more resistant to erosion. In addition, complex chemical reactions within the loose tephra within the island have gradually formed highly erosion resistant tuff material, in a process known as palagonitization [8]. Estimates of how long Surtsey will survive are based on the rate of erosion seen up to the present day. Assuming that the current rate does not change, the island will be mostly at or below sea level by 2100 [9].

China has few volcanic islands. They were mainly formed by volcanic eruptions and principally include the Penghu Islands, Pengjia Islet, Mianhua Islet and Huaping Islet in the north of Taiwan Island, Green Island, Orchid Island and Gueishan Island on the continental slope in the east of Taiwan, Diaoyu Islands in the East China Sea and so on.

2. Geothermal energy

In addition to the land resources introduced above, volcanoes provide us with clean energy - geothermal energy. So it has received the attention of all countries as a new energy source and the reserves are much more than the whole amount energy people are using currently. Its exploitation and utilization are developing rapidly. The definitions, distribution, present exploitation and utilization situation, and some typical examples would be introduced below.

Geothermal energy is the heat energy stored inside the Earth that originates from the melted magma of the Earth and the decay of radioactive substances. Most geothermal energy gathers around plate borders where most volcanic eruptions and earthquakes happen.

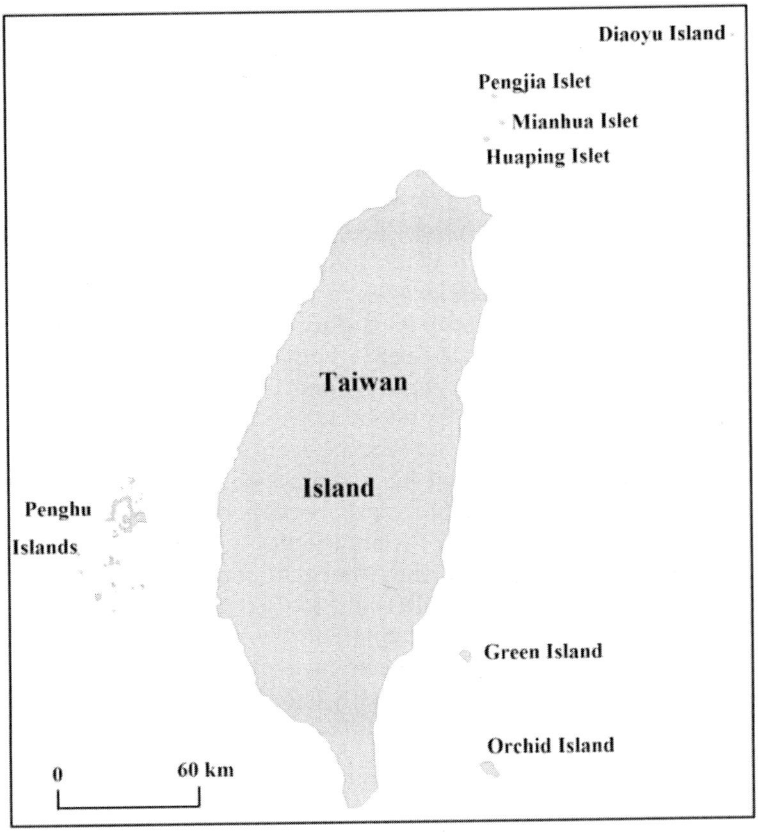

FIGURE 2.. Location of volcanic islands in China

Volcanically active areas normally have a background of high geothermal energy. Volcanic eruption is the most violent exhibition of the internal thermal energy of the Earth on the Earth's surface [10]. Areas with more volcano activities are generally areas with high geothermal flow of geothermal energy. This is because the hot magma chamber under the volcano may heat the circulating groundwater. The heated groundwater is either stored under the ground or spurts out of the ground surface to form hot springs, boiling springs (e.g., the Sirung Volcanic Boiling Spring, Lesser Sunda Islands, Indonesia and Great Boiling Spring, Nevada, United States), geysers (e.g., the Fly Ranch Geyser, Nevada, United States, the Strokkur Geyser, Iceland), fumaroles (e.g., the Valley of Ten Thousand Smokes, Katmai National Park, Alaska, United States) and boiling mud pots (Fountain Paint Pots, Yellowstone National Park, United States, the mud pool at Orakei Korako, north of Taupo, New Zealand and the mud pool at Hverir, Iceland)[10].

FIGURE 3. Penghu Island

The formation of a useful geothermal system needs to possess three essential conditions: an underground heat source (hot rocks), a heat transfer medium (groundwater) and a heat conducting channel (the fissures or boreholes communicating the underground heat reservoir and ground surface).

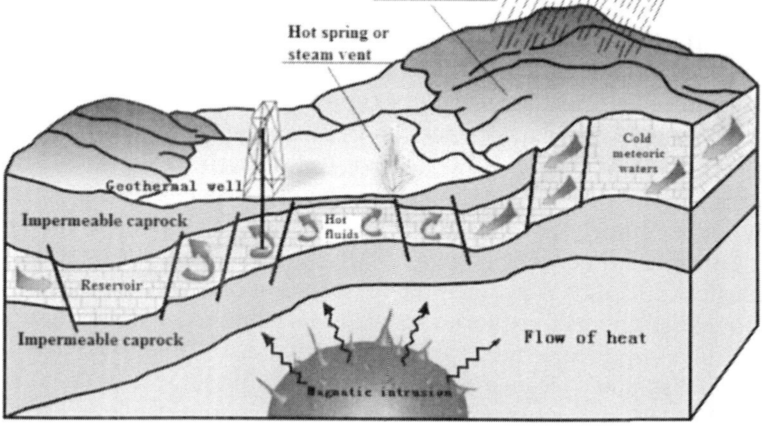

FIGURE 4. A natural geothermal system [11]

Geothermal energy may be classified into four categories [12]: (1) hydrothermal: the hot water or hydrothermal steam in the shallow layer with a depth of hundreds of metres ~ 2,000 metres; (2) geopressured geothermal energy: the high-temperature high-pressure fluid sealed thousands of metres under the ground in some large sedimentation basins; (3) magmatic thermal energy: the enormous thermal energy stored in magma pockets; (4) hot dry rock: it is a high-temperature rock mass stored deep underground without water or steam with the depth (thousands of metres) that can be reached with current drilling technology. Usually, the geothermal energy contained in hot dry rocks and magma pockets is much more than the geothermal energy of hydrothermal and geopressured heat reservoirs.

Hot dry rock is mainly metamorphic rock or crystalline rock. The key technology of heat collection is to build a heat exchange system in the body of hot dry rock without percolation. Generally, high-pressure water is injected into the rock stratum of 2,000~6,000 metres underground through a injection well which permeates the gaps of rock stratum and obtains geothermal energy; then the high-temperature water and steam in the gaps of rocks are picked up through a dedicated production well (at a distance of 200~600 metres) to the ground; the water and steam can be used to generate electricity after pouring into a heat exchange system; the water after refrigeration will be injected into the underground heat exchange system again via a high-pressure pump for recycling. The entire procedure is a closed system.

There are many problems in trying to tap Earth's internal heat as an alternative clean energy source. Earthquake risks and poorly understood geology are the two most important aspects. Domenico Giardini called for a better understanding of earthquake risk in pursuing deep geothermal energy using an enhanced geothermal system (EGS) [13]. In fact an EGS demonstration project in Geysers, north California, was halted by the geological anomalies. The California-based company AltaRock Energy was unable to penetrate the formation capping the hot rocks after months of drilling in 2009. Similar frustrations were encountered during EGS drilling projects at Paralana and the Cooper Basin, both in South Australia. In general, depths of 3–10 km are optimal for geothermal exploitation because they are extremely hot and accessible to modern drilling techniques. But this rule can be broken by geological surprises. In order to improve our geological understanding and enable us to find optimal drilling sites, China is launching the deep exploration technology and experimentation project, SinoProbe, to locate mineral resources and to find out more about earthquakes and volcanism. Meanwhile, the United States and China will inject US$150 million over the next five years into a joint Clean Energy Research Center [13].

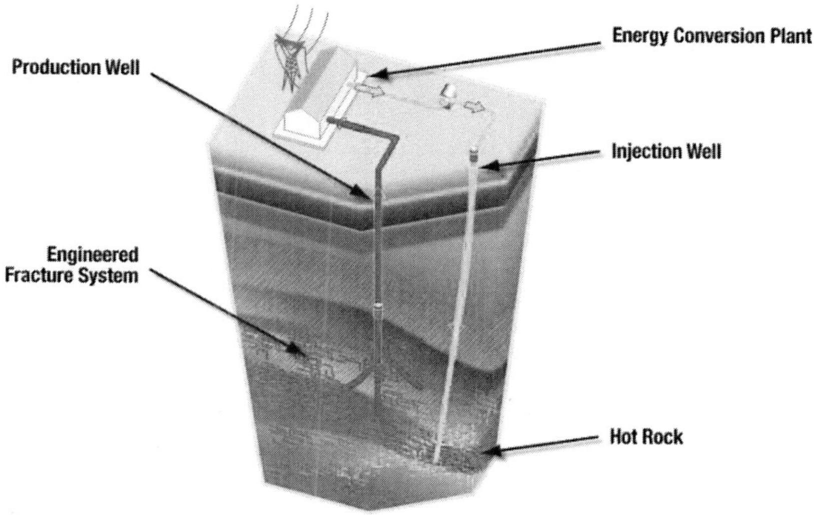

FIGURE 5. EGS Cutaway Diagram [14]

After the Second World War, countries around the world began to pay attention to exploiting and utilizing geothermal energy. Both the number of countries producing geothermal power and the total worldwide geothermal power capacity under development appear to be increasing significantly (Table 1and Table 2) [15].

Indonesia is a "volcano country". It owns 40% of the world's geothermal energy reserves [17]. According to the latest data released by the Indonesian Ministry of Energy and Mineral Resources, the potential installed capacity of geothermal power generation in the country is as much as 27,140 MW, equivalent to the power generated by 12 billion barrels of petroleum, twice the oil deposit of Indonesia and about 40% of the total reserves of geothermal resources in the world [18]. Indonesia ranks third in the world in terms of geothermal energy consumption, after the US and the Philippines. At present, the geothermal power generation capacity of Indonesia is 1,197 MW. It is also the third biggest emitter of greenhouse gases and aims to cut emissions by 16% by 2025 [15].

TABLE 1. Installed geothermal electric generating capacity worldwide from 1995 to 2000 in five-year intervals[16]

Country	1990 (MWe)	1995 (MWe)	2000 (MWe)	2005 (MWe)	2010 (MWe)
Argentina	0.7	0.6	0	0	0
Australia	0	0.2	0.2	0.2	1.1
Austria	0	0	0	1	1.4
China	19.2	28.8	29.2	28	24
Costa Rica	0	55	142.5	163	166
El Salvador	95	105	161	151	204
Ethiopia	0	0	8.5	7	7.3
France	4.2	4.2	4.2	15	16
Germany	0	0	0	0.2	6.6
Guatemala	0	33.4	33.4	33	52
Iceland	44.6	50	170	322	575
Indonesia	144.8	309.8	589.5	797	1197
Italy	545	631.7	785	790	843
Japan	214.6	413.7	546.9	535	536
Kenya	45	45	45	127	167
Mexico	700	753	755	953	958
New Zealand	283.2	286	437	435	628
Nicaragua	35	70	70	77	88
Papua New Guinea	0	0	0	39	56
Philippines	891	1227	1909	1931	1904
Portugal	3	5	16	16	29
Russia	11	11	23	79	82
Thailand	0.3	0.3	0.3	0.3	0.3
Turkey	20.6	20.4	20.4	20.4	82
USA	2774.6	2816.7	2228	2544	3093
Total	5831.7	6866.8	7974.1	9064.1	10716.7

TABLE 2. Development of generating geothermal power from 1990 to 2010

Country	1990-2010 (increase in MW.)	Increase(%)
Iceland	530.4	1189.24
Portugal	26	866.67
Indonesia	1052.2	726.66
Russia	71	645.45
Turkey	61.4	298.06
France	11.8	280.95
Kenya	122	271.11
Nicaragua	53	151.43
Japan	321.4	149.77
New Zealand	344.8	121.75
El Salvador	109	114.74
Philippines	1013	113.69
Italy	298	54.68
Mexico	258	36.86
China	4.8	25
USA	318.4	11.48
Thailand	0.3	0
Costa Rica	166	--
Papua New Guinea	56	--
Guatemala	52	--
Ethiopia	7.3	--
Germany	6.6	--
Austria	1.4	--
Australia	1.1	--
Argentina	-0.7	n/a
Total	4885	83.77

11.5% of the territory of Iceland is covered by modern glaciers. However, within the range of 103,000 km^2, it has at least 200 volcanoes formed in the last million years, including about 30 active volcanoes [19]. It has ample geothermal resources and more than 800 thermal fields. Called the "*Land of Ice and Fire*" it is one of the countries among a few with many thermal fields. The distribution of other thermal fields is closely related to the locations of volcanoes. In the volcanic zone running diagonally through the whole island from southwest to northeast, there are 26 high-temperature steam fields, about 250 low-temperature geothermal fields and more than 800 natural hot springs [19]. Within the range of the 0~10 km thick crust, the total amount of geothermal resources is 300 million TWh (1TWh is equivalent to 100 million kWh of electricity); within the range of the 0~3 km thick crust, the total amount is 30 million TWh; and the amount of geothermal resources technically exploitable is one million TWh [19]. Through long-term exploitation and utilization, Iceland has developed efficient geothermal utilization techniques and become the only country in the world almost non-dependent on petroleum. Heating and electric power both rely on natural geothermal energy [20].With a small population, the country is currently generating 100% of its power from renewable sources, deriving 25% of its electricity and 90% of its heating from geothermal resources [17]. Of course, the long-term exploitation and utilization of geothermal energy has also had some negative impacts on geothermal resources e.g., continual descent of the water level in some geothermal systems, temperature and chemical component changes because of the injection of cold water [21]. To address these impacts, Iceland has taken many measures, for example, according to Iceland's environmental law, any geothermal exploitation with a gross amount of above 25 MWe or a net amount of above 10 MWe must submit a detailed environmental impact assessment report [20]. This attitude of attaching importance both to resource exploitation and utilization, and to their conservation and future development is commendable.

New Zealand lies on the suture line between the southwestern margin of the Pacific Plate and the Indo-Australian Plate. This suture line extends from the sea southeast of North Island to the northwest of South Island and then goes further to the south along the western edge of South Island (see Figure 6). The Pacific Plate to its east subducts into the crust of North Island and forms the Taupo Volcanic Zone (see Figure 6), which is the home of the main active volcanoes and geothermal fields in New Zealand. Among the exploited geothermal fields, there are Wairakei, Broadlands, Rotokawa, Kawerau, Ohaaki and Mokai (see Figure 7). Wairakei Geothermal Power Station was the world's first geothermal power station which generates power from wet steam and also is the world's second largest geothermal power station, behind only Italian Larderello Geothermal Power Station [23]. This geothermal field is located in the central volcanic zone on North Island, New Zealand, and about 16 km to the northeast of Taupo Lake. It is the largest geothermal field in New Zealand. Wairakei Geothermal Power Station was developed in 1950 and started power generation in 1958. Now it has more than 100 gas wells, including 60 wells for power generation, with a total installed capacity of about 180 MW and annual power generation of 1501 GW h [23]. The "wet" steam extruded from the gas wells contains 80% water. Its temperature may reach 300°C at most. During power generation, white water vapour spurts out of the well mouth continuously up to the sky and turns into clouds shortly after. Against the blue sky and over the green pines it looks majestic (Figure 8).

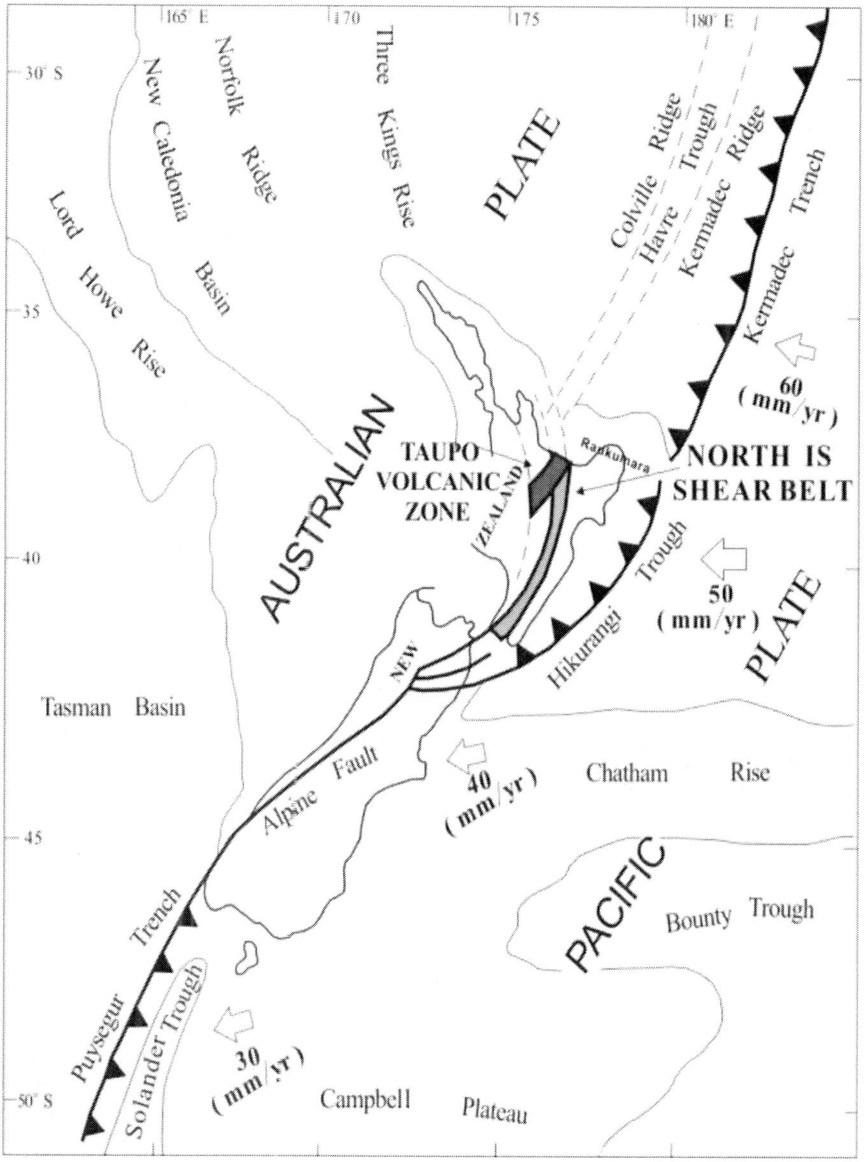

FIGURE 6. Location of Pacific-Australian plate boundary and Taupo Volcanic Zone[22]

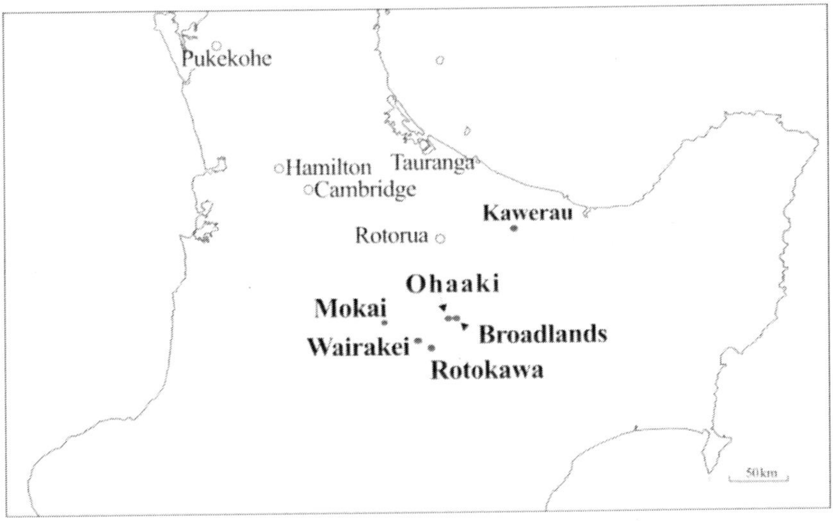

FIGURE 7. Location of the Wairakei Geothermal Power Station in the Taupo Volcanic Zone, New Zealand

FIGURE 8. Steam pipelines towards the Wairakei Geothermal Power Station [24]

3. HOT SPRINGS AND MINERAL SPRINGS

3.1. Hot Springs

A hot spring is a kind of spring belonging to ground water. It is called a hot spring if the temperature of the spring pouring out of the Earth's surface is higher than that of the local ground water. If the temperature of the spring is lower than that of the local ground water, it is called a cold spring. Hot springs are a display of geothermal energy.

A hot spring can be created in many ways; generally it is created by the ground water percolation cycling system effect. The average geothermal gradient of the Earth's near-suface is 3 degrees Celsius per 100 metres, the atmosphere penetrates into the underground, becomes aquifer and absorb heat from the deep underground rocks. The heated water can produce steam as well as the air included in the water expanding, that increases the pressure of the water-containing system, and then it pours out at the surface along cracks and gaps to become the hot spring.

Most hot springs are located in volcanic areas and are closely related to volcanic activities. No matter if the volcanoes are erupting or dormant (even extinct ones, under which magma pockets existing or magma activities always exist), lots of heat energy is accumulated, which heats the surrounding ground water and make it pour out as hot springs.

During the storing and moving process, due to the interaction of water and rocks, the hot spring includes many chemical components. Because of the different features of these components, the medical effects are also different.

For example [25], when people have a bath in hot spring containing carbonic acid gas, i.e., carbon dioxide gas, carbonic acid gas will adhere to the skin in bubbles and form a carbonic acid gas film which will keep exchanging with new small bubbles leading to a warm and cosy feeling. They also can stimulate the blood capillaries to expand and promote blood circulation. The carbon dioxide gas breathed into the lungs can strengthen gas metabolism and help balance acid-base organism.

Bicarbonate hot springs can soften and clean skin. People will feel smooth and cool after the bath. As the calcium in the spring can slightly dry the skin out, it is also a good medical treatment for trauma, chronic eczema and ulcers.

When people take a bath in a hot spring including hydrothion, sodium sulphide will be formed on the skin, which can stimulate the skin's blood circulation and nutrition metabolism, promote softer skin and dissolve cutin, reduce inflammation and increase immunity; it also can adjust blood pressure in two ways - improve the insufficiencies of coronary arteries. However, hydrothion is a poisonous gas; excessive hydrothion can lead to neurotoxicity, from headaches to dizziness to respiratory paralysis. Therefore, be aware when bathing in hydrothion hot springs.

Sulphate hot springs can be used for a bath and the water can be drunk; as a bath, the water stimulation from salt to skin can lead to the expansion of blood capillaries and promote the body's metabolism, it can also assist in the treatment of some skin diseases. Drinking sodium sulphate and magnesium sulphate in hot spring waters can promote the secretion and excretion of bile, clean the stagnancy of bile and prevent calculus forming, so it can act as a medical treatment for cholecystitis and gall-stones; drinking calcium sulphate hot spring water can help purine supersession and promote excretion of uric acid, so it can act as a medical treatment for gout and urethra inflammation.

A chloride hot spring is called a "nerve painkiller" as the excitement of the nerve can be reduced when bathing in chloride springs - a good medical treatment for neuralgia. The osmotic pressure of a chloride hot spring with medium concentration (content is above 5g/l) is close to a salt solution, therefore bathing in a hot spring of 36~38 degrees Celsius will help to treat trauma, haemorrhoids and skin diseases; chloride hot springs with a high concentration will help to promote the constitution, to recover the function of the ligament arthroclisis, muscle atrophy and dyskinesia.

Drinking from a hot spring with silicate (metasilicate) can help to adjust the metabolism and promote gastrointestinal motility and strength digestion; when bathing, silicate will adhere to the skin, clean the skin and skin mucosa.

In addition, hot springs with radon, arsenic, bromine, iodine and other microelements also have medically positive effects for humans.

There are hot springs on all continents and in many countries around the world. We list some of the famous hot springs with medical value around the world – see Table 3 and Figure 9.

3.2. Mineral Springs

Mineral springs are springs with lots of mineral substances. They may be hot, defined by Mazor as having a temperature of >6 degrees Celsius above that of the mean annual surface temperature, or cold [26]. Different stipulations had been set up by different organizations. The element content standards of the mineral spring water which has been adopted by the Codex Alimentarius Commission in 1993 can be seen in Table 4 [27].

TABLE 3. Some famous hot springs with medical value around the world

Country	Hot Spring
USA	Glenwood Springs
Chile	Chihuío
Ecuador	Baños de San Vicente
China	Tengchong Hot Spring Zone Taiwai (Beitou Hot Spring, Yangmingshan Hot Spring)
Indonesia	Maribaya Hot Spring
Japan	Kusatsu Onsen, Arima Onsen, Gero Onsen
Iceland	Blue Lagoon Spring, Geysir Hot Springs
Italy	Bormio, Sondrio Province, Lombardy (geothermal spa)
New Zealand	Rotorua Volcanic Hot Spring

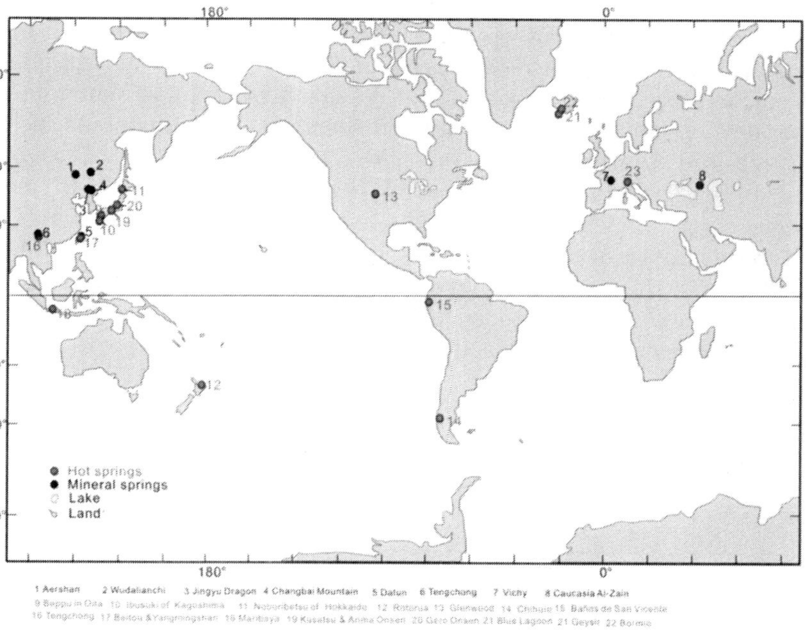

FIGURE 9. Location of famous hot springs and mineral springs around the world

But according to the stipulations of the International Mineral Spring Association, in natural water, if one or more microelements and mineral substances needed for humans, such as lithium, strontium, zinc, copper, barium, cadmium, selenium, arsenic, manganese, antimony, nickel, bromide, iodide, metasilicate, free carbon dioxide, etc. reaches the standard required, it can be called "mineral springs" [28]. The French Vichy spring contains four, the Russian Caucasia Al-Zain spring contains two and the Chinese Wudalianchi mineral spring contains more than six (Figure 9). However, the Russian Caucasia Al-Zain spring is created by the melting of Mount Elbrus' glaciers, while the French Vichy and Wudalianchi springs are closely related to volcanoes.

The French Vichy Mineral Spring was formed by a volcanic eruption and can be found at 103 locations. At present, only 15 mineral spring wells have been exploited and utilized. As none of the mineral springs is an artesian spring, the extraction is not easy and wells must be dug. In the downtown of Vichy there is a mineral spring called Célestins. It is the only mineral spring for daily drinking.

In China, Wudalianchi Yaoquan (Figure 10) (literally "Medicated Spring") Mountain in Heilongjiang province, the Aershan Wuli Spring in Inner Mongolia and the Jingyu Dragon Spring in Jilin province are all high-quality volcanic mineral springs (Figure 9). In Wudalianchi Yaoquan Mountain, the mineral water contains multiple desired elements and components such as carbonic acid, iron, zinc, strontium, lithium and germanium. The water features large reserves, is high quality and has medical value. It has an obvious curative effect on stomach disease, skin disease and arthritis. The mud there may be used for pelotherapy and many sanatoriums have been built there. The former wasteland around the area has become an emerging modern city - Wudalianchi today. It is probably the only city established by relying on volcanic resources and it has become a famous geopark.

TABLE 4. The elements content standards of the mineral spring water [27]

Zn	Cu	Ba	Cd	Cr_{Total}	Pb	Hg	Se	As_{Total}	Mn	Sulphide	Sb	Ni
-	-	0.7	0.003	0.05	0.01	-	-	0.01	0.5	-	0.05	0.02

3.3. Mineral Spring Culture

The historic culture of mineral springs can be traced back to two thousand years BC. In ancient Greek stories, the goddess who can cure human diseases lives in mineral springs, which made people desperate to attain the waters [29]. In the ancient Rome era third century AD, the development and utilization of mineral springs took shape, it is said that there were more than 860 mineral spring bathing places in the city of Rome.

After the 18th century, people began to study mineral spring as a science. In 1742, a German doctor called Hoffmann confirmed some components of mineral springs based on a predecessor's research, which laid a foundation for the development of mineral spring science [29].

The 20th century saw the development of theoretical research and applied research, and a specialized agency was founded in developed countries like Japan, France, Germany and the US to develop the research and train talent, even putting crenology on required courses of advanced medical schools.

Each country of the world has its own mineral spring historical cultural expression, especially Japan which is called the country of hot springs - thousands of hot springs and mineral springs are located across the Japanese islands. Every family has bathing equipment and it seems that Japanese people like to bathe in hot springs the most. Hot spring bathing can not only reduce tiredness, cure disease and strengthen the body, it can also be a place to communicate with fellow bathers.

Volcanoes have caused natural disasters in Japan, but have also created abundant hot spring resources across the country. Among the 2,200 natural hot springs, the most favoured hot springs are located in Oita, Kagoshima and Hokkaido with different features [29].

Beppu Hot Spring in Oita (Figure 9) has been known about in Japanese since ancient times, there are more than 3,800 spring water holes and the water inflow is more than 200 thousand tons every day [29]. Known as the city of spring, it is the biggest natural hot spring area in Japan and also a world-class hot spring city. Sand Steam Hot Spring (Figure 9) in Ibusuki, Kagoshima, attracts more women (Figure 10). It is the only sand steam hot spring in Japan. To "*sand steam*" is to bury the body except the head in hot sand, termed "*sand happiness*". It is similar to having a sauna to make you sweat - in less than five minutes your body will feel hot. The sand pressure and hot water promote blood circulation, sweating from the whole body and tiredness to reduce, which is a fantastic medical treatment for preventing rheumatism and nerve ache. There is "sodium" in the sand which makes skin fair, so it is favoured by women for cosmetic reasons. In Noboribetsu, Hokkaido (Figure 9), the hot spring has another fun aspect – the spring, rock, flowers and grass form extremely pleasant scenery. Hot spring hotels are located along the main street, the bathroom in the Noboribetsu International View Club is especially well-known - with length of 90 m and width of 20 m there are more than ten hot spring pools of different sizes and temperatures to choose from, and men and women can bathe together.

FIGURE 10. Sand steam hot spring in Ibusuki, Kagoshima [30]

Hot springs in New Zealand are located across the country. Rotorua (Figure 9) sitting on volcanic-prone area is called the "*New Zealand Hot Spring City*" and is home to the largest hot spring waterfall in the southern hemisphere and the only mud bath pool in New Zealand - "*Wai Ora Spa*". The mud contains abundant mineral substances, which have health benefits. In addition, the unique Maori culture all combine to make this area a thriving tourist attraction.

In China, there are fewer present-day volcanic eruptions, but the hot springs and mineral springs related to volcanoes widespread and some volcanic areas, such as Wudalianchi in Heilongjiang, Changbai Mountain in Jilin, Aershan in Inner Mongolia, Tengchong in Yunnan and Datun in Taiwan, produce hot springs and mineral springs (Figure 9). Aershan in Inner Mongolia in particular contains hot and cold springs, and mineral springs [31, 32]. Hot springs in Tengchong are not only located widely and with numerous spring holes, but it is also well-known that they have a high temperature and water flow [33, 34].

FIGURE 11. Landscape of Tengchong, China

4. VOLCANIC MATERIALS

Volcanic rocks forming volcanic edifices and surrounding volcanic ring plains consisting of volcanic ash, scoria, pumiceous deposits and the coherent volcanic rocks part of lava flows, lava domes or exposed subsurface facies of a core of a volcano such as dykes, sills, laccoliths are all good building materials widely used for construction such as paving and building houses. Scoria and the volcanic ash of basalt can be used directly as a filling in clinker-free cement to produce high-quality cement [35]; cast stone bricks, tiles and panels made of basalt are good fire-proof materials, which are not only heat-resistant, but also crush-resistant and corrosion-resistant, so much so that they are considered as a substitute for steel in machine tool and the chemical industry [36-38]. What attracts people's attention is basalt fibre, which takes basalt as the raw material. It has the advantage of good combination properties and cost effectiveness, and no poisonous substances, waste gas, waste water, residue or pollution in normal machining processes. It is applied widely in fire-fighting, environment protection, aviation, the arms industry, automobile and ship making, engineering plastics, the construction industry and the so-called "green industrial materials" of the 21st century. The classifications are continuous basalt fibre, thick fibre, narrow fibre, super-narrow fibre and subtle scale, among which the continuous basalt fibre has been previously studied and studies of the others are developing at present.

4.1. Development Background of Basalt Fibre

In 1840, the trial production of rock wool with basalt as the primary raw material succeeded in Wales, UK [39]. In the early 1950s, Germany, the then Czechoslovakia, Poland and Hungary produced basalt wool with an average fibre diameter of 25 μm~30μm from basalt using the centrifugal method [40]. In the early 1960s, the USA, the former Soviet Union and Germany vigorously developed a vertical blowing production process, resulting in the rapid growth of basalt wool output. The former Soviet Union introduced a German patent for producing mineral wool using the vertical blowing method. On the basis of digestion and absorption, it succeeded in applying this technology to the production of basalt wool. The production capacity is 38~40 tons of basalt wool a day [41].

Basalt fibre was successfully developed by the Russian Moscow Glass and Plastic Research Institute in 1953~1954. The first furnace for industrial production was built and put into operation in the Ukraine Fibre Laboratory (TZI) in 1985. The fibre-drawing process of a 200-hole bushing and combination furnace was adopted with an annual output of 260 t [40]. Then, they adopted a 400-hole tank furnace fibre-drawing process to produce CBF and its products. The annual output is about 700 t. In recent years, China, Japan, South Korea and Germany have also carried out relevant research and achieved some new research results [41]. At present, the USA, Germany, China, Japan and South Korea are also carrying out relevant research and are achieving some new research results [41].

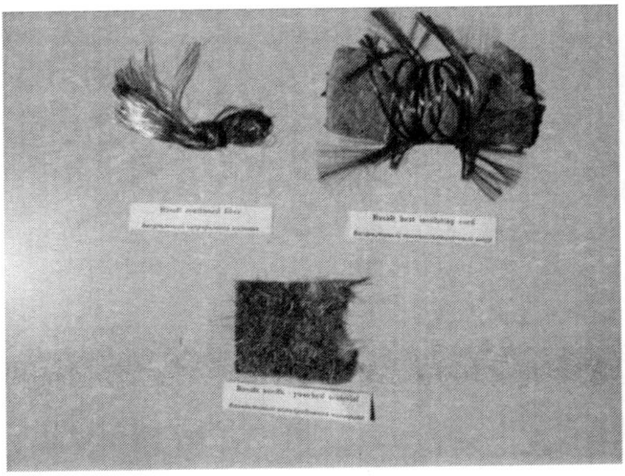

FIGURE 12. Basalt fibre

4.2. Properties and Use of Basalt Fibre

Compared with glass fibre, rock wool, asbestos and chemical fibre, basalt fibre and its products have extraordinary performance and have multiple application capabilities (Table 5):

1. Good thermal property and flame retardant property: basalt fibre is amorphous state inorganic silicate substance, with good temperature-resistance and heat insulation and without thermal contraction [42]. The temperature range of its usage is -269~700 degrees Celsius, the softening point is 60 degrees Celsius, which is much higher than that of glass fibre - 60~ 450 degrees Celsius and carbon fibre - 500 degrees Celsius. Under a temperature of 500 degrees Celsius, its stability of thermal shock resistance is unchanged. Initial mass fraction loss is less than 0.02 and it is 0.03 under 900 degrees Celsius. Under a temperature of 600 degrees Celsius, its breaking strength can still keep as 80% as the original intensity, and it will not shrink under 860 degrees Celsius while mineral wool with good temperature-resistance can only keep 50%~60%. Carbon fibre will produce CO and CO_2 under 300 degrees Celsius [43], while glass fibre is crushed completely. Therefore, basalt fibre can be used for the manufacture of flame retarding materials such as fire-proof suits, blankets and curtains, and high-temperature filters such as scrim, material and high-temperature resistant felt. In addition, with regard to low-temperature resistance, the intensity of basalt fibre will not change under the medium of low-temperature (-196 degrees Celsius) liquid nitrogen. So it is an effective low-temperature heat insulation material [41].

2. Good chemical stability and corrosion resistance: basalt fibre maintains favourable chemical stability and powerful acid and alkali resistance. With natural consistency and high stability in alkali medium such as cement, it can be used as the enhancement material for concrete building structures to replace rebar. Flake coating made of basalt can offer protection to building and objects under water, including warships, to strengthen the capacity and life of corrosion prevention.

3. Good stretch ability and modulus of elasticity: the density of CBF is 2.65~3.00g/cm3 and its hardness is very high, Mohs hardness scale 5~9, so it has excellent wear resistance and tensile reinforcement [43]. The tensile strength of CBF is 3800~4800 MPa, higher than that of large-tow carbon fibre, aramid fibre, PBI fibre, steel fibre, boron fibre and aluminium oxide fibre, and equivalent to S glass fibre. The strength of basalt fibre may be maintained for

1200 h in 70 degrees Celsius water, while ordinary glass fibre will lose its strength at 200 h. At 100~250 degrees Celsius, the tensile strength of basalt fibre may be increased by 30%, while the tensile strength of ordinary glass fibre will decrease by 23% [41]. Therefore, basalt fibre material has a huge advantage in bridge building as well as in sport material usage such as fishing rods, hockey sticks, antennae, skis, umbrella handles, poles, bows and arrows, and crossbows [43].

Meanwhile, basalt fibre keeps high modulus of elasticity: 9100 kg/mm2 ~11000 kg/mm2, higher than that of alkali-free glass fibre, asbestos, aramid fibre, PP fibre and silicon fibre, and close to that of expensive S glass fibre. It may replace S glass fibre in the making of heat insulating products and composite materials, for example, hard armour and GFRP products [40, 41].

1. Low coefficient of heat conduction, good heat-proof quality: the conductivity factor of basalt fibre is 0.031 W/m•K~0.038 W/m•K, lower than that of aramid fibre, aluminium silicate fibre, alkali-free glass fibre, rock wool, silicon fibre, carbon fibre and stainless steel [40]. It may be used for heat preservation of heat treatment equipment, heat insulation of automobiles and ships, and heat preservation of pipelines.
2. High acoustical absorption coefficient and good stealth performance: the acoustic absorption factor of CBF is 0.9~0.99, higher than that of alkali-free glass fibre and silicon fibre. Basalt fibre has good wave permeability, certain wave absorbability, excellent acoustical absorption and insulation. In addition, it may be used as a stealth material [42].
3. High dielectric coefficient and good insulating property: the specific volume resistance of basalt fibre is 1x1012Ω•m, much higher than that of alkali-free glass fibre and silicon fibre. By relying on its good dielectric performance, low hydroscopicity and good temperature resistance, basalt fibre may be made into high-quality PCB and blades [40]. After basalt fibre is treated with a special impregnating compound, its dielectric loss angle tangent is 50% lower than that of ordinary glass fibre, and it may be used to produce new-type heat-resistant dielectric materials [41].
4. Low hygroscopicity and good seepage-proof and anti-crack property: the hygroscopicity of basalt fibre is below 0.1%, lower than that of aramid fibre, rock wool and asbestos [40]. Compared to glass fibre, the hydroscopicity of basalt fibre is 6~8 times lower [44]. It has strong seepage control and crack resistance, and may be

widely used in expressways, runways, port terminals, hydroelectric engineering buildings and other infrastructure fields.

TABLE 5. Comparison of the performance of the basalt fibre with other fibres [43]

Properties	Performance	Basalt fibre	E-glass fibre	S-glass fibre	Carbon fibre-HS	Aramid fibre 1313
Physical properties	Density/g·m^{-3}	2.65-3.00	2.55-2.62	2.46-2.49	1.78	1.44
	Tensile strength/MPa	3000-4840	3100-3800	4590-4830	3100-5000	2758-3034
	Elasticity modulus/GPa	79.3-93.1	76-78	88-91	230-240	124-131
	Elongation after fracture/%	3.15	4.70	5.60	1.20	230
	Conductivity factor /w·(m·kg)$^{-1}$	0.031-0.038	0.034-0.04	0.036-0.04	5-185	Low
	Volume resistivity /Ω·m	1×10^{12}	1×10^{11}	1×10^{11}	2×10^{-5}	*/>1×10^{11}
	Acoustic absorption factor	0.9-0.99	0.8-0.93	0.8-0.93	Small	Small
Chemical properties	Softening point	960	850	1056	—	—
	Maximum operating temp/℃	700	380	300	2000	250
	Chemical resistance	Acid and alkali resistance	Moderate resistance to alkalis	Moderate resistance to alkalis	No effect	Moderate resistance to acids
	Price/yuan·kg^{-1}	38	17	20	300	200

The features of basalt decide the wide usage of basalt fibre, not only for aviation, the arms industry, fire-fighting, traffic, energy, environment protection and construction industry, but also espionage, communication and special materials under thermal shock. With abundant basalt resources, its future in industrial production and marketing promotion is bright.

5. GEMSTONES AND OTHER MINERAL RESOURCES

Many gemstones are related to or result from volcanic processes and therefore they are hosted in volcanic deposits and rocks. Gemstones in volcanic rocks include sapphire and ruby, and sometimes adamas can also be found; garnet, pyroxene and olivine with good crystals and bright colours can be found in basalt, which can be used directly as gemstones. Obsidian itself erupted from a volcano can be high quality pure volcanic glass with gem-quality. Crystal, agate and aragonite produced by volcano action are all valuable gemstones. Besides gemstones, many metal and non-metal minerals are related to volcano action, such as gold, silver, copper, lead, zinc and sulphur and diatomite.

5.1. Diamonds

Diamonds are commonly hosted in kimberlite pipes that are commonly looked upon as a specific type of maar diatreme volcano (Figure 13). Kimberlite is a type of potassic volcanic rock best known for sometimes containing diamonds. It is named after the town of Kimberley in South Africa, where the discovery of an 83.5 carat (16.7 g) diamond in 1871 spawned a diamond rush, eventually creating the Big Hole [45].

Diamonds form at a depth greater than 93 miles (150 kilometres) beneath the Earth's surface. After their formation, diamonds are carried to the surface of the Earth by volcanic activity. As this molten mixture of magma (molten rock), minerals, rock fragments and diamonds approaches the Earth's surface it begins to form an underground structure (pipe) that is shaped like a champagne flute. These pipes can lie directly underneath shallow lakes formed in the active volcanic calderas or craters [46].

Kimberlite occurs in the Earth's crust in vertical structures known as kimberlite pipes. According to the descriptions of Volker Lorenz [48], "the formations of maars and diatremes suggest a specific process. Magma rises along a fissure and contacts ground - or surface-derived water. The resulting phreatomagmatic eruptions give rise to base surge and air-fall deposits consisting of juvenile and wall-rock material. Spalling of the wall-rocks enlarges the fissure into an embryonic vent. At a critical diameter of the vent large-scale spalling at depth and slumping near the surface gives rise to a ring-fault of large diameter and subsidence of the enclosed wall-rocks and overlying pyroclastic debris. This subsidence leads to a maar crater at the surface. Various features of kimberlite diatremes seem to be consistent with this model. They extend into fissures along which hot kimberlite magma rise. The diatremes, however, indicate emplacement by a cool gas phase, probably steam. Indicators for subsidence along ring-faults may be diatremes with large diameter, slickenside on walls, saucer-shaped structures, subsided "floating reefs", concentration of xenoliths from specific horizons within certain areas, and zoning of diatreme rocks. It is suggested that formation of kimberlite diatremes may have been influenced by non-juvenile water." For example, the Premier kimberlite pipe and

Jagersfontein kimberlite pipe in South Africa, and the Mwadui kimberlite pipe in Tanzania are all famous diamond mines around the world.

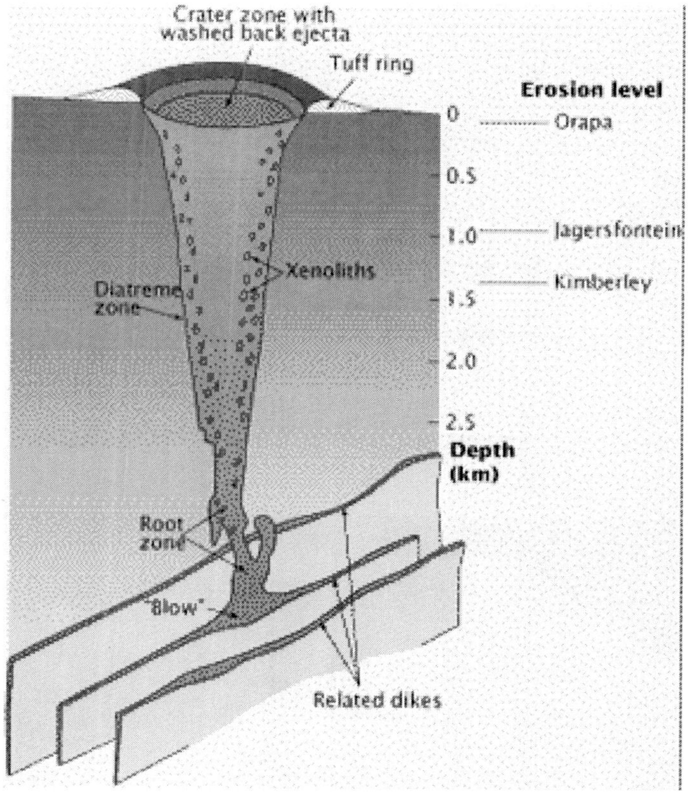

FIGURE 13. Kimberlite pipe [47]

Lamproite pipes produce diamonds to a lesser extent than kimberlite pipes. Lamproite pipes are created in a similar manner to kimberlite pipes, except that boiling water and volatile compounds contained in the magma act corrosively on the overlying rock, resulting in a broader cone of eviscerated rock at the surface. This results in a martini-glass shaped diamondiferous deposit as opposed to kimberlite's champagne flute shape [46].

The Argyle diamond mine in Western Australia is one of the first commercial open-cast diamond mines that is dug along an olivine lamproite pipe. The Argyle pipe is a diatreme, or breccia-filled volcanic pipe that is formed by gas or volatile explosive magma which has breached the surface to form a "tuff" cone [46].

The complex volcanic magmas that solidify into kimberlite and lamproite are not the source of diamonds, only the elevators that bring them with other minerals and mantle rocks to the Earth's surface. Although rising from much greater depths than other magmas, these pipes and volcanic cones are relatively small and rare, but they erupt in extraordinary supersonic explosions [49].

Kimberlite and lamproite are similar mixtures of rock material. Their important constituents include fragments of rock from the Earth's mantle, large crystals and the crystallized magma that glues the mixture together. The magmas are very rich in magnesium and volatile compounds such as water and carbon dioxide. As the volatiles dissolved in the magma change to gas near the Earth's surface, explosive eruptions create the characteristic carrot- or bowl-shaped pipes [49].

Diamonds also can be formed by subduction (Figure 14). When the ocean floor slides under the mantle, the basaltic rock becomes eclogite, and organic carbon in sediments may become diamonds [49].

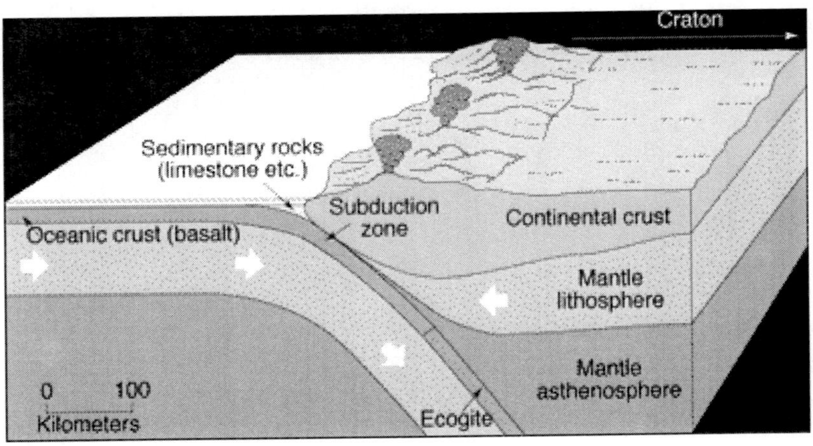

FIGURE 14. Diamonds can be formed by subduction [50]

5.2. Sapphires

Sapphires are commonly hosted in dykes, volcanic necks and lava flows which are composed by alkali basalts. Many sapphires have been found in Changle of Shandong, Liuhe of Jiangsu and Muling of Heilongjiang in China. Sapphires are also formed in basalts in Kanchanaburi and Banca in Thailand, Pailin in Cambodia and Anaky in Australia [51].

5.3. Garnets

Garnets are a group of silicate minerals that have been used since the Bronze Age as gemstones and abrasives. Garnets possess similar physical properties and crystal forms but different chemical compositions. The different species are pyrope, almandine, spessartine, grossular (varieties of which are hessonite or cinnamon-stone and tsavorite), uvarovite and andradite [52].In these species, only the pyrop has a relationship with the volcanic activities. They are mainly produced in kimberlite, basalt and mantle xenoliths. A famous pyrop, Bohemian garnet in the Czech Republic, is known to be associated with maar diatreme volcanoes of the Czech Republic.

In the abyssal lherzolite inclusions of the basalts, there are olivine, pyroxene, garnet, spinel and other minerals which have fine crystals and bright colours. They have been widely used as gems. This type of inclusion can be easily found in the countries which have volcanoes [51].

5.4. Opal and Others

Australia is the opal capital of the world. Opals are often host in fractures, voids and primary stomata of the volcanic rocks (basalts, andesites, rhyolites and tuffs). Besides Australia, opals have been found in the Czech Republic, Mexico, Honduras and other countries. The compositions and formations of the agate and chalcedony are similar to opal, but their distributions are much wider than opal. A lot of agates and chalcedonies have been produced in India, China, Brazil, the United States, Russia, Madagascar, Ireland, Namibia and Egypt [51].

Opal, agate and chalcedony are semi-precious stones that come under the category of quartz. These stones are characterized by fine granular texture and bright colours. They are processed and marketed in different shapes and sizes [53].

Gems with a variety of colours and textures can be formed by the aragonites that fill in the stomata of the basalts and andesites. The basalt of the Penghu Islands in China is rich in this precious gem. This type of gem has been produced in Spain, Italy, Austria, Chile, the United States and other countries [51].

5.5. Obsidian

Obsidian is a naturally occurring volcanic glass formed as an extrusive igneous rock. It is produced when felsic lava extruded from a volcano cools rapidly with minimum crystal growth. Obsidian is commonly found within the margins of rhyolitic lava flows known as obsidian flows, where the chemical composition (high silica content) induces a high viscosity and polymerization degree of the lava. Obsidian can be found in locations which have experienced rhyolitic eruptions. It can be found in Argentina, Armenia, Azerbaijan, Canada, Chile, Greece, El Salvador, Guatemala, Iceland, Italy, Japan, Kenya, Mexico, New Zealand, Peru, Scotland and the United States [54]. Anatolian sources of obsidian are known to have been the material used in the Levant in modern-day Iraqi Kurdistan from a time beginning sometime about 12,500 BC [55]. In Ubaid in the 5th millennium BC, blades were manufactured from obsidian mined in what is now Turkey [56]. Now obsidian has been used for blades in surgery. Obsidian is also used for ornamental purposes and as a gemstone. It possesses the property of presenting a different appearance according to the manner in which it is cut: when cut in one direction it is jet black, in another it is glistening grey. Plinths for audio turntables have been made of obsidian since the 1970s [54].

5.6. Gold Deposit

Gold deposits attract the constant attention of every country around the world. People have found that the formation of gold deposits has a close relationship with volcanic activities in terms of time and space. Hu et al. studied the classification of gold deposits relevant to volcano, sub-volcano, intrusion and hydrotermalism. He introduced classification methods of gold deposits according to the volcanic internal ore-controlling structures [57]: 1) gold deposits from bedrock, such as the super large gold bearing porphyry copper deposits in Bingham in the United States and Chuquicamata in Chile; 2) gold deposits from the contact zone between bedrock and volcanic caprock, such as the gold ore in Tuanjiegou, Heilongjiang, China; 3) gold deposits from the volcanic edifice of the caprock of volcanic rock series, such as Conrad in Kazakhstan; 4) gold deposits from hot spring accumulation near volcanic craters, such as Round Mountain in Nevada, California, the United States.

Jiang et al. [58] undertook an in-depth study on volcanism and gold deposits, and claimed that volcanism is one of the optimal geologic conditions for the formation of gold deposits. They classified gold deposits into three categories [58]. The first category is volcanic sediment deposits in tensioned structures which were formed by submarine volcanism. The second category is plutonic volcanic gold deposits formed due to plate activities and collisions during the orogenic and post-orogenic period. They are classified into three sub-categories: 1) gold deposits in intrusive rocks or contact zones including the vein deposits in intrusive rocks and porphyric gold deposits, e.g. Barrick (gold reserves: ~420t), Lihir Island (gold reserves: ~500t) and other huge sub-volcanic porphyry gold deposits in Papua New Guinea, which were formed along with the intrusive magmatism in the Cenozoic island-arc volcanic rock zone subducted by the Western Pacific Plate; 2) gold deposits from continental volcanic rocks; they are mainly developed in the Circum-Pacific Island Arc and related to Middle Cenozoic volcanic activities; 3) metasomatism and filling gold deposits related to plutonic volcanic magmatism, e.g., the Carlin type gold deposits in Nevada, the United States [58] and the Munmtau gold deposit belt in Uzbekistan [59]. The third category is the placer gold deposits formed due to weathering and sedimentation under hypergene conditions.

5.7. Diatomite

Diatomite is a naturally occurring, soft, siliceous sedimentary rock that is easily crumbled into a fine white to off-white powder. It has a particle size ranging from less than 1 micrometre to more than 1 millimetre, but typically 10 to 200 micrometres. Diatomite is formed by the accumulation of the amorphous silica (opal, $SiO_2\,nH_2O$) remains of dead diatoms (microscopic single-celled algae) in lacustrine or marine sediments [60]. So diatomite can be found in exposed maar crater lakes. Where there are maar crater lakes, there are diatomite (e.g., Germany, China, Hungary, Slovakia and so on). It is used as a filtration aid, mild abrasive, mechanical insecticide, absorbent for liquids, matting agent for coatings, reinforcing filler in plastics and rubber, anti-block in plastic films, porous support for chemical catalysts, cat litter, an activator in blood clotting studies and a stabilizing component of dynamite. As it is heat-resistant, it can also be used as a thermal insulator [60].

Most of these deposits are related to calc-alkali volcanic rock, mainly andesite. Some of these deposits are replacement and vein deposits. Stratiform lead and zinc deposits are found at Nong Phai and Song Thoin Kanchanaburi in middle Ordovician limestone. The zinc deposits at Pha Daeng, Mae Sot is the largest zinc deposit in Thailand. The ore are zinc carbonate and zinc silicate in the supergene enrichment in the Jurassic Kamawkala limestone near the Thai-Myanmar border.

6. VOLCANIC LANDSCAPES

Volcanoes are nature's sculptors, making numerous beautiful scenic spots and natural landscapes, which are not only tourist attractions but also ideal places for scientific research. Many famous landscapes and tourist attraction are also volcanic areas, and all of them are colourful and charming. Many of the world's geoparks and natural heritage sites are related to volcanoes.

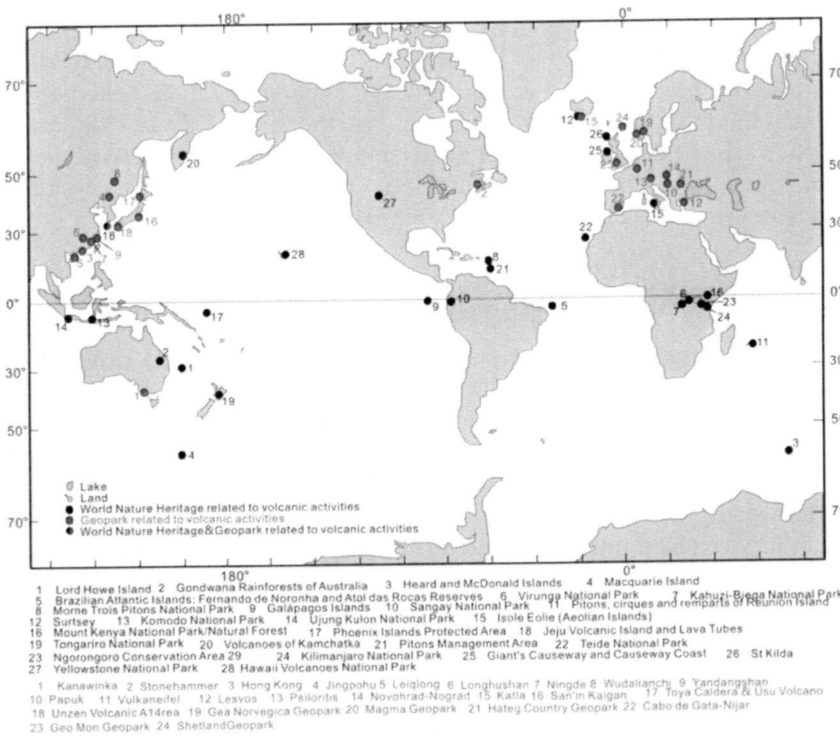

FIGURE 15. Location of the world nature heritage sites and geoparks which are related to volcanic activities

Heritage is our legacy from the past, what we live with today, and what we pass on to future generations. The United Nations Educational, Scientific and Cultural Organization (UNESCO) seeks to encourage the identification, protection and preservation of cultural and natural heritage around the world considered to be of outstanding value to humanity. This is embodied in an international treaty called the Convention concerning the Protection of the World Cultural and Natural Heritage, adopted by UNESCO in 1972 [61]. The World Heritage List includes 936 sites forming part of the cultural and natural heritage which the World Heritage Committee considers as having outstanding universal value [61].These include 725 cultural, 183 natural and 28 mixed properties in 153 states. Some of them are closely related to volcanic activities and the unique geological landscape and ecological systems in these places illustrate the charm of volcanic activities for humans.

A geopark is defined by UNESCO in its International Network of Geoparks' program as "a territory encompassing one or more sites of scientific importance, not only for geological reasons but also by virtue of its archaeological, ecological or cultural value" [62]. A global geopark is a unified area with geological heritage of international significance and where that heritage is being used to promote the sustainable development of the local communities that live there [63]. The key heritage sites within a geopark should be protected under local, regional or national legislation as appropriate.

We will now introduce some typical heritage sites and geoparks which are associated with volcanic activities (Table 6, Table 7).

TABLE 6. World nature heritage sites which are related to volcanic activities [61]

Country	World Nature Heritage Sites	
Australia	Lord Howe Island Group	Located in the South Pacific, 700 km north-east of Sydney, the property is included administratively in New South Wales. Lord Howe Island is the eroded remnant of a large shield volcano which erupted from the sea floor intermittently for about 500,000 years in the late Miocene (6.5-7 million years ago). The entire island group has remarkable volcanic exposures not known elsewhere.
	Gondwana Rainforests of Australia	The outstanding geological features displayed around shield volcanic craters and the high numbers of rare and threatened rainforest species are of international significance for science and conservation.
	Heard and McDonald Islands	Located in the Southern Ocean, approximately 1,700 km from the Antarctic continent and 4,100 km south-west of Perth. As the only volcanically active sub-Antarctic islands they 'open a window into the Earth', thus providing the opportunity to observe ongoing geomorphic processes and glacial dynamics.
	Macquarie Island	Macquarie Island is an oceanic island in the Southern Ocean, lying 1,500 km south-east of Tasmania and approximately halfway between Australia and the Antarctic continent. It is the only island in the world composed entirely of oceanic crust and rocks from the Earth's mantle deep below the surface.

Brazil	Brazilian Atlantic Islands: Fernando de Noronha and Atol das Rocas Reserves	The Fernando de Noronha Archipelago and Rocas Atoll represent the peaks of a large submarine mountain system of volcanic origin, which rises from the ocean floor some 4,000 m in depth. The Fernando de Noronha volcano is estimated to be between 1.8 million and 12.3 million years old.
Democratic Republic of the Congo	Virunga National Park	Virunga National Park is notable for its chain of active volcanoes and the greatest diversity of habitats of any park in Africa. Features include hot springs in the Rwindi plains and the Virunga Massif volcanoes, such as Nyamulagira (3,068 m) and Nyiragongo (3,470 m), are still active, which alone account for two-fifths of the historical volcanic eruptions on the African continent. They are especially notable because of their highly fluid alkaline lavas. The activity of Nyiragongo is globally significant for its demonstration of lava lake volcanism, with a quasi-permanent lava lake at the bottom of its crater, periodic draining of which has been catastrophic to the local communities.
	Kahuzi-Biega National Park	A vast area of primary tropical forest dominated by two spectacular extinct volcanoes, Kahuzi and Biega, the park has a diverse and abundant fauna.
Dominica	Morne Trois Pitons National Park	Luxuriant natural tropical forest blends with scenic volcanic features of great scientific interest in this national park centred on the 1,342-m-high volcano known as Morne Trois Pitons. With its precipitous slopes and deeply incised valleys, 50 fumaroles, hot springs, three freshwater lakes, a 'boiling lake' and five volcanoes, located on the park's nearly 7,000 ha.
Ecuador	Galápagos Islands	The islands were formed by volcanic processes and most represent the summit of a volcano, some of which rise over 3,000 m from the Pacific floor. Ongoing seismic and volcanic activities reflect the processes that formed the islands. The larger islands typically comprise one or more gently sloping shield volcano, culminating in collapsed craters or calderas. Other noteworthy landscape features include crater lakes, fumaroles, lava tubes, sulphur fields and a great variety of lava and other ejects such as pumice, ash and tuff.
	Sangay National Park	The site is situated in the Cordillera Oriental region of the Andes in central Ecuador. The park is dominated by three volcanoes, Tungurahua (5,016 m) and El Altar (5,139 m) to the north-west and Sangay (5,230 m) in the central section of the park.
France	Pitons, cirques and remparts of Reunion Island	The Pitons, cirques and remparts of Reunion Island site covers more than 100,000 ha or 40 % of La Réunion, an island comprising two adjoining volcanic massifs located in the south-west of the Indian Ocean. Dominated by two towering volcanic peaks, massive walls and three cliff-rimmed cirques, the property includes a great variety of rugged terrain and impressive escarpments, forested gorges and basins creating a visually striking landscape.
Iceland	Surtsey	Surtsey is a new island formed by volcanic eruptions in 1963-1967. It has been legally protected from its birth and provides the world with a pristine natural laboratory. There is a recent tendency to promote Surtsey Island as a geopark.

Indonesia	Komodo National Park	The generally steep and rugged topography reflects the position of the national park within the active volcanic 'shatter belt' between Australia and the Sunda shelf. These volcanic islands are inhabited by a population of around 5,700 giant lizards, whose appearance and aggressive behaviour have led to them being called 'Komodo dragons'. They exist nowhere else in the world.
	Ujung Kulon National Park	This national park, located in the extreme south-western tip of Java on the Sunda shelf, includes the Ujung Kulon peninsula and several offshore islands and encompasses the natural reserve of Krakatoa. In addition to its natural beauty and geological interest – particularly for the study of inland volcanoes – it contains the largest remaining area of lowland rainforests in the Java plain.
Italy	Isole Eolie (Aeolian Islands)	The Aeolian Islands provide an outstanding record of volcanic island-building and destruction, and ongoing volcanic phenomena. Studied since at least the 18th century, the islands have provided the science of vulcanology with examples of two types of eruption (Vulcanian and Strombolian).
Kenya	Mount Kenya National Park/Natural Forest	At 5,199 m, Mount Kenya is the second-highest peak in Africa. It is an ancient extinct volcano, during whose period of activity (3.1-2.6 million years ago) it is thought to have risen to 6,500 m.
Kiribati	Phoenix Islands Protected Area	The Phoenix Island Protected Area (PIPA) is a 408,250 sq.km expanse of marine and terrestrial habitats in the Southern Pacific Ocean. PIPA conserves one of the world's largest intact oceanic coral archipelago ecosystems, together with 14 known underwater sea mounts (presumed to be extinct volcanoes) and other deep-sea habitats.
Korea, Republic of	Jeju Volcanic Island and Lava Tubes	Jeju Volcanic Island and Lava Tubes together comprise three sites that make up 18,846 ha. It includes Geomunoreum, regarded as the finest lava tube system of caves anywhere, with its multicoloured carbonate roofs and floors, and dark-coloured lava walls; the fortress-like Seongsan Ilchulbong tuff cone, rising out of the ocean, a dramatic landscape; and Mount Halla, the highest in Korea, with its waterfalls, multi-shaped rock formations, and lake-filled crater.
New Zealand	Tongariro National Park	Tongariro National Park is situated on the central North Island volcanic plateau. The park's volcanoes, which are outstanding scenic features of the island, contain a complete range of volcanic features. The mountains at the heart of the park have cultural and religious significance for the Maori people and symbolize the spiritual links between this community and its environment.

Russian Federation	Volcanoes of Kamchatka	This is one of the most outstanding volcanic regions in the world, with a high density of active volcanoes, a variety of types, and a wide range of related features.
Saint Lucia	Pitons Management Area	Dominating the mountainous landscape of St Lucia are the Pitons, two steep-sided volcanic spires rising side by side from the sea. The Pitons are part of a volcanic complex, known to geologists as the Soufriere Volcanic Centre which is the remnant of one (or more) huge collapsed stratovolcano. The Pitons occur with a variety of other volcanic features including cumulo-domes, explosion craters, pyroclastic deposits (pumice and ash) and lava flows
Spain	Teide National Park	Teide National Park, dominated by the 3,781 m Teide-Pico Viejo stratovolcano, represents a rich and diverse assemblage of volcanic features and landscapes concentrated in a spectacular setting. Mount Teide is a striking volcanic landscape dominated by the jagged Las Cañadas escarpment and a central volcano that makes Tenerife the third tallest volcanic structure in the world. Teide National Park is an exceptional example of a relatively old, slow moving, geologically complex and mature volcanic system.
Tanzania, United Republic of	Ngorongoro Conservation Area	The Ngorongoro Conservation Area spans vast expanses of highland plains, savannah, savannah woodlands and forests. It includes the spectacular Ngorongoro Crater, the world's largest caldera. Ngorongoro Crater is one of the largest inactive unbroken calderas in the world which is unflooded. It has a mean diameter of 16-19 km, a crater floor of 26,400 ha, and a rim soaring to 400-610 m above the crater floor. The formation of the crater and other highlands are associated with the massive rifting which occurred to the west of the Gregory Rift Valley.
	Kilimanjaro National Park	Mount Kilimanjaro is one of the largest volcanoes in the world. It has three main volcanic peaks, Kibo, Mawenzi and Shira. With its snow-capped peak and glaciers, it is the highest mountain in Africa.
United Kingdom of Great Britain and Northern Ireland	Giant's Causeway and Causeway Coast	The Giant's Causeway lies at the foot of the basalt cliffs along the sea coast on the edge of the Antrim plateau in Northern Ireland. It is made up of some 40,000 massive black basalt columns sticking out of the sea. The Causeway Coast has an unparalleled display of geological formations representing volcanic activity during the early Tertiary period some 50-60 million years ago. The most characteristic and unique feature of the site is the exposure of a large number of regular polygonal columns of basalt in perfect horizontal sections forming a pavement. Tertiary lavas of the Antrim Plateau, covering some 3,800 km^2, represent the largest remaining lava plateau in Europe. The coastline is composed of a series of bays and headlands consisting of resistant lavas. The average height of the cliffs is 100 m, and has a stepped appearance due to the succession of five or six lava flows through geological time.

	St Kilda	This volcanic archipelago, with its spectacular landscapes, is situated off the coast of the Hebrides and comprises the islands of Hirta, Dun, Soay and Boreray. It has some of the highest cliffs in Europe, which have large colonies of rare and endangered species of birds, especially puffins and gannets.
United States of America	Yellowstone National Park	Yellowstone National Park, established in 1872, covers 9,000 km2 of a vast natural forest of the southern Rocky Mountains in the North American west. The park has a globally unparalleled assemblage of surficial geothermal activity, thousands of hot springs, mudpots and fumaroles, and more than half of the world's active geysers. The world's largest recognized caldera (45km by 75km – 27 miles by 45 miles) is contained within the park. It boasts an impressive array of geothermal phenomena, with more than 3,000 geysers, lava formations, fumaroles, hot springs and waterfalls, lakes and canyons. The park is part of the most seismically active region of the Rocky Mountains, a volcanic 'hot spot'. The latest eruptive cycle formed a caldera 45 km wide and 75 km long, when the active magma chambers erupted and collapsed. The crystallizing magma is the source of heat for hydrothermal features such as geysers, hot springs, mud pots and fumaroles.
	Hawaii Volcanoes National Park	Lies in the south-east part of the island of Hawaii (Big Island), the easternmost island of the State of Hawaii, and includes the summit and south-east slope of Mauna Loa and the summit and south-western, southern, and south-eastern slopes of the Kilauea Volcano. Mauna Loa and Kilauea are two of the world's most active and accessible volcanoes where ongoing geological processes are easily observed. This property serves as an excellent example of island building through volcanic processes. It represents the most recent activity in the continuing process of the geologic origin and change of the Hawaiian Archipelago.

There are many other volcanic landscapes around the world besides the heritage sites and geoparks which have not been listed in Table 6 and Table 7. Take the Llancanello Volcanic Field in Argentina as an example, together with the nearby Payun Matru Field, there are at least 800 scoria cones and voluminous lava fields that cover an extensive area behind the Andean volcanic arc, six volcanoes show evidence of explosive eruptions involving magma-water interaction. Tuff rings and tuff cones can also be seen in this field. The diversity of volcanic landforms is so well-preserved that some people are promoting it as a geopark. Changbai Mountain volcano area (Figure 17) in northeast China is also a famous resort around the world. In Changbai Mountain, the forest is boundless, waterfalls are plentiful, the mountain peaks poke into the clouds and steamy hot springs flow along the canyons. Tianchi is embraced by a group of peaks. It is a paradise for scientific research because of rich biological resources, forest resources, mineral resources and typical volcanic geomorphologic landscape. We believe that all volcanic heritage sites should be protected under local, regional or national legislation as appropriate.

TABLE 7. Global Geoparks which are related to volcanic activities [64, 65]

Country	Geopark	
Australia	Kanawinka Geopark	The Kanawinka Geopark is located in south-eastern Australia. The area covers 26,910 square kilometres, seven local government areas with 58 protected geosites within Australia's most extensive volcanic province. The surface geology of Kanawinka is a striking contrast of sweeping plains and spectacular cones which are largely the product of volcanic activity. The Blue Lake is one of the most significant volcanic sites within the Kanawinka Global Geopark. Mount Elephant is considered as "one of the highest and one of the major scoria cones in the largest homogenous volcanic plains on earth". The Geopark displays geological history and the nature of volcanoes in the region.
Canada	Stonehammer Geopark	Stonehammer Geopark is located in Southern New Brunswick on the East Coast of Canada. The Geopark encompasses 2500 square kilometres and extends from Lepreau Falls to Norton and from the Fundy Trail to the Kingston Peninsula. The landscape of the Stonehammer Geopark has been created by the collision of continents, the closing and opening of oceans, volcanoes, earthquakes, ice ages and climate change. It shows a billion years of Earth's history.
China	Hong Kong Geopark	Situated in close proximity to Hong Kong's world-famous financial and business centre, Hong Kong Global Geopark of China is lauded as a unique "geopark in the city". Hong Kong Geopark's major geo-attractions include well-outcropped acidic volcanic hexagonal rock columns, of which the average diameter is 1.2 metres and which are distributed over a land-and-sea area of 100 square kilometres. The Geopark also has comprehensive sedimentary rock formations formed about 400 to 55 million years ago. In addition, It is integrated with diverse ecological resources and historical relics, all combining to form the unique natural landscape of Hong Kong Geopark.
China	Jingpohu Geopark	The Jingpohu Geopark is located in the upper-middle reaches of the Mudanjiang River, Northeast China's Heilongjiang Province. Jingpo lake covers an area of 79 square kilometres with a total water volume of 1.62 billion cubic meters. It is the largest lava barrier lake in China, and the second largest in the world. About 1 million years ago, Lava as a result of volcanic eruption blocked the river bed of the Mudanjiang River forming a lava dam. As a result, the incipient Jingpo Lake began to take shape. In the late Glacial Period, larger volcanic eruptions helped build a more voluminous lava dam. Hence the Jingpo Lake as we see it today. The granite rocks after experiencing various earth movements are in the shape of cliffs which are called a "geological corridor" by visitors. The Diaoshuilou Waterfall is the only large collapsed-lava waterfall of China which is also rarely seen in the world.

China	Leiqiong Geopark	Leiqiong Global Geopark is located in the southern margin of Chinese Mainland, straddling Qiongzhou Strait. More than 100 volcanoes are densely distributed across the Geopark. These include examples of nearly all volcanic types. Judging from the number, variety and completeness of the volcanoes, the park is considered topmost among the Quaternary volcanic belts of China. Two districts of Leiqiong Geopark feature several 'maar' craters – broad, low-relief craters caused by groundwater coming into contact with hot lava. The Geopark is extremely rich in volcanic landscapes and lava structures, such as different kinds of lava flows and tunnels. The territory of the park spans Haikou City and Zhanjiang City, which are famous cultural centres. Also, the Geopark is positioned in an ecological transition area characterized by a rich diversity of flora and fauna.
	Longhushan Geopark	Located in China's Jiangxi Province, Longhushan Geopark holds geological, geomorphological, human and natural ecological landscapes. The major landform in the park is the Danxia landform and others but minor are volcanic landforms.
	Ningde Geopark	Ningde Geopark is located in the southeastern part of the Eurasian Plate, belonging to the continental marginal active zone adjacent to the Pacific Ocean. Ningde Geopark is an integration of various landforms such as miarolite landform, volcanic landform, erosion riverbed landform, and erosion coastal landform. The occurrence of many landforms in a geopark suggests that the area has a complicated geological history. At the same time, the park landform landscape also has the very high ornamental value.
	Wudalianchi Geopark [66] (Figure 16)	Wudalianchi Geopark is located in the north central Heilongjiang Province, China. Wudalianchi World Geopark is such a precious legacy left by the volcanic actions in the Quaternary period. Fourteen young and old volcanoes, the world's top volcanic resources, stand in the 1,060 km^2 geopark, with their eruption ages ranging from 2.07 million years prehistory to 280 years before now. They are the world's best-conserved volcanic landforms with the most varieties and most typical forms. The 14 volcano cones that sprung from the ground add beauty to the mountains and rivers, composing a remarkable picture; the micro-topographic landscapes in various forms such as lava platform, lava sea, lava cascade, lava postern, lava stalactite, lava Vortex, trunk lava, flower-like lava, fumarolic cone and disc, lapilli and volcanic bomb are named "Natural Museum of Volcano" and "An Open Textbook on Volcano". The five connected lakes, like a string of beads, are formed by the latest volcanic magma filling the ancient Baihe valley, thus named Wudalianchi (five connected lakes). The mineral water here is renowned as "The World's Three Cold Springs" together with Vichy mineral water of France and Ciscaucasia mineral water of Russia.

	Yandangshan National Geopark	Yandangshan National Geopark is located in Zhejiang Province in China, which covers an area of 294.6 square kilometres. The Mount Yandang Shan is a natural park based on its geological landform of Cretaceous volcanic rhyolite. The geoheritage in the Park is referred to as a typical example of the formation and evolvement mode of reviviscent caldera on the edge of Asian Plates in late Mesozoic. From the Geoheritage, we can see the complete geo-evolving process of the eruption, subsiding and reviviscent apophysis of volcanoes. The Geoheritage provides a permanent sample for studying the volcanoes in Mesozoic.
Croatia	Papuk Geopark	Papuk Geopark is located in Slavonia, in eastern Croatia. Mount Papuk has a "stormy" geological history going as far back as the Precambrian with metamorphic rocks that date back as far as the Precambrian, more than 540 million years ago. More recent geological features result from mountain building with Mesozoic sediments 260 and 65 million years old, including carbonate rocks, mostly dolomites, as well as the Cretaceous volcanic sedimentary complex of Rupnica. The Rupnica geosite is famous due to a characteristic exposure of the columnar jointing developed in the albite rhyolite volcanic rock.
Germany	Vulkaneifel Geopark	Located in the middle of Central Europe, at the northwestern part of the 'Rheinish Slate Mountains', the rolling Eifel highlands are a hilly landscape with deep, glacially carved valleys cut into old Devonian sediments (360-415 million years old). Volcanoes dot the landscape, with 350 known eruption centres, and give the area its name – Vulkaneifel. In some craters, bogs and lakes have formed while others remain dry. Known as 'maar' craters, these bodies reveal a nearly uninterrupted stack of sediments dating back to 150,000 years ago that provides data for the reconstruction of past climate, vegetation and ecology. Similarly, fossils found in 43 million years old sediments of Eckfeld Maar are of worldwide importance, since they contained an archetypal horse and the oldest known honey bee. The Vulkaneifel has attracted geo-scientists for 200 years and many international research projects have been conducted here.
Greece	Lesvos Petrified Forest Global Geopark	Located in the western part of the Greek island of Lesvos, north-east Aegean Sea, the Lesvos Petrified Forest Geopark features rare and impressive fossilised tree-trunks. Formed some 15 to 20 million years ago, due to intense volcanic activity, the trees were covered by lava, ashes and other materials that were spewed into the atmosphere. The Lesvos Petrified Forest Global Geopark brings visitors to an ancient forest preserved by a massive volcanic eruption 20 million years ago.
	Psiloritis Geopark	Psiloritis Monts rose up through the sea a few million years ago when the African continent encroached on Europe. Psiloritis Geopark is characterized by its superb geodiversity. This is reflected by a great variety of volcanic, sedimentary and metamorphic rocks aging from Permian to Pleistocene (300 to 1 million year ago), outstanding folds and faults, fascinating caves and deep gorges with rich biodiversity.

Hungary-Slovakia	Novohrad-Nograd	Novohrad-Nograd Geopark lies at the border of Hungary and Slovakia. Being transnational, the Geopark's name comes from the Slovak and Hungarian names of the very county, where the Novohrad - Nograd Geopark is located. The geology of the Geopark includes diverse past volcanic events and a geological history dating back the last 30 million years from the birth of the Pannonian basin. Within a small area, the Geopark contains a wide spectrum of volcanic sites of spectacular sights, and several landscape protection areas and other territories. The area is also recognized as an important centre for the Palóc ethnic group's folk art and living traditions. Recently people pay more attention to the Maar diatreme volcanism and Miocene andesite volcanism in the region.
Iceland	Katla Geopark	The Katla Geopark is named after the volcano Katla that has for centuries had great impact on Icelandic nature and people living in the area. Katla is one of the largest central volcanoes in Iceland, covered by the glacier Mýrdalsjökull. Katla Geopark is located in the southern part of Iceland. The Geopark consists of three municipalities: Skaftárhreppur, furthest east; Mýrdalshreppur, in the middle; and Rangárþing eystra, in the west. The volcanic activity of Eyjafjallajökull, Katla, Grímsvötn, Lakagígar and Eldgjá and its widespread effect on the landscape in the area provide the geological basis for the Geopark. Apart from the ice-capped volcanoes and lava streams, sandur plains with their black beaches and rootless vents (pseudocraters) are prominent features in the landscape.
Japan	San'in Kaigan Geopark	The San'in Kaigan Geopark is located in the west of Japan and stretches from the eastern Kyogamisaki Cape, Kyoto to the western Hakuto Kaigan Coast, Tottori. The San'in Kaigan Geopark is home to a diversity of geological sites related to the formation of the Sea of Japan, including granite outcrops formed when Japan was part of the Asian continent (70 million years ago), as well as sedimentary and volcanic rocks accumulated when Japan rifted away from Asia (25 to 15 million years ago) to form the Sea of Japan, a geological process still on-going today. It also contains geographical features, such as ria type coasts, sand dunes, sandbars, volcanoes and valleys. Thanks to such diversity, the Geopark is home to rare plants like Pseudolysimachion ornatum and Ranunculus nipponicus, as well as Ciconia boyciana (Oriental White Storks) - a symbol of biodiversity.

Japan	Toya Caldera and Usu Volcano Global Geopark	The Toya Caldera and Usu Volcano Global Geopark is located in Hokkaido, northern Japan and displays a unique showcase of active volcanism on the Pacific Rim. This is a unique region which has a wealth of characteristic geological relics in a relatively compact area, from the 110,000-year-old Toya Caldera to the 20,000 to 10,000-year-old Nakajima domes and a strato-volcano Usu, as well as the recent history of eruptions. The recent eruptive stage of Usu volcano started in 1663, and repeated nine times, creating lava domes, such as "Showa-Shinzan" that was born in a wheat field during 1943-1945. There are also precious fauna and flora living in the conserved thick forest and abundant water resources.
	Unzen Volcanic Area Global Geopark	Unzen volcano gives us a lot of gifts: outstanding landscapes, hot springs, spring water, fertile agricultural soil, and so on. On the other hand, Unzen volcano erupts repeatedly and causes serious disasters. In 1792, about 15,000 people were killed by the tsunami derived from the strong earthquakes due to sector collapse of an old lava dome, one of the worst volcanic disasters in Japan. In 1991-1995, a part of lava dome at the summit of the mountain collapsed. The pyroclastic surges containing giant hot ash clouds took 44 lives. Many buildings, houses, and school were also completely burned or buried by the flows and many residents lost their property. The burned school building and destroyed houses of the last eruption have been preserved just as they were. Now, these facilities are the main geosites of the Geopark and have been utilized widely for local disaster prevention educational programs in Japan. The theme of Unzen Volcanic Area Global Geopark is "the coexistence of an active volcano and human beings".
Korea	Jeju Geopark	See Table 6.
Norway	Gea Norvegica Geopark	Gea Norvegica Geopark is located in southeastern Norway, in the counties of Vestfold and Telemark. The story told in this geopark is a 1.5 billion year-long journey, from old mountain chains, the tropical sea, strange volcanoes, rifting of a continent and a glaciated surface -and how we all depend on this geological diversity and natural resources.
	Magma Geopark	Magma Geopark is situated in southwest Norway. The story began as early as 1.5 billion years ago when red-hot magma and sky-high mountains characterized the region. Through millions of years, glaciers helped to form today's characteristic landscape. Although the magma has cooled down and solidified and the mountains have been worn away, the area offers a glimpse into the roots of an ancient mountain chain. Here is a rock type called anorthosite that is more common on the moon than on Earth.
Romania	Hateg Country Geopark	The Hateg Country Geopark is located in the central part of Romania, in a very fertile region surrounded by mountains. The region is world famous for its dwarf dinosaurs from the end of Cretaceous, 65 million years ago. Also well documented at the Geopark are the volcanic rocks-tuffs, lavas and craters that mark eruptions that took place during the age of the dinosaurs.

Spain	Cabo de Gata-Nijar Global Geopark	The Geopark's geodiversity derives from ancient Miocene volcanic substrata emplaced between 16 and 8 million years ago. In fact, the Geopark represents the most extensive and complex calco-alcaline fossil volcanism in the Iberian Peninsula. Visitors can walk through an open air geological museum with lava flows, volcanic domes, calderas, columnar joints, and fossilized sand beaches with tropical fossil reefs. This semi-arid climate and poor soil supports a surprising richness of plant species which ranks among the most diverse in Europe.
United Kingdom	Geo Mon Geopark (Wales)	The beautiful Isle of Anglesey lies off the west coast of Wales, UK. The island is renowned for its diverse tectonic geology. Brilliantly coloured Precambrian (about 800 Ma) 'pillow' lavas (erupted on the deep ocean floor with a characteristic shape) and deep ocean sediments are exposed at the western end whilst on the north coast is the world type locality for melange containing blocks of limestone with 800 million year old fossils. The remarkable folds and faults of South Stack date from the Cambrian period (520 Ma). Carboniferous limestones deposited about 350 Ma, crowded with fossil coral and shells show how ancient ice ages, sea level changes and plate movements affected the world long ago.
	Shetland Geopark	This tiny windswept archipelago has played host to tropical seas, volcanoes, deserts, ice ages and ancient rivers. The islands can boast the best section through the flank of a volcano in the UK, the best exposure of one of Europe's major tectonic faults, and are one of the best places in the world to see a compact vertical section through ancient oceanic crust.

FIGURE 16. Landscape of Wudalianchi, China

FIGURE 17. Landscape of Changbai Mountain, China

7. CONCLUSION

Volcanoes have provided us with material and spiritual wealth. The land resources enlarge our habitats; geothermal activity is a clean and regenerative energy source; hot springs and mineral springs are beneficial to our health; volcanic materials had become new and popular materials of the 21st century; gemstones and mineral resources are symbols of wealth; volcano landscapes provide us with an opportunity to experience the rich and extraordinary natural world of the volcanic zones. It is said that volcanoes have had a far-reaching impact upon our lives and participated in the progress of our society. So we should protect volcanic resources, exploit them reasonably and appreciate all the gifts which volcanoes give us.

REFERENCES

1. Allen J. A., Ewel K. C., Jack J. Patterns of natural and anthropogenic disturbance of the mangroveson the Pacific Island of Kosrae. Wetlands Ecology and Management 2001; 9: 279-289.
2. Louvat P., Allègre C. J. Riverine erosion rates on Sao Miguel volcanic island, Azores archipelago. Chemical Geology 1998; 148: 177-200.
3. Fridriksson S., Magnússon B. Development of the Ecosystem on Surtsey with References to Anak Krakatau. GeoJournal 1992; 28(2):287-291.
4. Partomihardjo T., Mirmanto E., Whittaker R. J. Anak Krakatau's Vegetation and Flora circa 1991, with Observations on a Decade of Development and Change. GeoJournal 1992; 28(2):233-248.
5. Klumpp D.W., Burdon-Jones C. Investigations of the potential of bivalve molluscs as indicators of heavy metal levels in tropical marine waters. Australian Journal of Marine and Freshwater Research 1982; 33(2):285-300.

6. Baidu. Baike: Volcanic island. http://baike.baidu.com/view/15872.htm (accessed 16 June 2012).
7. The Surtsey Research Society. Surtsey-The Surtsey Research Society: Geology: Erosion of the Island. http://www.surtsey.is/pp_ens/geo_2.htm (accessed 16 June 2012).
8. The Surtsey Research Society. Surtsey-The Surtsey Research Society: Geology: Formation of Palagonite Tuffs. http://www.surtsey.is/pp_ens/geo_4.htm (accessed 16 July 2012).
9. Garvin J. B., Williams J., R.S., Frawley J. J., Krabill W. B. Volumetric evolution of Surtsey, Iceland, from topographic maps and scanning airborne laser altimetry: 1968-1998. In: Surtsey Research 11. Reykjavil: Gutenberg; 2000. p127–134. Available from http://www. surtsey.is/pp_ens/report/report_XI.htm (accessed 16 July 2012).
10. Liu R. X., Li N. Review on Volcano Resources. Bulletin of Mineralogy Petrology and Geochemistry 2000; 19(3): 172-174.
11. INFORSE - International Network for Sustainable Energy. Geothermal Energy: Geothermal System. http://www.inforse.dk/europe/dieret/Geothermal/ geotermal. html (accessed 12 May 2012).
12. Wang J. Y., Liu S. B., Zhu H. Z. Development Strategy of China's Geothermal Energy in 21st Century. Electric Power 2000; 33(9):85-94.
13. Huang S. P., Liu J. Q. Geothermal energy stuck between a rock and a hot place. Nature 2010; 463: 293.
14. Department of Energy U.S. An evaluation of enhanced geothermal systems technology. http:// www1.eere.energy.gov/geothermal/enhanced_systems.html (accessed 9 May 2012)
15. Alison H., Leslie B., Dan J., Karl G. Geothermal Energy: International Market Update. Geothermal Energy Association. http://www.geo-energy.org/pdf/reports/ gea_international_ market_ report_final_may_2010.pdf (accessed 12 June 2012).
16. International Geothermal Association. Geothermal in the world: Electricity Generation: Installed Generating Capacity. http://www.geothermal-energy.org/ 226,installed_generating_capacity.html (accessed 12 June 2012).
17. Desy N., Alfian. Indonesia set to become No.1 user of geothermal. The Jakarta Post. http://www.thejakartapost.com/news/2010/04/27/indonesia-set-become-no-1-user-geothermal.html (accessed 12 June 2012).
18. Si W. Development and utilization of geothermal resources in Indonesia. Geothermal Energy 2010; 5: 31-32.
19. Wu, X. M., Bo, Q., Yuan, H. M., Wan, X. Z., Zhao, X. J. Present situation of the development and utilization of geothermal resources in Iceland. Hydrogeology and Engineering Geology 2007; (5): 129-130.
20. Chu J. The country which using geothermal energy instead of the oil. Life & Disaster 2009; (4): 43.
21. Ragnarsson A., Du S P., Zhao P. Geothermal energy in Iceland. Geothermal Energy 1999; (4):15-21.
22. Cole J.W. Structural control and origin of volcanism in the Taupo volcanic zone, New Zealand. Bulletin of Volcanology 1990; 52:445-459.
23. Zhang, D. Z. & Yan, Y. H. The coexistence of the high temperature geothermal power generation and volcanic geothermal landscape in New Zealand. In: The proceedings of the Meeting of Development and Protection of Geothermal Resources in China: the National Protection Research Seminar on the Exploitation and Protection of the Geothermal Resources: 42-44, October 2007, Tengchong, Yunnan Province, China.
24. Wikipedia. Steam pipelines towards Wairakei geothermal power station.jpg. http://en. wikipedia.org/wiki/File:Steam_pipelines_towards_Wairakei_geothermal_ power_station.jpg (accessed 16 June 2012).
25. Yang X. China Daily: The type and effect of hot spring. Http://www.chinadaily. com.cn//hqylss/ 2007-01/15/content_783931.htm (accessed 20 May 2012).

26. Cartwright I., Weaver T., Tweed S., Ahearne D., Cooper M., Czapnik K., Tranter J. Stable isotope geochemistry of cold CO2-bearing mineral spring waters, Daylesford, Victoria, Australia: sources of gas and water and links with waning volcanism. Chemical Geology 2002; 185: 71-91.
27. Baidu. Baike: Mineral spring water. http://baike.baidu.com/view/35726.htm (accessed 16 June 2012).
28. Baidu. Baike: The mineral spring water of Wudalianchi. http://baike.baidu.com/view/877619.htm (accessed 16 June 2012).
29. An K. S. Health Preserving and Culture of mineral spring. In: International Forum on hot spring culture of China conference proceedings, 27-28 May, 2008, Xi'an, China.
30. Japan-i. Explore Japan: Kyushu: Kagoshima: Ibusuki: Ibusuki Hot Spring. http://www.japan-i.jp/explorejapan/kyushu/kagoshima/ibusuki/d8jk7l00000 2rrsa. html (accessed 16 June 2012).
31. Chen Q. Q., Du G. Y. & Yang Y. Q. Discussion on designing of tourism products in the Aershan volcano and warm spring national geopark. Resources & Industries 2006; 8(4): 85-88.
32. Han X. J., Jin X. & Sun, C. H. Geothermal structure of Aershan hot springs, Inner Mongolia. Acta Geoscientica Sinica 2001; 22(3): 259-264.
33. Li S. X., Bin D. Z. & Chen J. P. Development and protection suggestion of the volcanic warm spring resources in Tengchong. Geothermal Energy 2007; (6): 20-22.
34. Dong W., Zhang M. T., editor. Geothermal Energy in Tengchong. Beijing: Science Press;1989.
35. Li X. H. Volcanic ash used as cement admixture and its effect on the cement performance. Wealth Today 2011; (12): 268-269.
36. Su X. H. A Study on rapid annealing technology and thermo-instrument in the control system in industrial test for energy saving tunnel furnace in cast basalt production. China Building Materials Science 1992; (2): 40-43, 50.
37. Xiao G. P., Li J. X., Wang Z. D. Application of Casting Basalt Technology in the Sluiceway of BF. Shandong Metallurgy 2003; (6): 80.
38. Bai S. L., Zhao L., Su X. H. Basalt cast stone annealing process parameters and yield. China Building Materials Science & Technology 2000; (3):37, 40.
39. Qi F. J., Li J. W., Li C. X., Wei H. X. & Gao Y. Z. Summary on the research of continuous basalt fiber. Hi-Tech Fiber & Application 2006; 31(2): 42-46.
40. Wen, F. Basalt fiber: the new environmentally friendly fiber in 21st century. China Building Materials News: (2009, April 16), pp. A3.
41. Lei J., Dang X. A. & Li J. J. Characteristic, application and development of basalt fiber. New Chemical Materials 2007; 35(3): 9-11.
42. Guo H., Ma Y. & Chen Z. N. Development and application prospect of continuous basalt fiber. China Fiber Inspection 2010; (5): 76-79.
43. Wang R. S. Discuss of Basalt Filament's Property and Application. Progress in Textile Science & Technology 2010; (1): 40-42.
44. Xie E. G. Characteristics of basalt fiber and its application prospects in China. Fiber Glass 2005; (5): 44-48.
45. Wikipedia.Kimberlite. http://en.wikipedia.org/wiki/Kimberlite (accessed 24 June 2012).
46. AllAboutGemstones. Diamonds: Diamond Geology & Kimberlites. http://www.allaboutgemstones.com/diamond_geology.html (accessed 24 June 2012).
47. Diamond Source of Virginia, Inc. Education: Diamonds Education: What is a diamond? http://www.diamondsourceva.com/Education/diamonds-what-is-a-diamond.asp (accessed 24 June 2012).
48. Lorenz V. Formation of phreatomagmatic maar-diatreme volcanoes and its relevance to kimberlite diatremes. Physics and Chemistry of the Earth 1975; (9):17-27.
49. American Museum of Natural History. Origins: The nature of diamond: Kimberlite and lamproite. http://www.amnh.org/ exhibitions/ diamonds/kimberlite.html (accessed 24 June 2012).

50. American Museum of Natural History. Origins: The nature of diamond: What does the carbon come from? http://www.amnh.org/exhibitions/ diamonds/ kimberlite .html (accessed 24 June 2012).
51. Luan B. A. Gem. Beijing: Metallurgical Industry Press; 1993.
52. Wikipedia.Garnet. http://en.wikipedia.org/wiki/Garnet (accessed 24 June 2012).
53. India mart. Company Directory: Gems & Jewelry: Precious Gemstones. http://www. indiamart. com/gem-world/precious-gemstones.html (accessed 24 June 2012).
54. Wikipedia.Obsidian. http://en.wikipedia.org/wiki/Obsidian (accessed 24 June 2012).
55. Pollard A. M., Heron C. History of obsidian. In: Archaeological Chemistry. Royal Society of Chemistry. 2008; 91.
56. Wilford J. N. In Syria, a Prologue for Cities. The New York Times. http://www.nytimes. com/2010/04/06/science/06archeo.html?pagewanted=all (accessed 24 June 2012).
57. Hu, S. X., Wang, W. B. Inquiring into the classification of gold deposits related to volcanism-subvolcanism-intrusion-hydrothermal processes. Gold Geology 1997; 3(3): 25-29.
58. Jiang, F. Z., Wang, Y. W. Volcanism and gold deposit. Chinese Geology 2003; 30(1):84-92.
59. Dru, L. J., Zhang Q. M. Geological and tectonic evolution of Muruntau gold deposit in Uzbekistan. Geological Science and Technology Abroad 1997; (2): 28-38.
60. Wikipedia. Diatomaceous earth. http://en.wikipedia.org/wiki/Diatomite (accessed 24 June 2012).
61. UNESCO. UNESCO World Heritage Centre: World Heritage List. http://whc.unesco.org/en/list (accessed 20 May 2012).
62. Great Arc Initiative. Niagara Escarpment Resource Network. http:// www. escarpmentnetwork.org/ great-arc-initiative.php (accessed 16 June 2012).
63. UNESCO. Natural Sciences: Environment: EARTH SCIENCES: Geoparks: What is a Global Geopark? http://www.unesco.org/new/en/natural-sciences/environment /earth-sciences/geoparks/some-questions-about-geoparks/what-is-a-global-geopark/ (accessed 16 June 2012).
64. UNESCO. Natural Sciences: Environment: EARTH SCIENCES: Geoparks: Members. http://www.unesco.org/new/en/natural-sciences/environment/earth-sciences/geoparks/members/ (accessed 16 June 2012).
65. Geoparks. Nature Park Eisenwurzen. http://www.geoparks.it/node/123 (accessed 16 June 2012).
66. Wudalianchi. About us: instruction. http://english.wdlc.com.cn/yyweb /show. asp?scid=7 (accessed 18 June 2012).

CHAPTER 9

Satellite-Based Thermophysical Analysis of Volcaniclastic Deposits: A Terrestrial Analog for Mantled Lava Flows on Mars

Mark A. Price [1,*], Michael S. Ramsey [1] and David A. Crown [2]

[1]*Department of Geology and Planetary Science, University of Pittsburgh, 4107 O'Hara Street, Pittsburgh, PA 15260, USA,*

[2]*Planetary Science Institute, 1700 East Ft. Lowell Rd., Suite 106, Tucson, AZ 85719, USA*

ABSTRACT

Orbital thermal infrared (TIR) remote sensing is an important tool for characterizing geologic surfaces on Earth and Mars. However, deposition of material from volcanic or eolian activity results in bedrock surfaces becoming significantly mantled over time, hindering the accuracy of TIR compositional analysis. Moreover, interplay between particle size, albedo, composition and surface roughness add complexity to these interpretations. Apparent Thermal Inertia (ATI) is the measure of the resistance to temperature change and has been used to determine parameters such as grain/block size, density/mantling, and the presence of subsurface soil moisture/ice. Our objective is to document the quantitative relationship between ATI derived from orbital visible/near infrared (VNIR) and thermal infrared (TIR) data and tephra fall mantling of the Mono Craters and Domes (MCD) in California, which were chosen as an analog for partially mantled flows observed at Arsia Mons volcano on Mars. The ATI data were created from two images collected ~12 h apart by the Advanced Spaceborne Thermal Emission and Reflection Radiometer (ASTER) instrument. The results were validated with a quantitative framework developed using fieldwork that was conducted at 13 pre-chosen sites. These sites ranged in grain size from ash-sized to meter-scale blocks and were all rhyolitic in composition. Block size and mantling were directly correlated with ATI. Areas with ATI under 2.3×10^{-2} were well-mantled with average grain size below 4 cm; whereas values greater than 3.0×10^{-2} corresponded to mantle-free surfaces. Correlation was less accurate where checkerboard-style mixing between mantled and non-mantled surfaces occurred below the pixel scale as well as in locations where strong shadowing occurred. However, the results validate that the approach is viable for a large majority of mantled surfaces on Earth and Mars. This is relevant for determining the volcanic history of Mars, for example. Accurate identification of non-mantled lava surfaces within an apparently well-mantled flow field on either planet provides locations to extract important mineralogical constraints on the individual flows using TIR data.

Keywords: Apparent Thermal Inertia (ATI); ASTER; volcanic terrain; Mono Craters; Arsia Mons; Mars

1. INTRODUCTION

The erosion, transport and deposition of fine-grained particulates over bedrock surfaces serve as indicators of the past and more recent volcanic and eolian processes on Earth and Mars. For example, lava flows and other recently emplaced volcanic terrains on the Earth commonly become covered by remobilized ash, tephra fall and/or material from pyroclastic density currents (PDC's). In comparison, the movement of sand and dust over time on the surface of Mars significantly modifies the exposure of any bedrock surfaces [1]. Identification of the major rock-forming minerals that comprise the underlying rocks is important to understand the geologic and petrologic history of the planet and is possible with thermal infrared (TIR) remote sensing data [2,3,4]. However, the pervasive mantling typically masks these spectral signatures with the relatively featureless TIR spectrum of the fine-grained material [5,6,7]. Therefore, the higher albedo regions (e.g., regions with more dust mantling) on Mars are generally avoided during TIR spectral analyses.

Similarly, surfaces on Earth mantled to any degree must be treated cautiously during TIR spectral analysis. Fine-grained mantling of bedrock exposures on Earth is not as pervasive as on Mars due mainly to the higher rates of weathering. Consistent erosion on shorter time scales (e.g., years to centuries) and tectonic processes over much longer time scales (e.g., millions of years) remove this material. However, mantling is found on younger surfaces and/or those that experience higher rates of new mantling due to a variety of processes (e.g., explosive volcanic eruptions, eolian deposition, and desertification). Knowledge of the presence of these mantling deposits and, perhaps more importantly, identification of their composition and that of the underlying rocks, becomes invaluable in active volcanic environments, for example. Mantling in these locations may arise from a recent eruption, with the direction and thickness of the tephra fall deposits indicating the magnitude of the eruption and wind direction at the time, whereas the composition of the pyroclastic material reveals information about any evolution of the eruption or magmatic center over time. Analysis of the composition of active dunes can also relay similar information about the underlying substrate as well as the paleo-environmental conditions that produced the dunes [8,9,10].

Some regions of Mars have become heavily mantled by deposits that were derived locally, and thus retain the chemical and spectral signatures of the underlying rocks [11,12]. More generally, however, fine-grained surface materials have become globally-homogenized and therefore hinder accurate characterization of the underlying bedrock from orbit [13,14]. This mantling is observed to be spatially variable in areas such as the lava flow fields south of Arsia Mons. This field contains flows that are visibly similar and in close proximity but have very different thermophysical properties (Figure 1). This region was examined in prior studies and shown to contain some of the roughest terrain on Mars with flows that are geologically young (~100 My) for Mars [4,15,16,17,18,19].

The Arsia flows display unusual thermophysical properties such as variable and distinct diurnal temperature changes that vary from flow to flow and even within a given flow. This behavior motivated the objective of the current study, which is to assess the applicability of using a thermophysical analysis approach to determine the presence and degree of mantling on these flows by first validating it on an Earth analog site. Although deriving information from mantled surfaces using TIR data is difficult, orbital thermophysical measurements have been used to approximate the thickness of such deposits overlying bedrock layers on Mars and Earth [9,10,20]. Furthermore, locating specific bedrock regions within mantled flows that either protrude through the fine-grained covering or are only thinly-mantled is one possible approach that could be used to determine the composition of those regions [4,17,18].

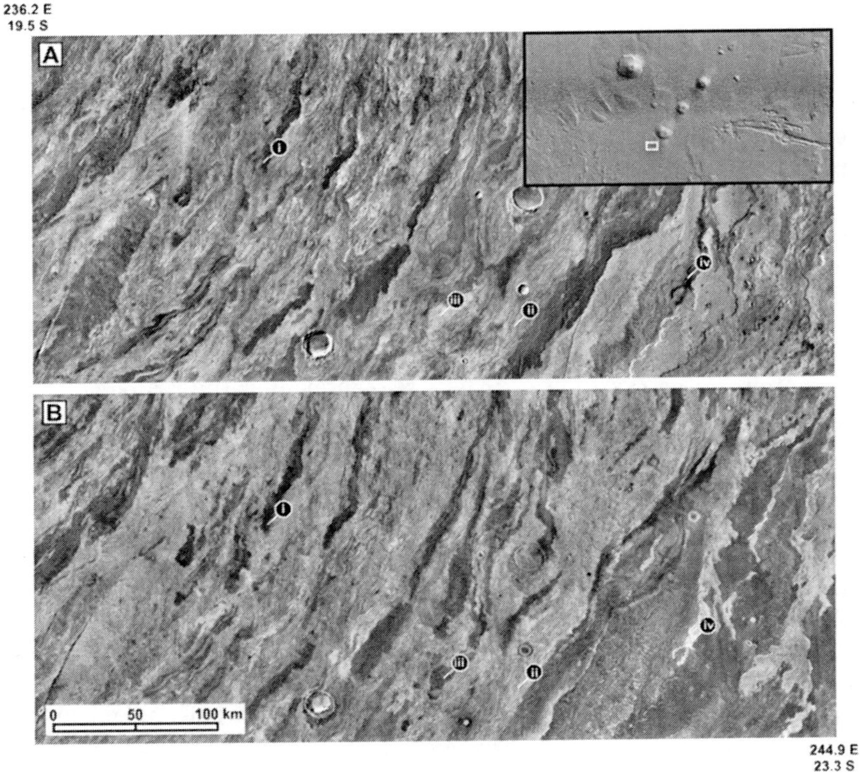

Figure 1. THEMIS IR global composite (100 pixel/degree) of a portion of the southern Arsia Mons flow field. (**A**) Day; (**B**) Night. Locator pins identify common examples of the four different thermophysical behaviors exhibited by individual lava flows over the diurnal cycle: (i) unchanging temperature, always cool; (ii) unchanging temperature, always warm; (iii) changing temperature; warm during the day and cool at night, indicative of a low thermal inertia surface; (iv) changing temperature, cool during the day and warm at night, indicative of a high thermal inertia surface. Inset is a MOLA-derived shaded relief image with the study area denoted by the white box.

The Mono Craters and Domes (MCD) in east-central California were chosen because many of the obsidian-rich, rhyolitic composition lava flows are mantled by material with much finer grain sizes (Figure 2).

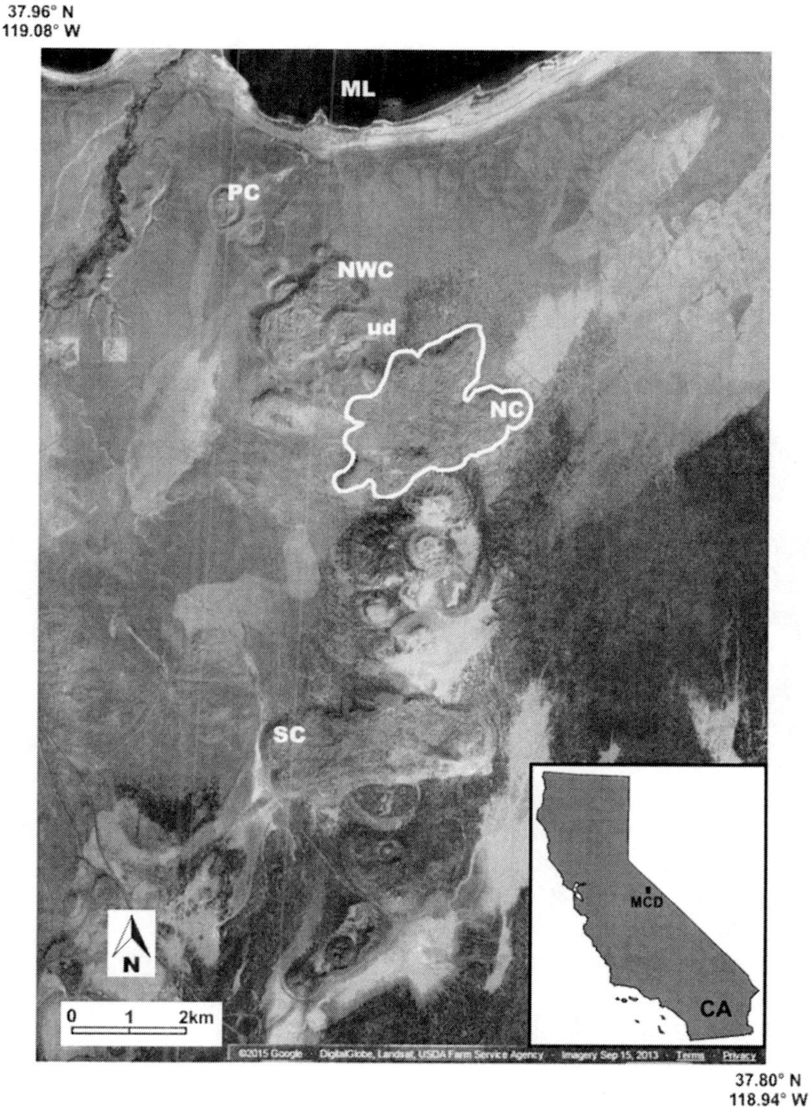

Figure 2. High-resolution visible satellite image from Digital Globe (within Google Earth) the Mono Craters and Domes with North Coulee (NC) outlined in white. Other features include: Mono Lake (ML), Panum Crater (PC), Northwest Coulee (NWC), South Coulee (SC), and the informally-named Upper Dome (UD). Inset shows the state of California with the study area (MCD) denoted by the black box.

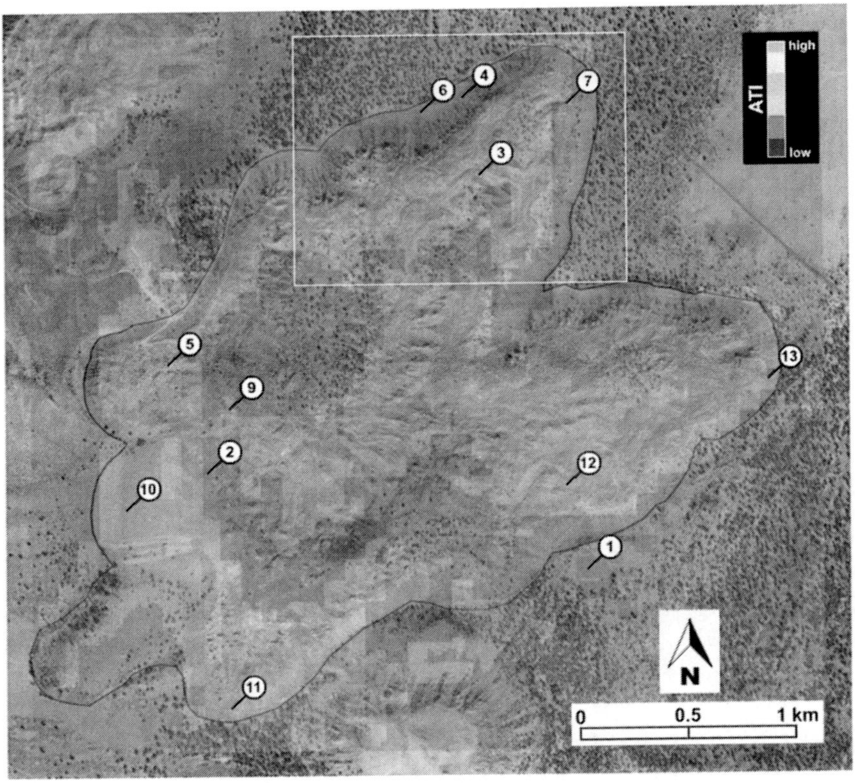

Figure 3. Relative apparent thermal inertia image of North Coulee (outlined in black) derived from 10 July 2011 ASTER data (created by Ramsey and Crown, 2010) [4]. The colorized scale image is overlain on a Digital Globe (within Google Earth) high-resolution color image for detail. This image was used as a guide for later fieldwork and the numbered pins refer to visited field sites. The white box denotes the area shown in Figure 6. Data courtesy of NASA/GSFC/METI/Japan Space Systems. U.S./Japan ASTER Science Team.

Although not a direct compositional analog for Mars, the mantled flows and domes of the MCD provide an excellent process and technique analog [14,21,22,23]. The rough, blocky flows and domes display a wide range of tephra mantling depth on their surfaces, from heavily mantled (e.g., >10 cm) by tephra to areas that are non-mantled. These surfaces span many dozens of TIR 90 m scale pixels or can be within meters of each other, therefore providing all scales of spatial mixing and mantling depths. The similar albedo and composition of the lava flows to the mantling deposits hinders discrimination using purely spectral data such as VNIR reflectance or TIR emissivity. This is analogous to the case on Mars, however with mainly basaltic compositions. Furthermore, the ability to validate the satellite data using ground and laboratory data (Figure 3) provides the opportunity to test the accuracy of using a thermophysical rather than a spectral approach. This study uses spaceborne-derived data to first identify the mantled regions on domes of the MCD followed by field validation of those regions and later quantification of field samples using laboratory spectral and petrographic techniques. This tiered approach allows the image data to guide the field mapping and the field and laboratory data to later quantify the initial results from image analysis, thus producing a particle/block size map from the ATI data.

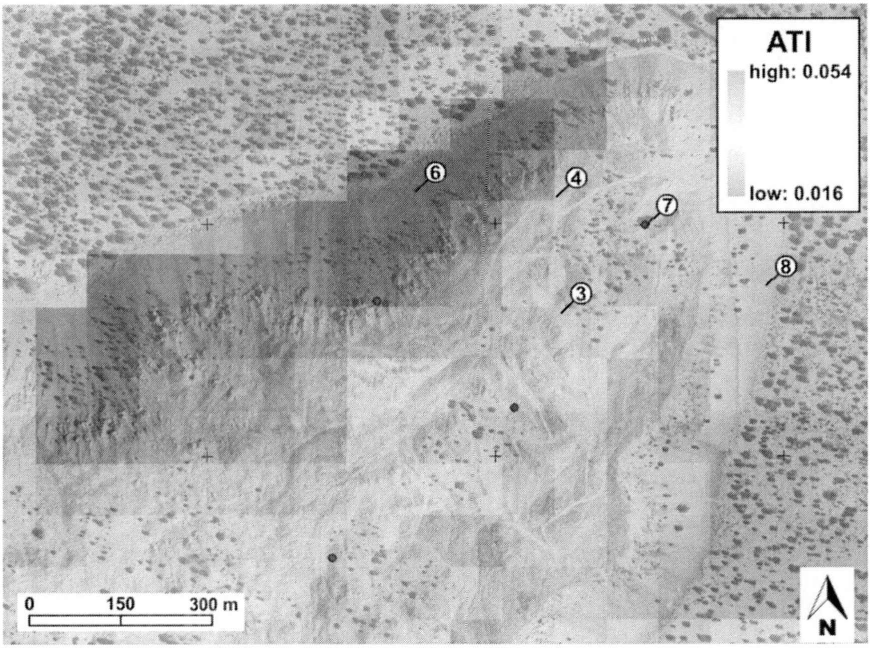

Figure 6. ATI draped over a high-resolution Digital Globe image (within Google Earth) of the northeast lobe of North Coulee. Blue dots indicate proposed field sites prior to the fieldwork being conducted. The numbered pins denote visited sites with associated field data. Note the very different ATI behavior on the northwest versus the southeast facing slopes. Figure created in ArcGIS 10 using Bing Maps layer.

2. BACKGROUND

2.1. Terrestrial TIR Data

The Advanced Spaceborne Thermal Emission and Reflection Radiometer (ASTER) instrument is one of five on the Terra platform [24] and has acquired data since its launch in December 1999. ASTER data have improved local- and regional-scale understanding of various processes, including surface-atmosphere interactions, geology, and volcanic activity [25]. ASTER has three spectral channels in the VNIR wavelength region, six in the shortwave infrared (SWIR) wavelength region, and five in the TIR wavelength region, with resolutions of 15 m/pixel, 30 m/pixel, and 90 m/pixel, respectively [25]. The SWIR data are no longer acquired, however, due to a failed detector cooling system. The spectral range and moderately high spatial resolution of the VNIR and TIR allow a variety of surface materials to be distinguished and a better understanding of compositional mixing within a given pixel or region of pixels [10,25].

2.2. Mars Thermophysical Data

Thermal Inertia (TI) is an important remote sensing tool for identifying certain physical properties of a surface such as the particle size, cementation state, and/or soil moisture content [9,26,27]. It can be thought of as a measure of the resistance of a surface material to temperature change over some time scale [28] and is defined as

$$TI = (K\rho c)^{1/2} \tag{1}$$

where K is the thermal conductivity ($W \cdot m^{-1} \cdot K^{-1}$), ρ is the density ($kg \cdot m^{-3}$), and c is the specific heat ($J \cdot kg^{-1} \cdot K^{-1}$) giving TI units of $J \cdot m^{-2} \cdot K^{-1} \cdot s^{-1/2}$. TI nor the terms that comprise it can be determined directly from satellite measurements and therefore it must either be approximated or modeled where only remote sensing is available.

The relative simplicity of the Martian atmosphere and surface compared to Earth has allowed well-described thermal models to be developed, which predict the surface temperature at any time during the day in different seasons and various latitudes [29,30,31]. Modeled diurnal temperature curves are then fit to measurements of two or more temperatures from orbital instruments and used to estimate physical parameters that influence TI such as the presence and amount of bedrock and blocks on the surface, particle size of the sediments, layering, *etc.* For example, Kieffer *et al.* (1977) [32] and Palluconi and Kieffer (1981) [29] use this approach to differentiate between different surface types on the planet. Edgett and Christensen (1991) [33] expand on the work of Christensen (1982) [34] to determine the particle size of unconsolidated materials in Martian dunes. Mellon *et al.* (2000) [35] compare thermal inertia to thermal emission spectrometer (TES) observations of the Martian surface to determine the small-scale variability on the planet's surface by mapping particle size, rock abundance, and exposures of bedrock. Christensen *et al.* (2001) [2] apply TES-derived thermal inertia to interpret the composition of Martian surface materials, rocks, ices, and the thermal properties of the surface. Putzig *et al.* (2005) [36] create a new thermal inertia map of Mars and use it to quantify new surface units of low and high thermal inertia. The work of Rogers *et al.* (2005) [37] use Thermal Emission Imaging System (THEMIS) multispectral and thermophysical data combined with TES hyperspectral and albedo readings to determine the mineralogical heterogeneity of Martian bedrock exposed at Ares Valles. Edwards *et al.* (2009) [38] expand upon that work to identify numerous other bedrock outcrops on Mars using thermal inertia.

2.3. Apparent Thermal Inertia

Similar quantitative thermophysical modeling for Earth surfaces is typically much more difficult with complications that include (but are not limited to) the effects of soil moisture, especially if there is a change between the day and night data collection; atmospheric thickness, condensation and gas conductivity, all of which make modeling of the grain to grain heat transfer more complex. Earth's thick heterogeneous atmosphere absorbs and scatters energy in the visible to thermal infrared wavelength regions. Additionally, it hinders the efficiency of radiant transfer of energy between the grains of a particulate surface. Diurnal temperature variations are also more sensitive to atmospheric humidity and precipitation, which reduces the accuracy of thermal inertia estimates [39,40,41,42]. Thermal inertia investigations of Earth must also account for vegetative cover, which obscures geologic surfaces in addition to effecting albedo and surface temperature. Despite these factors, thermal inertia has been used successfully on Earth, most commonly to map the moisture content of soils and for geologic interpretations [9,39,40,41,42]. Cracknell and Xue (1996) [43] use data from the Advanced Very High Resolution Radiometer (AVHRR) sensor to create thermal inertia images to estimate surface heat flux after consulting local weather data. A study by Sobrino and El Kharraz (1999a) [44] expands on this work by successfully applying a derived thermal inertia proxy to AVHRR images of a region of Niger. Additional maps were created of the Iberian Peninsula and Morocco [45]. More recently,

Scheidt *et al.* (2010) [9] use ASTER-derived thermophysical data to create high-spatial resolution soil moisture maps of the White Sands dune field in New Mexico and successfully identify an extreme drying period that later led to a large-scale dust emission event from the region.

The complexity of thermal inertia modeling on Earth has led to the creation of Apparent Thermal Inertia (ATI) as a proxy. In the terrestrial literature, ATI is commonly referred to the semi-quantitative approximation for the actual thermophysical response (*i.e.*, thermal inertia) using data from the VNIR, SWIR and TIR regions. It is defined as:

$$ATI = (1 - a)/(\Delta T) \tag{2}$$

where a is the broadband VNIR/SWIR albedo and ΔT is the temperature difference between a day and night pair of coregistered TIR images. Therefore, ATI is a unitless measure of the reflected solar albedo relative to the difference in emitted brightness temperature over the diurnal cycle [9,28,46,47].

The use of apparent thermal inertia in the Martian literature is quite different [9,46,48]. In past studies, ATI simply denotes apparent seasonal or yearly differences in thermal inertia for the same surface. For example, Putzig and Mellon (2007) [48] investigate surface heterogeneity and the effects of diurnal and seasonal change on thermal inertia models, calling these differences "apparent" thermal inertia. In this study, we use the traditional semi-quantitative terrestrial definition of ATI.

2.4. The Mono Craters and Domes

The MCD, 25–30 km north of the town of Mammoth Lakes, California, is an arcuate chain of craters and high-silica lava flows (Figure 2). The most recent phase of activity occurred over the past 10,000 years and therefore represents some of the youngest activity in the Long Valley Caldera region [49]. The chain consists of an alignment of approximately 28 domes, craters, and coulees. The coulees have relief greater than 600 m above the surrounding terrain and a maximum elevation of 2800 m [50]. The vents also have associated tephra rings and extensive aprons of pumiceous fall, flow, and reworked deposits [49,50,51]. Beginning in the Holocene, the Mono system spread both north and south from the current central part of the dome chain, producing dike-fed domes of crystal-poor, high-silica rhyolite [49]. A volumetric eruption rate of ~0.2 km^3/ka in the early Holocene increased to about 0.8 km^3/ka during the last 3000 years [52]. This volumetric surge coincided with a switch from crystal-poor to virtually aphyric rhyolite [52,53]. Composition of the

MCD lavas is similar for all younger effusive deposits in the region. All lavas but one (an undated crystal-rich rhyodacite at 68 wt.% SiO_2) at the Mono Craters is high-silica (75.4 wt.%–77 wt.% SiO_2) rhyolite [49]. Major element (and most trace-element) contents are quite similar to those of Glass Mountain and the early Bishop Tuff [49].

The youngest eruptions in the MCD chain date to approximately the fourteenth century [49,54,55]. These eruptions produced 0.2 km^3 of pyroclastic fall, flow, and surge deposits and 0.4 to 1.0 km^3 of lava domes and flows [53,54]. These rhyolitic lavas originated from five aligned vents in the northwestern zone of the Mono chain. According to Bailey (1989) [53], the modern exposures at Northwestern Coulee, Panum Crater, Upper Dome, Crater Dome, and North Coulee were all directly formed during this eruptive phase (Figure 2). Eruption sequences generally consisted of explosive sub-Plinian events that formed thick deposits of tephra fall on older domes within a few kilometers of the vent followed by pyroclastic flows and surges and finally extrusion of the viscous dome-forming lavas [53,54].

2.5. The Arsia Mons Flow Field

Arsia Mons, the southernmost of the three prominent Tharsis shield volcanoes, has a well-developed summit caldera and extensive lava flow fields [56,57,58,59,60]. There are two large flow aprons with a multitude of individual flows and flow lobes that extend from alcoves on the lower NE and SW flanks of the volcano, which postdate its main shield [22,23,61,62]. Previous work in the area includes series of 1:2M scale lava flow maps of the Tharsis region by Scott (1981) [63] and Scott *et al.* (1981a,b) [64,65] using Viking Orbiter images. More recent studies use Mars Express High Resolution Stereo Camera (HRSC) images to examine stages in effusive volcanism in the Tharsis region, including the Arsia Mons shield [66,67]. Lang *et al.* (2009) [23] examine this flow field further south and found it to be basaltic with a possible affinity to the basaltic shergottites class of SNC meteorites.

Like the Mono Craters, the Tharsis region has been imaged by high-resolution VNIR and TIR datasets and is shown to contain aerially extensive lava flow fields with variable amounts of mantling derived from both eolian infilling of the flow surfaces and airfall dust. These flows have a Thermal Emission Spectrometer (TES) derived albedo of between 0.22 and 0.24 and a dust cover index of between 0.95 and 0.97 [14,62], which generally denotes a less than desirable, region for thermal infrared spectral analyses due to the presence of dust.

3. METHODOLOGY

This investigation is designed to validate the assumption that thermophysical variability is directly related to the degree of mantling and its particle size on volcanic surfaces through the combined use of image, field, and laboratory data. Therefore, it serves as a terrestrial analog calibration/validation study to ongoing thermal inertia studies of the mantled flow fields on Mars.

A preliminary study by Ramsey and Crown (2010) [4,19] noted statistically significant variations in ATI for the MCD chain (Figure 3). However, the qualitative study merely showed that ATI did correlate to the mantled regions of the MCD shown in high-resolution visible images. Therefore, the methodological approach used here was to recreate the results of Ramsey and Crown (2010) [4] and then develop a more quantitative validation of the ATI product. New ATI images were created showing pixel-by-pixel difference in ATI as compared to the regional differences seen in the 2010 study. These data were compared with high-resolution visible satellite images, and then used to preselect field site locations. These field sites represent the full range of ATI values, mantled and unmantled surfaces as well as those at different elevations and slopes. Data were acquired at 13 field sites including samples collected for later laboratory analysis. Analysis of the field and laboratory results allow quantitative constraints to be placed on the ATI results, including extrapolating to larger block sizes than would normally not be resolved by ATI alone.

After an extensive examination of the various flows within the MCD, North Coulee was chosen as the field study site. It is one of the largest (~0.4 km^3) of the recent lava flows, comprising over half the total volume of all the fourteenth century eruptions [54,68]. The coulee contains a wide variety of mantled surfaces that have identical composition to the volcanic flows found in the area. These mantled surfaces originated from the deposition of tephra produced during younger eruptions to the north.

Prior studies showed that the calculation of ATI provides a reasonable proxy for identification of the mantled zones on the domes [17,18,62]. For this study, four ASTER image pairs of the Mono Craters spanning several years were chosen and ATI was calculated for each. Cloud free, chronologically-similar (within 12–30 h of each other) day and night level 1B ASTER images were first identified. The day and night images were as close in time to one another as possible considering the latitude, acquisition, and geometry limitations of the ASTER instrument. The data pair from July 2011 was chosen because the day and night images were obtained less than 12 h apart and the weather conditions were ideal.

Average relative humidity on 10 July 2011 (day of the ASTER overpasses used for this work) was 33% (Table 1) (Weather Underground, 2013). Precipitation records from Sonora Junction, CA (~60 km from MCD) and Bridgeport, CA (~45 km from MCD) were used to verify that there were no large rain events either before or within the image pair acquisition that could affect the ATI results. Records from these stations indicate a scarcity in precipitation in the month prior to the ASTER overpass and none during the overpass interval [69]. Furthermore, the low humidity, high surface porosity, and an average temperature in July of ~29 °C [69,70] minimizes the possibility of any surface moisture.

Table 1. Precipitation records for the closest weather stations to the Mono Craters (Sonora Junction, CA and Bridgeport, CA). Note the lack of rain before the ASTER overflights (10 July 2011) from which the ATI data are based.

Bridgeport, CA		Sonora Junction, CA	
Elevation: 598.4 m	Location: 38.25° N, 119.22° W	Elevation: 639.7 m	Location: 38.4° N, 199.5° W
Date	Precipitation (cm)	Date	Precipitation (cm)
28 June 2011	0	28 June 2011	1.02
29 June 2011	0	29 June 2011	0
30 June 2011	0	30 June 2011	0
1 July 2011	trace	1 July 2011	trace
2 July 2011	0	2 July 2011	0
3 July 2011	0	3 July 2011	0
4 July 2011	0	4 July 2011	0
5 July 2011	0	5 July 2011	0
6 July 2011	0	6 July 2011	0
7 July 2011	0	7 July 2011	0
8 July 2011	0	8 July 2011	0
9 July 2011	0	9 July 2011	0
10 July 2011	0	10 July 2011	0
11 July 2011	0	11 July 2011	0
12 July 2011	0	12 July 2011	0
13 July 2011	0	13 July 2011	0

3.1. ASTER Data Processing

With the constraints mentioned, the ASTER diurnal pair dated from 10 July 2011 was used to create the ATI product for this study. The 10 July (day) image (Scene ID: AST_09T_00307102011185047) was collected at 18:50:47 GMT (11:50:47 local time) with a solar azimuth angle of 131.06° and a solar elevation angle of 68.11°. The 11 July (night) image

(Scene ID: AST_09T_00307112011055433) was acquired at 05:54:33 GMT (22:54:33 local time on 10 July). Therefore, there was a 12:03:46 absolute time difference between the pair.

All data were obtained from the ASTER Level 2 (L2) on-demand processing pathway at the Land Processes Distributed Active Archive Center (LP DAAC) [71]. The ASTER L2 atmospherically-corrected TIR radiance data serve as the input product to derive the kinetic temperature images. An emissivity of 0.96 was used as a maximum assumed value in order to separate temperature and emissivity from the ASTER thermal infrared radiance images based upon previous spectral studies of glassy rhyolite [72,73]. The temperature difference product was created from these data (Figure 4). The standard L2 atmospherically corrected VNIR reflectance product was converted to albedo assuming no preferential directional scattering (generally valid on the rough and moderate albedo rhyolite surfaces) and calculated by averaging the three ASTER VNIR reflectance images. All VNIR data were down-sampled from 15 to 90 m/pixel to match the TIR resolution. The $(1 - a)$ product, where a is the albedo, was calculated and the image pair geographically aligned by rotating each to north and cropping the images to identical UTM coordinates. Precise pixel alignment was maintained throughout to ensure accurate comparison of the same pixel area. Once aligned and spatially matched, the temperature difference and complement of the albedo products were used to create the final ATI image using Equation (2) (Figure 5).

It should be noted that no SWIR data were used in the albedo calculation because of the failed system two years prior. The lack of SWIR for calculation of broadband albedo creates only minor errors for this analysis. The SWIR reflectance of glassy rhyolite lava and the rhyolitic mantling material are spectrally-similar across the SWIR wavelength region [74]. Therefore, no non-uniform spectral differences are expected between the lava and the mantling material with the lack of the SWIR data. Including the SWIR (if it were available) would have therefore only slightly lowered the $(1 - a)$ term thus making the overall ATI calculation marginally lower throughout the image.

3.2. Field Approach

The ATI image created for this study, as well as the product created by Ramsey and Crown (2010) [4] as a proof-of-concept, were both used to help identify potential field targets (Figure 3). It was considered important to select a variety of sites with a diverse range of ATI values in order to validate surface roughness, surface fragment size, surface temperature, mantling thickness, and collect samples. The ASTER TIR

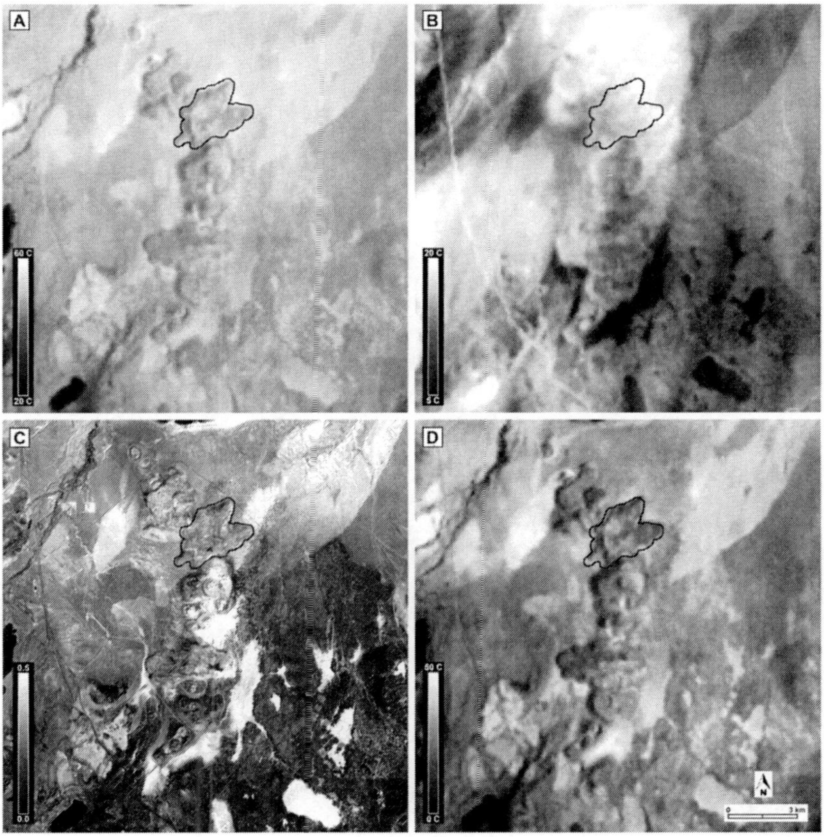

Figure 4. ASTER data products used to compile the apparent thermal inertia image. (**A**) Atmospherically-corrected, daytime surface kinetic temperature (ASTER level 2 AST_08 product) acquired on 10 July 2011 at 11:50:47 PDT; (**B**) Atmospherically-corrected nighttime surface kinetic temperature (ASTER level 2 AST_08 product) acquired on 10 July 2011 at 22:54:33 PDT; (**C**) Visible/Near Infrared surface albedo acquired at the same time as (**A**); (**D**) Temperature difference product (day – night).

emissivity data were also used independently of the ATI to select areas of interest for field sampling. The day and night infrared emissivity data were compared with the ATI image to select areas with different micron-scale roughness [8] and identify any potential compositional variability. However, this comparison showed that the emissivity data were generally unable to resolve the differences identified by the ATI approach. A detailed description of the field work is given in Section 9, "Supplementary Information".

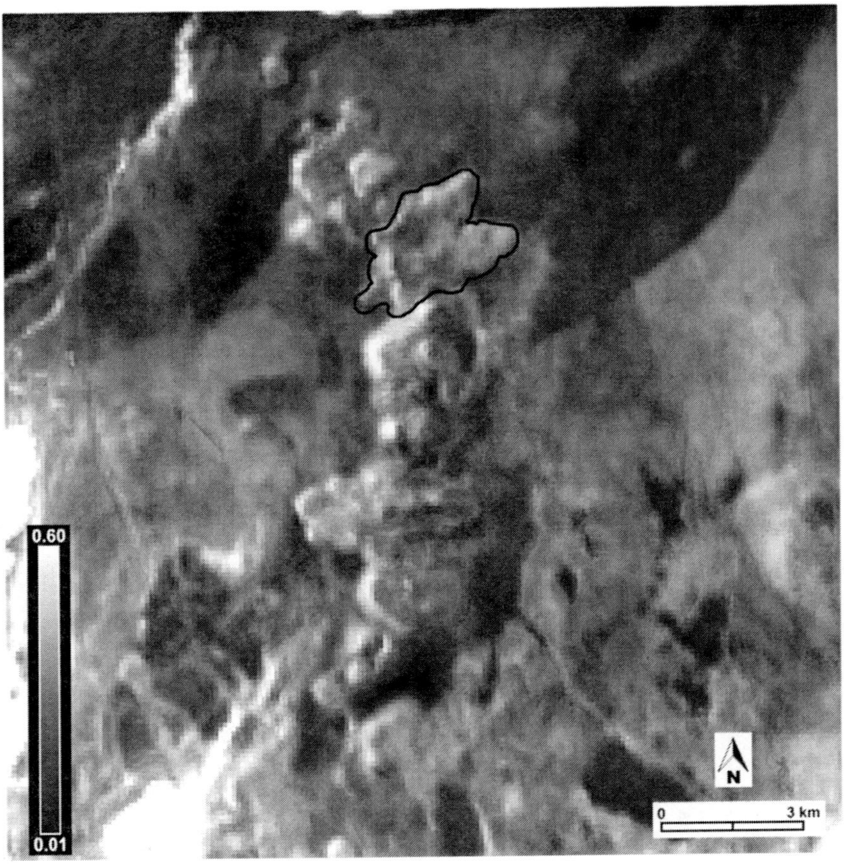

Figure 5. Apparent thermal inertia image derived for this study from the data shown in Figure 4. June Lake (**lower left**) and northwest-facing steep slopes have very high ATI values, whereas large, non-vegetated pumice deposits and old fire scars (**upper right**) (exposing the underlying pumice) have the lowest ATI values. Note that North Coulee has a range of ATI values 942 consistent with the preliminary work of Ramsey and Crown (2010) (see Figure 3).

3.3. Laboratory Analyses

Because it was important to describe a variety of textures and particle sizes, a broad selection of samples from areas with various degrees of mantling were analyzed to validate the ATI results. High-resolution laboratory thermal emission spectra of all samples were acquired at the

Image Visualization and Infrared Spectroscopy (IVIS) facility at the University of Pittsburgh following the fieldwork. A petrographic study of the samples was also conducted. A detailed description of the laboratory analyses is given in Section 9, "Supplementary Materials".

4. RESULTS

4.1. Image Results

Ramsey and Crown (2010) [4] hypothesized that ASTER-derived ATI values correlate with the presence of mantling and therefore perhaps variations in mantling thickness (Figure 3). Figure 5 shows the calibrated ATI results for this study. Our new analysis improved upon the earlier study by focusing on specific small-scale ATI targets, comparing results to the emissivity variation (both from ASTER and the laboratory), as well as tracking any changes in those values with time. For the date range described here, statistically significant differences in ATI were found. For example, initial comparison of lowest ATI regions with the high-resolution image data showed that these areas correlate precisely to heavily mantled surfaces on the dome (e.g., smooth morphology, lack of shadowing and the presence of sparse vegetation). In comparison, blocky surfaces had higher ATI values, topographic variability indicated by the presence of shadows and no vegetation. Furthermore, the ATI results indicate a similar correlation on the other domes and flows along the chain, although these were not visited during the field campaign.

4.2. Field Results

The calibrated ATI image was first used to identify the areas of interest for the field validation, with a wide variety of potential sites chosen that had different values of ATI, spanning the full range in ATI calculated. A subset of the selected field sites can be found in Figure 6 with data from all sites presented in Table 2. Comparison of high-resolution orbital images and the ATI image also helped to further constrain the choice of sites.

Table 2. Characterization of the 13 targeted field sites in the study. Sorting and grain size determined after analysis using a petrographic microscope.

Site 1	
Average fragment size (cm):	0.5–1.0
Mantling depth (cm):	8.0–16.0
ATI value (× 100):	**1.960**
General composition:	rhyolitic *(80% pumice/20% obsidian)*
Sorting degree:	well-sorted
Surface temperature (°C):	17.0
Local time:	10:15 a.m.
Field description:	*Course, angular grains, sand to gravel-sized. Well-sorted.*
Site 2	
Average fragment size (cm):	1.0–2.0
Mantling depth (cm):	<10
ATI value (× 100):	**2.317**
General composition:	rhyolitic
Sorting degree:	moderately-sorted
Surface temperature (°C):	30.0
Local time:	15:00 p.m.
Field description:	*Well-sorted pyroclastic deposits. Course, angular, sand to gravel-sized grains. Predominantly pumice in composition.*
Site 3	
Average fragment size (cm):	bimodal *(0.03 and 0.50)*
Mantling depth (cm):	not measured *(inaccessible)*
ATI value (× 100):	**2.576**
General composition:	rhyolitic
Sorting degree:	very poorly sorted
Surface temperature (°C):	not measured *(inaccessible)*
Local time:	n/a
Field description:	*Poorly-sorted mixture of pumice, obsidian, and other dome rocks. Large variation in size, from small sand-sized grains to 0.5m blocks.*
Site 4	
Average fragment size (cm):	40.0
Mantling depth (cm):	0.0
ATI value (× 100):	**2.860**
General composition:	rhyolitic *(pumice and obsidian boulders)*
Sorting degree:	well-sorted
Surface temperature (°C):	not measured
Local time:	n/a
Field description:	*Well-sorted large volcanic debris consisting of pumice obsidian, and other dome rocks. Large, angular blocks ~1 m.*
Site 5	
Average fragment size (cm):	<0.5
Mantling depth (cm):	<30
ATI value (× 100):	**2.645**
General composition:	rhyolitic
Sorting degree:	well-sorted
Surface temperature (°C):	46.2
Local time:	~2:00 p.m.
Field description:	*Well-sorted pumice dune consisting of fine sand. Wind-deposited. Surrounded by slopes of larger volcanic fragments.*

Site 6	
Average fragment size (cm):	3.0–6.0
Mantling depth (cm):	none
ATI value (×100):	**2.850**
General composition:	rhyolitic
Sorting degree:	poorly-sorted
Surface temperature (°C):	n/a
Local time:	n/a
Field description:	*Poorly-sorted weathered slope debris. Fragments range from smaller rocks a few cm in diameter to large meter-sized boulders. Minimal mantling.*
Site 7	
Average fragment size (cm):	1.0–3.0
Mantling depth (cm):	20.0
ATI value (×100):	**2.777**
General composition:	rhyolitic *(pumice and obsidian boulders)*
Sorting degree:	poorly-sorted
Surface temperature (°C):	not measured
Local time:	n/a
Field description:	*Poorly-sorted sand- and gravel-sized grains. About 30% is rhyolitic pumice boulders. Average boulder size ~0.3 m.*
Site 8	
Average fragment size (cm):	4.0
Mantling depth (cm):	none
ATI value (×100):	**2.461**
General composition:	rhyolitic
Sorting degree:	poorly-sorted
Surface temperature (°C):	n/a
Local time:	n/a
Field description:	*Poorly-sorted debris weathered off the east slope. 70% pumice 30% obsidian. Most fragments ~3–5 cm in size. Larger fragments ~15–20 cm.*
Average fragment size (cm):	4.0
Mantling depth (cm):	none
ATI value (×100):	**2.461**
General composition:	rhyolitic
Sorting degree:	poorly-sorted
Surface temperature (°C):	n/a
Local time:	n/a
Field description:	*Poorly-sorted debris weathered off the east slope. 70% pumice 30% obsidian. Most fragments ~3–5 cm in size. Larger fragments ~15–20 cm.*
Site 9	
Average fragment size (cm):	1.0
Mantling depth (cm):	n/a
ATI value (×100):	**2.555**
General composition:	rhyolitic
Sorting degree:	well-sorted
Surface temperature (°C):	n/a
Local time:	n/a
Field description:	*Well-sorted wind-deposited pumice sediment. Some outcrops of larger volcanic debris.*
Site 10	
Average fragment size (cm):	<3
Mantling depth (cm):	None
ATI value (×100):	**2.731**
General composition:	rhyolitic *(pumice boulders)*
Sorting degree:	Intermediate
Surface temperature (°C):	22.5
Local time:	n/a
Field description:	*Slope with larger pumice fragments. Weathering of fragments varied. 80% pumice and 20% obsidian. Average grain size ~3 cm, varied.*

Site 11	
Average fragment size (cm):	0.5
Mantling depth (cm):	<30
ATI value (× 100):	**3.340**
General composition:	rhyolitic *(90% pumice/10% obsidian)*
Sorting degree:	well-sorted
Surface temperature (°C):	55.7
Local time:	11:45 a.m.
Field description:	*Very well sorted rhyolitic grains. Some obsidian grains. High ATI value probably the result of sun angle/shadowing on western slopes.*
Site 12	
Average fragment size (cm):	40.0
Mantling depth (cm):	None
ATI value (× 100):	**2.943**
General composition:	rhyolitic *(pumice and obsidian boulders)*
Sorting degree:	well-sorted
Surface temperature (°C):	not measured
Local time:	n/a
Field description:	*Well-sorted rhyolitic boulders with minimal mantling.*
Site 13	
Average fragment size (cm):	4.0–6.0
Mantling depth (cm):	n/a
ATI value (× 100):	**2.697**
General composition:	rhyolitic *(70% pumice/30% obsidian)*
Sorting degree:	poorly-sorted
Surface temperature (°C):	22.5
Local time:	9:00 a.m.
Field description:	*Poorly sorted slope debris. All grains and blocks very angular. Debris ranges from fine sand to >0.3 m fragments.*

Field data were acquired at a total of 13 sites both on the dome surface as well as around the perimeter. The results are described in detail for four of these sites to give a representative description of the variability encountered as well as the measurement approach employed. In general, the results (Table 2, above) from the field validation at North Coulee correlate very well with the image results (Figure 7 and Figure 8). Areas of low ATI (Figure 7, Site 1) predominantly consisted of tephra deposits; whereas areas of higher ATI were comprised exclusively of non-mantled larger (>10 cm) rhyolite pumice and obsidian blocks (Figure 8, Site 4). Zones with intermediate ATI commonly consisted of either homogenous intermediate-sized particles (1–4 cm) (e.g., Figure 7, Site 2) or a bimodal distribution of block sizes from fine to coarse (e.g., Figure 8, Site 3). A detailed explanation of the field sites is given in Section 9, "Supplementary Materials".

Figure 7. Field photographs of Site 1 (**upper row**) and Site 2 (**lower row**) with context images shown in the left hand column and close-up images in the right hand column. Site 1 is a large pumice plain at the base of the coulee (black circle shows an approximate 30 cm high pine tree sapling for scale). The particle size for this site is 1 cm or less. Site 2 is a mantled location on the coulee (white circle shows an approximate 12 cm camera case for scale). Note the bimodal distribution of particle sizes at this site with the largest particles exceeding 12 cm.

Figure 8. Field photographs of Site 3 (**upper row**) and Site 4 (**lower row**) with context images shown in the left hand column and close-up images in the right hand column. Site 3 has a strongly bimodal particle size consisting of mantling similar to Site 2 as well as a non-mantled primary lava flow surface (blocks). Site 4 is a completely non-mantled location on the coulee. The block size distribution measured along a transect for this site is shown in Figure 9.

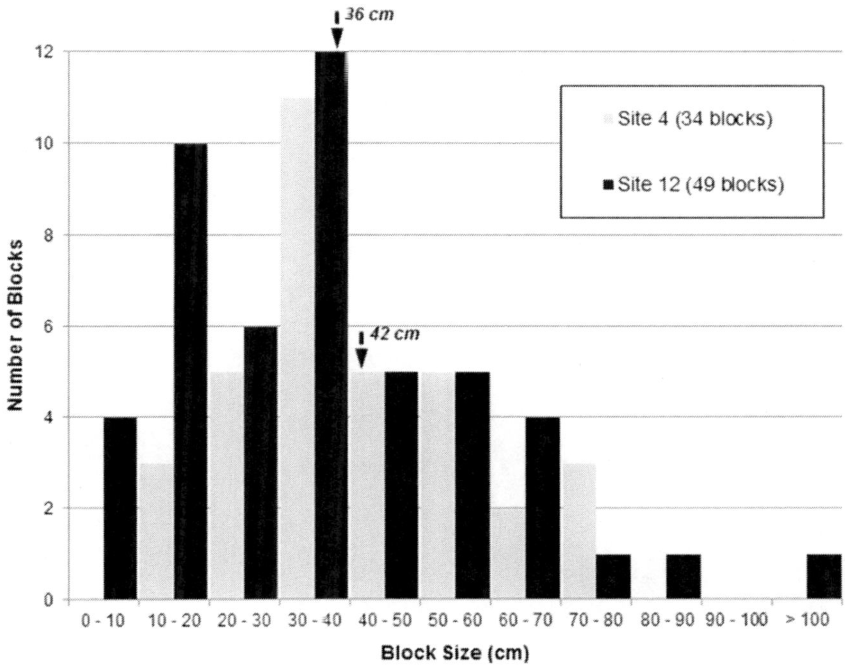

Figure 9. Block size distributions for the non-mantled Site 4 and Site 12. The average block size for each is denoted with the arrows.

4.3. Laboratory Results

Laboratory thermal infrared emissivity spectra were acquired of the field samples. Nearly all had identical spectral morphology with a prominent absorption feature at 9.1 microns (ASTER TIR band 12) indicative of high-silica glass [75]. However, one sample spectrum had a much deeper feature; likely caused either by fewer highly-vesicular fragments and/or the larger particle size of the sample (~3–5 cm), each of which would minimize photon scattering and produce deeper spectral features [73]. The 9.1 micron emissivity values for each sample were plotted against the corresponding ATI value extracted for the same pixel, with no correlation found (Figure 10). This result is not surprising for a surface consisting of approximately the same composition (rhyolite) and widely varying particle size (mm-scale fragments to m-scale blocks), varying textures (*i.e.*, glassy obsidian and highly vesicular pumice), and/or micron scale roughness values [73]. This homogeneity in the composition and heterogeneity in the textures and particle sizes limits the use of emissivity alone to determine details of the surface properties and highlights the importance of ATI for mantled surfaces such as these.

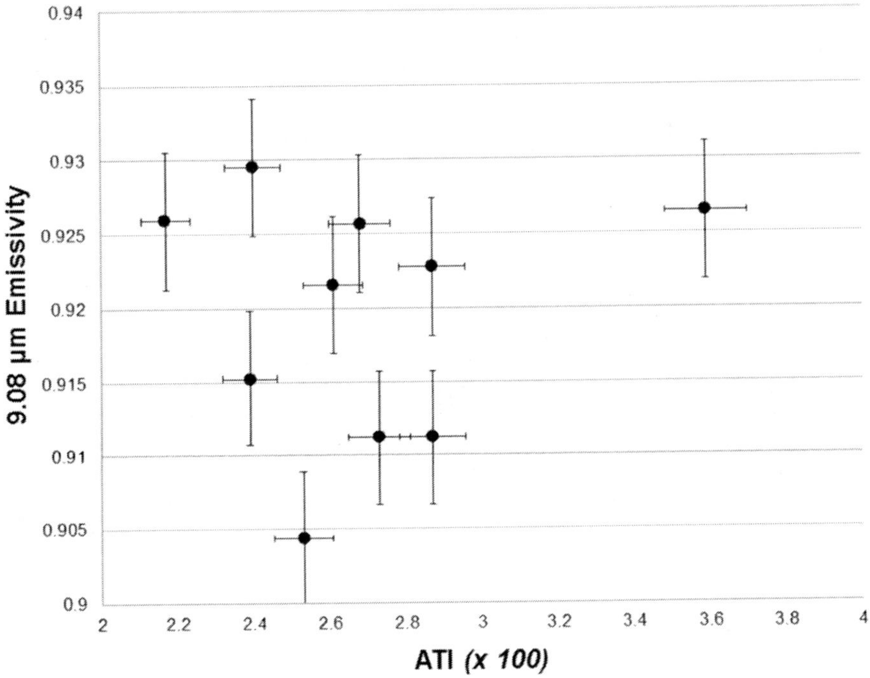

Figure 10. ASTER TIR band 12 (9.1 micrometers) surface emissivity (Level 2, AST_05) versus ATI for various the field sites. The lack of correlation is a direct result of very similar composition (pumiceous rhyolite lava and air fall pumice) reflected in the emissivity with very different particle sizes (pyroclastic air fall to blocks) reflected in the ATI. Error bars denote the standard error associated with the AST_05 product and the composite error in the derived ATI.

The petrographic analysis of the same samples confirmed the findings from the satellite and field analyses. Particulate samples were prepared on a slide and point counts conducted on a 5 by 5 cm grid using a petrographic microscope [49,50]. The composition was found to be very similar for most samples with ratios of ~3:1 rhyolitic pumice tephra to obsidian glass fragments with visible vesicles seen in two-thirds of the samples. The majority of the samples from the low ATI sites contained well-sorted particles with an average grain size of ~25 mm. Several samples contained larger (~150 mm) well-sorted grains, but these are less common. Four samples are considered poorly sorted, with grain sizes ranging from 6 mm to as large as 30 cm.

4.4. Data Integration

To further explore the variations observed in the thermophysical properties, the ATI data were compared to field and laboratory determined composition and particle size by creating a dome-wide particle size map (Figure 11). This is the first terrestrial application to confirm the relationship between ATI and particle size in volcanic environments. ATI values less than 2.3×10^{-2} at North Coulee were confirmed to be associated with the pyroclastic tephra deposits. For example, Site 1 had an ATI value of 1.9×10^{-2} ATI with an average measured grain size between 5–10 mm, and a mantling thickness of ~16 cm. Ground validation also confirmed sites with ATI values greater than 2.3×10^{-2} are not entirely mantled (e.g., Site 4). These are either associated with homogeneous smaller block size surfaces or are an aerial mixture of mantled surfaces and larger boulders (e.g., Site 3). However, the average particle size increases and correlates well with ATI values above 2.8×10^{-2}. Pixels with values greater than this threshold are entirely associated with mantle-free flow surfaces consisting of larger lava blocks. Surfaces with average block sizes >~20 cm all result in very similar ATI, which indicates this block size is near the upper limit of the ATI sensitivity.

As noted previously, this ATI approach results in inaccurate values for pixels containing steep topography and hence, shadowing. Large-scale (greater than the pixel-scale) shadowing is found primarily on the north and northwestern flow fronts. These shadows lower the amount of VNIR reflectance and daytime temperature of the surface thereby producing much higher ATI values. Because the main focus of this study was the mantled flow surfaces themselves and not the flow fronts, we did not correct for this shadowing. Work by Simurda and Ramsey (2014) [76] is addressing these shadowing errors using image and ground based digital elevation models (DEMs) to calculate and remove the effects of both large-scale and sub-pixel roughness elements.

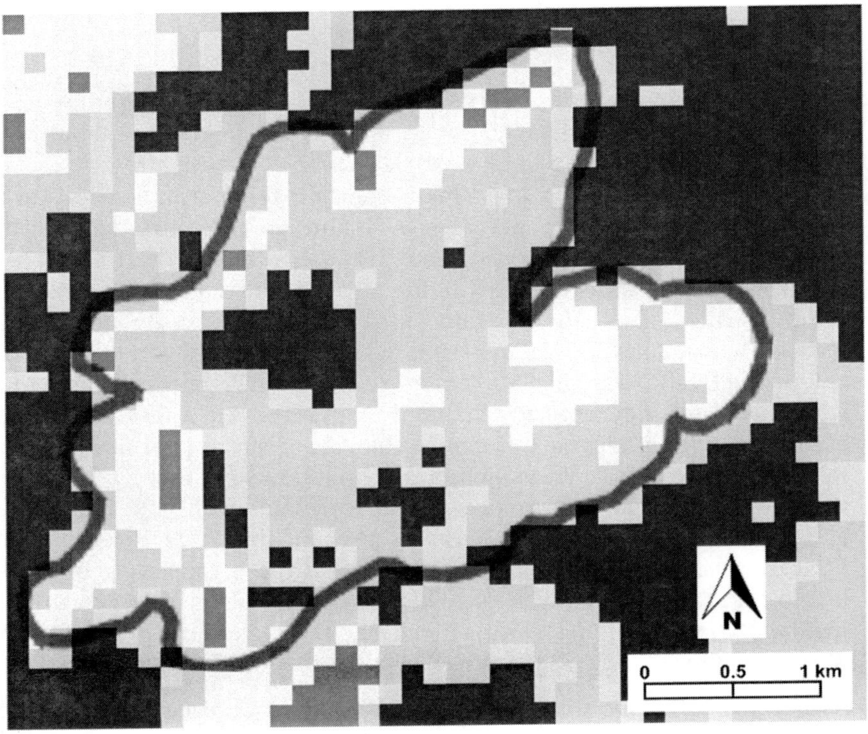

Figure 11. Average pixel-integrated particle size map of the North Coulee 972 flow derived from extensive field measurements and related back to the calculated ATI, with blue (<1 cm), cyan (1 cm–3 cm); green (3 cm–500 cm); yellow (500 cm–1.0 m); and red (>1 m). The approximate flow boundary is shown by the black outline. Map created from ATI image.

5. DISCUSSION

5.1. Terrestrial Applications

The distribution of grain and block sizes on the surface of North Coulee is related to both the emplacement of the high-silica rhyolite flow itself as well as subsequent volcanic activity in the region. Post-emplacement mechanical weathering would have had little effect on the block size distribution due to the young age of these flows. However, emplacement of the high-viscosity lavas and stress conditions during flow have been shown to break larger blocks into smaller, sub-meter fragments with distance from the vent [77]. Vent regions generally contain the largest average block size, whereas ridged zones further from the vent will have the smallest average block size due to compressional stresses. The non-mantled surfaces of North Coulee follow a similar pattern in the ATI result. Sites 1 and 2 (lower ATI) are situated near southern ridges, and contain homogenous, smaller block sizes. Whereas sites 3 and 4 (higher ATI) are further north on the dome and closer to the vent region [54].

During dome emplacement the higher temperature lava below the surface inhibits fracturing and also promotes the formation of distinct flow stratigraphy and volcanic breccia through annealing [78]. The process of breccia formation and flow stratigraphy formation are both evident at various locations around North Coulee where large breccia blocks and obsidian outcrops are visible. Areas where these outcrops occur in non-shadowed regions produced some of the highest measured ATI values. Weathering and gravitational collapse of these blocks has produced fragmental material that also has a higher ATI than the detrital material.

Complicating the direct relationship between ATI, particle size, and dome emplacement processes at North Coulee is the presence of the overlying mantling deposits, which was the focus of this study. Emplaced during eruptions of younger domes, these deposits mask the flow surface in many areas but also create an ideal test location for the Mars flows. High ATI values ($>3.0 \times 10^{-2}$) are generally found in areas dominated by large obsidian and rhyolite boulders. Conversely, lower ATI values ($<2.3 \times 10^{-2}$) are indicative of extensive finer-grained pyroclastic mantling. Pixels with ATI values intermediate to this range generally can be divided into two categories: homogenous, moderately-sized particles (Figure 2, Site 2) or heterogeneous mixing of fine-grained mantling deposits with unmantled larger blocks (Figure 2, Site 3). Within this first group, ATI values between 2.3×10^{-2} and 2.5×10^{-2} generally corresponded to surfaces with particles ranging from 1 to 4 cm, whereas larger ATI values corresponded to blocks greater than 5 cm.

However, the correlation between ATI and emissivity is poor; a result similar to the studies of mantled flows on Mars [4,17,62]. This lack of correlation can be explained by the composition and texture of the surface materials. The domes and mantling deposits at North Coulee all consist of high silica rhyolite of various block and particle sizes and surface textures. Therefore, emissivity is not a reliable indicator to distinguish mantled from non-mantled surfaces, nor extract the fragment size or vesicular texture because of similar composition and vesicular state [73].

Distinguishing non-mantled lava flows from flows that contain some level of volcanic and/or eolian mantling on Earth is possible with the appropriate caution and validation using this approach. ATI relies on diurnal temperature measurements, which are affected by atmospheric conditions such as changes in humidity, precipitation, surface cooling by winds, as well as degassing and heat loss on active flows and domes. Accurate atmospheric correction is important and considered to be relatively well understood for terrestrial TIR corrections. However, active domes on Earth will produce gases that are not commonly corrected in standard atmospheric models such as MODTRAN and have temperatures well above the average background. This internal heat would complicate the normal diurnal heating and cooling process, thereby making ATI more complex to interpret. Finally, the presence of vegetation on older domes (if extensive) could lead to erroneous results. Future work using this approach in terrestrial applications should therefore be focused mainly on younger flows that have cooled and have sparse vegetation cover. In addition, an accurate knowledge of the recent precipitation and climate conditions prior to the data collection is important.

It should be noted that more robust ATI calculations can be implemented using various additions to the basic ATI equation. For example, the inclusion of simple scaling factors [9] through a much more rigorous implementation to convert ATI to approximated actual thermal inertia [41]. Scheidt *et al.* (2010) [9] includes two variables (n and c) in the calculation of ATI using: $ATI = nc\,(1 - a)/\Delta T$, where n and c are correction factors to account for variations in solar flux with latitude and solar declination [79]. The effects of these scaling factors at this latitude and time of year are generally quite small and would have no effect on reducing the larger scale shadowing errors. Therefore, their addition is not deemed critical to this study, but may be important for applications at other locations.

5.2. Applications to Mars

The application of this type of approach for mantled lava flows on the SW apron of Arsia Mons and within the plains of Daedalia Planum in the southern part of the Tharsis region of Mars was initially explored by Ramsey and Crown (2010) [4] and Crown *et al.* (2012) [16]. Finding an approach to accurately discriminate regions of mantling on volcanic surfaces from the underlying lava flows using a thermophysical methodology has the potential to greatly improve the understanding of the compositional and volcanic history of these regions. However, the ATI approach should also be compared to model-based theoretical thermal inertia results in order to determine its applicability for Mars. For example, ATI derived from THEMIS could be used as simpler alternative to model-derived thermal inertia with the results being used to quickly determine inter-pixel mixing and homogeneous cover of coarser grained particles [34]. The success of using ATI to determine grain and block size of the North Coulee flow indicate that an ATI analysis of the flow fields of Arsia Mons and surrounding regions could be useful. The similarity between the two regions and the success of ATI analysis on Earth warrant an analogous study on Mars.

6. CONCLUSIONS

Initial applications were successful in utilizing apparent thermal inertia to distinguish mantled from non-mantled terrestrial volcanic surfaces and directly link ATI to particle size for the Mono Craters and Domes. For this study, mantled regions had a calculated ATI of $<2.3 \times 10^{-2}$, whereas blocky non-mantled deposits had an ATI value above 3.0×10^{-2}. ATI findings were able to distinguish the average grain size of these mantling deposits adding further fidelity to the study. A mantled pixel with a very low (1.6×10^{-2}) value correlated with an average grain size of ~0.5 cm in the field, whereas another mantled pixel with an ATI value of 2.3×10^{-2} had an average grain size of 1–2 cm. The sensitivity of this ATI approach is obvious and therefore valuable for mapping flows on Earth that are partially-mantled or that contain a wide range of particle/block sizes due to flow emplacement conditions. The use of ATI is also an asset in hazardous or inaccessible terrain where ground validation becomes impossible or dangerous. However, pixel mixing and larger-scale shadowing can hinder accurate interpretation of ATI results depending on image resolution and caution should be taken where interpreting these results without direct field observations or validation using other high-resolution images. For example, moderate ATI values at the MCD indicate either heterogeneous (mixed) particle sizes or a homogenous pixel of an intermediate particle size. These cannot be uniquely separated with the ATI data alone. In these cases, ground validation (either on-site or from higher-resolution aerial or satellite images) is required until the development of a more nuanced ATI model. Finally, very high ATI values in steep topography (particularly on slopes opposite the solar azimuth angle) are likely the result in a large amount of pixel-integrated shadowing. These are typically easy to identify and ongoing work is addressing these errors through incorporation of a DEM into the ATI calculation. Future work using ATI should also address mantling thickness using a layered model to estimate thicknesses and attempt to decouple the effects of layering and or accurately identify areas of the lava that protrude through the mantling deposit.

The success of this terrestrial ATI study bodes well for its inclusion in research identifying mantled surfaces in volcanic regions on Mars. Using ATI in addition to modeled thermal inertia allows areas of mantling on Mars to be properly classified, mantling thickness to be constrained, and the percentage of checkerboard mixing between mantled deposits and the underlying lava to be determined. This will then allow spectral analysis to be concentrated on just those regions of exposed lava with minimal to no mantling, which enables textural and compositional analysis both down flow from the vent as well as across different flow units comprising the larger flow field. Where combined with geologic mapping and relative age dating, this integrated analysis becomes an important tool for interpreting the larger volcanic history of the flow field under study (as opposed to the eolian history of the mantling). Ideal study sites include the lava flow fields of southern Arsia, which display a wide range of thermophysical behavior with some flows exhibiting a thermophysical response consistent with a complete dust cover and others appearing to be dust-free. Even more confounding are flows that do not show any consistent thermophysical behavior throughout the day by staying consistently cool or warm [4]. Therefore, identifying these mantling deposits, their thickness, their separation from the underlying rocks, and the eventual mapping of the underlying lava composition should provide a much greater insight into this relatively young volcanic field on Mars and the subsequent eolian modification of these regions.

SUPPLEMENTARY MATERIALS

Details of Field Approach

Fieldwork was conducted on the North Coulee flow from 13–17 July 2012, and included detailed measurements at 13 sample sites across the dome surface and around the perimeter (Table 2, above). Sites were selected on the North Coulee flow with high values of ATI as well as at numerous low ATI targets assumed to be mantled (Figure 6). Data collection included photographs, GPS location, surface roughness measurements, sediment descriptions, direct contact temperature measurements using a thermocouple at the surface and three depths where possible (1 cm, 15 cm, and 25 cm), measurement of the large block sizes, as well as sample collection for later TIR laboratory spectral analysis. Samples and locations were selected specifically to be representative of the deposits within the 90-meter ASTER TIR pixel and the dome as a whole. If a layer of mantling material covered a site, samples of the mantling deposit and underlying material (buried mantled deposits or dome rock) were obtained. The depth of the mantling layer was also measured (where possible). If a site had no mantling deposits, the block sizes were measured along transect lines and a representative sample was collected for later laboratory analysis. Contiguous rock-mantle contacts at the surface were mapped by walking the contact using a Trimble XRS real-time differential Global Position System (GPS) unit with an accuracy of ~30 cm in order to later calculate the areal proportion of mantling within a given pixel. In addition, a complete and detailed survey of an ASTER TIR pixel (90 m^2) was conducted at one site to assess the degree of surface variability and improve the confidence in the sampling strategy at the other sites. This pixel calibration survey consisted of visual observations taken at one meter intervals in a crosswise transect from one corner to the opposite corner. Surface components (e.g., percent vegetation, vegetation type, vegetation height, mantling percentage, average particle size of the mantling material, composition, *etc.*) were all logged and later averaged over the entire pixel to gain a thorough understanding of the sub-pixel diversity.

Details of Laboratory Analyses

The Nicolet Nexus 670 spectrometer used in laboratory analyses has a 2 wavenumber spectral resolution between 2000 and 400 wavenumber (~5–25 μm). A thorough account of the measurement approach and data calibration processing can be found in Ruff *et al.* (1997) [80]. Several grams of particulate sample were placed in 2.5 cm diameter cups and heated to 80 °C in a temperature-controlled oven for a 24-h period. Each sample was then placed on a controlled heating stage within a CO_2 and H_2O purged glovebox. The spectral spot size of approximately 2 cm assured a well-represented spectrum of each mixed sample. Each analysis took ~5 min resulting in a spectrum with a signal to noise of ~800. The spectra were also compared to the corresponding five-point TIR emissivity spectra obtained from the ASTER emissivity of the pixel that contained that sample. Both sets of spectra (laboratory and ASTER) were deconvolved using the approach of Ramsey and Christensen (1998) [7] with obsidian and blackbody end-members to determine surface vesicularity [7,73]. In addition to the spectral analysis, all samples were viewed under a petrographic microscope and point counts made on a five by five centimeter grid to constrain both texture and grain size.

Detailed Results of Field Sites

Site 1 is located at the base of the flow front on the southeastern side of North Coulee (see Figure 3 for location, Figure 7for ground images). A 90 m-pixel survey was conducted at this site to validate the ATI results, which had the lowest ATI in the MCD scene (1.6×10^{-2}). Considering this location had an ATI value less than the maximum (ATI = 2.3×10^{-2}) for a mantled surface on the dome, it was assumed that this pixel contained a very fine-grained deposit with a thick mantling depth. This was verified by the pixel survey, with the site having an approximate mantling depth of 16 cm and an average grain size of 0.5 to 1.0 cm with grains as small as 13 mm. Field investigation found Site 1 to be relatively homogenous in composition and particle size across the surveyed region, with no large blocks, vegetation or slopes. It was therefore considered the ideal calibration pixel for the later analyses of other sites.

The location of Site 2 (Figure 7) is on the eastern section of a large pumice deposit on the western slope of North Coulee. The ATI of 2.3×10^{-2} is considered to be at the upper end of the range of a well-mantled surface. Ground analysis here revealed the average grain size to be larger (1–2 cm) than Site 1, with grains as small as ~2 mm. This site also is fairly homogenous, but closer to large boulder piles and with no measureable vegetation.

Site 3 (Figure 8) is located on the northeastern lobe of North Coulee and is an example of bimodal particle-size areal mixing that resulted in a pixel of moderate ATI (2.6×10^{-2}). Within this pixel area, large (~0.5 m) blocks dominated one half, whereas finer-grained pyroclastic material covered the remainder. Several moderately high pine trees (~10 m) and small shrubs were found growing in the fine-grained material. However, this vegetation comprised no more than 2%–3% of the total area and therefore did not affect the ATI result in a measurable way. The average grain size was not measured here due to rough terrain but was observed to fall between the size of the blocks (~0.15 m) and pyroclastic particles (1–4 cm).

Site 4 (Figure 8) is also located on the northeastern lobe of North Coulee, about 100 m north of Site 3 near the edge of the northern flow front. This region had one of the higher calculated ATI values (2.9×10^{-2}) of all surfaces on North Coulee. The blocks in this area are larger than previous sites. Each block was measured along a 50 m transect and resulted in an average size of 40 cm, with the largest measuring 720 cm and the smallest being 170 cm (see Figure 9 for block size measurements).

Other sites included an outlying dome on the southwest edge of the coulee with a large pumice deposit (Figure 3, Site 10) and a high ATI value, as well as plains of fine-grained pumice deposited near the western edge of North Coulee (Figure 3, Site 9). The large pumice deposit was examined specifically because the higher ATI should be indicative of larger blocks or well-cemented deposits, counterintuitive to what was presumed to be at this site. The deposit was found to be comprised of loosely-consolidated, coarse-grained pyroclastic granules (~1–5 cm) deposited on a steep slope. Because of the position of this deposit on the western side of the dome (and the fact that other western and northern slopes have higher ATI), the elevated ATI was presumed to be caused mainly by shadowing produced from the slope face. Site 9 was located on a region of the dome with the most pervasive mantling. It contained very well-sorted, finer-grained, wind-deposited pumice with outcrops of larger volcanic debris protruding through. The larger outcrops were ~10% of the pixel area and raised the ATI result into the intermediate range (2.6×10^{-3}). Locations such as these were seen as the most similar Mars analog found on the North Coulee flow.

ACKNOWLEDGMENTS

Funding for this research was made possible through the NASA Planetary Geology and Geophysics (PGG) Program (grant NNX11AP17G). The quality and clarity of this manuscript were greatly improved by reviews from Gaetana Ganci, Gabor Kereszturi and David Pieri. The authors would also like to thank Anna Downey (PGG Undergraduate Research Program student) for her assistance during the fieldwork and research preparation.

AUTHOR CONTRIBUTIONS

Michael Ramsey and David Crown conceived and designed the experiments. Mark Price performed the experiments. Mark Price and Michael Ramsey analyzed the data. Michael Ramsey contributed materials and laboratory equipment. Mark Price wrote the paper.

CONFLICTS OF INTEREST

The authors declare no conflict of interest.

REFERENCES AND NOTES

1. Malin, M.C.; Edgett, K.S. Mars Global Surveyor Mars Orbiter Camera: Interplanetary cruise through primary mission. J. Geophys. Res. 2001, 106, 23429–23570.
2. Christensen, P.R.; Bandfield, J.L.; Hamilton, V.E.; Ruff, S.W.; Kieffer, H.H.; Titus, T.N.; Malin, M.C.; Morris, R.V.; Lane, D.M.; Clark, R.N.; et al. Mars Global Surveyor Thermal Emission Spectrometer experiment: Investigation description and surface science results. J. Geophys. Res. 2001, 106, 23823–23871.
3. Bandfield, J.L.; Hamilton, V.E.; Christensen, P.R.; McSween, H.Y. Identification of quartzofeldspathic materials on Mars. J. Geophys. Res. 2004, 109.
4. Ramsey, M.S.; Crown, D.A. Thermophysical and spectral variability of Arsia Mons lava flows. In Proceedings of the 41st Lunar and Planetary Science Conference, The Woodlands, TX, USA, 1–5 March 2010.
5. Hunt, G.R.; Vincent, R.K. The Behavior of Spectral Features in the Infrared Emission from Particulate Surfaces of Various Grain Sizes. J. Geophys. Res. 1968, 73, 6039–6046.
6. Mustard, J.F.; Hays, J.E. Effects of hyperfine particles on reflectance spectra from 0.3 to 25 μm. Icarus 1997, 125, 145–163.
7. Ramsey, M.S.; Christensen, P.R. Mineral abundance determination: Quantitative deconvolution of thermal emission spectra. J. Geophys. Res. 1998, 103, 577–596.
8. Ramsey, M.S.; Fink, J. Estimating silicic lava vesicularity with thermal remote sensing: A new technique for volcanic mapping and monitoring. Bull. Volcanol. 1999, 61, 32–39.
9. Scheidt, S.; Ramsey, M.; Lancaste, N. Determining soil moisture and sediment availability at White Sands Dune Field, New Mexico, from apparent thermal inertia data. J. Geophys. Res. 2010, 115.
10. Scheidt, S.; Lancaster, N.; Ramsey, M. Eolian dynamics and sediment mixing in the Gran Desierto, Mexico, determined from thermal infrared spectroscopy and remote-sensing data. Geol. Soc. Am. Bull. 2011, 123, 1628–1644.
11. Edgett, K.S.; Lancaster, N. Volcaniclastic aeolian dunes: Terrestrial examples and application to Martian sands. J. Arid Environ. 1993, 25, 271–297.
12. Fenton, L.K.; Bandfield, J.L. Aeolian processes in Proctor Crater on Mars: Sedimentary history as analyzed from multiple data sets. J. Geophys. Res. 2003, 108.
13. Johnson, J.R.; Christensen, P.R.; Lucey, P.G. Dust coatings on basaltic rocks and implications for thermal infrared spectroscopy of Mars. J. Geophys. Res. 2002, 107.
14. Ruff, S.W.; Christensen, P.R. Bright and dark regions on Mars: Particle size and mineralogical characteristics based on thermal emission spectrometer data. J. Geophys. Res. 2002, 107.
15. Bandfield, J.L. Effects of surface roughness and graybody emissivity on martian thermal infrared spectra. Icarus 2009, 202, 414–428.
16. Crown, D.A.; Ramsey, M.S.; Berman, D.C. Morphologic and chronologic studies of lava flow fields in the southern Tharsis region of Mars. In Proceedings of the 43rd Lunar and Planetary Science Conference, The Woodlands, TX, USA, 19–23 March 2012.
17. Ramsey, M.S.; Crown, D.A.; Price, M.A. Decoupling lava flow composition and emplacement processes from eolian mantling deposits using thermal infrared data. In Proceedings of the 43rd Lunar and Planetary Science Conference, The Woodlands, TX, USA, 19–23 March 2012.
18. Price, M.A.; Ramsey, M.S.; Crown, D.A. Thermophysical characteristics of mantled terrestrial volcanic surfaces: Infrared analogs to the Arsia Mons flows. In Proceedings of the 44th Lunar and Planetary Science Conference, The Woodlands, TX, USA, 18–22 March 2013.
19. Crown, D.A.; Ramsey, M.S. Morphologic and thermophysical characteristics of lava flows southwest of Arsia Mons, Mars. J. Volc. Geotherm. Res. 2016. under review.
20. Hardgrove, C.; Moersch, J.; Whisner, S. Thermal imaging of alluvial fans: A new technique for remote classification of sedimentary features. Earth Planet. Sci. Lett. 2009, 285, 124–130.

21. Greeley, R. The Snake River Plain, Idaho: Representative of a new category of volcanism. J. Geophys. Res. 1982, 87, 2705–2712.
22. Plescia, J.B. Morphometric properties of martian volcanoes. J. Geophys. Res. 2004, 109.
23. Lang, N.P.; Tornabene, L.L.; McSween, H.Y., Jr.; Christensen, P.R. Tharsis-sourced relatively dust-free lavas and their possible relationship to Martian meteorites. J. Volcanol. Geotherm. Res. 2009, 185, 103–115.
24. Terra|The EOS Flagship. Available online: terra.nasa.gov (accessed on 7 October 2015).
25. Yamaguchi, Y.; Kahle, A.; Tsu, H.; Kawakami, T.; Pniel, M. Overview of Advanced Spaceborne Thermal Emission and Reflectance Radiometer (ASTER). IEEE Trans. Geosci. Remote Sens. 1998, 36, 1062–1071.
26. Edgett, K.S.; Malin, M.C. Martian sedimentary rock stratigraphy: Outcrops and interbedded craters of northwest Sinus Meridiani and southwest Arabia Terra. Geophys. Res. Lett. 2002, 29.
27. Bandfield, J.L.; Feldman, W.C. Martian high latitude permafrost depth and surface cover thermal inertia distributions. J. Geophys. Res. 2008, 113.
28. Price, J.C. Thermal inertia mapping: A new view of the earth. J. Geophys. Res. 1977, 82, 2582–2590.
29. Palluconi, F.D.; Kieffer, H.H. Thermal inertia mapping of Mars from 60 S to 60 N. Icarus 1981, 45, 415–426.
30. Jakosky, B.M.; Mellon, M.T.; Kieffer, H.H.; Christensen, P.R.; Varnes, E.S.; Lee, S.W. The thermal inertia of Mars from the Mars global surveyor thermal emission spectrometer. J. Geophys. Res. 2000, 105, 9643–9652.
31. Fergason, R.L.; Christensen, P.R.; Kieffer, H.H. High-resolution thermal inertia derived from the Thermal Emission Imaging System (THEMIS): Thermal model and applications. J. Geophys. Res. 2006, 111.
32. Kieffer, H.H.; Martin, T.Z.; Peterfreund, A.R.; Jakosky, B.M.; Miner, E.D.; Palluconi, F.D. Thermal and albedo mapping of Mars during the Viking primary mission. J. Geophys. Res. 1977, 82, 4249–4291.
33. Edgett, K.S.; Christensen, P.R. The particle size of Martian aeolian dunes. J. Geophys. Res. 1991, 96, 22765–22776.
34. Christensen, P.R. Martian dust mantling and surface composition: Interpretation of thermophysical properties. J. Geophys. Res. 1982, 87, 9985–9998.
35. Fink, J.H.; Manley, C.R. Origin of pumiceous and glassy textures in rhyolite flows and domes. In The Emplacement of Silicic Domes and Lava Flows; Fink, J.H., Ed.; Geological Society of America Special Paper; Geological Society of America: Boulder, CO, USA, 1987; Volume 212, pp. 77–88.
36. Putzig, N.E.; Mellon, M.T.; Kretke, K.A.; Arvidson, R.E. Global thermal inertia and surface properties of Mars from the MGS mapping mission. Icarus 2005, 173, 325–341.
37. Rogers, A.D.; Christensen, P.R.; Bandfield, J.L. Compositional heterogeneity of the ancient Martian crust: Analysis of Ares Vallis bedrock with THEMIS and TES data. J. Geophys. Res. 2005, 110.
38. Edwards, C.S.; Bandfield, J.L.; Christensen, P.R.; Fergason, R.L. Global distribution of bedrock exposures on Mars using THEMIS high-resolution thermal inertia. J. Geophys. Res. Planets 2009, 114.
39. Van de Griend, A.; Camillo, P.; Gurney, R. Discrimination of soil physical parameters, thermal inertia, and soil moisture from diurnal surf. Water Resour. Res. 1985, 21, 997–1009.
40. Rosema, A.; Fiselier, J. Meteosat-based evapotranspiration and thermal inertia mapping for monitoring transgression in the Lake Chad region and Niger Delta. Int. J. Remote Sens. 1990, 11, 741–752.
41. Cai, G.; Xue, Y.; Hu, Y.; Wang, Y.; Guo, J.; Luo, Y.; Wu, C.; Zhong, S.; Qi, S. Soil moisture retrieval from MODIS data in Northern China Plain using thermal inertia model. Int. J. Remote Sens. 2007, 28, 3567–3581.
42. Matsushima, D.; Kimura, R.; Shinoda, M. A Study on Soil Moisture Estimation using Thermal Inertia. J. Jpn. Soc. Civ. Eng. B1 2012, 67.

43. Cracknell, A.P.; Xue, Y. Estimation of ground heat flux using AVHRR data and an advanced thermal inertia model (SoA-TI model). Int. J. Remote Sens. 1996, 17, 637–642.
44. Sobrino, J.A.; El Kharraz, M.H. Combining afternoon and morning NOAA satellites for thermal inertia estimation 1: Algorithm and its testing with Hydrologic Atmospheric Pilot Experiment-Sahel data. J. Geophys. Res. 1999, 104, 9445–9453.
45. Sobrino, J.A.; El Kharraz, M.H. Combining afternoon and morning NOAA satellites for thermal inertia estimation 2. Methodology and application. J. Geophys. Res. 1999, 104, 9455–9465.
46. Price, J.C. On the analysis of thermal infrared imagery: The limited utility of apparent thermal inertia. Remote Sens. Environ. 1985, 18, 59–73.
47. Kahle, A.B. Surface emittance, temperature, and thermal inertia derived from Thermal Infrared Multispectral Scanner (TIMS) data for Death Valley, California. Geophysics 1987, 52, 858–874.
48. Putzig, N.E.; Mellon, M.T. Apparent thermal inertia and the surface heterogeneity of Mars. Icarus 2007, 191, 68–94.
49. Hildreth, W. Volcanological perspectives on Long Valley, Mammoth Mountain, and Mono Craters: Several contiguous but discrete systems. J. Volcanol. Geotherm. Res. 2004, 136, 169–198.
50. Bailey, R.A. Geologic Map of Long Valley Caldera, Mono-Inyo Craters Volcanic Chain, and Vicinity, Eastern California; U.S. Geological Survey Map I-1933. U.S. Geological Survey: Reston, VA, USA, 1989.
51. Bursik, M.; Sieh, K. Range front faulting and volcanism in the Mono Basin, Eastern California. J. Geophys. Res. 1989, 94, 15587–15609.
52. Wood, S. Chronology of Late Pleistocene and Holocene Volcanics, Long Valley and Mono Basin Geothermal Areas, Eastern California; United States Geological Survey Open File Report. U.S. Geological Survey: Reston, VA, USA, 1983.
53. Bailey, R.A. Quaternary volcanism of Long Valley Caldera and Mono-Inyo Craters, Eastern California. In Field Trip Guidebook T313; American Geophysical Union: Washington, DC, USA, 1989.
54. Sieh, K.; Bursik, M. Most recent eruption of the Mono Craters, Eastern Central California. J. Geophys. Res. 1986, 91, 12539–12571.
55. Hill, D.P. Response Plan for Volcano Hazards in the Long Valley Caldera and Mono Craters Region, California; U.S. Geological Survey: Reston, VA, USA, 2002; Volume 2185.
56. Crumpler, L.S.; Aubele, J.C. Structural evolution of Arsia Mons, Pavonis Mons, and Ascraeus Mons: Tharsis region of Mars. Icarus 1978, 34, 496–511.
57. Crumpler, L.S.; Head, J.W.; Aubele, J.C. Calderas on Mars: Characteristics, structure and associated flank deformation. In Volcano Instability on the Earth and Other Planets; McGuire, W.J., Jones, A.P., Neuberg, J., Eds.; Geological Society Special Publication: London, UK, 1996; Volume 110, pp. 307–348.
58. Head, J.W.; Siebert, N.; Pratt, S.; Smith, D.; Zuber, M.; Solomon, S.; McGovern, P.J.; Garvin, J.B.; The MOLA Science Team. Characterization of major volcanic edifices on Mars using Mars Orbiter Laser Altimeter (MOLA) data. In Proceedings of the 29th Lunar and Planetary Science Conference, The Woodlands, TX, USA, 16–20 March 1998.
59. Head, J.W.; Siebert, N.; Pratt, S.; Smith, D.; Zuber, M.; Garvin, J.B.; McGovern, P.J.; The MOLA Science Team. Volcanic calderas on Mars: Initial views using Mars Orbiter Laser Altimeter (MOLA) data. In Proceedings of the 29th Lunar and Planetary Science Conference, The Woodlands, TX, USA, 16–20 March 1998.
60. Mouginis-Mark, P.J. Prodigious ash deposits near the summit of Arsia Mons volcano, Mars. Geophys. Res. Lett. 2002, 29.
61. Scott, D.; Zimbelman, J. Geologic Map of Arsia Mons Volcano, Mars; U.S. Geological Survey: Reston, VA, USA, 1995.
62. Ramsey, M.S.; Crown, D.A. Mantled lava flow surfaces on Earth: Thermo-physical analogs for Martian volcanism? In Proceedings of the American Geophysical Union, Fall Meeting, San Francisco, CA, USA, 5–9 December 2011.
63. Scott, D. Map Showing Lava Flows in the Southeast Part of the Phoenicis Lacus Quadrangle of Mars; U.S. Geological Survey: Reston, VA, USA, 1981.

64. Scott, D.; Schaber, G.; Dial, A. Map Showing Lava Flows in the Southwest Part of the Phoenicis Lacus Quadrangle of Mars; U.S. Geological Survey: Reston, VA, USA, 1981.
65. Scott, D.; Schaber, G.; Horstman K.; Dial, A.; Tanaka, K. Map Showing Lava Flows in the Northwest Part of the Phoenicis Lacus Quadrangle of Mars; U.S. Geological Survey: Reston, VA, USA, 1981.
66. Bleacher, J.E.; Greeley, R.; Williams, D.A.; Cave, S.R.; Neukum, G. Trends in effusive style at the Tharsis Montes, Mars, and implications for the development of the Tharsis province. J. Geophys. Res. 2007, 112.
67. Garry, W.B.; Williams, D.A.; Bleacher, J.E.; Dapremont, A.M. Geologic mapping of Olympus Mons and Tharsis Montes, Mars. In Proceedings of the 46th Lunar and Planetary Science Conference, The Woodlands, TX, USA, 16–20 March 2015.
68. Russell, I.C. Quaternary History of the Mono Valley, California; U.S. Geological Survey Eighth Annual Report. U.S. Geological Survey: Reston, VA, USA, 1889; pp. 267–394.
69. National Climatic Data Center (NCDC). NOAA, n.d. Web. 24 March 2013.
70. Google Earth. 37°53′42.34″ N, 119°0′16.07″ W. Available online: https://www.google.com/earth/ (accessed on 15 November 2012).
71. Abrams, M. The Advanced Spaceborne Thermal Emission and Reflection Radiometer (ASTER): Data products for the high spatial resolution imager on NASA's Terra platform. Int. J. Remote Sens. 2000, 21, 847–859.
72. Realmuto, V.J. Separating the effects of temperature and emissivity: Emissivity spectrum normalization. In Proceedings of the 2nd TIMS Workshop, Pasadena, CA, USA, 6 June 1990; Jet Propulsion Laboratory Publication: Pasadena, CA, USA, 1990; pp. 90–155.
73. Ramsey, M.S.; Fink, J.H. Estimating lava vesicularity: A new technique using thermal infrared remote sensing data. Am. Geophys. Union Eos Trans. 1996, 77, F803.
74. Mars, J.C.; Rowan, L.C. Spectral assessment of new ASTER SWIR surface reflectance data products for spectroscopic mapping of rocks and minerals. Remote Sens. Environ. 2010, 114, 2011–2025.
75. Lee, R.J.; King, P.L.; Ramsey, M.S. Spectral analysis of synthetic quartzofeldspathic glasses using laboratory thermal infrared spectroscopic methods. J. Geophys. Res. 2010, 115.
76. Simurda, C.M.; Ramsey, M.S. Correcting topographic shadowing errors in apparent thermal inertia images. In Proceedings of the 45th Lunar and Planetary Science Conference, The Woodlands, TX, USA, 17–21 March 2014.
77. Anderson, S.W.; Stofan, E.R.; Plaut, J.J.; Crown, D.A. Block size distributions on silicic lava flow surfaces: Implications for emplacement conditions. Geol. Soc. Am. Bull. 1998, 110, 1258–1267.
78. Vaughan, R.A., Ed.; Remote Sensing Applications in Meteorology and Climatology; Springer Science & Business Media: Berlin, Germany, 1987; Volume 201.
79. Mellon, M.T.; Jakosky, B.M.; Kieffer, H.H.; Christensen, P.R. High-resolution thermal inertia mapping from the Mars global surveyor thermal emission spectrometer. Icarus 2000, 148, 437–455.
80. Ruff, S.W.; Christensen, P.R.; Barbera, P.W.; Anderson, D.L. Quantitative thermal emission spectroscopy of minerals: A laboratory technique for measurement and calibration. J. Geophys. Res. 1997, 102, 14399–14913.

CHAPTER 10

Sensitivity to Volcanic Field Boundary

Melody G. Runge[1]*, Mark S. Bebbington[2], Shane J. Cronin[1], Jan M. Lindsay[1] and Mohammed Rashad Moufti[3]

[1] *University of Auckland, Private Bag 92019, Auckland 1142, New Zealand.*

[2] *Massey University, Palmerston North, New Zealand.*

[3] *King Abdulaziz University, Jeddah, Kingdom of Saudi Arabia.*

ABSTRACT

Volcanic hazard analyses are desirable where there is potential for future volcanic activity to affect a proximal population. This is frequently the case for volcanic fields (regions of distributed volcanism) where low eruption rates, fertile soil, and attractive landscapes draw populations to live close by. Forecasting future activity in volcanic fields almost invariably uses spatial or spatio-temporal point processes with model selection and development based on exploratory analyses of previous eruption data. For identifiability reasons, spatio-temporal processes, and practically also spatial processes, the definition of a spatial region is required to which volcanism is confined. However, due to the complex and predominantly unknown sub-surface processes driving volcanic eruptions, definition of a region based solely on geological information is currently impossible. Thus, the current approach is to fit a shape to the known previous eruption sites. The class of boundary shape is an unavoidable subjective decision taken by the forecaster that is often overlooked during subsequent analysis of results. This study shows the substantial effect that this choice may have on even the simplest exploratory methods for hazard forecasting, illustrated using four commonly used exploratory statistical methods and two very different regions: the Auckland Volcanic Field, New Zealand, and Harrat Rahat, Kingdom of Saudi Arabia. For Harrat Rahat, sensitivity of results to boundary definition is substantial. For the Auckland Volcanic Field, the range of options resulted in similar shapes, nevertheless, some of the statistical tests still showed substantial variation in results. This work highlights the fact that when carrying out any hazard analysis on volcanic fields, it is vital to specify how the volcanic field boundary has been defined, assess the sensitivity of boundary choice, and to carry these assumptions and related uncertainties through to estimates of future activity and hazard analyses.

Keywords: Probabilistic Hazard forecasting Harrat Rahat Auckland volcanic field

BACKGROUND

Regions of distributed volcanism are often referred to as volcanic fields. As each eruption tends to occur in a different location, they can be further classified as monogenetic volcanic fields and are frequently modelled as spatial or spatio-temporal point processes (Connor and Hill 1995; Marzocchi and Bebbington 2012; Bebbington 2013). Volcanic activity within these regions tends to be predominantly basaltic, with low flux rates. Consequently, volcanic fields tend to grow laterally rather than build up a single edifice in one location (Wood and Shoan 1981), and eruptions typically form cinder cones, tuff cones and rings, maars, lava domes, shield volcanoes, and lava flows (Connor and Conway 2000).

Determination of a distributed volcanic field's spatio-temporal behaviour is key to understanding its underpinning magmatic processes, evolution, and hazard potential. As these subsurface processes are predominantly unknown, and highly complex, this determination is accomplished via approximation of the known spatio-temporal evolution of the field, and then extrapolated into the future to obtain hazard forecasts. This almost invariably uses spatial or spatio-temporal point processes with model selection and development based on analysis of previous eruption data. The location of future eruptions substantially influences the areas at risk from eruption related hazards. The specific location may also affect the eruption style, for example, the presence of ground-water may lead to an explosive phreatomagmatic eruption (Ross et al. 2011). Thus, the probable location is one of the most important pieces of information for volcanic hazard estimation.

To obtain probable future locations, the first step is to assess the existence of patterns in previous eruption locations and timings. This is accomplished using various exploratory statistics including lineament identification (Wadge and Cross 1988; Lutz and Gutmann 1995) and cluster analyses (Clark and Evans 1954; Hopkins and Skellam 1954; Ripley 1979). For these, a surface region representing the volcanic field must be defined, both for area calculations and Monte Carlo point placement steps. A region must also be defined for the evaluation of spatio-temporal models (Connor and Hill 1995; Cronin et al. 2001), and for assessment of all spatial and spatio-temporal models during verification and validation steps.

The spatial boundaries of a volcanic field are conventionally defined as the extent of the existing lava flows, scoria cones, or other volcanic edifices. These are usually well constrained via geological mapping. This provides a present-day view of the location of past volcanic material from a resource-mapping perspective. However, for assessment of future volcanic activity, this definition is not sufficient as it does not account for the locations of new eruption vents and may confound the locations of previous volcanic eruptions with factors controlling lava distribution. Depending on rheology and underlying topography, lava may extend tens of kilometres from a source vent and accumulate in depressions giving false weight to these regions far from potential eruption locations.

Volcanic fields form above mantle regions where partial melting produces magma, which may be related to low rates of crustal extension (Takada 1994). In these sites, magma supply is insufficient to maintain a single conduit, such that each new magma batch has to find its own path to the surface (Fedotov 1981). Due to the variety of geological conditions under which mantle materials melt to form magmas (e.g., within subduction zones, in areas of weak crustal thinning, or hot-spot upwelling), boundary parameters remain unknown, thus a well-informed geological approach to boundary definition is currently impossible. For forecasting purposes however, the volcanic field boundary has a significant influence on estimates of event recurrence rate and spatial intensity. Where a hard boundary is required (e.g., for clustering statistics, or spatio-temporal models), it is assumed that there is zero probability of an eruption occurring outside of this region. Often, even when a soft boundary is feasible, a hard boundary is imposed for convenience in spatial models (Lindsay et al. 2010; Sandri et al. 2012).

Estimation of the region of potential volcanic vent breakout in a field represents a typical bias-variance trade-off. Underestimating the area may result in future eruptions occurring outside the area of analysis and an overestimate of average vent density, which may in turn bias statistical diagnostics (Clark and Evans 1954; Hopkins and Skellam 1954; Ripley 1979). Overestimation ensures a conservative 'safe' solution, but potentially underestimates average vent-density and consequently spatial intensity. It also increases the coverage area over which supplementary data is required for a hazard analysis, thus either increasing costs, or lowering the data resolution.

In this work, various approaches for the definition of volcanic field boundaries are evaluated for spatial intensity estimates at two quite different volcanic fields: Harrat Rahat, Kingdom of Saudi Arabia, and Auckland Volcanic Field (AVF), New Zealand. This paper begins with an introduction to these two volcanic fields selected as case studies. Five boundary options are then outlined which comprise of three commonly applied fitting methods (an ellipse, a rectangle, and a convex hill), alongside two vent-density based suggestions (isotropic and anisotropic kernel smoothing methods). Five geological considerations are then discussed for the two volcanic fields that then form the underlying assumptions and related uncertainties for each fitted boundary. These considerations combined with the five boundary options result in ten potential boundary options for the volcanic field case studies. These provide the basis for the sensitivity analyses for each field and the affect of boundary definition is assessed using exploratory methods (summary statistics) and spatial intensity estimates.

HARRAT RAHAT

Harrat Rahat is one of at least nineteen volcanic fields within the western Arabian Plate (Brown 1970). The harrat ('volcanic-field') has a total erupted volume of ~2000 km^3 (Blank and Sadek 1983) from at least 968 identified vents (Runge et al. 2014, Fig. 1a) and is predominantly basaltic with some more evolved compositions. The cause of volcanism across the Arabian Shield is still very much under discussion (Moufti et al. 2013; Rolandone et al. 2013), and although there are other volcanic fields, and a major zone of crustal rifting (the Red Sea rift) proximal to Harrat Rahat, little is known about their respective influences on magma generation within the harrat.

Erupted products include extensive lava flows (Murcia et al. 2014), as well as scoria and spatter cones, shield volcanoes, tuff rings, and lava domes (Camp and Roobol 1989). Eruptions in Harrat Rahat have been dated between 10 Ma and 1256 AD, with age data broadly suggesting a northward focussing of volcanism with time (Camp and Roobol 1989). A caveat on this, however, is that definitive ages are only available for 3 % of vents, and recent $^{40}Ar/^{39}Ar$ ages suggest there are also older flows in the north, and younger flows in the centre of the field (Moufti et al. 2013).

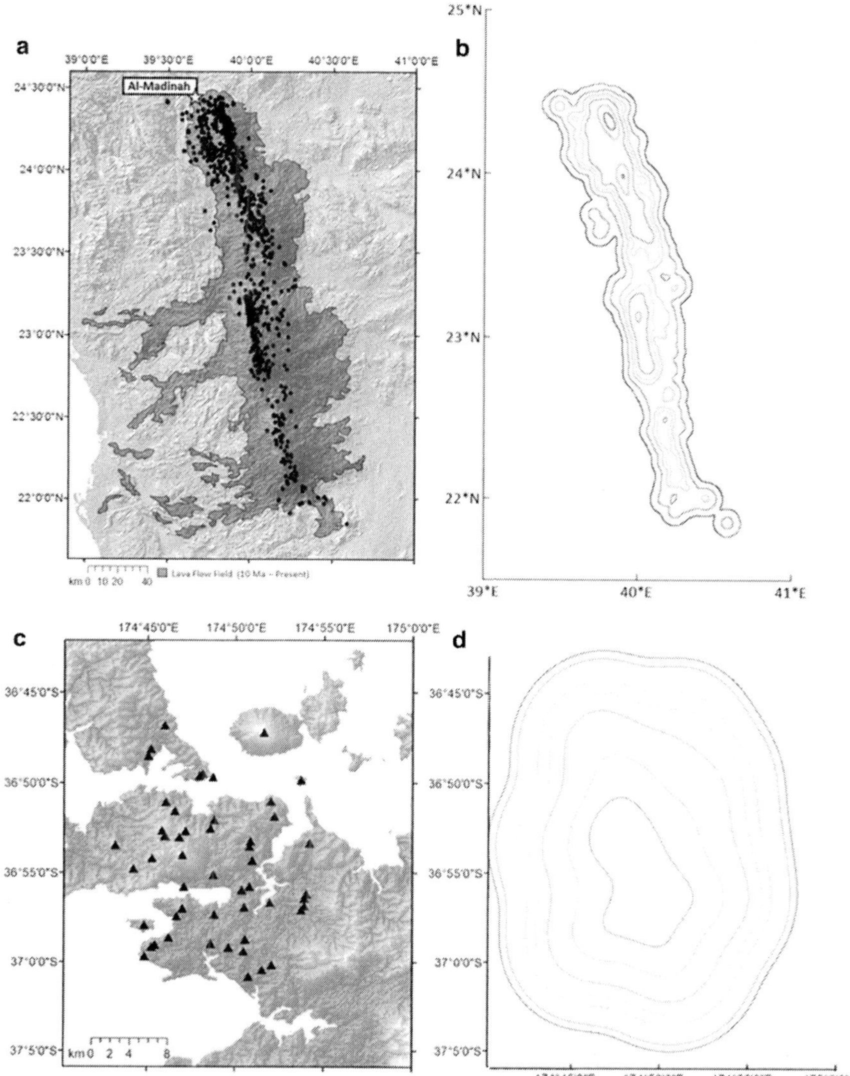

Figure 1. Volcanic field examples. **a** Harrat Rahat vent locations, shaded area represents lava flow field; **b** Harrat Rahat vent density distribution; **c** Auckland Volcanic Field vent locations; **d** Auckland Volcanic Field vent density distribution. Vents are shown as black marks, North is up. Vent density contours at [0.05, 0.5, 1, 5, 10, 20, 50, 80] % of maximum density

Previous boundary definitions for Harrat Rahat have included the mapped limits of lava flows (Moufti et al. 2010), a convex hull of the previous known eruption locations (Runge et al. 2014), and a rectangular area of interest within the northern region known as Harrat Al-Madinah (Moufti et al. 2010; El Difrawy et al. 2013).

Auckland volcanic field

The Auckland Volcanic Field (AVF) is a basaltic field located on the North Island of New Zealand. The field has a total erupted volume of at least 1.7 km^3 (Kereszturi et al. 2013) from approximately 50 volcanoes (Fig. 1c), Hayward et al. 2011). Volcanic features include lava flows, scoria cones, maars, and tuff rings (Allen and Smith 1994). Eruption order and locations are relatively well constrained (Bebbington and Cronin 2011) with events dated between ~250 and 0.6 ka (Lindsay et al. 2011).

The AVF boundary has been defined as a minimum area ellipse fitted to the outer vent locations (Spörli and Eastwood 1997), and was used for evacuation planning (Tomsen et al. 2014). An elliptical boundary was also used by Sandri et al. (2012). This approach was expanded by Bebbington (2013) to include an outer ellipse with an exponentially decaying intensity. Rectangular bounds were employed by Lindsay et al. (2010), and a convex hull with a ~1 km buffer zone was employed for exploratory analyses by Le Corvec et al. (2013a).

Boundary options

While the area is an integral part of the description of a volcanic field, very few studies define how the boundary is represented, thus comparison of exploratory statistics (e.g., average vent densities) between fields is often meaningless, and repeatability of the study impossible. Here, three of the most commonly applied volcanic field boundary definitions will be investigated: (1) convex hull, (2) minimum area ellipse, and (3) minimum area rectangle. Two additional boundary options based on isotropic and anisotropic kernels are also explored. A description of how each of these shapes were fitted to the data is also provided.

Convex hull

For a volcanic field, a convex hull is the smallest convex set that contains all of the vents (any straight line between any two vents must be entirely contained within the hull). Various algorithms can be used to obtain this shape, such as gift-wrapping (Chand and Kapur 1970), or divide-and-conquer (Preparata and Hong 1977), and the majority of statistical packages (e.g., MATLAB, or R) have inbuilt functions for this.

For this reason, and because the fitting of a convex hull is quick, and always renders the same answer, a convex hull is often used as the default shape when applying initial exploratory methods on volcanic fields. One of the first uses of a convex hull in describing volcanic systems was by Zhang and Lutz (1989), for the assessment of the spatial extent in kimberlites. Since then it has been frequently used in volcanic hazard evaluations, e.g., Springerville volcanic field, Arizona (Condit and Connor 1996), Mount Gambier, Newer Volcanic Province, Australia (Bishop 2007) and Auckland Volcanic Field (Le Corvec et al. 2013c). In this study, the convex hull shape was determined via the inbuilt MATLAB function ('convhull').

Minimum area ellipse

The main rationale for an elliptical, or curved boundary to a volcanic field is based primarily on thermodynamic arguments (Fedotov 1981). A mantle source zone below a field is likely to be susceptible to partial melting from a relatively stable (in terms of location) heat source (Condit and Connor 1996). These tend to be semi-circular in plan-view, because heat radiates equally in all directions as long as the lithology is broadly homogeneous.

A minimum area ellipse can be fitted to a set of points (in this case, volcanic vents) via a variety of methods such as non-linear least squares (Angell and Barber 1977), maximum likelihood (Kanatani 2005), or bootstrapping (Cabrera and Meer 1996). Ellipse fitting algorithms are slightly more complex to run than a convex hull algorithm because there are five parameters to vary within the chosen optimisation process: the centre of the ellipse (x_o, y_o), the major and minor radii (r_a, r_b), and the orientation angle (θ). In this work, the minimum area ellipse method of Khachiyan (1996) was adopted which greatly reduces the computational time by finding the bounding convex hull first, reducing the number of points required to fit an ellipse.

Minimum area rectangle

A minimum bounding rectangle can be fitted to the points of a convex hull from a data set, similarly to an ellipse. Several relatively straightforward methods are available such as rotating callipers (Toussaint 1983), or minimum perimeter (Freeman and Shapira 1975). Due to their simplicity both in terms of proportional area calculations, and presentation of results, rectangular shapes are often used when analysis of a sub-region of a field is required for example, around an area being assessed as a nuclear waste facility (Connor and Hill 1995).

Although rectangular volcanic field bounds are difficult to justify geologically, similar shapes have been used in areas where there is thought to be some geological structural control. Kear (1964) fitted oblong structures to the Bay of Islands Volcanic Zone, NZ, and a rift-subparallel structure known as the Mekkah-Madinah line was inferred in the Kingdom of Saudi Arabia (Camp and Roobol 1989). This shape also has several advantages over an ellipse: it is easier to fit a minimum area rectangle, and it tends to truncate the field at one or more ends which minimises empty space. In this work, the minimum area rectangle was calculated from the convex hull of the vents, and a rectangle fitted based on the orientations of each edge, the best-fit is then the rectangle corresponding to the minimum area.

Isotropic and anisotropic kernels (Vent Density Contours)

The final two boundary methods included in this work are based on Gaussian isotropic (Eq. 1), and anisotropic (Eq. 2) kernels:

$$\lambda_s = \frac{1}{N}\sum_{i=1}^{N}\frac{1}{2\pi h}\exp\left[\frac{\|x_i - x\|^2}{2h^2}\right] \quad (1)$$

$$\lambda_s(x) = \frac{1}{2N\pi\sqrt{|H|}}\sum_{i=1}^{N}\exp\left[-0.5(x_i - x)^T H^{-1}(x_i - x)\right] \quad (2)$$

Where N: total number of eruptions, h: 1D kernel bandwidth, H: 2D kernel bandwidth matrix, |H|: is the determinant of H, and $\|x_i - x\|^2$: distance between ith vent at x_i and location x.

These methods have been frequently applied for spatial intensity estimates (Connor and Hill 1995; Martin et al. 2004; Connor and Connor 2009; Kiyosugi et al. 2010; Bebbington and Cronin 2011; Germa et al. 2013; Bebbington 2013) or to represent vent density variation (Connor 1990; Lutz and Gutmann 1995). Despite this potential reflection of spatial variation in magma fertility or persistent melt columns (as represented by high vent density regions, Germa et al. 2013), kernel methods have been suggested (Germa et al. 2013) but are not yet widely used to define a volcanic field boundary. In this study, bandwidth estimation (h, H) is accomplished via SAMSE ('Squared-asymptotic-mean squared error', Duong and Hazelton 2003) using the R package 'ks' (Duong 2007).

The use of kernel density estimation is consistent with the probabilistic approach taken for intensity estimates by assuming that future volcanic activity is strongly linked to past activity, and thus more fertile regions of the fields (as represented by a greater number of vents) are more highly weighted, pushing the boundary out farther from them. As the exponential decay function (with distance) has infinite support, a user-defined cut-off value is required to define a boundary. Here, this truncation was assumed as the maximum isocontour that includes all vents within the same shape, without containing any empty regions (simply connected).

GEOLOGICAL CONSIDERATIONS

Regions of distributed volcanism are the surface representation of highly complex subsurface systems which vary from system to system. Thus, each volcanic field must be considered independently with respect to any potential constraints or known system properties (such as major crustal lithological or fault boundaries). There are several assumptions that need to be made when imposing a boundary and since some may be subjective, it is important that their accompanying uncertainties are propagated through to intensity estimates and hazard forecasts. Assumptions need to address the following:

- Temporal invariance (i)
- Unobserved (hidden) vents (ii)
- Anomalous vents (iii)
- Geological constraints (iv)
- Field maturity (v)

These are considered here one by one.

Temporal invariance

In the case of volcanic field boundaries, temporal invariance suggests a fixed region within which all eruptions have, could have, or will occur. Questions that need to be asked include whether the eruption locations have shifted in a systematic way with time, or if the locus (or loci) of activity has drifted towards, or away from, a particular area.

Shifts in the locus of activity within a volcanic field have been noted by Tanaka et al. (1986) from west to east in the San Francisco volcanic field over the last 2.5 My, and Connor and Hill (1995) found substantial movement in volcanic activity (east to west) within the Yucca Mountain region, Nevada. Kiyosugi et al. (2010) noted a constant location of volcanism within the Abu Monogenetic Volcano Group, Japan over the last 0.46 My, and Condit and Connor (1996) observed continuous waxing and waning of certain areas across Springerville volcanic field, Arizona.

If it is thought that the locus of activity is moving in a specific direction, then a volcanic field boundary may need to be extended, or widened, in that direction to accommodate future eruptions. Or, if no young eruptions have been identified in a region of the volcanic field, this area could be removed from the current or future boundary region by shortening, or narrowing, the boundary in these regions. Recognising that tectonic stress regimes, upward mantle flow (driving partial melting), and plate motions all vary with time, a time-invariant volcanic field boundary is unlikely. However, for young fields, or if the influence of these on local volcanism is minor, then assuming temporal invariance may have little effect on subsequent analysis.

For Harrat Rahat, the two most recent eruptions occurred in the northern most part of the field. However, due to the high levels of uncertainty regarding the ages of the older eruptions, northward migration of the source region cannot be conclusively demonstrated. Thus, temporal invariance will be assumed. While this may be a conservative estimate for the boundary definition, it may lead to underestimation of the hazard, especially in the northern areas of the harrat.

Spörli and Eastwood (1997) found no shift in activity with time for the AVF, neither did more recent work by Bebbington and Cronin (2011) nor Bebbington (2013). Given that this field is very young (<0.25 Ma), temporal invariance is a justifiable assumption.

Unobserved (Hidden) vents

For the majority of volcanic fields, the eruption record is typically incomplete as eruptive centres can be both hidden (e.g., due to surface water, vegetation, or severe erosion), or buried (e.g., by sedimentation or subsequent eruptions). Geological, geophysical, seismic, and remote sensing studies can be used to reduce the number of unidentified vents/eruptions (Perry et al. 2005), but this is not always feasible or affordable. The main problem with an incomplete eruption record is that of underestimating temporal recurrence rates and spatial intensities (Runge et al. 2014). Further, hidden vents may lie significantly outside boundaries derived from observed vents, leading to underestimation of the field area. This argument can be used to justify the inclusion of a lava flow field's extent within a volcanic field boundary, although surface topography and magma rheology are likely uncorrelated with the spatial location of source areas and magmas.

For Harrat Rahat, Runge et al. (2014) used the ages and extents of lava flows with an erosion function to estimate that two thirds of eruptive vents are missing from the record. This estimate correlated well with estimates of lava flow volumes (Camp and Roobol 1989) and volcanic pile thickness (Blank and Sadek 1983). The maximum thickness of the eruptive sequence across the harrat commonly reaches > 200 m and along the centre of the field can be up to 400 m (Blank and Sadek 1983). More than 50 % of observed cones and related vent structures are < 50 m in height, hence they are easily buried by lava in central regions of the field. Since total lava thickness dramatically reduces with distance from the central spine of the field (Fig. 1a), Blank and Sadek 1983), the likelihood of vent burial at its edges is low. Thus it is assumed that no unobserved vents are located outside of the region enclosed by those observed.

While the number of known eruption centres within the Auckland Volcanic Field has recently increased (e.g., Hayward et al. 2011), these were rediscovered and were present on an existing map of the field (Von Hochstetter 1864). The development of Auckland city almost entirely on top of this field suggests that there may be a few hidden eruption centres towards the centre of the field but the decrease in urbanisation towards the external vents means that it is assumed here that no unobserved vents are located outside of the region enclosed by those observed.

Anomalous vents

This section considers whether atypical volcanic vents within or near a volcanic field should be included or ignored. Anomalous eruptions might have distinct geochemistry, erupted volume, or age. There are also regions where polygenetic volcanoes coexist within, or overlap, the region of a volcanic field, for example, the Cascade Range (Guffanti and Weaver 1988), and the Eastern Volcanic Zone, Iceland (Jakobsson 1979).

Condit and Connor (1996) discounted three outlying vents from the Springerville volcanic field on the basis of previous cluster analysis work (Connor et al. 1992). Lutz and Gutmann (1995) discounted any distal vents deemed to be located at the edges of the Pinacate volcanic field when performing lineament analysis using a circular region. For island-based distributed volcanism (Jeju, Brenna et al. 2012; Ambae Island, Vanuatu, Németh and Cronin 2009), offshore eruptions are often discounted when forming a volcanic field boundary for spatial intensity estimation (Tenerife, Martí et al. 2009; El Hierro, Canaries, Melían et al. 2014), although they were included in work by Becerril et al. (2013). This may considerably underestimate the hazard, especially where magma-water interaction may cause the most explosive eruptions from such fields (Németh and Cronin 2009).

The most widely dispersed volcanic vents of Harrat Rahat occur at the most north-western, and south-eastern extents of the field (Fig. 1a). The vents in the north-west are among the most recent eruptions (641 AD), and those in the south-east among the oldest (though to be 8.5–2.5 Ma, Camp and Roobol 1989; Runge et al. 2014). Since there is no evidence other than their location to suggest these vents are anomalous, they must be included in the field boundary estimations.

Evidence of more complex eruption behaviour is also present in Harrat Rahat in the form of more evolved (trachytic) eruptions within a small region around 24°5′N. These eruptions have erupted through older more mafic volcanic structures. While anomalous in eruption style, these vents and their magma compositions are related to the basaltic lavas in this field, and are common to other volcanic fields on the Arabian Shield (Khaybar, Camp et al. 1991), as well as other similar volcanic fields in other parts of the world (Jeju, Brenna et al. 2012). The co-genetic relationship of these eruptions of differing composition leads to the decision to include all observed vents within the volcanic field boundary. However, more sophisticated modelling techniques that consider the behaviour as two separate processes could lead to multiple (likely overlapping) boundary definitions for the same field.

In the Auckland Volcanic Field, the most recent eruption and volcanic structure, Rangitoto, is anomalous in terms of size (Kereszturi et al. 2013), potential polygenetic eruption style, and composition (Needham et al. 2011). However, the most recent eruption may be the best indicator of future eruptions, and it is not uncommon for volcanic fields to contain large, shield-like volcanic structures (Mount Gambier, Bishop 2007; Jeju, Brenna et al. 2012), which may represent slightly different manifestations of volcanism from the same source area. Thus, in the AVF, all vents are included in the field boundary estimations.

Geological considerations

Knowledge of any field-limiting factors should be incorporated into the volcanic field boundary definition, such as crustal/fault structure, geological boundaries, or specific impenetrable lithologies. For example, the Uinkaret volcanic field, Colorado is bounded to the east and west by the Toroweap and Hurricane faults respectively (Karlstrom et al. 2007), and in the Garrotxa volcanic field, Spain, structural control on magma at depth was inferred and used to provide field-limiting constraints (Barde-Cabusson et al. 2014).

Another consideration may be the presence of 'holes' within a volcanic field representing localised regions of depleted mantle or crustal conditions that hinder magma reaching the surface. Stalled, or failed eruptions, where magma is intruded into the shallow crust may also provide additional boundary information. At Harrat Rahat, a 1999 AD earthquake swarm, was interpreted as a site of potential dyke intrusion (Lindsay and Moufti 2014), similar to that of the 2009 AD intrusion in nearby Harrat Lunayyir (Pallister et al. 2010). The location of these inferred intrusions could either be treated as sites where eruptions are now less likely, or conversely, where magma may be stalled, and instead erupt as a more evolved, and therefore potentially more explosive magma in the future. The incorporation of 'holes' in volcanic field boundary definitions would require a substantial change to how boundaries are construed beyond a simply connected polygon. However, with evidence for localised magma depletion, an initial attempt could begin with holes located at previous eruption sites.

Observed structures across and proximal to Harrat Rahat include suspected shear zones, terrane boundaries, and Precambrian sutures and faults (Camp and Roobol 1989). There are also volcanic fields nearby, which have been identified as separate systems based on the variation in vent clustering within and between regions, and relative geochemistry (Coleman et al. 1983). Although identified structures may control the orientations of vent lineaments (Connor 1990; Magill et al. 2005), little is known about their influence on magma pathways for Harrat Rahat, so no definitive field-limiting factors can be assumed.

The eastern boundary of the Auckland Volcanic Field was hypothesised at a contact with up-faulted greywacke basement rocks (Allen and Smith 1994). Consequently, Lindsay et al. (2010) extended their AVF boundary to the North where no outcropping of basement is observed. However, this geological bound was not considered here during boundary construction in the interest of a conservative approach, but this assumption should be revisited if more substantial support for this idea comes to light.

Field maturity

Fitting a boundary to the observed vents based on a minimum constraint, e.g., a convex hull, or minimum area polygon, is used to obtain a unique solution to a mathematical construct. However, this process means that those outermost vents are likely to lie on, or very close to, the volcanic field edge. This assumes that all future eruptions will occur in the currently confined area, an assumption that may not be valid for a very young field, or one with evidence of persistent vent migration, or where the most recent activity is focussed on a sub-area.

For the Auckland Volcanic Field, Spörli and Eastwood (1997) argued that vents at the outer limit of the current field extent varied substantially with age, potentially defining a constant magma source area. In many other settings however, this argument is hard to justify. Specifics of magma transport through the crust below volcanic fields are debated (Lister and Kerr 1991; Buck 2006; Tait and Taisne 2013), but generally it is thought to move via dyke propagation (Valentine and Hirano 2010). The geometry and orientation of dykes may be influenced by existing tectonic or geological boundary structures within the crust (Le Corvec et al. 2013b). The degree of lateral forcing of such structures is little known, however, if dyke propagation is forced too close to horizontal (e.g., via reaching a neutral buoyancy within the upper crust), a sill will form and magma is unlikely to reach the surface (Kavanagh et al. 2006).

As a conservative compromise to cover source uncertainty, a 5 km buffer zone was added to each fitted boundary. This value comes from Runge et al. (2014) and represents the > 95 % upper limit of dyke length from posterior distributions adapted to Harrat Rahat, with initial prior distributions taken from an eroded analogue volcanic field (San Rafael, Utah, Delaney and Gartner 1997), and expert elicitation. Although adapted specifically for Harrat Rahat, the same distance was applied to the Auckland Volcanic Field, while Le Corvec et al. (2013a) used ~1 km. In this work, Harrat Rahat was used to illustrate the sensitivity of spatial intensity results to the inclusion of a buffer. Thus, the buffer distance is applied to the AVF is purely to show the variation in scale (Fig. 2) and is not used in any sensitivity studies.

For the minimum area ellipse and rectangle boundaries, a buffer zone is straightforward to implement, with 10 km added to both the major and minor axes of the shape. For a convex hull, each of the hull points is moved 5 km further away from the polygon's centroid. For the contour boundaries, which already include a form of buffer zone, the boundary cut-off criteria was set to 10 % of the optimally determined value ('no holes' criterion) of the spatial density, thus increasing the bounds in all directions, but not necessarily by the same distance (Fig. 2).

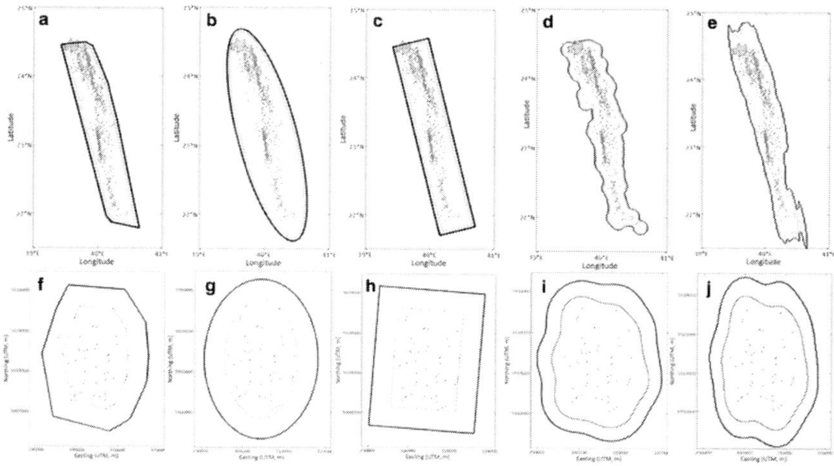

Figure 2. Volcanic field boundary options, solid lines represent buffered boundaries, dashed lines represent minimum fitted shapes (see text for shape definitions), North is up. Harrat Rahat: convex hull (**a**), ellipse (**b**), rectangle (**c**), isocontour (**d**), anisocontour (**e**), red shaded area represents the extent of Al-Madinah city (Fig. 6a). Auckland Volcanic field: convex hull (**f**), ellipse (**g**), rectangle (**h**), isocontour (**i**), anisocontour (**j**)

RESULTS – SUMMARY STATISTICS

Five volcanic field boundaries for the two volcanic fields were constructed based on the above assumptions (Fig. 2). This section assesses the sensitivity of four commonly applied exploratory methods (analysing vent clustering and preferred orientations) to the choice of boundary definition. First-order statistics are based on a single measurable variable (e.g., the mean of a set of values). The methods of Clark and Evans (1954) and Hopkins and Skellam (1954) have been used in previous volcanic field work (Connor and Hill 1995; Martin et al. 2004) to determine the existence of clustering. Second-order statistics are based on variation, or distribution, of a measured value (e.g., the standard deviation). Here, the K-function (Ripley 1979) is applied to identify variations in clustering behaviour at different scales (Fig. 3) which has been used previously for the Auckland Volcanic Field (Magill et al. 2005).

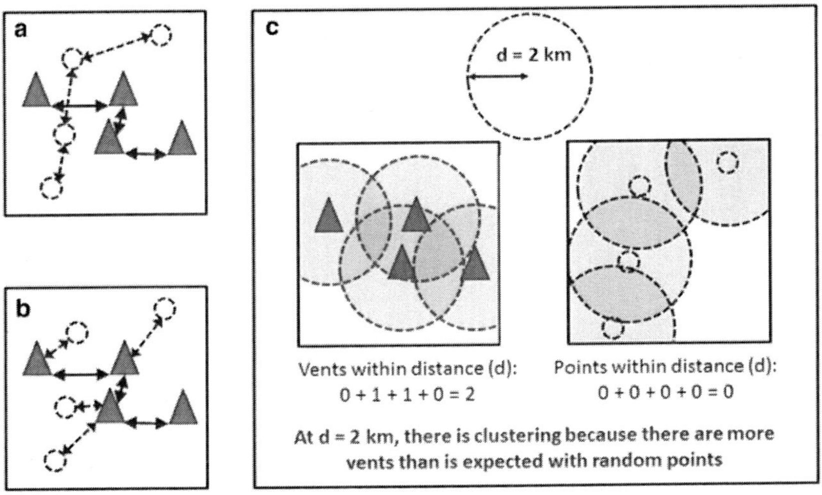

Figure 3. Visualisation of clustering statistics. Red triangles represent vents. Small blue circles represent randomly placed points. **a**Clark-Evans, **b** Hop-F, **c** K-function. See text for mathematical formula and explanation of individual statistics

The final exploratory method considered here is designed for lineament identification, the two-point azimuth (TPA) method of Lutz and Gutmann (1995) who applied it to the Pinacate volcanic field, Mexico. Connor (1990) also used TPA within and across inferred clusters of the TransMexican volcanic belt. All the exploratory methods were carried out before a buffer zone was applied. A buffer zone would cause an increase in area, a corresponding decrease in average vent density, and stronger clustering results.

Clark-Evans: 1st order randomness statistic

The Clark-Evans test compares the mean distance between neighbouring samples (r¯¯ar¯a) to the typical distance of an equivalent random (homogeneous Poisson) distribution (r¯¯er¯e, Fig. 3a). The random distribution (r¯¯er¯e) requires an average vent density and therefore relies heavily on the estimated volcanic field area (A):

$$\bar{r}_a = \frac{\sum_{i=1}^{N} r_i}{N}, \quad \bar{r}_e = \frac{1}{2\sqrt{\rho}}, \quad where\ \rho = \frac{N}{A}$$

where N is the number of vents, ρ is average vent density, r_i is the distance between the ith vent and nearest neighbour vent, and A is the area. r¯¯er¯e can also be estimated by placing the same number of points within the defined bounds via a Monte Carlo (MC) approach.

$$R = \frac{\bar{r}_a}{\bar{r}_e} \tag{3}$$

A ratio (R) of the two of R = 1 suggests a random distribution, while R < 1 indicates clustering, and R > 1, regular spacing or dispersion (Clark and Evans 1954). The statistical significance can be assessed with the standard normal variate (c)

$$c = \frac{\bar{r}_a - \bar{r}_e}{\sigma_{\bar{r}_e}}, \quad where\ \sigma_{\bar{r}_e} = \frac{0.26136}{\sqrt{N\rho}} \tag{4}$$

where σr¯¯¯eσr¯e is the standard deviation for the equivalent random distribution (r¯er¯e). The constant 0.26136 was derived from the probability of the number of points found within an area of specified size given a mean population density using a Poisson exponential function (derivation found in the appendix of Clark and Evans 1954). Where *c* lies outside of the range ± 2.58, the result can be assumed more than 99 % significant (Clark and Evans 1954), although Hamilton et al. (2010) suggest this critical value of *c* should be increased for analysis of low sample sizes.

All Clark-Evans results (R, Table 1) suggest significant clustering (R < 1) in Harrat Rahat. The elliptical volcanic field boundary suggests much stronger clustering than the other three due to the substantially larger area encompassed, much of which is empty. The Auckland Volcanic Field results (Table 2) are highly dependent on the choice of volcanic field boundary. The contour based boundaries suggest clustering, the ellipse and rectangular boundaries suggest random dispersion, and the convex hull suggests a more regularly spaced vent distribution. However, none of these results appear to be statistically significant (−2.58 < c < 2.58).

Table 1. Harrat Rahat boundary summary statistics. Clark-Evans (R) values are reported alongside the standard variate (c), Hop-F (A_g) values are reported with the standard error on the mean from 1000 simulations

Boundary	Area (km^2)	Vent density (km^{-2})	R (Clark-Evans)	A_g (Hop-F)
Convex hull	12,535	0.077	0.652, c = −20.77	1.014 ± 0.047
Ellipse	22,430	0.043	0.487, c = −30.58	1.006 ± 0.043
Rectangle	14,717	0.066	0.601, c = −23.77	0.978 ± 0.048
Isocontour	13,496	0.072	0.628, c = −22.18	0.914 ± 0.038
Anisocontour	15,292	0.063	0.5899, c = −24.45	0.956 ± 0.047

Table 2. Auckland Volcanic Field boundary summary statistics. Clark-Evans (R) values are reported alongside the standard variate (c), Hop-F (A_g) values are reported with the standard error on the mean from 1000 simulations

Boundary	Area (km^2)	Vent density (km^{-2})	R (Clark-Evans)	A_g (Hop-F)
Convex hull	324.5	0.1527	1.134, c = 1.829	0.887 ± 0.160
Ellipse	385.7	0.129	1.04, c = 0.547	1.191 ± 0.242
Rectangle	400.2	0.127	1.021, c = 0.287	1.0582 ± 0.226
Isocontour	585.7	0.087	0.844, c = −2.132	0.793 ± 0.141
Anisocontour	579.4	0.088	0.849, c = −2.069	0.896 ± 0.173

Hop-F: 1st order randomness statistic

The Hop-F statistic compares the distance between vent I and its nearest neighbour vent (r_i), and the distance between a random point and the nearest vent (d_i) (Fig. 3b).

$$A_g = \frac{\sum_{i=1}^{N} d_i^2}{\sum_{i=1}^{N} r_i^2} \tag{5}$$

where N is the number of vents. The resulting coefficient of aggregation (A_g) indicates randomness if $A_g = 1$, clustering if $A_g > 1$, and regular spacing or dispersion if $A_g < 1$ (Hopkins and Skellam 1954). The statistic A_g is most often reported with the standard error on the mean using the results of at least 100 simulations (Cressie 1991). It is this simulation step that requires definition of the specific field boundary.

The Hop-F results from 1000 simulations (A_g, Table 1) for the ellipse, rectangle, and convex hull suggest random dispersal of vents across Harrat Rahat with the standard error bounds all including unity (random). The isocontour boundary suggests slightly dispersed vent locations ($A_g < 1$). For the Auckland Volcanic Field, results vary dramatically as with the Clark-Evans results, however, similar to the Harrat Rahat, almost all results include the null hypothesis of randomness. The exception is again that of the isocontour, which suggests dispersed vent locations, and the ellipse boundary (although the confidence bound includes one) that more strongly suggests clustering than any of the other bounds.

While the Clark-Evans statistic tests the degree of departure from a Poisson distribution (Fig. 3a), the Hop-F statistic suggests clustering (Fig. 3b). These results suggest that the Clark-Evans statistic is more sensitive to departures from randomness other than clustering, such as lineaments and anisotropy.

The presence of multiple-vent events (Runge et al. 2014) may have more influence on the Hop-F statistic when using a more closely fitting boundary within which random points can fall. Since contour boundaries tailor the boundary to the vent locations, they effectively discard empty space furthest from vents and the false-positive clustering results are avoided. However, without knowledge of the actual volcanic field boundary, it is difficult to determine if this makes the contour results more reliable for awkward volcanic field shapes, such as that of Harrat Rahat, or instead, overly biased towards observed eruption locations.

K-Function: 2nd order randomness statistic

This second order test compares the number of vents within distance d of another vent, with the number expected for a random distribution with the same density (Fig. 3c)

$$\hat{K}(d) = \frac{A}{N^2}\sum_{j=1}^{N}\sum_{i\neq j}^{N}\frac{H(d-d_{i,j})}{w_{i,j}} \tag{6}$$

where N is the number of vents, A is the area of the field (requires a boundary), $d_{i,j}$ is the distance between ith and jth vents, H is the Heaviside function, and $w_{i,j}$ is an edge correction (Martinez and Martinez 2001). This statistic is most easily interpreted as a graph of the L-function

$$\hat{L}(a) = \sqrt{\frac{\hat{K}(d)}{\pi}}-d \tag{7}$$

where L^(d)=0L^(d)=0 suggests randomness, L^(d)>0L^(d)>0implies clustering, and L^(d)<0L^(d)<0, regular spacing (Ripley 1979). Significance levels are determined via random Poisson (MC) simulations run over the defined area.

K-function (or L-function) results are often diluted or smoothed if the method is applied over larger areas containing highly varying vent densities. The strong spine-type structure of Harrat Rahat (Fig. 1a) thus produces unusual results when this method is applied across the whole field and, in practice, much smaller regions should be assessed to look for specific clustering distances. All the volcanic field boundaries fitted to Harrat Rahat show clustering (L^(d)>0L^(d)>0) at all distances (Fig. 4a), with similar shapes following similar trends (e.g., both contour boundaries follow similar lines, as do the convex hull and rectangular bounds).

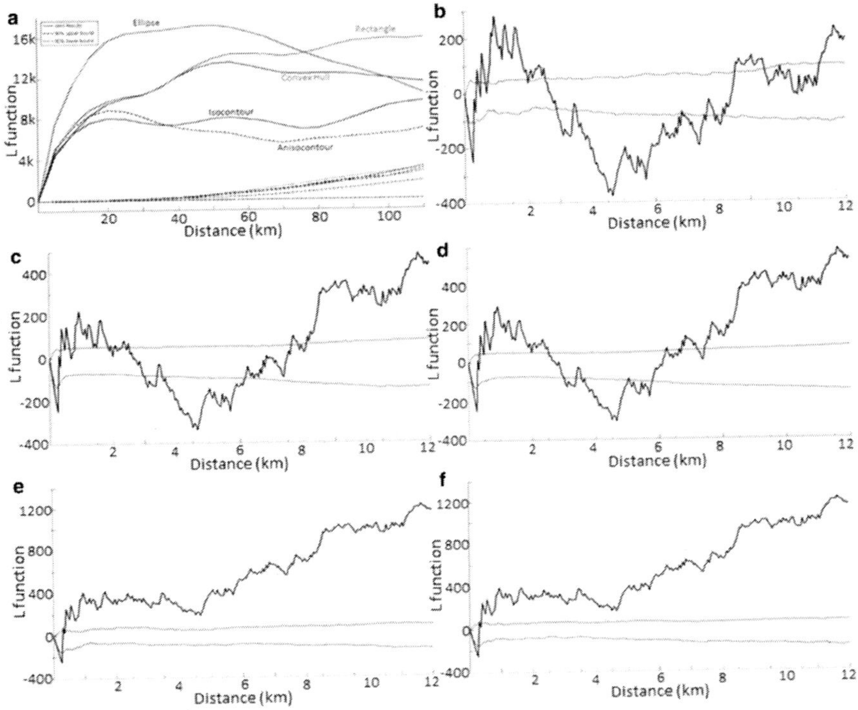

Figure 4. L-function (L^(d)L^(d)) statistic for Harrat Rahat (HR) and Auckland Volcanic field (AVF) vent locations within various field boundaries. Thin lines represent upper and lower 95 % significance levels from 1000 random (Poisson) simulations. Thick black lines represent vent location based values. **a** HR All boundaries, **b** AVF convex hull, **c** AVF ellipse, **d** AVF rectangle, **e** AVF isocontour, **f** AVF anisocontour. See Fig. 2 for boundary illustrations, and text for L-function definition

For the Auckland Volcanic Field, results are more informative for actual clustering distances due to the more evenly dispersed eruption locations, and lower vent numbers (Fig. 4). All results show the minimum vent distance at ~ 300 m (first trough), and suggest clustering at distances > 11 km (vent results lie above the upper significance bound), which likely reflects the maximum/regular vent-spacing, and consequently, field dimensions. The convex hull, ellipse, and rectangular bounds suggest clustering between ~ 0.5 and 2.5 km (above upper bound), and then maximum/regular spacing between 3 and 6 km (below lower bound, Fig. 4c and d) (3 and 8 km for the convex hull, Fig. 4b), and then clustering at larger distances. The two contour boundaries however, suggest clustering at all distances above ~ 0.5 km. Previous results from application of the K-function (Ripley 1979) to 49 vents of the Auckland Volcanic Field by Magill et al. (2005) found clustering between 0.9 and 1.6 km, and then maximum spacing at 4.6 km, which closely resembles the convex hull results presented here. The significant variation in results between boundaries is important. Although the boundary shapes of the Auckland Volcanic Field are broadly similar, the extra space incorporated around the external vents in the contour methods dilutes the second-order statistics by effectively forcing a cluster in the centre of the region (Fig. 2 and Fig. 4e-f).

TPA: lineament identification method

Lutz and Gutmann (1995)'s method for lineament identification (Two Point Azimuth: TPA) is based on the azimuth between each sample and every other sample and provides a distribution of the frequency of alignment angles, usually presented within groups of 10° intervals (Connor 1990; Wadge and Cross 1988). To accommodate any non-circular field shape, the results are then compared to Monte Carlo simulations with the same number of samples placed within the defined area, thus the results are dependent on the specific field boundary definition.

Harrat Rahat TPA results for the five volcanic field boundaries show approximately the same results, but with, in some cases, substantially varying proportions (Fig. 5). For example, the convex hull results show that vent alignments are only slightly more likely to be between 160° and 170° than randomly placed points within the defined field area.

In contrast, the elliptical boundary results show ~ 50 % more alignments than would be expected for random points within the elliptical field (Fig. 5b, the difference in vent-vent alignment frequency for the thick black line representing vents, and the grey lines representing significance bounds). Had only a convex hull boundary been used, these apparent NNW/SSE vent alignments may have been discarded as an artefact of overall volcanic field shape. However, the fact that there are consistently more vents than would be expected aligned between 160° and 170° may suggest similar structural controls on at both large (the region of volcanism) and small (vent lineaments/structures) scales. The only variation is that of the anisocontour boundary which does not pick up the peak that the other four bounds do (160° and 170°). This is a result of the base kernel (as defined by H, Eqn. 2) being pre-aligned (via SAMSE) to the overall dominant alignment direction, i.e., this alignment direction of vents is not anomalous within this boundary definition, as all the simulations are based on, and therefore reproduce, it.

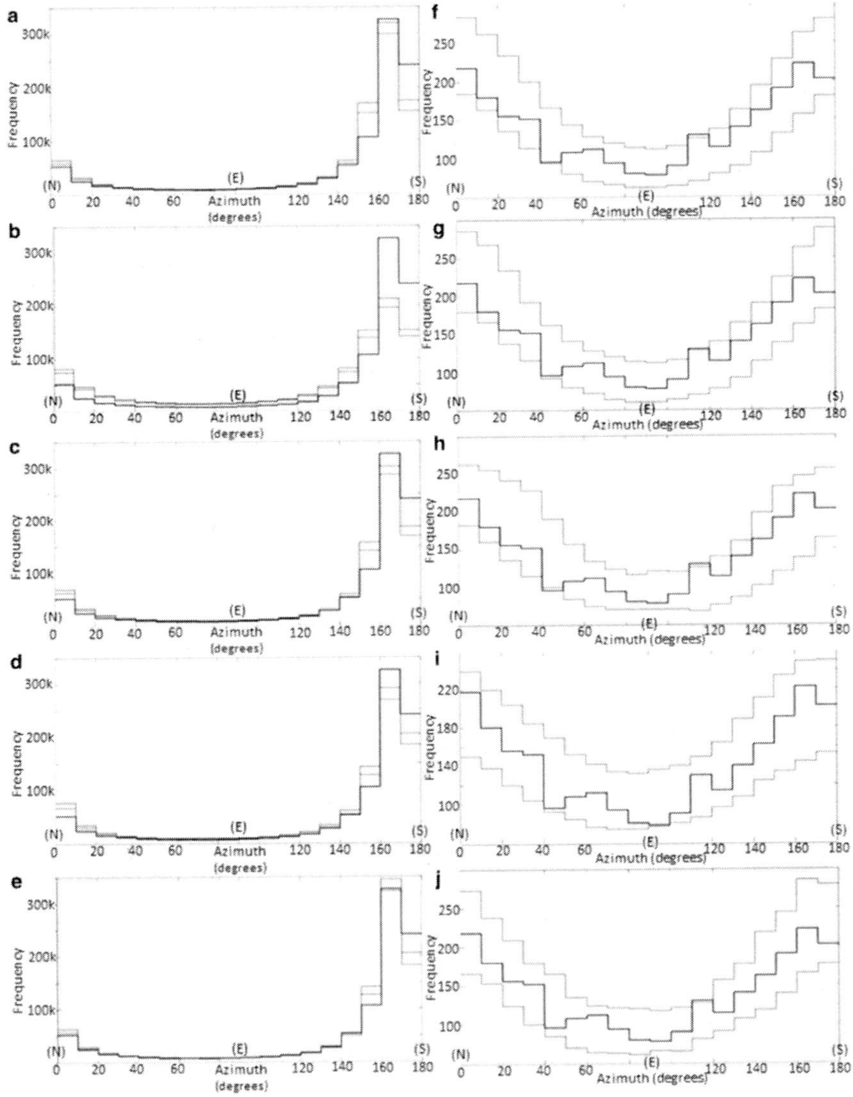

Figure 5. TPA frequency histograms for Harrat Rahat (HR) and Auckland Volcanic Field (AVF) vent locations within various field boundaries. Grey lines represent the 95 % significance levels from 1000 random (Poisson) simulations. Black lines represent vent location based values. **a** HR convex hull, **b** HR ellipse, **c** HR rectangle, **d** HR isocontour, **e** HR anisocontour, **f** AVF convex hull, **g** AVF ellipse, **h** AVF rectangle, **i** AVF isocontour, **j** AVF anisocontour

Across the Auckland Volcanic Field, all boundaries (except the isocontour) suggest a slight tendency for alignments between 110° and 120° (Fig. 5). This matches the work of Von Veh and Németh (2009) who used the Hough Transform (Wadge and Cross 1988) and an elliptical boundary. The rectangular bounds suggest a negative preference for alignments between 40° and 50°. Although several of the other results get close, none breach the 95 % significance levels. Unlike the previous clustering results (Table 2, Fig. 4), the TPA results are insensitive to the volcanic field boundary of the Auckland Volcanic Field because this analysis is based on field shape (relative proportions) rather than absolute dimensions.

RESULTS – SPATIAL INTENSITY ESTIMATES

For exploratory methods, a buffer is superfluous as it is the existing locations and spacings of eruptions that are assessed. However, for forecasts, field maturity and potential eruption migration need to be considered, with a buffer zone included in the definition of a boundary as a conservative measure. However, the addition of a buffer zone is non-trivial, especially where specific spatial densities are important as excessively increasing the volcanic field area will result in an underestimate of average vent-density and consequently spatial intensity. This is illustrated here using the city of Al-Madinah, located immediately to the north-west of Harrat Rahat (Fig. 1a) which is spread over an area of ~30 km^2 (Fig. 6a), with a population of 1.5 million.

A large number of methods can be employed to estimate spatial density within volcanic fields (Marzocchi and Bebbington 2012; Bebbington 2013). Here, a uniform spatial intensity is assumed (the simplest of the available methods), whereby all regions within the volcanic field boundary are assumed equally likely to host future eruptions. This allows the sensitivity of the spatial intensity results to boundary choice to be assessed without assumptions inherent to more complex methods confounding the comparison. For a uniform spatial density model, given an eruption occurs (probability across the whole region must sum to one), the spatial density at a point ($\gamma(x)$) is equal to $1/A$ where A is the total field area. The probability of an eruption occurring within the area of interest then becomes the ratio between the proportion of the area of interest (in this case, the city of Al-Madinah) within the volcanic field boundary and the total area of the field.

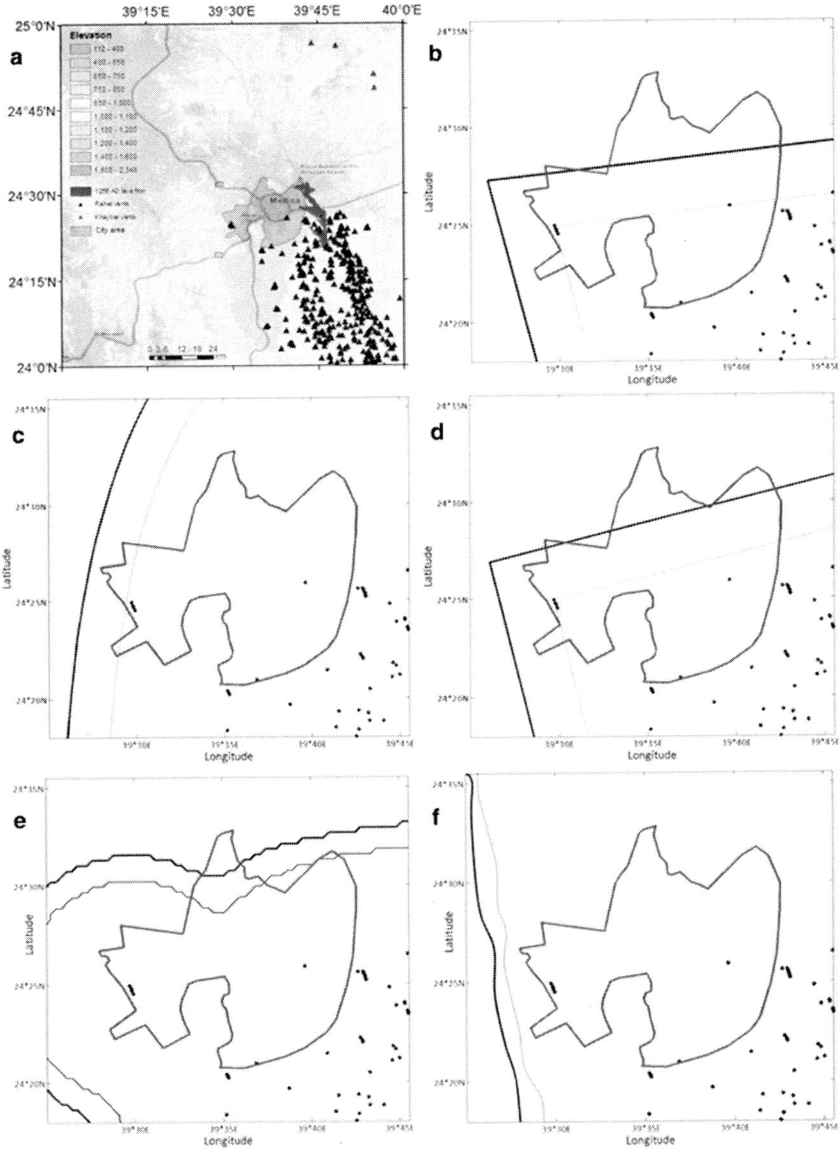

Figure 6. Harrat Rahat boundary variations overlain by the city limits of Al-Madinah (*red*). Thin black lines represent fitted shapes, thick black lines represent boundaries with buffer, North is up. **a** 1256 AD lava flow extent shown in red, Al-Madinah city boundary shown in blue. **b** convex hull, **c** ellipse, **d** rectangle, **e** isocontour, **f** anisocontour

Four buffer variations were assessed for each boundary type: (1) a fitted boundary (no buffer), (2) an equal intensity buffered boundary, and (3),(4) two inner-outer boundary approaches. For these latter two, the uniform spatial density estimate is imposed across the region, but with the values in the outer boundary set as a proportion ($p \leq 1$) (50 and 10 % here) of that of the inner boundary.

$$\gamma(x) = \begin{cases} \frac{1}{\eta_o A_1} x \in A_1(inner) \\ \frac{p}{\eta_o A_1} x \in A_2 x \notin A_1(outer) \end{cases}, \quad \eta_o = \left[p\left(\frac{A_2}{A_1} - 1\right) + 1\right]^{-1} \tag{8}$$

where A_1 is the area within the inner boundary, A_2 is the area within the outer boundary (i.e., $A_1 \subseteq A_2$, and $A_2 \geq A_1$). This is a generalised form of that suggested in Bebbington (2013) in which the inner area was assumed to have twice the probability of the outer.

As the city of Al-Madinah borders the current flow field extents, the effect of volcanic field boundary definition on the area susceptible to future eruptions is substantial (Fig. 6). For the fitted boundaries (without a buffer) the anisocontour boundary already includes the whole of Al-Madinah, and both the isocontour and elliptical bounds incorporate almost the entire extent of Al-Madinah within the area deemed susceptible to a future eruption. Thus, the addition of the buffer zones results in a decrease in eruption probability within the city limits as the spatial density is spread over a larger region (Table 3). Conversely, the convex hull and rectangular boundaries enclose less than half of Al-Madinah without a buffer zone, and thus when the buffer is added, the proportion of the city within the boundary increases sufficiently to also increase the eruption probability, even though the spatial intensity at any specific point is reduced (Table 3).

Table 3. Eruption probability estimates for an eruption within the area of interest, the city of Al-Madinah. Estimates are based on a uniform distribution of eruption probability, given that an eruption occurs within the region defined for Harrat Rahat

Boundary	No buffer	Equal intensity buffer	50 % intensity buffer	10 % intensity buffer
Convex hull	0.0092	0.0147	0.01227	0.0099
Ellipse	0.0132	0.0119	0.0125	0.0131
Rectangle	0.0100	0.0146	0.0125	0.0106
Isocontour	0.0206	0.0190	0.0198	0.0204
Anisocontour	0.0200	0.0184	0.0192	0.0199

The estimate of the intensity within Al-Madinah therefore is a balance between the extent of the city enclosed by the bound, and the spread of the intensity across the whole field. Comparing across boundaries, the isocontour boundary with no buffer resulted in the highest probability of eruption within the area of Al-Madinah city, with values more than twice that of the corresponding unbuffered convex hull and rectangular boundaries (Table 3). The probability of an eruption within the city limits (given that an eruption occurs) was assessed as 0.0206, or a 2.1 % chance. OF the five boundary shapes tested, the convex hull results were most sensitive to the addition of a buffer zone, varying up to ~160 % (Table 3), predominantly due to the change in proportion of the area of interest falling within a susceptible region (Fig. 6b). This is closely followed by the rectangular volcanic field boundary with a variation of ~140 % for the same reason (Fig. 6d).

DISCUSSION

The results above show the significant effect that the definition of a volcanic field boundary has on even the simplest of exploratory methods (Tables 1 and 2). The effect on spatial density estimation is also substantial, with the estimate of eruption probability within Al-Madinah varying significantly even between the few boundaries assessed here (Table 3). The use of a constant (fixed) boundary is also questionable, especially in those cases (such as Harrat Rahat), where a possible shift in eruption locations with time is noted. In these cases, a temporally varying boundary may be more appropriate by allowing the boundary definition to expand with time, or even move with time for those fields where regions of older activity are considered no longer active – a limited-memory boundary could be imposed which is fitted only to the eruptions from the most recent *x* years.

For forecasting purposes, a buffer zone is recommended as a conservative approach. It allows some accommodation of spatio-temporal migration or focussing (e.g., Harrat Rahat), and on-going field development (e.g., Auckland Volcanic Field). However, the size of the buffer zone was not investigated here and while the 5 km buffer zone imposed for Harrat Rahat is a relatively small increase in total area, this value has a substantial effect on the area designated to the Auckland Volcanic Field (Fig. 2). A buffer zone distance based on the dimensions of a volcanic field may be more appropriate (e.g., the ~1 km buffer zone imposed by Le Corvec et al. 2013a).

The 5 km buffer zone for Harrat Rahat was taken from previous estimates of dyke length (Runge et al. 2014), however it is likely that the distribution of dyke lengths, and subsequently the distance along which magma may erupt from, may vary with field thus, at the very least, the tectonic regime should be considered (Pinel and Jaupart 2004). The presence of fissure eruptions (as noted in Harrat Rahat) may also justify an increase in buffer zone distance, and, if multiple fissures show similar alignments, anisotropic buffer zones may be more appropriate. Evidence of waning may also suggest a decrease (or even exclusion) of a buffer zone as the future eruptions may be smaller, and from shorter subsurface structures (Valentine and Perry 2006).

Smaller scale structures should also be considered during volcanic field definition. For example, in Harrat Rahat, where there is strong anisotropic structure (Runge et al. 2014), or dominant alignment directions, the anisotropic kernel based boundary added a large amount of empty space in the directions determined by the SAMSE optimised base kernel, which in turn is strongly influenced by regional vent alignments. The isotropic kernel based contour resulted in a much more irregular volcanic field boundary than that of the anisotropic kernel (Fig. 1), mainly due to the smaller optimum bandwidth (also reflected in the TPA results for Harrat Rahat, Fig. 5). This is consistent with previous work that suggests vent-specific locations are influenced by smaller-scale local, rather than regional, structures (Connor et al. 1992).

For Harrat Rahat, none of the boundary methods would have included the 641 AD eruption (the most north-westerly vent area) should this analysis have been carried out in 640 AD. This suggests an evolving field boundary and a buffer zone should be imposed for Harrat Rahat, and other volcanic fields exhibiting strong spatio-temporal migration.

For the Auckland Volcanic Field, with the exception of the rectangular boundary, the range of fitting processes resulted in similar shapes. However, had the bounds been fitted prior to the most recent eruption (Rangitoto), only the elliptical and rectangular bounds would have included it, reflecting the unusual number of vents within a short distance of the elliptical boundary (Spörli and Eastwood 1997; Bebbington 2015). While the field is still relatively young (Lindsay et al. 2010), previous work has suggested that the spatial distribution of eruptions within the AVF has been relatively stable since the formation of the 15th eruption (Bebbington 2013; Le Corvec et al. 2013a). This may suggest the addition of a buffer zone is unnecessary. However, as the most recent eruption was anomalous both in terms of polygenetic behaviour (Needham et al. 2011) and large erupted volume (Kereszturi et al. 2013), a buffer zone is advised to conservatively accommodate, at least in part, these potential changes in eruption characteristics.

CONCLUSIONS

Forecasts of long-term volcanic activity should be accompanied by an estimate of their precision and uncertainties imposed by the assumptions made. However, hazard analyses are required for incompletely understood volcanic regions threatened by future eruptions. This requires parameterisation of the unknown. This study showed the substantial effect that the definition of a volcanic field boundary may have on even the simplest exploratory methods for hazard forecasting.

As research on volcanic fields and geophysical methods progress, boundary fitting may become better informed by more extensive knowledge of specific field constraints, however, a region definition will always be vital for probabilistic approaches. When carrying out any analysis on volcanic fields, it is vital to specify how the volcanic field boundary has been defined, assess the sensitivity of boundary choice, and to carry these assumptions and related uncertainties through to estimates of future activity and related hazard analyses.

For Harrat Rahat, sensitivity of results to boundary definition is substantial and a buffer zone is highly recommended due to the spatio-temporal behaviour of the field. For the Auckland Volcanic Field, the range of options resulted in similar shapes, nevertheless, some of the statistical tests still showed substantial variation in results.

This work highlights the fact that when carrying out any hazard analysis on volcanic fields, it is vital to specify how the volcanic field boundary has been defined, assess the sensitivity of boundary choice, and to carry these assumptions and related uncertainties through to estimates of future activity and hazard analyses.

ACKNOWLEDGEMENTS

This work is part of the VORiSA (Volcanic Risk in Saudi Arabia) collaborative project between King Abdulaziz University and The University of Auckland. Valuable comments were also provided by Laura Sandri and an anonymous reviewer.

AUTHORS' CONTRIBUTIONS

MR wrote and implemented the code for the statistical analyses and drafted the manuscript. MB advised on methods and suggested the comparison with the AVF. SC, JM, and MM helped define the research question and commented on the manuscript. All authors read and approved the final manuscript.

REFERENCES

1. Allen S, Smith I. Eruption styles and volcanic hazard in the Auckland Volcanic Field, New Zealand. Geosci Rep Shizuoka Univ. 1994;20:5–14.
2. Angell I, Barber J. An algorithm for fitting circles and ellipses to megalithic stone rings. Sci Archaeol. 1977;20:11–6.
3. Barde-Cabusson S, Gottsmann J, Martí J, Bolós X, Camacho A, Geyer A, et al. Structural control of monogenetic volcanism in the Garrotxa volcanic field (Northeastern Spain) from gravity and self-potential measurements. Bull Volcanol. 2014;76(1):1–13.
4. Bebbington M. Assessing spatio-temporal eruption forecasts in a monogenetic volcanic field. J Volcanol Geotherm Res. 2013;252:14–28.
5. Bebbington MS. Spatio-volumetric hazard estimation in the Auckland Volcanic Field. Bull Volcanol. 2015;77:39.
6. Bebbington M, Cronin S. Spatio-temporal hazard estimation in the Auckland Volcanic Field, New Zealand, with a new event-order model. Bull Volcanol. 2011;73:55–72.
7. Becerril L, Cappello A, Galindo I, Neri M, Del Negro C. Spatial probability distribution of future volcanic eruptions at El Hierro Island (Canary Islands, Spain). J Volcanol Geotherm Res. 2013;257:21–30.
8. Bishop M. Point pattern analysis of eruption points for the Mount Gambier volcanic sub-province: a quantitative geographical approach to the understanding of volcano distribution. Area. 2007;39(2):230–41.
9. Blank H, Sadek H. Spectral analysis of the 1976 aeromagnetic survey of Harrat Rahat, Kingdom of Saudi Arabia. 1983. Tech Report USGS-OF-03-67, USGS.
10. Brenna M, Cronin S, Smith I, Sohn Y, Maas R. Spatio-temporal evolution of a dispersed magmatic system and its implications for volcano growth, Jeju Island Volcanic Field, Korea. Lithos. 2012;148:337–52.
11. Brown G. Eastern margin of the Red Sea and the coastal structures in Saudi Arabia. Phil Trans R Soc Lond Ser A Math Phys Sci. 1970;267:75–87.
12. Buck W. The role of magma in the development of the Afro-Arabian Rift System. Geol Soc Lond Spec Pub. 2006;259(1):43–54.
13. Cabrera J, Meer P. Unbiased estimation of ellipses by bootstrapping, pattern analysis and machine intelligence. IEEE Trans. 1996;18(7):752–6.
14. Camp VE, Roobol MJ. The Arabian continental alkali basalt province: Part I. Evolution of Harrat Rahat, Kingdom of Saudi Arabia. Geol Soc Am Bull. 1989;101(1):71–95.
15. Camp VE, Roobol MJ, Hooper PR. The Arabian continental alkali basalt province: Part II. Evolution of Harrats Khaybar, Ithnayn, and Kura, Kingdom of Saudi Arabia. Geol Soc Am Bull. 1991;103(3):363–91.
16. Chand D, Kapur S. An algorithm for convex polytypes. J ACM. 1970;17(1):78–86.
17. Clark P, Evans F. Distance to nearest neighbour as a measure of spatial relationships in populations. Ecology. 1954;35:s–53.
18. Coleman RG, Gregory RT, Brown GF. Cenozoic volcanic rocks of Saudi Arabia. 1983. Tech Rep Open File Report USGS-OF93, Saudi Arabian Deputy Minister of Mineral Resources.

19. Condit C, Connor C. Recurrence rates of volcanism in basaltic volcanic fields, an example from the Springerville volcanic field, Arizona. Geol Soc Am Bull. 1996;108:1225–41.
20. Connor C. Cinder cone clustering in the TransMexican volcanic belt: implications for structural and petrologic models. J Geophys Res Solud Earth (1978–2012). 1990;95(B12):19,395–405.
21. Connor C, Connor L. Estimating spatial density with kernel methods. Volcanic and tectonic hazard assessment for nuclear facilities. Cambridge: Cambridge University Press; 2009. p. 346–68.
22. Connor C, Conway F. Basaltic volcanic fields. In: Encyclopedia of volcanoes. New York: Academic; 2000. p. 331–43.
23. Connor C, Hill B. Three nonhomogeneous Poisson models for the probability of basaltic volcanism: application to the Yucca Mountain region, Nevada. J Geophys Res Solid Earth (1978–2012). 1995;100(B6):10,107–25.
24. Connor C, Condit C, Crumpler L, Aubele J. Evidence of regional structural controls on vent distribution: Springerville volcanic field, Arizona. J Geophys Res Solid Earth (1978–2012). 1992;97(B9):12,349–59.
25. Cressie N. Statistics for spatial data, Wiley Series in Probability and Mathematical Statistics: Applied Probability and Statistics. New York: Wiley; 1991.
26. Cronin S, Bebbington M, Lai C. A probabilistic assessment of eruption recurrence on Taveuni volcano, Fiji. Bull Volcanol. 2001;63(4):274–88.
27. Delaney P, Gartner A. Physical processes of shallow mafic dike emplacement near the San Rafael Swell, Utah. Geol Soc Am Bull. 1997;109(9):1177–92.
28. Duong T. ks: Kernel density estimation and kernel discriminant analysis for multivariate data in R. J Stat Softw. 2007;21(7):1–16.
29. Duong T, Hazelton ML. Plug-in bandwidth selectors for bivariate kernel density estimation. J Nonparametric Stat. 2003;15:17–30.
30. El Difrawy M, Runge M, Moufti MR, Cronin S, Bebbington M. A first hazard analysis of the Quaternary Harrat Al-Madinah volcanic field, Saudi Arabia. J Volcanol Geotherm Res. 2013;267:39–46.
31. Fedotov S. Magma rates in feeding conduits of different volcanic centres. J Volcanol Geotherm Res. 1981;9(4):379–94.
32. Freeman H, Shapira R. Determining the minimum-area encasing rectangle for an arbitrary closed curve. Commun ACM. 1975;18(7):409–13.
33. Germa A, Connor L, Canon-Tapia E, Le Corvec N. Tectonic and magmatic controls on the location of post-subduction monogenetic volcanoes in Baja California, Mexico, revealed through spatial analysis of eruptive vents. Bull Volcanol. 2013;75(12):1–14.
34. Guffanti M, Weaver C. Distribution of late Cenozoic volcanic vents in the Cascade Range: volcanic arc segmentation and regional tectonic considerations. J Geophys Res Solid Earth (1978–2012). 1988;93(B6):6,513–29.
35. Hamilton CW, Fagents SA, Thordarson T. Explosive lava-water interactions ii: self-organization processes among volcanic rootless eruption sites in the 1783–1784 laki lava flow, Iceland. Bull Volcanol. 2010;72(4):469–85.
36. Hayward BW, Murdoch G, Maitland G. Volcanoes of Auckland: the essential guide. N Z Sci Rev. 2011;68:123.
37. Hopkins B, Skellam J. A new method for determining the type of distribution of plant individuals. Ann Bot. 1954;18(2):213–27.
38. Jakobsson S. Petrology of recent basalts of the Eastern Volcanic Zone, Iceland. Acta Nat Islandica. 1979;26:1–103.
39. Kanatani K. Statistical optimization for geometric computation: theory and practice. 2005. Courier Dover Publications.
40. Karlstrom K, Crow R, Peters L, McIntosh W, Raucci J, Crossey L, et al. 40Ar/39Ar and field studies of Quaternary basalts in Grand Canyon and model for carving Grand Canyon: Quantifying the interaction of river incision and normal faulting across the western edge of the Colorado Plateau. Geol Soc Am Bull. 2007;119(11–12):1,283–1312.
41. Kavanagh J, Menand T, Sparks R. An experimental of sill formation and propogation in layered elastic media. Earth Planet Sci Lett. 2006;245(3):799–813.

42. Kear D. Volcanic alignments north and west of New Zealand's central volcanic region. N Z J Geol Geophys. 1964;7(1):24–44.
43. Kereszturi G, Németh K, Cronin S, Agustín-Flores J, Smith I, Lindsay J. A model for calculating eruptive volumes for monogenetic volcanoes-implication for the Quaternary Auckland Volcanic Field, New Zealand. J Volcanol Geotherm Res. 2013;266:16–33.
44. Khachiyan L. Rounding of polytopes in the real number model of computation. Math Oper Res. 1996;21(2):307–20.
45. Kiyosugi K, Connor C, Zhao D, Connor L, Tanaka K. Relationships between volcano distribution, crustal structure, and P-wave tomography: an example from the Abu Monogenetic Volcano Group, SW Japan. Bull Volcanol. 2010;72(3):331–40.
46. Le Corvec N, Bebbington M, Lindsay J, McGee L. Age, distance, and geochemical evolution within a monogenetic volcanic field: analyzing patterns in the Auckland Volcanic Field eruption sequence. Geochem Geophys Geosyst. 2013a;14(9):3,648–65.
47. Le Corvec N, Menand T, Lindsay J. Interaction of ascending magma with pre-existing crustal fractures in monogenetic basaltic volcanism: an experimental approach. J Geophys Res Solid Earth. 2013b;118(3):968–84.
48. Le Corvec N, Spörli K, Rowland J, Lindsay J. Spatial distribution and alignments of volcanic centers: clues to the formation of mono- genetic volcanic fields. Earth Sci Rev. 2013c;124:96–114.
49. Lindsay J, Moufti MR. Assessing volcanic risk in Saudi Arabia. Eos Trans Am Geophys Union. 2014;95(31):277–8.
50. Lindsay J, Marzocchi W, Jolly G, Constantinescu R, Selva J, Sandri L. Towards real-time eruption forecasting in the Auckland Volcanic Field: application of BET-EF during the New Zealand national disaster exercise 'Ruaumoko'. Bull Volcanol. 2010;72(2):185–204.
51. Lindsay J, Leonard G, Smid E, Hayward B. Age of the Auckland Volcanic Field: a review of existing data. N Z J Geol Geophys. 2011;54(4):379–401.
52. Lister J, Kerr R. Fluid-mechanical models of crack propagation and their application to magma transport in dykes. J Geophys Res Solid Earth (1978–2012). 1991;96(B6):10,049–77.
53. Lutz T, Gutmann J. An improved method for determining and characterizing alignments of pointlike features and its implications for the Pinacate volcanic field, Sonora, Mexico. J Geophys Res Solid Earth (1978–2012). 1995;100(B9):17,659–70.
54. Magill C, McAneney K, Smith I. Probabilistic assessment of vent locations for the next Auckland Volcanic Field event. Math Geol. 2005;37(3):227–42.
55. Martí J, Ortiz R, Gottsmann J, García A, De La Cruz-Reyna S. Characterising unrest during the reawakening of the central volcanic complex on Tenerife, Canary Islands, 2004–2005, and implications for assessing hazards and risk mitigation. J Volcanol Geotherm Res. 2009;182(1):23–33.
56. Martin A, Umeda K, Connor C, Weller J, Zhao D, Takahashi M. Modeling long-term volcanic hazards through Bayesian inference: An example from the Tohoku volcanic arc, Japan. J Geophys Res Solid Earth (1978–2012). 2004;109:B10.
57. Martinez W, Martinez A. Computational statistics handbook with MATLAB. Boca Raton: CRC press; 2001.
58. Marzocchi W, Bebbington MS. Probabilistic eruption forecasting at short and long time scales. Bull Volcanol. 2012;74(8):1,777–805.
59. Melían G, Hernández P, Padrón E, Pérez NM, Barrancos J, Padilla G, et al. Spatial and temporal variations of diffuse CO2 degassing at El Hierro volcanic system: Relation to the 2011–2012 submarine eruption. J Geophys Res Solid Earth. 2014;119(9):6,976–91.
60. Moufti MR, Matsah MI, Soliman MA, Moghazi A. Arabian plume dynamics beneath Al-Madinah Al-Munawwarah region and its related geohazards, final report. 2010. Technical Report ARP-26-79, King Abdulaziz City for Science and Technology.
61. Moufti MR, Moghazi A, Ali K. 40Ar/39Ar geochronology of the Neogene-Quaternary Harrat Al-Madinah intercontinental volcanic field, Saudi Arabia: implications for duration and migration of volcanic activity. J Asian Earth Sci. 2013;62:253–68.

62. Murcia H, Németh K, Moufti MR, Lindsay J, El-Masry N, Cronin S, et al. Late Holocene lava flow morphotypes of northern Harrat Rahat, Kingdom of Saudi Arabia: Implications for the description of continental lava fields. J Asian Earth Sci. 2014;84:131–45.
63. Needham A, Lindsay J, Smith I, Augustinus P, Shane P. Sequential eruption of alkaline and sub-alkaline magmas from a small monogenetic volcano in the Auckland Volcanic Field, New Zealand. J Volcanol Geotherm Res. 2011;201(1):126–42.
64. Németh K, Cronin S. Phreatomagmatic volcanic hazards where rift-systems meet the sea, a study from Ambae Island, Vanuatu. J Volcanol Geotherm Res. 2009;180(2):246–58.
65. Pallister J, McCausland W, Jónsson S, Lu Z, Zahran H, El Hadidy S, et al. Broad accommodation of rift-related extension recorded by dyke intrusion in Saudi Arabia. Nat Geosci. 2010;3(10):705–12.
66. Perry F, Cogbill A, Kelley R. Uncovering buried volcanoes at Yucca Mountain. Eos Trans Am Geophys Union. 2005;86(47):485–8.
67. Pinel V, Jaupart C. Magma storage and horizontal dyke injection beneath a volcanic edifice. Earth Planet Sci Lett. 2004;221(1):245–62.
68. Preparata F, Hong SJ. Convex hulls of finite sets of points in two and three dimensions. Commun ACM. 1977;20(2):87–93.
69. Ripley B. Tests of randomness for spatial point patterns. J R Stat Soc Ser B. 1979;41:368–74.
70. Rolandone F, Lucazeau F, Leroy S, Mareschal JC, Jorand R, Goutorbe B, et al. New heat flow measurements in Oman and the thermal state of the Arabian Shield and Platform. Tectonophysics. 2013;589:77–89.
71. Ross PS, Delpit S, Haller M, Németh K, Corbella H. Influence of the substrate on maar–diatreme volcanoes—an example of a mixed setting from the Pali Aike volcanic field, Argentina. J Volcanol Geotherm Res. 2011;201(1):253–71.
72. Runge M, Bebbington M, Cronin S, Lindsay J, Kenedi C, Moufti MR. Vents to events: determining an eruption event record from volcanic vent structures for the Harrat Rahat, Saudi Arabia. Bull Volcanol. 2014;76(3):1–16.
73. Sandri L, Jolly G, Lindsay J, Howe T, Marzocchi W. Combining long-and short-term probabilistic volcanic hazard assessment with cost-benefit analysis to support decision making in a volcanic crisis from the Auckland Volcanic Field, New Zealand. Bull Volcanol. 2012;74(3):705–23.
74. Spörli K, Eastwood V. Elliptical boundary of an intraplate volcanic field, Auckland, New Zealand. J Volcanol Geotherm Res. 1997;79(3):169–79.
75. Tait S, Taisne B. The dynamics of dike propagation. In: Fagents S, Gregg T, Lopes R, editors. Modeling volcanic processes: the physics and mathematics of volcanism. 2013. p. 33–54.
76. Takada A. The influence of regional stress and magmatic input on styles of monogenetic and polygenetic volcanism. J Geophys Res Solid Earth (1978–2012). 1994;99(B7):13,563–73.
77. Tanaka K, Shoemaker E, Ulrich G, Wolfe E. Migration of volcanism in the San Francisco volcanic field, Arizona. Geol Soc Am Bull. 1986;97(2):129–41.
78. Tomsen E, Lindsay J, Gahegan M, Wilson T, Blake D. Evacuation planning in the Auckland Volcanic Field, New Zealand: a spatio- temporal approach for emergency management and transportation network decisions. J Appl Volcanol. 2014;3(1):6.
79. Toussaint G. Solving geometric problems with the rotating calipers. Proc IEEE Melecon. 1983;83:A10.
80. Valentine G, Hirano N. Mechanisms of low-flux intraplate volcanic fields—Basin and Range (North America) and northwest Pacific Ocean. Geology. 2010;38(1):55–8.
81. Valentine G, Perry F. Decreasing magmatic footprints of individual volcanoes in a waning basaltic field. Geophys Res Lett. 2006;33:14.
82. von Hochstetter F. Geology of New Zealand. In: Delattre T, editor. Explanation of the geographical & topographical atlas of New Zealand. 1864.
83. Von Veh MW, Németh K. An assessment of the alignments of vents based on geostatistical analysis in the Auckland Volcanic Field, New Zealand. Geomorphology. 2009;3/2009:175–86.

84. Wadge G, Cross A. Quantitative methods for detecting aligned points: an application to the volcanic vents of the Michoacan-Guanajuato volcanic field, Mexico. Geology. 1988;16(9):815–8.
85. Wood C, Shoan W. Growth patterns of monogenetic volcano fields. Eos Trans Am Geophys Union. 1981;62:1061.
86. Zhang D, Lutz T. Structural control of igneous complexes and kimberlites: a new statistical method. Tectonophysics. 1989;159(1):137–48.

CHAPTER11

Volcanic Rock-Hosted Natural Hydrocarbon Resources: A Review

Jiaqi Liu[1], Pujun Wang[1], Yan Zhang[1], Weihua Bian[1], Yulong Huang[1], Huafeng Tang[1] and Xiaoyu Chen[2]

[1] *College of Earth Sciences, Jilin University, Changchun, China, Institute of Geology and Geophysics, Chinese Academy of Sciences, Beijing, China*

[2] *Institute of Geology and Geophysics, Chinese Academy of Sciences, Beijing, China*

1. INTRODUCTION

Evolution and the hydrocarbon bearing capacity of basins are closely related to volcanic activity, and not only source rock maturity, but also hydrocarbon trapping are influenced by volcanism within a basin. Volcanic rocks act as important basin filling material in different types of basins, for instance, rift basins, epicontinental basins, basins in a trench-arc system, back-arc foreland basins, etc. [1]. Volcanic accumulation of oil and gas is a new global field of hydrocarbon exploration and has been proved in more than 300 basins in 20 countries and regions [2]. The Cenozoic volcanic rocks, especially Jurassic Cretaceous and Tertiary, contribute about 70% of the total preservation globally [3-7].

Derivations of hydrocarbon in volcanic accumulation have organic as well as inorganic sources [8-10]. Volcanic rocks could act as a reservoir or cover within hydrocarbon traps, whose thermal effects could accelerate the maturity of source rocks or destroy preserved hydrocarbon [11-13]. Primary hydrocarbon accumulations could be reformed or destroyed during tectonic and volcanic processes, the preserved hydrocarbon remobilized to other traps or the ground surface [14]. Effective reservoirs have been found in most lithology [15-17]. Lithofacies, including deposits and rocks formed by explosive, effusive, extrusive and subvolcanic processes, could bear hydrocarbon, and the facies combination close to a volcanic conduit shows better porosity and permeability due to an increased number of fractures and reservoir spaces, or an elevated volume of coarse-grained fragmented rocks [7, 18]. Reservoir spaces within volcanic rocks are composed of primary pores, secondary pores and fissures with significant heterogeneity [19]. Tectonism, weathering and fluid saturation and/or movement could modify reservoir space [6, 20-23]. Upward cover and lateral seal of volcanic rocks could form hydrocarbon traps [24]. Lateral distribution of volcanic rocks can be mapped based on aeromagnetic and gravity data [25]. The reflective seismic features of volcanic rocks are summarized [26] and visualized [12] based on qualified seismic data. Volcanic edifice is identified by trend surface analysis and spectrum imaging methods [27]. A volcanic reservoir has been predicted based on seismic wave impedance [28].

Over the last half a century the Songliao Basin (Figure 1) has been the most productive basin in China for hydrocarbon and the Xujiaweizi fault depression has been proved as a typical volcanic accumulation. Based mainly on the achievements of hydrocarbon exploration in volcanic rocks in the Songliao Basin, the authors reviewed the hydrocarbon-related volcanic impacts, volcanic lithofacies and key geophysical techniques for volcanic accumulation exploration.

2. VOLCANISM IMPACTS ON THE FORMATION OF OIL AND GAS ACCUMULATION

2.1. Volcanic Activity Provides A Catalyst For The Evolution Of Organic Matter

During the transformation from organic matter to hydrocarbon, the role of volcanic is mainly to supply a catalyst and thermal energy. Volcanogenic zeolite and olivine can be a catalyst in turning organic matter into hydrocarbon [29]. Hydrothermal liquid contains many transition metals, such as Ni, Co, Cu, Mn, Zn, Ti, V etc. [30]. The transition metals are catalysts for organic matter thermal degradation [31]. Studies have shown that some volcanic minerals undergo catalysis and hydrogenation which can produce more oil and gas source rocks at lower temperature and pressure. Jin [32] performed a catalysis and hydrogenation experiment on volcanic minerals and source rocks. He used zeolite as a catalyst collected from volcanic rocks, olivine as intermediates of accelerating hydrogen generation and type II and type III organic matter as source rocks. The experimental results show

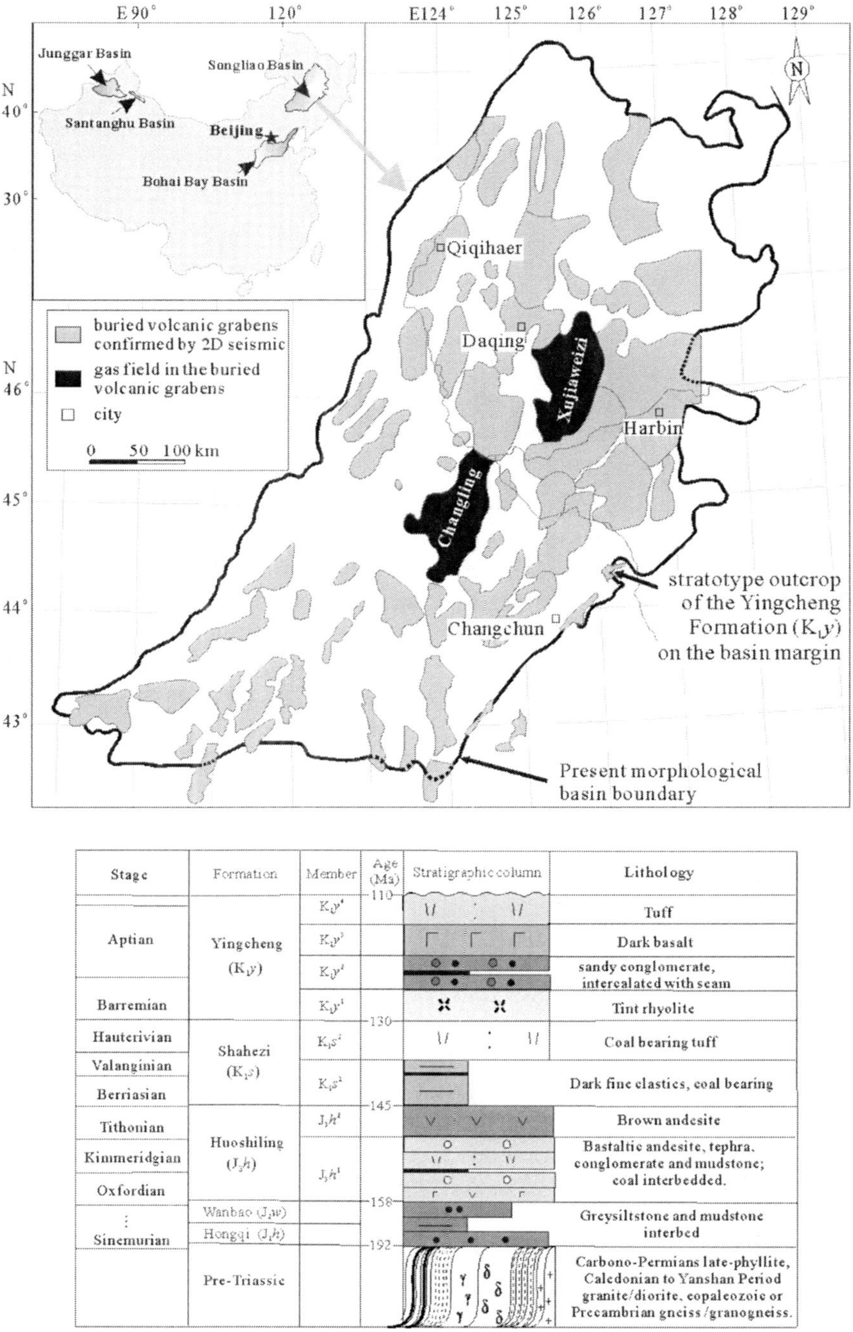

FIGURE 1. Geological structure and stratigraphic framework of the fault-depression sequence of the Songliao Basin, China

that the hydrogen production rate increased after olivine addition, while when adding zeolite and olivine hydrogen, the production rate still improved. This is due to olivine alteration occuring and reacting with water to produce hydrogen in the organic matter into hydrocarbon conversion process. The reaction is as follows:

$$6(Mg_{1.5}Fe_{0.5})SiO_4 + 13H_2O \rightarrow 3Mg_3SiO_2O_5(OH)_4 + Fe_3O_4 + 7H_2$$

The results show that after the source rocks interact with zeolite and olivine, the production rate of methane improved 2 to 3 times, which is related to the hydrogen increasing. The results also show that the better the organic matter or kerogen types, the higher the production rate of hydrogen and methane. However, the catalytic minerals are not only the clay minerals such as zeolite, but also pyrite.

Pyrite can be found in kerogen commonly, whose mass fraction is closely related to the type of kerogen. The data analysis of the Songliao Basin from Zhang et al. [33] shows that pyrite mass fraction is up to 38% ~ 76% in type I kerogen, which is 10% to 30% in type II kerogen and 2.5% to 3.5% in type III kerogen, i.e., the better the kerogen type, the higher the pyrite mass fraction. Electron microscopy reveals that type I and type II kerogen often closely coexist with pyrite and kerogen, exhibiting obvious zoning around pyrite [34]. Kerogen forms have some connection with pyrite forms, different kerogen forms and different pyrite forms: type III kerogen is a mainly contour shape without microcrystal pyrite inclusion; type III kerogen is mainly amorphous with rich microcrystal pyrite inclusion; type II kerogen form is a mixed type whose relative proportion of components change greatly.

Pyrite is the most widely distributed sulphide in the crust, which can be formed in a variety of different geological conditions [35]. Copper containing a pyrite layer hosted in volcanic rock series is the largest pyrite mass fraction deposit, formed by volcanic sedimentation and hydrothermal processes [36]. Although most pyrite in the kerogen is usually of organic origin, pyrite mass fraction in kerogen will undoubtedly be affected by volcanism. Sulphur-rich material provided by the volcanism can increase the sulphur content in an aqueous medium participating in the formation of pyrite related to kerogen, while volcanic rocks or pyroclastic rocks can form pyrite directly [37].

2.2. Volcanic Activity Provide Thermal Energy For The Evolution Of Organic Matter

The thermal effect of volcanic activity has a dual function on organic matter hydrocarbon generation, which can accelerate the maturation of immature source rocks and hydrocarbon generation [38], and also can make mature source rocks over-mature or destroy oil and gas reservoirs formed earlier [39]. The temperature of magma can be more than 1,000 degrees Celsius [40-41], which of hydrothermal fluid can be up to 300-400 Celsius degree, making it a carrier with large amounts of heat energy. The heat will accelerate the maturity of organic matter [42-44]. Studies at the Illinois Basin by Schimmelmann et al. [45] have shown that R_o values increased from 0.62% to 5.03% within 5 metres at the coal contact to large intrusion, while R_o values increased from 0.63% to 3.71% within 1 metre at the coal contact to small intrusion. George [46] found that intrusion made R_o rise from 0.55% to 6.55% when he investigated the maturity of the Scottish Midland Valley oil shale. Raymond and Murchison [47] found that vitrinite reflectance was significantly higher around bedrock in the carboniferous strata, Midland Valley, Scotland.

The research data show that the igneous body's effect on organic matter maturity incidence is relevant to the igneous body's size. Carslaw and Jaeger [48] thought that the sphere of influence of the intrusion is in the range of 1-1.5 times rock mass thickness. Through the vitrinite reflectance analysis of rocks around intrusion, Dow [49] concluded that influence scope can be up to twice the thickness of the intrusive body. According to the study of sill and dike on the east Siberia platform, Galushkin [50] considered the scope of sill and dike to be in general within 30-50% sill or dike thickness, rarely more than the thickness. Chen [51] thought that influence range of a sill to organic matter is from less than the sill thickness to more than double thickness, even reaching four times the thickness of the sill. Galushkin [50] reached the conclusion that the intensity of intrusion alternation was in the range of 50-90% sill or dike thickness through many analysed examples. Zhu et al. [52] thought that only within a 15m scope, was organic matter obviously affected by sill. When Raymond and Murchison [47] studied the sediments in in carboniferous strata, Midland Valley, he found that vitrinite reflectance of the organic matter in the tuff close to the volcanic neck is significantly higher than that in the sedimentary rocks. So the magmatic intrusion effect on the evolution extent of organic matter requires further study.

3. BIOGENIC AND ABIOGENIC HYDROCARBON RELATED TO VOLCANIC RESERVOIRS

3.1. Organic Hydrocarbon Generation

The origin of oil and gas has been a long debated theoretical issue. There are two opposing points of view: 1) the organic origin theory and 2) the inorganic origin theory. Organic origin theory considers oil and gas to come from biological processes. Inorganic origin theory explains the origin of oil and gas through inorganic synthesis and mantle degassing. The earliest organic origin theory was proposed by Lomonosov in 1763 [53]. He thought that fertile substances underground, such as oil shale, carbon, asphalt, petroleum and amber, originated in plants. The hydrocarbon formation theory of kerogen thermal degradation proposed by Tissot and Welte [54] and Hunt [55] are the representatives of the organic hydrocarbon generation theory.

The hydrocarbon formation theory of kerogen thermal degradation is based on the diagenesis of organic matter resulting from biopolymers into geopolymers, then kerogen. Kerogen is the main precursor material of oil compounds during the process of hydrocarbon generation, when thermal degradation plays a major role [54]. For sufficient hydrocarbon class and commercial oil gathering, sedimentary rocks must experience the hydrocarbon generation and temperature threshold. Mass hydrocarbons are formed at temperatures from 60 to 150°C by heated organic matter [55]. According to this theoretical model, the sedimentary organic matter maturity, especially for kerogen, becomes the key factor for evaluating hydrocarbon potential. When the threshold burial depth reaches, kerogen will be changed from immature to mature. Oil and gas generates by series of thermal degradation.

Thus the organic origin theory of petroleum has been completely established - it is consistent with the object geological facts, especially the basic law of sedimentary organic matter evolution. The theory has been accepted gradually by the majority of petroleum geologists and plays a major role in oil and gas exploration [56].

3.2. Inorganic Hydrocarbon Generation

Although organic origin theory has been the guiding theory of modern oil exploration, with foundation of immature oil and ultra-deep liquid hydrocarbon, inorganic origin theory has aroused much attention among geologists. Take the Xujiaweizi area where the Daqing oilfield as an example

(Figure 1). Here many reservoirs have been found to contain a lot of alkane gas and non-hydrocarbon gas with inorganic origin, such as CH_4 and CO_2 [57-60]. Carbon isotope of carbon dioxide ($\delta^{13}C_{CO2}$) is an important indicator to identify the carbon dioxide origin, and many domestic and foreign scholars have undertaken research on this [58, 61-64]. Dai [65] pointed out that the $\delta^{13}C_{CO2}$ value is from +7 ‰ to - 39 ‰ in China, in which the organic origin $\delta^{13}C_{CO2}$ value is from - 10 ‰ to - 39.14 ‰, with the main frequency scope of -12 ‰ ~ -17 ‰; the inorganic origin $\delta^{13}C_{CO2}$ value is from +10 ‰ to -8 ‰ with the main frequency scope of -3 ‰ ~ -8 ‰ (Figure 2). Inorganic origin CO_2 can be divided into mantle-derived, carbonate pyrolysis, magma degassing and so on. The $\delta^{13}C$ value of mantle-derived CO_2 is around - 6‰, which of carbonate pyrolysis is from +3‰ to -3‰. CO_2 volume fraction of Well FS9 in the Xujiaweizi area, Songliao Basin, is 89.73%, and the $\delta^{13}C$ value is from -4.06‰ to -5.46‰. The $\delta^{13}C_{CO2}$ value is -6.61‰ of Well FS6, which confirms that CO_2 of the Xujiaweizi region belongs to the mantle-derived category.

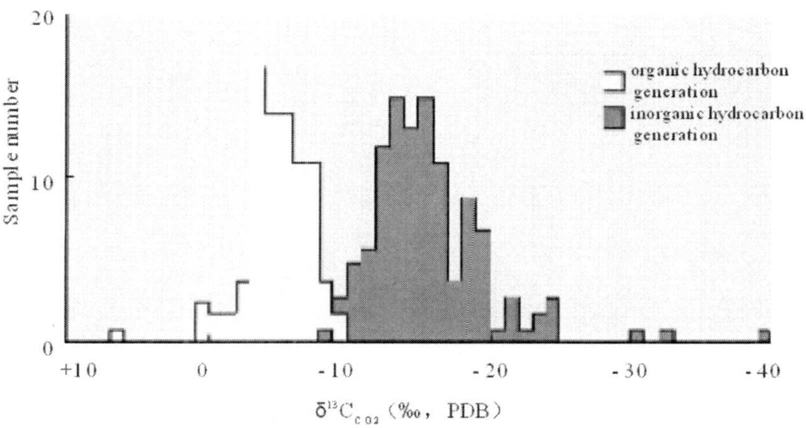

FIGURE 2. Organic and inorganic origin$\delta^{13}C_{CO2}$ frequency (Dai [62, 65])

4. VOLCANIC LITHOFACIES RELATED TO OIL AND GAS ACCUMULATION

Current classification of volcanic lithofacies is mainly based on the style of volcanism or eruptive and/or pyroclast/volcaniclast transport, and corresponds to a modern volcano architecture that forms volcanic rocks

[66-70]. The diagenetic process significantly influences the porosity and permeability of volcanic deposits turning to volcanic rocks during burial in a sedimentary basin. According to the characteristics of volcanic rocks and the hydrocarbon exploration situation in the Songliao Basin (Figure 1), Wang et al. [71] introduced a classification system on volcanic lithofacies (Table 1). Volcanic lithofacies are classified as "*5 facies and 15 sub-facies*", and distinguished with representative features. These features include the transportation mechanism and origin, diagenesis style, representative lithology and structure, facies sequence and rhythm, and potential reservoir spaces. The classification emphasizes the relationship between lithofacies and reservoir capability, and will facilitate the recognition of volcanic lithofacies in various scales of outcrop section, well core, well cutting and thin section. Lithofacies can be identified with seismic and logging data by proper correlation. The reservoir capability of volcanic rocks can be primarily evaluated based on their facies and sub-facies.

TABLE 1. Classification of volcanic facies and corresponding characteristics for each sub-facies

Facies	Sub Facies	Material and Transport	Diagenesis	Lithology	Structure	Texture	Sequence	Reservoir Space
	V3 Coal-bearing tuffaceous sediment	Tuffaceous pyroclast, plant-enriched turf		Interbedded tuff and coal seam	Pyroclastic/ clastic structure	Rhythmic bedding, horizontal bedding	Swamp, close to volcanic dome	
V Volcanogenic Sedimentary Facies	V_2 Reworked volcanogenic sediment	Pyroclast reworked by fluid	Compaction and consolidation	Layered pyroclastic rocks/ tuff	Pyroclastic structure with rounded gravel, no epi-clast.	Cross bedding, trench bedding, graded bedding, massive	Depression between volcanic domes, conduit-close facies of large volcanic edifice	Intergranular pore, primary and secondary pore, fissure, porosity and permeability similar to sedimentary rock
	V_1 Epiclast-bearing volcanogenic sediment	Pyroclast dominating with epiclast		Epiclast-bearing tuff (tuffaceous sandy conglomerate)	Pyroclastic / clastic structure with rounded gravel. Few epi-clast		Depression between volcanic domes	
	IV3 Outer extrusive sub-facies	Lava front condense, deform, scrap and wrap new and old rocks. The mixture wriggles under internal force	Condensing lava weld new and old rock fragments	Breccia lave with deformed fluidal structure	Welded breccia and welded tuff structure	Deformed fluidal structure	Outer part of extrusive facie, transition to effusive facies	Inter-breccia fissure, micro-fissure, fissure between fluidal structure
IV Extrusive								

Facies (late stage of a cycle)	IV2 Middle extrusive sub-facies	Lava with high viscosity flows under internal force, domed near crater	Lava condense and consolidate (quench)	Massive pearlite and cryptocrystalline rhyolite	Vitric structure, perlitic structure, mortar structure	Massive, layered, lenticular and wrapping	Middle part of extrusive facie	Primary micro-fissure, tectonic fissure, inter-crystal pore
	IV 1 Inner extrusive sub-facies			Pillow or sphericity shaped perlite		Spherical, pillow, dome	Core of extrusive facies	Inter-perlite sphere space, pores within loosely packing perlite, micro-fissure, inter-crystal pore
III Effusive Facies (middle stage of a cycle)	III3 Upper effusive sub-facies	Crystals and syn-eruption braccia bearing lava flows on surface under gravity and propelling of subsequent lava	Lava condense and consolidate	Vesicular rhyolite	Spherulitic structure, cryptocrystalline structure, microcrystalline structure	Vesicular, amygdaloidal, lithophysae	Upper part of flow unit	Vesicular, inner space of lithophysae, inner space of amygdaloidal
	III 2 Middle effusive sub-facies			Rhyolite with fluidal structure	cryptocrystalline structure, microcrystalline structure, porphyritic structure	Fluidal structure, few Vesicular-amygdaloidal structure	Middle part of flow unit	fissure between fluidal structure, vesicular, tectonic fissure
	III 1 Lower effusive sub-facies			Cryptocrystalline rhyolite, syn-genetic braccia bearing rhyolite	Vitric structure, cryptocrystalline structure, porphyritic structure, breccia structure	Massive, dashed deformed fluidal structure	Lower part of flow unit	Slaty and wedge joint, tectonic fissure
II Explosive Facies (early stage of a cycle)	Ii3 Pyroclastic flow deposti	Volatile bearing hot mixture of pyroclast and magma slurry flows on surface under gravity and propelling of subsequent lava	Lava condensation and consolidation as well as compaction	Crystal fragment, vitric fragment and magma slurry and lithic fragment bearing welded tuff (lava); lava cemented polymictic conglomerate	Welded tuff structure, pyroclastic structure	Massive, normal grading, inversed grading, orientated vesicular and vitric, matrix support	Early stage of cycle, Upper part of explosive facies, transition to effusive facies	Intergranular pore, vesicular, loose deposit of condensing unit
	II 2 Base surge deposit	Air ejecting multiphase turbidity of gas-solid-liquid flows rapidly under gravity on surface (maximum velocity: 240km/hour)	Compaction	Crystal fragment, vitric fragment and magma slurry bearing tuff	Pyroclastic structure (crystal fragment bearing tuff structure dominating)	Parallel bedding, cross bedding, regressive sand wave bedding	Middle-lower part of explosive facies, interbedded with air fall deposit, normal grading, thinning bedding upward, cover the palaeo-slope	Loose deposit within volcanic body, intergranular pore, inter-breccia fissure
	II 1 Air fall deposit	Free fall of air ejecting solid and plastic material (under influence of wind)	Compaction	Bomb and pumice bearing agglomerate, breccia, crystal fragment bearing tuff	Agglomerate structure, breccia structure, tuff structure	Granular support, normal grading, trajectory falling blocks	Lower part of explosive facies, normal grading, thinning bedding upward, intercalated bedding	

I Conduit Facies (lower part of a edifice)	I3 Crypto-explosive breccia	Volatile-enriched magma intrudes and explodes within surrounding rocks underground. Surrounding rocks were in-situ broken and cemented by magma. Explosion and cementation function simultaneously.	Cement by condensation of hydrothermal or fine pyroclast	Cyrpto-explosive breccia	Crypto-explosive breccia structure, self-mortar structure, cataclastic structure	Column, layered, dike, branch, fissure filling	Near crater, top of sub-volcanic rock, fingered in surrounding rock	Inter-breccia pore, filled primary micro fissure
	I2 Dikes and sills	Syn or post magma intrusion	Lava condensation and crystallization	Sub-volcanic rock, porphyrite and porphyry	Porphyritic structure, holocrystalline structure	Chilled border, flow plane, flow line, columnar and tabular, xenolith	Several to fifteen hundreds of meters beneath volcanic edifice, fingered with other lithofacies and surrounding rocks	Slaty and columnar joint, fissure between dike and surrounding rock
	I1 diatreme	Detained lave in conduit, Collapse of crater	Lava condensation, lava welding pyroclast, compaction	Lava, welded breccia/tuff lave, tuff/breccia	Porphyritic structure, welded structure, breccia/tuff structure	Packing structure, ring or radial joint, belted lithology	Diameter of hundreds of meters, vertical occurrence, penetrate other facies	Inter-breccia pore, ring and radial fissure

Sequences of facies and sub-facies assemblage follow certain principles. In the Songliao Basin, the felsic sequences are explosive facies → effusive facies/extrusive facies (probability: 50%±), conduit facies → effusive facies/extrusive facies (probability: 30%±) and explosive facies → conduit facies →extrusive facies/explosive facies (probability: 20%±). The intermediate – basic sequences are effusive facies → explosive facies (probability: 50%±), effusive facies → volcanogenic sedimentary facies (probability: 30%±), effusive facies → explosive facies → volcanogenic sedimentary facies (probability: 20%±). The sequences of felsic rocks inter-bedded with intermediate – basic rocks are more complex and mainly include effusive facies → explosive facies → volcanogenic sedimentary facies (probability: 30%±), explosive facies → volcanogenic sedimentary facies (probability: 20%±), explosive facies → effusive facies → volcanogenic sedimentary facies (probability: 20%±) and effusive facies → conduit facies → extrusive facies (probability: 10%±). Sequences of facies are the basis of volcanic lithofacies modelling, geological interpretation of seismic data and prediction of volcanic reservoir.

According to facies assemblage in drill cores and outcrops, lava of explosive facies and effusive facies can be directly linked to the pyroclastic rocks of volcanogenic sedimentary facies, especially in the proximity of a volcanic conduit, while most volcanogenic sedimentary facies form along volcanic edifice flanks. In general, felsic eruptive sequences start with explosive facies, while in conduit-close regions, they starts with conduit facies (Figure 3).

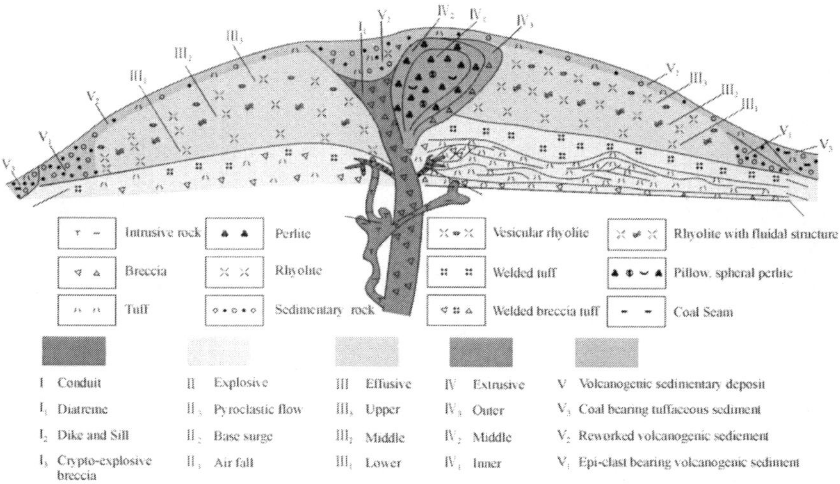

FIGURE 3. Model of the facies of Mesozoic acidic volcanic rocks in the Songliao Basin, China

Since diagenesis of volcanic lava is condensing consolidation-dominated, its porosity-change influenced by the burial is less pronounced than for sedimentary rocks, thus, volcanic rock will contribute more reservoirs when burying depth exceeds a threshold value. In the Songliao Basin, the threshold burying depth is about 3,500 m. Beneath this depth, sandstone is densely compacted and loses reservoir capability, the reservoir is volcanic rock-dominated. Reservoir spaces within volcanic rocks show complex structures and strongly heterogeneous distribution. Based on observation of well core, cutting and analysis of micro-structure, reservoir spaces of volcanic rocks in the Songliao Basin are classified as primary pores, secondary pores and fissures which include 13 types of elemental components (Figure 4). In general, the reservoir space of volcanic rocks has a dual-component of pore and fissure.

Primary vesicular pores and tectonic joints well develop within upper sub-facies of effusive facies, effectively connected intergranular pores are found within loose deposits between sub-facies of explosive facies, primary fissures and intergranular pores well develop within inner sub-facies of extrusive facies. Exploration of hydrocarbon accumulation in volcanic rocks should target these sub-facies in case of the existence of source rocks and traps.

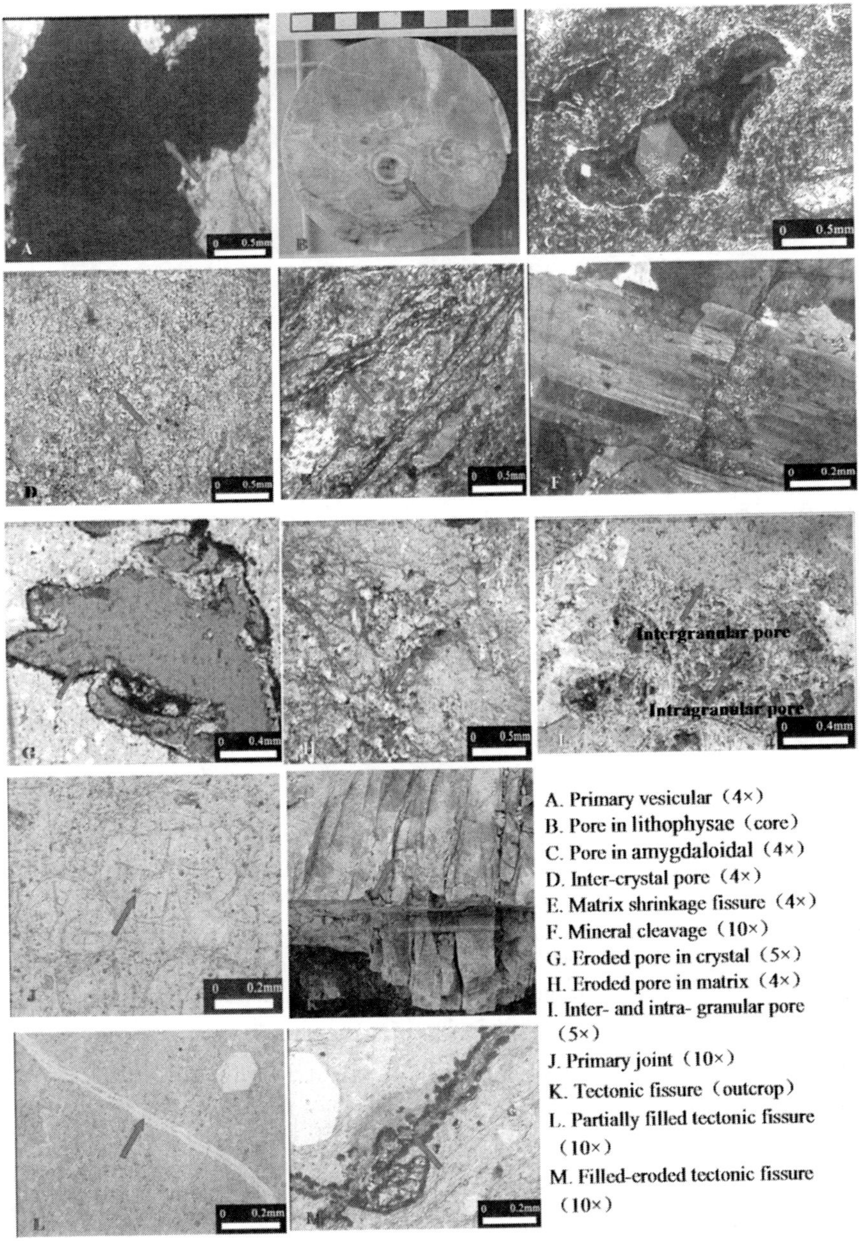

A. Primary vesicular (4×)
B. Pore in lithophysae (core)
C. Pore in amygdaloidal (4×)
D. Inter-crystal pore (4×)
E. Matrix shrinkage fissure (4×)
F. Mineral cleavage (10×)
G. Eroded pore in crystal (5×)
H. Eroded pore in matrix (4×)
I. Inter- and intra- granular pore (5×)
J. Primary joint (10×)
K. Tectonic fissure (outcrop)
L. Partially filled tectonic fissure (10×)
M. Filled-eroded tectonic fissure (10×)

FIGURE 4. Volcanic reservoir space types

5. IDENTIFICATION OF VOLCANIC RESERVOIRS BY WELL-LOGGINGS

Identification of volcanic lithology and lithofacies by logging is primarily based on calibration with drilling cores and cuttings, then logging parameters are used to make cross-plots and frequency distribution histograms. In addtion, logging facies' analysis and FMI image interpretation is used so as to discriminate volcanic rocks as well as their textures and structures.

5.1. Discrimination Analysis of Volcanic Lithology and Lithofacies by Cross-Plots of Logging Parameters

Cross-plots of logging parameters are simple and effective methods which are generally used to discriminate volcanic lithology and lithofacies in drilled wells [72-73]. Primary logging parameters includ natural gamma (GR), natural gamma-ray spectral logging (U, Th and K), electrical resistivity (RT), NPHI porosity, RHOB density, acoustic log (DT), photoelectric absorption coefficient (Pe) as well as compound parameters M and N. Two of theses parameters are plotted in a X and Y coordinate system, different regions are divided by the concentration of data points, then will be assigned with corresponding geological information. Generally, this method is used firstly on well sections with known lithology and lithofacies, so as to make master plates which are then applied to the other unknown sections in the same area. Applications in the Songliao Basin show that GR-Th, Pe-Th and M-N cross-plots are the most effective methods for discriminations of volcanic lithologies (Figure 5). Moreover, logging facies' analysis and FMI image interpretation are used to identify the textures and structures of volcanic rocks, and then finally determine the discrimination of volcanic lithology and lithofacies in detail.

5.2. Logging Facies' Analysis Of Volcanic Rocks

Comparative analysis between the volcanic facies and logging facies of drilling core sections is aimed at revealing and summarizing the relationship between geologic properties and logging responses, so as to solve the multiplicity of interpretation by logging parameters, and then set up identification standards of logging facies in the study area. Identification of logging facies is by means of configuration analysis of logging curves including SP, GR, RT, ML, RHOB, as well as dip logging interpretation. Moreover, the standard logging facies could be interpreted as lithofacies on the basis of geologic data.

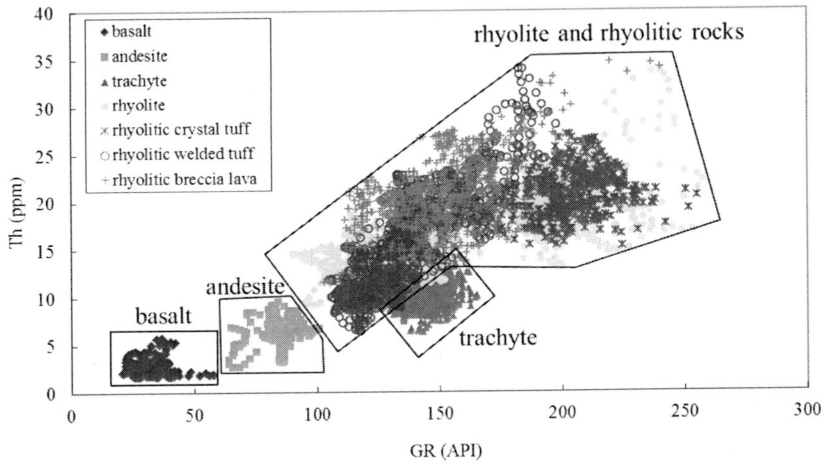

FIGURE 5. Cross-plot of GR versus Th for lithological identification of volcanic rocks of the Lower Cretaceous Yingcheng Formation in the Xujiaweizi depression, Songliao Basin, China

Electrical conductivity of volcanic reservoirs is mainly influenced by lithology, porosity and permeability, saturation, content of metal elements and also burial depth. Occurrence of hydrocarbons will greatly increase the resistivity, while it will obviously decrease with water. The shape of logging curves and their assemblages are related closely to volcanic lithologies as well as their textures and structures which have become good markers for discrimination of volcanic lithofacies. For massive volcanic rocks, the framework is the main medium of conduction. Under this circumstance, lithology, lithofacies and burial depth are the main controlling factors to the conduction of rocks and changes of logging curve shapes. For example, intermediate-felsic volcanic rocks of vent facies are characterized with high-GR and mid-RT, and their logging curves appear as a high amplitude dentiform and peak shape. While basalts of volcanic vent facies show low-GR and the tuff displays low-RT.

The Mesozoic volcanic rocks are the most important gas reservoirs in the northern Songliao Basin. Five lithofacies and 15 sub-facies have been recognized in the volcanic rocks. The best reservoirs were generally found in three of the 15 sub-facies including pyroclastic bearing lava flow, upper effusive and inner extrusive sub-facies. The corresponding logging characteristics are as follows. The pyroclastic rock-bearing lava flow sub-facies show high-GR values with high amplitude dentiform and medium to mid-high RT with low frequency, low amplitude dentiform. The upper effusive sub-facies show high GR with high amplitude dentiform and mid-high RT with finger and peak shapes. The inner extrusive sub-facies show high GR with medium amplitude dentiform and mid-high to high RT with medium amplitude dentiform. In addition, crypto-explosive and outer extrusive sub-facies may also be good reservoirs. The occurrence of hydrocarbons will cause a remarkable increase of resistivity, while water does the contrary. The changing of resistivity without influence of fluids from low to high are respectively followed as volcanogenic sedimentary facies, extrusive facies, explosive facies, volcanic conduit facies and effusive facies [74].

5.3. Identification of Volcanic Textures and Structures by Fmi Image Interpretation

With the characteristics of high resolution, total borehole coverage and visibility, FMI image interpretation may reveal continuous geologic information such as lithology, textures and structures, as well as pores and fractures by means of calibrations with drilling core sections. Sizes and shapes of volcanic breccia and conglomerates, as well as features of rock structures and beddings, can give much geologic information on volcanic lithofacies and pore spaces, especially for well sections lacking drilled cores [75-77].

Features displayed by FMI images of volcanic rocks are the synthesized effects of logging response units including volcanic fragments, framework, fractures and pores. On the FMI images, bright tone corresponds to high resistivity, while dark tone relates to low resistivity, and warm colours, such as yellow and orange, indicate medium resistivity (Table 2).

TABLE 2. Image interpretation of volcanic imaging logging (Li et al. [72])

Image type	Tone	Resistivity	Geological interpretation
static	bright (white)	high	massive structure
	dark (brown, black)	low	fractures and pores
	mottle	heterogeneous	heterogeneous rock mass
dynamic	bright (white)	high	volcanic breccias, rock fragments, crystal fragments, magma fragments, amygdala
	yellow, orange	middle	rock matrix or framework
	dark (brown, black)	low	fractures and pores

Comparatively, rock fragments generally display bright tones due to high resistivity, while matrix shows dark as a result of low resistivity, and these featuresare common in pyroclastic lava rocks and pyroclastic rocks. The transformation of bright and dark zones on a FMI image not only indicates resistivity changes, but also reflects the contact relations among different parts of rocks. Descriptions of drilled cores reveal that there is great difference between volcanic fragments (breccia, conglomerates, rock fragments, crystal fragments and magma fragments) and their surrounding matrix (lava framework or volcanic ash) due to distinguishing colour, content and shape, which may result in colour diversities of the FMI images according to resistivity changes. Standard interpretation models of volcanic textures and structures which are used to identify lithofacies have been summarized through calibrations of FMI images with geologic information, for example, lava texture, welded texture, tuff texture, breccia texture and massive structure, vesicular-amygdaloidal structure, flow structure (Figure 6).

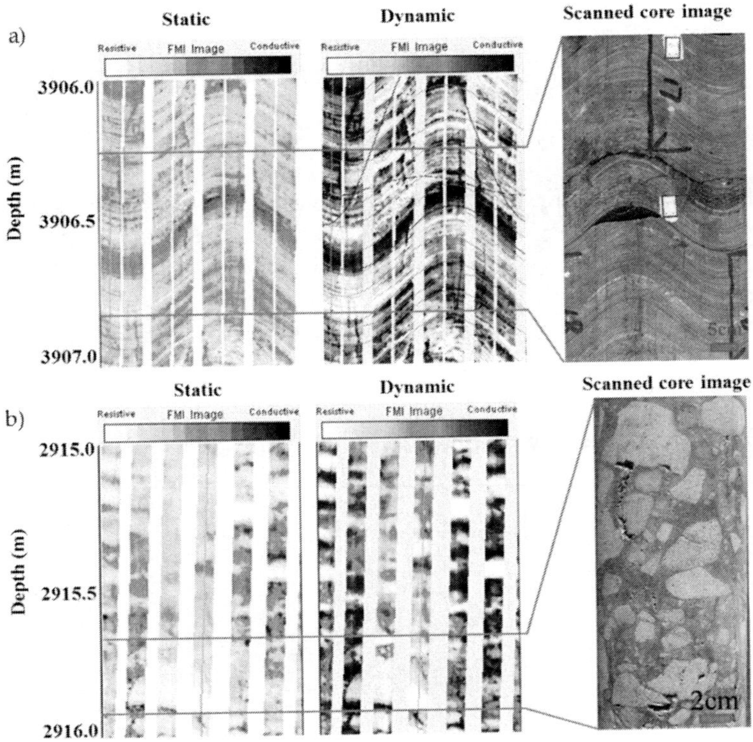

FIGURE 6. Typical lithological structures on FMI logging image. a) Flow bandings of extrusive rhyolite; b) Volcanic breccia structure of explosive facies.

6. IDENTIFICATION OF VOLCANIC RESERVOIRS BY SEISMIC DATA

At the exploration stage, the volcanic facies mapping relies mainly on the artificial seismic facies' analysis and is under the control of well facies or according to an experience template, converted into a volcanic facies map. At the development stage, the volcanic facies planar prediction relies mainly on the waveform classification method for obtaining the seismic facies map. The volcanic facies is interpreted under the control of facies of wells and the volcanic edifice belts [78-79]. This method can identify the volcanic facies and its combination. Now it is widely used in the volcanic exploration of the Songliao Basin. The waveform classification method of volcanic facies' identifying is explained in this paper.

The actual seismic facies were calculated by combining different amplitude, frequency, phase and time intervals. A seismic facies map can be obtained by the waveform classification calculation. Volcanic facies is predicted through observing the combination and distribution characteristics of the model trace in the seismic facies map. The number of waveform classifications (model trace) can be ensured by the seismic characteristics of volcanic facies. After that, we conduct the waveform classification experiment by selecting multiple waveform classifications or by using different thickness of time intervals. Lastly, the stability of the calculation results need be checked. The optimal time interval of the waveform classification calculation is between half a wavelength to two wavelengths.

Taking the volcanic rocks of the upper Yingcheng Formation in the Changling rift YYT work area of the southern Songliao Basin as an example (Figure 1), we introduce the prediction method of the volcanic facies plane. The volcanic facies is predicted by selecting 7 or 15, respectively, as the number of waveform classification. The prediction results of volcanic facies show good consistency. Volcanic facies is predicted by number waveform classification being set to 7. Firstly, overlap the volcanic seismic facies maps with the structure maps. The results show that, in some high tectonic areas, the waveforms characteristics of the seismic facies have unorganized distribution, but in the relatively flat tectonic area, the waveforms characteristics of seismic facies show continuous distribution. Next, the waveform classification characteristics of the seismic facies can be calibrated with well facies, the waveform characteristics of seismic facies in the central region of volcanic edifices are multi-waveform clutter distribution, while the waveform characteristics of the far-source area far away from the centre of volcanic edifices are continuous distribution (Figure 7). In this way, the centre's facies belt (volcanic conduit / effusive facies) distribution of volcanic edifices can be predicted. The different waveforms' seismic facies are calibrated by volcanic facies revealed by the well. A waveform classification planar map should be converted to the volcanic facies map. The seismic characteristics of volcanic conduit facies and its combination are rounded, massive, banded and messy reflection, in the edge is ring banded. The seismic characteristics of explosive facies and their combination are banded, messy, good continuous reflection. The seismic characteristics of effusive facies and their combination are mottled massive, middle-bad continuum reflection. According to the seismic characteristics, volcanic facies planar distribution is identified in the YYT area. The effusive facies distribution is dominating, and the explosive facies distribution is less. The effusive facies distribute mainly on both sides of the central fault. Explosive facies distribute mainly in the southeast far away the central fault. There are two volcanic facies sequences, one is the volcanic conduit facies-effusive facies, the other is the volcanic conduit facies-explosive facie/volcanic sedimentary facies.

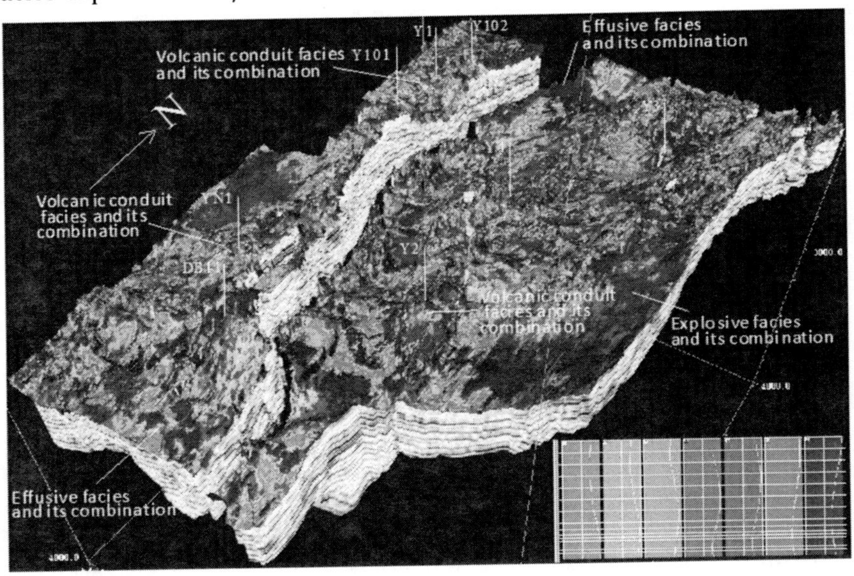

FIGURE 7. Waveform classification (seismic facies) map of volcanic rocks of the upper Yingcheng Formation in the YYT work area

7. MECHANISM AND GEOLOGICAL OCCURRENCE OF OIL AND GAS ACCUMULATION

Volcanic oil and gas reservoirs are mainly not only accumulations in volcanic rocks, but also those hydrocarbon reservoirs with volcanic rocks as seals or forming traps. The formation and distribution of hydrocarbon accumulations in volcanic rocks is different from non-volcanic (silici-)clastic rocks. Since volcanic rocks cannot produce hydrocarbons, neighbouring source rocks are essential to the formation of oil and gas accumulations in volcanic rocks, thus it will be more favourable for better matching relationships between volcanic reservoirs and source rocks [80-81].

7.1. Volcanic Rocks As Oil And Gas Reservoirs

Since the porosity and permeability of volcanic rocks do not decrease remarkably according to the increase of burial depth, they are more favourable for hydrocarbon accumulations compared to sedimentary rocks in the deep part of the basin. So far, hundreds of volcanic reservoirs, such as the Niigata Basin in the Honshu Island of Japan [82], Austral and Neuquen Basins of Argentina [7] and the Bohai Bay Basin, Songliao Basin and Junggar Basin in China have been found [5, 24, 71, 81]. As a whole, the volcanic reservoirs are mostly Cenozoic and Mesozoic, especially Jurassic, Cretaceous and Tertiary, which may be related to the frequent global volcanism in these epochs. As revealed by petroleum explorations, occurrences of hydrocarbons have been found in almost all types of volcanic rocks, and in detail, basalts have the largest proportion while the rest follow as andesite, rhyolite and pyroclastic rocks [83]. Prolific volcanic reservoirs have been found in explosive, extrusive and volcanic-sedimentary facies, while considering inside a volcanic edifice, they have generally the best reservoir properties in volcanic vents and near vent facies [71, 84]. Vertically, favourable reservoirs are developed in the upper part of volcanic sequences due to post-eruption weathering, leaching and dissolving [85].

7.2. Formation Of Oil Traps Related To Volcanic Activities

Besides being reservoirs, volcanic rocks can also be good cap rocks. After volcanic ash falls into water, it will inflate and form layers of bentonite or mudstone with bentonite which may become excellent cap rocks [86]. While mudstone lacks sealing abilities in the mid-shallow part of basins, unaltered massive basalts are generally rather better cap rocks, taking the Scott Reef oil field in the Browse Basin of Australia and Eastern Sag in the Liaohe Basin of China as examples [2, 24]. Some layered intrusive rocks may also be good cap rocks, for example, the Lin 8 oil trap in the Huimin depression of the Bohai Bay Basin, north China [87]. In the deep part of the Xujiaweizi depression of the Songliao Basin, two types of volcanic rocks have been found to be cap rocks including the pyroclastic type and lava type. Due to better sealing abilities, the pyroclastic type cap rocks control the regional accumulation and distribution of gas in volcanic reservoirs, while the lava type only control the local accumulation and distribution of gas in volcanic rock bodies [88]. Besides, diverse oil and gas traps can be formed by the local structures of intrusions as well as their matching with sedimentary layers, commonly forming arched uplifts and lateral barriers [24, 89].

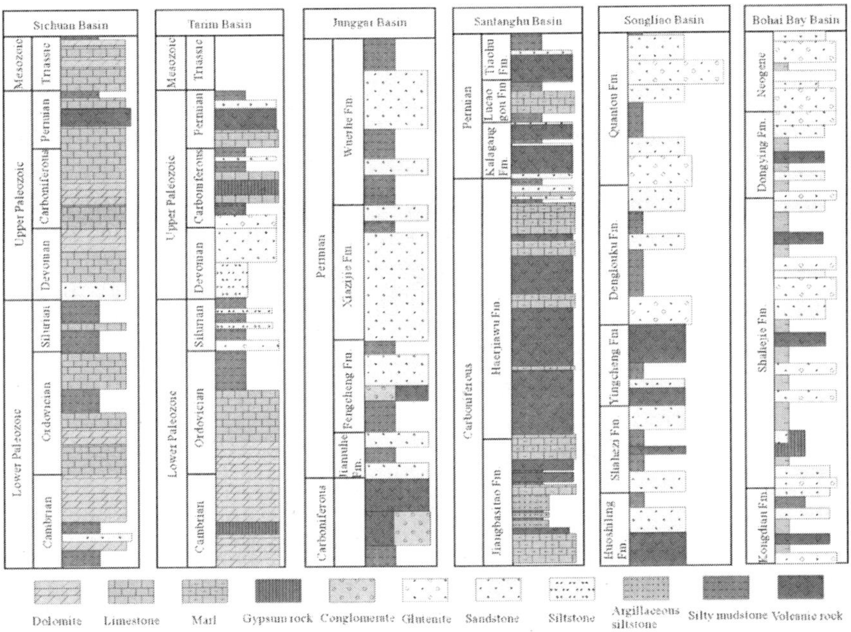

FIGURE 8. Source-reservoir-cap assemblages of the volcanic oil and gas accumulations in main petroliferous basins of China (Zou et al. [**81**])

All aspects of the common hydrocarbon accumulating conditions and their favourable matching relationships are also necessary to the volcanic oil and gas reservoirs. So far, most of the discovered volcanic oil and gas reservoirs are structural-lithologic and stratigraphic. Near-source accumulations are formed when volcanic rocks emplace close to source rocks, developing concentration zones of volcanic oil and gas reservoirs. While there is a long distance between source rocks and volcanic rocks, certain accumulations may also form due to communications of faults and unconformities. There are two accumulation patterns divided by volcanic reservoir forming conditions such as near-source play and distal play [81]. Near-source plays are mostly discovered in eastern Chinese depressions, for example, the

Paleogene mafic volcanic rocks in the Bohai Bay Basin and the Lower Cretaceous felsic volcanic rocks which developed prolific oil and gas accumulations emplaced right on the top of high-quality source rocks. Both near-source and distal plays are found in western Chinese basins, for instance, volcanic rocks and source rocks have developed together in carboniferous-Permian of the Junggar and Santanghu Basins which formed near-source plays, and the source rocks mainly developed in the Lower Paleozoic while the reservoir volcanic rocks emplaced in the Permian, thus forming distal plays in the Sichuan and Tarim Basins (Figure 8).

8. EXPLORATION OF VOLCANIC ACCUMULATION OF OIL AND GAS

8.1. The Characteristics Of Volcanic Oil And Gas Accumulation

The characteristics of volcanic oil and gas accumulation mainly include volcanic reservoirs and reservoir forming elements. Taking the characteristics of volcanic oil and gas accumulation in the Songliao Basin as an example, volcanic gas accumulation can be classified into acid type and intermediate-basic type by lithology. By the characteristics of volcanic edifices, these two types mentioned above can be furthe rclassified into six sub-types, including acid pyroclastic sub-type, lava sub-type, complex sub-type and intermediate-basic pyroclastic sub-type, lava sub-type, complex sub-type. Great differences have been discovered in developmental degrees among the types in the volcanic gas accumulation (Figure 9). The acid and intermediate-basic lava sub-types account for 72% of volcanic gas accumulation of the Yingcheng Formation in the north of the Songliao Basin, and the contribution degree of acid lava sub-type can reach 50%. The acid type accounts for 92% of volcanic gas accumulation in the south of the Songliao Basin, while only the intermediate-basic lava sub-type gains industrial gas. The highest deliverability in a single well is gained in the acid complex sub-type; the deliverability of the intermediate-basic type is lower than the acid type. There are few differences among the intermediate-basic pyroclastic sub-type, lava sub-type and complex sub-type. However, there are great differences among the acid pyroclastic sub-type, lava sub-type and complex sub-type.

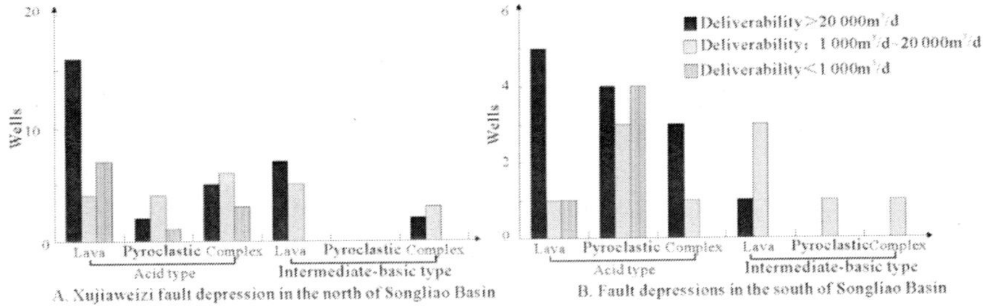

FIGURE 9. The relationship between hydrocarbon accumulations and volcanic edifices in the Songliao basin

By analysing the relationship between reservoirs and reservoir forming structures, most of the volcanic gas accumulations are structural-lithologic gas accumulations. The gas mainly originated from the mud and coal-bearing strata of the early Cretaceous Shahezi Formation and Huoshiling Formation. The fluid transforming system is made up by the faults, joints and high porosity-permeability transforming zones. The distribution range of gas layers is not absolutely controlled by the structural trap. When above the water-gas contact (WGC), the high porosity-permeability zone forms a gas layer, the medium-low porosity-permeability zone forms a poor gas layer and the zone which has fewer pores and fractures forms a dense layer or baffle layer. The proportion of poor gas layers increases gradually from the acid complex type to the acid pyroclastic sub-type to intermediate-basic lava sub-type (Figure 10). WGC is an uneven contact surface caused by the peculiarity of the stratigraphic construction of volcanic edifices. There are great differences in shape among the different volcanic edifices. The gas thickness of acid pyroclastic sub-type and acid complex sub-type changes little, forming a tabular and sill-like shape. There are great changes in the gas thickness of intermediate-basic lava sub-type, the maximum thickness is 2~3 times thicker than the minimum thickness and forms a mound or wedge shape. Moreover, in the same gas layer, the deliverability is also different in the different locations.

8.2. The Reservoir Forming Elements Of Volcanic Oil And Gas Accumulation

By the comparison between industrial gas wells and other wells, the advantages of forming high production gas include reservoir space diversity, high porosity, good source rocks, anticlinal /faulted anticline traps and the vertical migration pathway. The reservoir forming effects will be poor once one of the above conditions is absent.

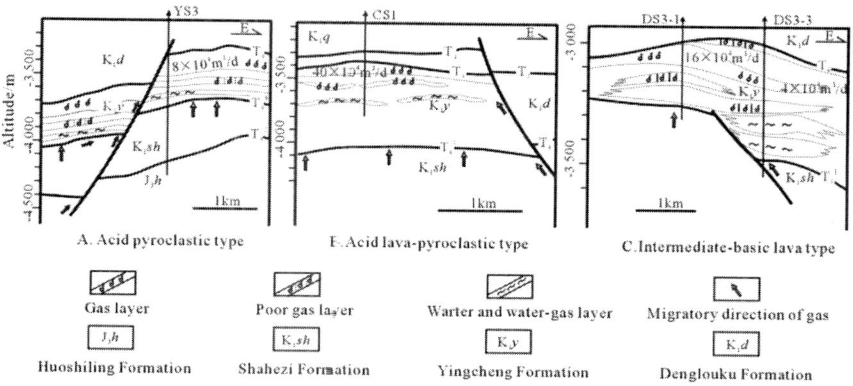

FIGURE 10. Forming pattern of gas pools of volcanic edifices in the Songliao Basin

TABLE 3. The characteristics of gas pool forming of volcanic edifices of faulted sequences in the Songliao Basin

Volcanic gas accumulation type		Primary pore		Secondary pore		Trap type	Migration type	Relations among source, reservoir and cap	Deliverability $10^4 m^3/d$	Typical well (Formation)
Type	Sub-type	Pore	Fracture	Pore	Fracture					
Acid	Pyroclastic	Intergranular pore, Intercrystalline pore, Intercentric pore (breccia)	Crash fractures, width : 0.1 mm , filling degree: 95%	/	High-angle fractures, width : 1~2 mm, upper part with no fill, lower part filled by calcite filling degree: 100%	faulted anticline trap	vertical	self generation and self preservation, lower generation and upper preservation	8	YS3 (k_1y)
		Interbreccial pore, Corrosion pore	Intercrystalline crash fractures	Dissolved pore	High-angle fracture	fault nose trap	lateral, vertical	self generation and self preservation, lower generation and upper preservation	gas show	SS1 (k_1y)
	Lava	Intercentric pore	Crash fractures	/	Joints and high-angle fracturesfilling degree: 80%	fault nose trap	lateral	self generation and self preservation	gas show	YN1 (k_1y)
		Vesicle, Intercentric pore	Intercrystalline crash fractures	/	High-angle fractures and oblique crossing fractures, reticular partly, width: 0.5~2 mm	anticline trap	lateral vertical	self generation and self preservation, lower generation and upper preservation	30	YS2 (k_1y)
		Vesicle, Interbreccial pore, Corrosion pore	Intercrystalline crash fractures	Dissolved pore	high-angle fractures and Joints	anticline trap	lateral vertical	self generation and self preservation, lower generation and upper preservation	40	YS1 (k_1y)

	Complex	/	Crash fractures, filled by calcite	/	Branched fractures, width:1 cm, filled by red magma, filling degree: 30%~100%	fault nose trap	lateral vertical	self generation and self preservation, lower generation and upper preservation	gas show	YS4 (k_1y)
intermediate-basic	Pyroclastic	Interbreccial pore, vesicle (breccia)	Crash fractures	Dissolved pore	High-angle structural fractures	anticline trap	vertical	lower generation and upper preservation	5.6	DS3 (k_1y)
	Lava	Micro-vesicle	/	/	High-angle fractures width: 2~30 mm, filled by calcite, filling degree: 10%~100%	fault nose trap	lateral vertical	self generation and self preservation, lower generation and upper preservation	0.4	DB11 (k_1y)
		Amygdale, vesicle, Interbreccial pore	Crash fractures	/	Reticular fractures, width:2~50 mm, filled by calcite, filling degree: 10%~100%	anticline trap	lateral vertical	lower generation and upper preservation	4.2	DX5,DS3-1 (k_1y)
	Complex	Amygdale, vesicle,	/	/	Reticular fractures, width:2~5 cm, filled by calcite, filling degree 10%~100%	anticline trap	lateral vertical	lower generation and upper preservation	5.0	DS4 (k_1y)

Notes:"/"shows not discovered by the cores/debris. k_1y is Yingcheng Formation ofLower Cretaceous

For example, although having good source rocks, a DB10 well only gains low producing gas because of unitary reservoir spaces and poor porosity-permeability. A YN1 well does not even gain industrial gas because of poor source rocks and porosity-permeability despite having diversiform reservoir spaces. An SS1 well has good porosity-permeability and diversiform reservoir spaces, while its source rock is poor, it does not gain industrial gas. The wells show that overlying strata on the volcanic rocks of the Yingcheng Formation can be regional caps in the Songliao Basin and there are a wealth of high angle joints and faults in the volcanic rocks of the Yingcheng Formation. So the main reservoir forming elements of volcanic gas accumulation include effective source rocks, faults connecting to the source rocks and reservoir porosity-permeability.

9. SUMMARY

Hydrocarbon exploration in volcanic rocks is a relatively new and important topic today. As a typical example in China, the Songliao Basin has been introduced here.

Natural transformation from organic matter to hydrocarbon is a slow process. This slow process can be accelerated by volcanic heat that is the thermal effect of volcanic activity. In addition, it can also be catalyzed by volcanogenic minerals, such as zeolite and olivine, and transition metals in hydrothermal liquid, such as Ni, Co, Cu, Mn, Zn, Ti, V, etc.

The origin of natural gas hosted in volcanic reservoirs can be both biogenic and abiogenic. In the Songliao Basin of China, most of hydrocarbon has been proved as having organic derivation, a few alkane gas and non-hydrocarbon gas have inorganic derivation. The origin of hydrocarbon can be distinguished with isotopes such as C and He.

Volcanic lithofacies can be classified into "*5 facies and 15 sub-facies*" with respect to the lithification mechanism and reservoir significance. In general, a felsic sequence starts with explosive facies, and an intermediate – basic sequence starts with effusive facies. The reservoir spaces of volcanic rocks are composed of primary pores, secondary pores and fissures. The upper part of effusive facies, loose intercalation in between explosive facies, and inner sub-facies of extrusive facies are the main targets for hydrocarbon exploration in the Songliao Basin because of their good porosity and permeability.

Volcanic lithology and lithofacies in drilled wells are effectively discriminated with cross-plots of different logging parameters. The most effective methods are GR-Th, Pe-Th and M-N cross-plots for lithological discrimination. Lithofacies are characterized by GR and RT with respect to amplitude, outer shape and frequency. The texture and structure information of volcanic rocks can be depicted with FMI images.

The spatial distribution of lithological and lithofacies associations can be characterized by seismic parameters, such as amplitude, frequency, phase and time intervals. Seismic facies are mapped with waveform classification between half a wavelength to two wavelengths. Geological facies are interpreted from seismic facies coupled with core section description and well-log information.

Volcanic rocks mainly act as reservoir or seal rocks in hydrocarbon accumulations. Most of the discovered volcanic oil and gas reservoirs are structural-lithologic and stratigraphic. Source rocks are essential to the formation of oil and gas. Although plays of proximal facies are predominant, distal facies have also been discovered to be productive in the Songliao Basin.

Porosity and permeability in volcanic rocks are more heterogeneous than sedimentary rocks. High resolution data are necessary for hydrocarbon exploration in volcanic rocks. Furthermore, diagenesis of volcanic rocks is one of the most important topics in the future because it is the controlling factor on the porosity and permeability of volcanic reservoirs.

10. ACKNOWLEDGEMENTS

This research was supported by a grant from the Major State Basic Research Development Programme of China (no. 2009CB219300), Key laboratory of Evolution of Past Life and Environment in Northeast Asia (Jilin University), Ministry of Education, China.

REFERENCES

1. G Einsele, Sedimentary Basins: Evolution, Facies, and Sediment Budget. Berlin: Springer 2000
2. S. R Schutter, Occurrences of Hydrocarbons in and around Igneous Rocks. Geological Society, London, Special Publications 20032143568
3. B Benyamin, Facies distribution approach from log and seismic to identification hydrocarbon distribution in volcanic fracture. In: 9th SPWLA Japan Formation Evaluation SYMP 20072526
4. A Dutkiewicz, H Volk, J Ridley, S. C George, Geochemistry of Oil in Fluid Inclusions in a Middle Proterozoic Igneous Intrusion: Implications for the Source of Hydrocarbons in Crystalline Rocks. Organic Geochemistry 2004358937957
5. Z. Q Feng, Volcanic Rocks as Prolific Gas Reservoir: A Case Study from the Qingshen Gas Field in the Songliao Basin, NE China. Marine and Petroleum Geology 2008
6. T Kawamoto, Distribution and Alteration of the Volcanic Reservoir in the Minami-Nagaoka Gas Field. Journal of the Japanese Association for Petroleum 20016614655
7. P Sruoga, N Rubinstein, Processes Controlling Porosity and Permeability in Volcanic Reservoirs from the Austral and Neuquen Basins, Argentina. AAPG Bulletin 2007911115129
8. B Beeskow, P. J Treloar, A. H Rankin, et alA Reassessment of Models for Hydrocarbon Generation in the Khibiny Nepheline Syenite Complex, Kola Peninsula, Russia. Lithos 2006
9. J. X Dai, X Shi, Y. Z Wei, Summary of the Abiogenic Origin Theory and the Abiogenic Gas Pools (Fields). Acta Petrolei Sinica 2001in Chinese with English abstract).
10. W. H Liu, M. J Chen, P Guan, et alTernary Geochemical-Tracing System in Natural Gas Accumulation. Science in China (Series D) 2007a501014941503
11. R Othman, K. R Arouri, C. R Ward, et alOil Generation by Igneous Intrusions in the Northern Gunnedah Basin, Australia. Organic Geochemistry 2001321012191232
12. K Thomson, Volcanic Features of the North Rockall Trough: Application of Visualisation Techniques on 3D Seismic Reflection Data. Bulletin of Volcanology 2005672116128
13. X Wang, I Lerche, C Walter, Effect of Igneous Intrusive Bodies on Sedimentary Thermal Maturity. AAPG Bulletin 198973911771178
14. Q Jin, L Xu, C Wan, et alInteractions between Basalts and Oil Source Rocks in Rift Basins: CO2 Generation. Chinese Journal of Geochemistry 20072615865
15. L. X Gu, Z. W Ren, C. Z Wu, et alHydrocarbon Reservoirs in a Trachyte Porphyry Intrusion in the Eastern Depression of the Liaohe Basin, Northeast China. AAPG Bulletin 2002861018211832
16. L. E Levin, Volcanogenic and Volcaniclastic Reservoir Rocks in Mesozoic- Cenozoic Island Arcs: Examples from the Caucasus and the Nw Pacific. Journal of Petroleum Geology 1995183267288

17. L Vernik, A New Type of Reservoir Rock in Volcaniclastic Sequences. AAPG Bulletin 1990746830836
18. P. J Wang, S. M Chen, W. Z Liu, et alRelationship between Volcanic Facies and Volcanic Reservoirs in Songliao Basin. Oil & Gas Geology 2003in Chinese with English abstract)
19. S. C Bergman, J. P Talbot, P. R Thompson, The Kora Miocene submarine andesite stratovolcano hydrocarbon reservoir, northern Taranaki Basin, New Zealand. In: New Zealand Oil Exploration Conference, Wellington, New Zealand, 1992178206
20. W. Z Liu, Y. M Pang, H. Y Wu, et alRelationship between Pore-Space Evolution and Deeply Buried Alteration during Diagenetic Stage of the Volcanic Fragments in Reservoir Greywacke from the Songliao Basin, NE China. Journal of Jilin University (Earth Science Edition) 2007in Chinese with English abstract).
21. J. L Luo, C. L Zhang, Z. H Qu, Volcanic Reservoir Rocks: A Case Study of the Cretaceous Fenghuadian Suite, Huanghua Basin, Eastern China. Journal of Petroleum Geology 1999224397416
22. K. L Rogers, P. S Neuhoff, A. K Pedersen, D. K Bird, CO2 Metasomatism in a Basalt-Hosted Petroleum Reservoir, Nuussuaq, West Greenland. Lithos 2006
23. H. L Zhao, Z. W Liu, J Li, et alPetrologic Characteristics of Igneous Rock Reservoirs and their Research Orientation. Oil & Gas Geology 2004in Chinese with English abstract).
24. Z. Y Chen, Y Huo, J. S Li, et alRelationship Between Tertiary Volcanic Rocks and Hydrocarbons in the Liaohe Basin, People's Republic of China. AAPG Bulletin 199983610041014
25. H Yang, Y Zhang, C. N Zou, et alVolcanic Rock Distribution and Gas Abundance Regularity in Xujiaweizi Faulted Depression, Songliao Basin. Chinese Journal of Geophysics, 49(4): 1136-1143 (in Chinese with English abstract).
26. S. A Stewart, P. J Allen, D Seismic Reflection Mapping of the Silverpit Multi-Ringed Crater, North Sea. Geological Society of America Bulletin 2005
27. C. J Jiang, X. Y Feng, Y. J Zhan, et alNew Methodology to Explore Gasbearing Volcanic Reservoir in Xujiaweizi Fault Depression of the Northern Songliao Basin. Petroleum Geology & Oilfield Development in Daqing 2007in Chinese with English abstract).
28. D. M Hansen, J. A Cartwright, D Thomas, D Seismic Analysis of the Geometry of Igneous Sills and Sill Junction Relationships. Geological Society, London, Memoirs 2004291199208
29. C. L Wan, Q Jin, B. J Fan, Current Research Situation on Hydrocarbon-Generating Evolution of Volcanic Minerals upon Hydrocarbon Source Rocks. Petroleum Geology and Recovery Efficiency 2001in Chinese with English abstract).
30. J. H Reuter, E. M Perdue, Importance of Heavy Metal-Organic Matter Interactions in Natural Waters. Geochimica Et Cosmochimica Acta 1977412325334
31. F. D Mango, Transition Metal Catalysis in the Generation of Petroleum and Natural Gas. Geochimica Et Cosmochimica Acta 1992561553555
32. Q Jin, Hydrocarbon Generation in Rift Basins, Eastern China: Catalysis and Hydrogenation-Interaction between Volcanic Minerals and Organic Matter. Advance in Earth Sciences 1998in Chinese with English abstract).
33. J. L Zhang, P. Z Zhang, A Discussion of Pyrite Catalysis on The Hydrocarbon Generation Process. Advance in Earth Sciences 1996in Chinese with English abstract).
34. Q Lu, H. F Liu, Studies of Kerogen from Baize Basin, Guangxi-Also on Relationship of Evolution of Kerogen and Clay Minerals. Acta Sedimentologica Sinica 1993in Chinese with English abstract).
35. Y. N Wei, D. D Zhang, Mineral Petrology. Beijing: China Coal Industry Publishing House; 2007
36. G. F Luo, Crystallography and mineralogy. Nanjing: Nanjing University Press; 1993
37. R. H Cheng, P. J Wang, W. Z Liu, et alInfluence of Tectonic Activity and Volcanism on Hydrocarbon Generation of Kerogen. Geological Science and Technology Information 2003in Chinese with English abstract).
38. W Fjeldskaar, H. M Helset, H Johansen, et alThermal Modeling of Magmatic Intrusions in the Gjallar Ridge, Norwegian Sea: Implications for Vitrinite Reflectance and Hydrocarbon Maturation. Basin Research 2008201143159

39. Q. H Zhou, Z. H Feng, G. T Men, Present Geothermal Features and Relationship with Gas in Xujiaweizi Fault Depression, Songliao Basin. Science in China (Series D: Earth Sciences) 2007SII): 177-188 (in Chinese with English abstract).
40. C. Q Wang, X. R Du, J. X Liu, Keluo-Wudalianchi-Erke Volcanic Cluster, Chapter 17, Characteristics of Melt Inclusions and Diagenesis Temperatures of Volcanic Rocks. Journal of East China College of Geology (Natural Science Edition) 1987in Chinese).
41. J. F Wang, A Preliminary Study on the Geochemical Behaviors of Trace Elements and the Origin of Mesozoic Volcanic Rocks in North- Western Zhejiang Province. Geochimica 1992in Chinese with English abstract).
42. N. B Çiftçi, R. Ö Temel, Y. H Iztan, Hydrocarbon Occurrences in the Western Anatolian (Aegean) Grabens, Turkey: Is there a Working Petroleum System? AAPG Bulletin 2010941218271857
43. P Farrimond, J. C Bevan, A. N Bishop, Hopanoid Hydrocarbon Maturation by an Igneous Intrusion. Organic Geochemistry 1996
44. J. Q Liu, F. C Meng, Hydrocarbon Generation, Migration and Accumulation Related to Igneous Activity. Natural Gas Industry 2009in Chinese with English abstract).
45. A Schimmelmann, M Mastalerz, L Gao, et alDike Intrusions Into Bituminous Coal, Illinois Basin: H, C, N, O Isotopic Responses to Rapid and Brief Heating. Geochimica Et Cosmochimica Acta 2009732062646281
46. S. C George, Effect of Igneous Intrusion on the Organic Geochemistry of a Siltstone and an Oil Shale Horizon in the Midland Valley of Scotland. Organic Geochemistry 1992185705723
47. A. C Raymond, D. G Murchison, Development of Organic Maturation in the Thermal Aureoles of Sills and its Relation to Sediment Compaction. Fuel 1988671215991608
48. H. S Carslaw, J. C Jaeger, Conduction of Heat in Solids. New York: Oxford University Press;1959
49. W. G Dow, Kerogen Studies and Geological Interpretations. Journal of Geochemical Exploration 1977707999
50. Y. I Galushkin, Thermal Effects of Igneous Intrusions on Maturity of Organic Matter: A Possible Mechanism of Intrusion. Organic Geochemistry 1997
51. R. S Chen, S He, Q. L Wang, et alA Preliminary Discussion of Magma Activity on the Maturation of Organic Matter-Taking Geyucheng-Wenan Area of Hebei Province as an Example. Petroleum Exploration and Development1989in Chinese with English abstract).
52. D. Y Zhu, Z. J Jin, W. X Hu, et alEffect of Igneous Activity on Hydrocarbon Source Rocks in Jiyang Sub-Basin, Eastern China. Journal of Petroleum Science and Engineering 2007
53. M Lomonosov, Ma WJ (translator). Stratum. Beijing: Science Press; 1958
54. B. P Tissot, D. H Welte, Petroleum Formation and Occurrence. Berling Heidelberg New York Tokyo: Springer; 1978
55. J. M Hunt, Petroleum Geochemistry and Geology. San Francisco: W H Freeman; 1979
56. W. H Liu, D. F Huang, C. W Xiong, et alAdvances in Hydrocarbon Generation Theory and Immature and Low Mature Oil and Gas Distribution and Research. Natural Gas Geoscience 1999in Chinese with English abstract).
57. Z. Q Guo, X. B Wang, A Discussion of Abiogenic Natural Gas in the Songliao Basin. Science in China (Series B) 1994in Chinese).
58. X. B Wang, C. Y Li, J. F Chen, et alA Discussion of Abiogenic Natural Gas. Chinese Science Bulletin 1997in Chinese)
59. Y. F Yang, Q Zhang, H. P Huang, et alAbiogenic Natural Gases and their Accumulation Model in Xujiaweizi Area, Songliao Basin, Northeast China. Earth Science Frontiers 2000in Chinese with English abstract).
60. P. J Wang, Q. J Hou, K. Y Wang, et alDiscovery and Significance of High CH4 Primary Fluid Inclusions in Reservoir Volcanic Rocks of the Songliao Basin, NE China. Acta Geologica Sinica 2007811113120
61. D. M Kerrick, K Caldeira, C. O Metamorphic, Degassing and Early Cenozoic Paleoclimate. GSA Today 199445765
62. J. X Dai, Abiogenic Gas in Oil-Gas Bearing Basins in China and Its Reservoirs. Natural Gas Industry 1995in Chinese with English abstract).

63. G. C Tu, The Discussion on Some CO2 Problems. Earth Science Frontiers 1996in Chinese with English abstract).
64. J. B Yun, Q. S Pang, E. C Xu, et alResearch on the Formation Condition of CO2 Gas Pool in the Southern Songliao Basin. Journal of Daqing Petroleum Institute 2000in Chinese with English abstract).
65. J. X Dai, Composition Characteristics and Origin of Carbon Isotope of Liuhuangtang Natural Gas in Tengchong County, Yunnan Province. Chinese Science Bulletin 1989341210271030
66. Cas RAFWright JV. Volcanic Successions: Ancient and Modern. London: Allen and Unwin; 1987
67. R. V Fisher, H. U Schmincke, Pyroclastic Rocks. Berlin-Heidelberg-New York: Springer; 1984
68. J Lajoie, Facies Models 15: Volcaniclastic Rocks. Geoscience Canada 197963129139
69. J. X Qiu, Volcanic Lithofacies and their Characteristics. Geological Science and Technology Information 1984in Chinese with English abstract).
70. D. Z Wang, X. M Zhou, Volcanic Petrology. Beijing: Science Press; 1982
71. P. J Wang, Y. L Chi, W. Z Liu, et alVolcanic Facies of the Songliao Basin: Classification, Characteristics and Reservoir Significance. Journal of Jilin University (Earth Sciences Edition) 2003in Chinese with English abstract).
72. N Li, D. X Qiao, Q. F Li, et alTheory on Logging Interpretation of Igneous Rocks and Its Application. Petroleum Exploration and Development 2009in Chinese with English abstract).
73. B. Z Pan, Z. B Li, Y. S Fu, et alApplication of Logging Data in Lithology Identification and Reservoir Evaluation of Igneous Rock in Songliao Basin. Geophysical Prospecting for Petroleum 2009in Chinese with English abstract).
74. Z. H Guo, P. J Wang, C. H Yin, et alRelationship between Lithofacies and Logging Facies of the Volcanic Reservoir Rocks in Songliao Basin. Journal of Jilin University (Earth Science Edition) 2006in Chinese with English abstract).
75. M Wang, L. F Xue, B. Z Pan, Lithology Identification of Igneous Rock Using FMI Texture Analysis. Well Logging Technology 2009in Chinese with English abstract).
76. R. S Yao, P. J Wang, L. Z Song, et alImaging Logging Response to Volcanic Pores and Fractures of Yingcheng Formation in the Songliao Basin. Progress in Geophysics 2011in Chinese with English abstract)
77. Y Zhang, B. Z Pan, C. H Yin, et alApplication of Imaging Logging Maps in Lithologic Identification of Volcanic. Geophysical Prospecting for Petroleum 2007in Chinese with English abstract).
78. H. F Tang, P. J Wang, C. J Jiang, et alApplication of Waveform Classification to Identify Volcanic Facies in Songliao Basin. Oil Geophysical Prospecting 2007in Chinese with English abstract).
79. Z. S Xu, Y. M Wang, Y. M Pang, et alIdentification and Evaluation of Xushen Volcanic Gas Reservoirs in Daqing. Petroleum Exploration and Development 2006in Chinese with English abstract).
80. Z. Y Chen, J. S Li, G Zhang, et alRelationship between Volcanic Rocks and Hydrocarbon within Liaohe Depression of Bohai Gulf Basin, China. Petroleum Exploration and Development 1996in Chinese with English abstract).
81. C. N Zou, W. Z Zhao, C. Z Jia, et alFormation and Distribution of Volcanic Hydrocarbon Reservoirs in Sedimentary Basins of China. Petroleum Exploration and Development 2008in Chinese with English abstract).
82. K Magara, Volcanic Reservoir Rocks of Northwestern Honshu Island, Japan. Geological Society, London, Special Publications 200321416981
83. N Petford, K Mccaffrey, Hydrocarbons in Crystalline Rocks: An Introduction. Geological Society, London, Special Publications 2003214115
84. H. F Tang, Y. M Pang, W. H Bian, et alQuantitative Analysis on Reservoirs in Volcanic Edifice of Early Cretaceous Yingcheng Formation in Songliao Basin. Acta Petrolei Sinica 2008in Chinese with English abstract).

85. Y. L Huang, P. J Wang, R Shao, Porosity and Permeability of Pyroclastic Rocks of the Yingcheng Formation in Songliao Basin. Journal of Jilin University (Earth Science Edition) 2010in Chinese with English abstract).
86. Z. Q Guo, Volcanic Activity versus Formation and Distribution of Oil and Gas Fields. Xinjiang Petroleum Geology 2002in Chinese with English abstract).
87. C. G Li, Oil and Gas Related to the Volcanic Rocks in Dongying and Huimin Depressions. Explorationist 1997in Chinese with English abstract).
88. G Fu, M Hu, D Yu, Volcanic Cap Rock Type and Evaluation of Sealing Gas Ability: An Example of Xujiaweizi Depression. Journal of Jilin University (Earth Science Edition) 2010in Chinese with English abstract)
89. G. H Lee, Y. I Kwon, C. S Yoon, et alIgneous Complexes in the Eastern Northern South Yellow Sea Basin and their Implications for Hydrocarbon Systems. Marine and Petroleum Geology 2006236631645

CHAPTER 12

Stereoscopic Estimation of Volcanic Ash Cloud-Top Height from Two Geostationary Satellites

Luca Merucci [1,*], Klemen Zakšek [2,3], Elisa Carboni [4] and Stefano Corradini [1]

[1]*Istituto Nazionale di Geofisica e Vulcanologia (INGV), CNT, Via di Vigna Murata 605, 00143 Rome, Italy*

[2]*Institute of Geophysics, Center for Earth System Research and Sustainability, University of Hamburg, Bundesstr. 55, D-20146 Hamburg, Germany*

[3]*ZRC SAZU, Institute of Anthropological and Spatial Studies, Novi trg 2, 1000 Ljubljana, Slovenia*

[4]*COMET, Atmospheric Oceanic and Planetary Physics, University of Oxford, Clarendon Laboratory, Parks Road, Oxford OX1 3PU, UK*

ABSTRACT

The characterization of volcanic ash clouds released into the atmosphere during explosive eruptions includes cloud height as a fundamental physical parameter. A novel application is proposed of a method based on parallax data acquired from two geostationary instruments for estimating ash cloud-top height (ACTH). An improved version of the method with a detailed discussion of height retrieval accuracy was applied to estimate ACTH from two datasets acquired by two satellites in favorable positions to fully exploit the parallax effect. A combination of

MSG SEVIRI (HRV band; 1000 m nadir spatial resolution, 5 min temporal resolution) and Meteosat-7 MVIRI (VIS band, 2500 m nadir spatial resolution, 30 min temporal resolution) was implemented. Since MVIRI does not acquire data at exactly the same time as SEVIRI, a correction procedure enables compensation for wind advection in the atmosphere. The method was applied to the Mt. Etna, Sicily, Italy, eruption of 23 November 2013. The height of the volcanic cloud was tracked with a top height of ~8.5 km. The ash cloud estimate was applied to the visible channels to show the potential accuracy that will soon be achievable also in the infrared range using the next generation of multispectral imagers. The new constellation of geostationary meteorological satellites will enable full exploitation of this technique for continuous global ACTH monitoring.

Keywords: volcanic ash; cloud top height; parallax; satellite; stereo view

1. INTRODUCTION

An estimation of ash cloud-top height (ACTH) affects estimation of cloud location, the resulting height of the most dangerous zone for air traffic, the forecast accuracy of transport and deposition models [1], the quantitative retrieval of the ash cloud burden [2,3,4], the estimation of mass eruption rate [5,6], *etc*. Many techniques have been proposed to measure the volcanic ACTH from ground-based instruments, including those based on ground weather radar [7,8,9] and volcano surveillance cameras [10]. These methods typically offer very accurate estimates of the height of the eruption column, but are only applicable to a few better-monitored volcanoes. Continuous ACTH measurements of volcanic ash layers can also be obtained from ground-based LiDARs, either near the volcano source where these instruments are deployed [11], or on the nodes of ceilometer and LiDAR networks [12,13,14,15].

Satellite data offers the best way to extend the ability to estimate ACTH when volcanic eruptions occur, providing the continuous global coverage required for effective monitoring [16]. Table 1 briefly reviews quantitative satellite remote sensing techniques for ash/meteorological cloud height retrieval. The accuracy of the listed methods depends on the spatial resolution of the sensors and the cloud height [17]. The best estimates using LiDAR from a space-based platform offer an accuracy within less than 200 m, and can provide either daytime and night-time measurements of a volcanic cloud height and thickness with a revisit time of 16 days [18]. When available, ash cloud height measured by LiDARs or LiDAR networks are very valuable for comparison validation with other observations in integrated and synergistic approaches [6,19,20]. Operational height estimates based on CO_2 absorption [21] are available several times a day but with significantly lower accuracy [22]: bias = −1.4 km; std = 2.9 km. A recent method based on the parallax between volcanic ash cloud images recorded by the satellite multispectral sensors MODIS (Moderate Resolution Imaging Spectroradiometer) and SEVIRI (Spinning Enhanced Visible and InfraRed Imager) achieves a higher accuracy of ~600 m. It was developed and successfully applied to the large 2010 Eyjafjallajökull eruption by Zakšek *et al.* [23]. This method exploits the different points of view of the SEVIRI instrument on board the MSG geostationary platform, and the MODIS instrument on board the polar-orbiting AQUA and TERRA platforms, for the observation of volcanic ash clouds produced during eruptions. The use of these two sensors makes the most of the better spatial resolution of MODIS, but is conversely limited by only a few MODIS acquisition per day. Moreover, the position of MODIS changes continuously, reducing the number of MODIS and SEVIRI image couples of the same eruption that can be effectively used to estimate the height of a volcanic cloud.

Table 1. Overview of satellite methods for cloud top height retrieval.

Methodology	Pros/Cons
LiDAR and radar [6,19,24]	+ very high vertical resolution and accuracy – excessive revisit time (16 days) and only nadir observations from currently-operational instruments (LiDAR CALIOP, radar CPR)
Radio occultation [25,26]	+ high resolution in lower troposphere – globally available only about 2000 times per day
Backward trajectory modeling [27,28]	+ estimate possible even for clouds drifted away from the source – requires wind field data for a large area and a reliable trajectory model (e.g., turbulence not easy to handle); homogenous wind field results with high uncertainty of the source height
Brightness temperature [3,17,27]	+ easy to apply, possible with instruments with a short revisit time – requires atmospheric profile and emissivity of the cloud; assumption of thermal equilibrium; problems around tropopause
O_2 A-band absorption [29]	+ high accuracy – requires high spectral resolution data (not available on many satellites, long revisit time); good performance only over dark surfaces; requires radiative transfer modeling; daytime only
CO_2 absorption [21,30,31]	+ good performance also with semi-transparent clouds – accurate only in the high levels of the troposphere; problems around tropopause
Shadow length [3,32]	+ easy to apply; requires no additional data – possible only during daytime; retrieves the height of the cloud horizontal edge and not its top
Stereoscopy [17,23,33–37]	+ high accuracy; requires no additional data; based on geometry → no problems in the case of ash reaching the stratosphere – requires simultaneous data from two different viewpoints
Optimal estimation [38–40]	+ includes error estimate – requires atmospheric profiles, ash optical properties, and radiative transfer

The use of two sensors, both onboard geostationary platforms, would allow maximization of the number of pairs of images available for ACTH estimation, making it possible to follow the evolution of a volcanic ash cloud height. The current EUMETSAT policy for the use of geostationary weather satellites is to position the newest satellite in the best nominal position (0° longitude) with an Earth full-disk acquisition every 15 min, moving the previous satellite in a back-up position (9.5° E longitude) in rapid scan mode service (RSS) with acquisition focused on much of the northern hemisphere every 5 min.

Attempts to use pairs of images acquired by these two SEVIRI sensors to estimate the height of volcanic clouds have been unsuccessful because the baseline, *i.e.*, the difference in angle of view between the two sensors, is too small and, consequently, the error in the estimated height is too large at the current spatial resolution of SEVIRI (1000 m in nadir for the HRV channel). However, one of the available satellites previously launched was subsequently moved by EUMETSAT to the position 57.5° E degrees over the Indian Ocean, supporting the Indian Ocean Data Coverage (IODC) service. The difference in angle of view of the IODC satellite relative to both the main satellite (0° longitude) and back-up satellite (9.5° E longitude) allows better exploitation of the parallax method for ACTH estimation.

Currently, the satellite located on the Indian Ocean is Meteosat First Generation (MFG) carrying aboard the Meteosat Visible and InfraRed Imager (MVIRI), which is the predecessor of SEVIRI. Although the characteristics of SEVIRI are superior to those of MVIRI, the data acquired by this pair of sensors can be exploited to illustrate the potential of the parallax method when applied to two sensors on geostationary platforms placed in optimal geometric positions [41]. The short term development plans of all the major space agencies include an increased number of meteorological geostationary platforms with multispectral sensors capable of capturing images with improved spatial and temporal resolution, such the Advanced Himawari Imager recently launched on platform HIMAWARI-8, the Flexible Combine Imager (FCI) on Meteosat Third Generation (MTG-I) platform in the Sentinel-4 mission, and the Advanced Baseline Imager (ABI) on the Geostationary Operational Environmental Satellite-R Series (GOES-R). The method developed and described in this article will then be able to provide even more accurate ACTH estimates.

The next sections will describe the sensors and data used for the Mt. Etna (Sicily, Italy) eruption that occurred on 23 November 2013. This eruption was a short-lived lava fountain episode, and was chosen because it was short and difficult to characterize, therefore suitable to reveal any problems of the proposed approach. Moreover, clear sequences of images were captured by both SEVIRI and MVIRI (Section 2). The application and description of the parallax method (Section 3) refers to this selected Mt. Etna case study. The results are discussed in Section 4.

2. DATASETS

Several details are crucial to correctly synchronize the stereo couples from two different sensors and improve the accuracy of ACTH estimation. The importance of precise geographic positioning and exact time determination for each pixel will be discussed in Section 3. However, it is first necessary to establish the exact position of the satellite itself. Although the satellites in question are called "geostationary", in reality the projection of their orbit onto the Earth's surface is not a single point but instead could be represented approximately as a number "8". The position of each satellite can vary by several degrees from its nominal position; thus, it is necessary to determine the position as a function of time. Although such an orbit is not easy to parameterize, it can be closely approximated in a short period of time (about one hour) using a third-degree polynomial from the metadata provided in the EUMETSAT data product (level 1.5 in HDF format was applied for both products).

In order to exploit the spatial resolution of the two instruments to the maximum, the datasets used where the broad visible channels of both MVIRI (VIS 0.50–0.90 μm) and SEVIRI (HRV 0.6–0.9 μm). This choice offered the best possible ACTH estimation accuracy.

2.1. MVIRI

On the 5th of December 2006 EUMETSAT relocated the Meteosat-7 satellite over longitude 57.5° E for the IODC service. Meteosat-7 is the last operational unit of the Meteosat First Generation (MFG) program carrying aboard the MVIRI sensor. The satellite is spin-stabilized and the MVIRI data are acquired at the rate of one image line per satellite rotation. Thus, the Earth disk is scanned horizontally from east to west at each rotation and when the line is completed the next is collected with a stepping tilt mirror operated from south to north. At 100 rpm a whole image of 2500 scan lines is acquired every 25 min over IR (thermal infrared wavelengths in the range 10.5 to 12.5 μm) and WV (radiation from the water vapour absorption band in the range 5.70 to 7.10 μm) spectral bands. At the same time a third composite image is acquired in the visible to near infrared range (0.5 to 0.9 μm, VIS) by two detectors with a twice finer resolution than the IR and WV detectors. A 5 min retrace and stand-by period prepares the sensor for the next image acquisition and completes the 30 min repetition cycle of MVIRI, which is 48 data acquisitions per day. The full disk complete images are 2500 × 2500 8-bit pixels for the IR and WV bands, doubled resolution (5000 lines of 5000 pixels) for the VIS band, with a ground resolution at the sub-satellite point of 5 km for the IR and WV images, and 2.5 km for the VIS image.

2.2. SEVIRI

The primary instrument of the Meteosat Second Generation (MSG) is the SEVIRI radiometer characterized by 12 spectral channels from visible to thermal infrared and a repetition cycle of full disk acquisition of 15 min for a nominal satellite located at 0° longitude. The backup satellite located at 9.5° E is operated in rapid scan service mode (RSS) and provides one image of the northern part of the hemisphere every 5 min. The high-resolution visible channel (HRV) has a ground resolution of 1 km at nadir, while the other 11 channels have a resolution of 3 km.

The acquisition of SEVIRI is similar to MVIRI in that it exploits a 100 rpm counter-clockwise spin used to stabilize the satellite platform and to align its longitudinal rotation axis with the Earth's rotational axis. However, MVIRI has a single detector per channel (two for the VIS channel) while SEVIRI has three detectors per channel scanning three lines simultaneously (nine for HRV), which results in larger images: 3712 × 3712 pixels (11,136 × 11,136 for HRV channel). Every repeat cycle of 15 min is required for a complete full disk image acquisition and transmission is composed of eight data image segments (24 segments for the HRV channel). Each segment has 464 lines of image data, with acquisition starting at the lower right pixel and ending at the top left pixel.

3. THE PARALLAX METHOD

The proposed ACTH estimate is based on the apparent ash cloud displacement due to parallax, using data collected at nearly the same time by two multispectral radiometers on board different geostationary satellites. The parallax can be measured as the distance between the apparent positions of the same cloud observed from two different observation points, and it depends on the height of the cloud (see Figure 1). Since the known position of the satellites and the georeferenced images define the viewing geometry, the ACTH can easily be estimated in three steps. The MVIRI data is first projected onto the SEVIRI spatial grid.

Then the images are co-registered in an automatic matching procedure. Finally, the lines of sight connecting each observed point from the two satellites are generated and their intersections are used to estimate the ACTH.

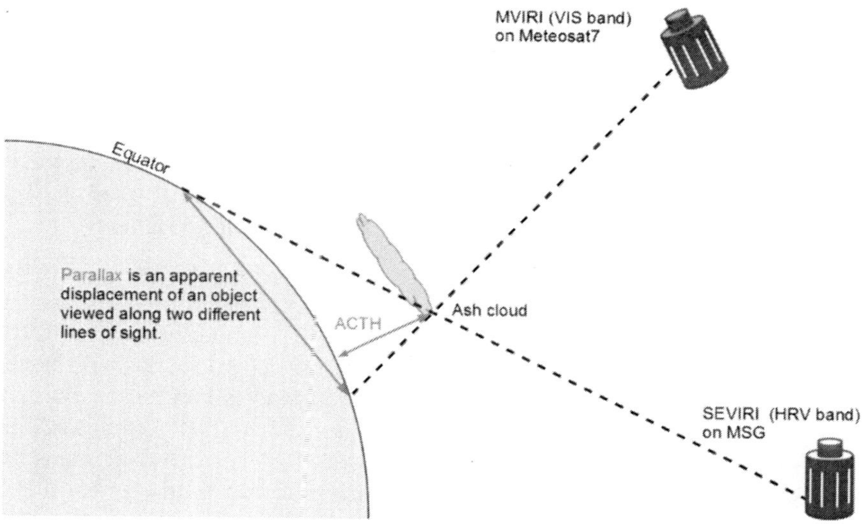

Figure 1. The scheme of observations as seen by an observer above the North pole; the lines of sight from two satellites cross at one point. Its coordinates are easy to determine within the Cartesian geocentric coordinates system. Conversion to geodetic coordinates gives us ACTH.

3.1. Projection of MVIRI Data onto the SEVIRI Grid

To detect parallax, the images retrieved from different satellites first have to be referenced within the same image coordinate system. This means that all data must have the same georeference, *i.e.*, the same map projection, geodetic datum, and raster resolution. Once this is achieved, it can be observed that all objects with elevation zero (coastlines) are aligned in all images. If an object is elevated, then its position in the image coordinate system depends on its height and on the position of the satellite. All satellite data are projected in a general near-side perspective projection, which is not a standard map projection. Projections of geostationary satellites differ basically only in sub-satellite longitude. The image coordinate system also depends on the size of the image and the spatial resolution of the instrument. These data are retrieved from EUMETSAT metadata delivered as level 1.5 data. Using bilinear interpolation, the MVIRI data is reprojected from its native projection to the SEVIRI projection grid. The SEVIRI grid was chosen for reference because it is centred over the area of interest (Europe) and the SEVIRI repetition cycle is used to estimate the influence of wind.

3.2. Automatic Image Matching

The image matching identifies point pairs between two satellite images. If the images are not acquired by the same instrument, as in this case, results might be of poor quality because of differences in spatial resolution, viewing geometry, instrument response functions, and solar illumination conditions. The results of image matching are, however, usually good enough to estimate ACTH. The method used is that described by e.g., Scambos *et al.* [42], and Zakšek *et al.* [23].

Image matching estimates image shifts between pairs of images. However, in the proposed procedure, three images are always used to account for the influence of wind: one MVIRI image, and two SEVIRI images acquired just before and after the MVIRI image was retrieved. Starting from MVIRI coordinates, image matching is conducted twice to find matching points between the MVIRI and both SEVIRI images. Since the coordinates of the matched points were not retrieved at the same time as the MVIRI was collected, the position of the cloud was estimated as it would have appeared in a SEVIRI image collected at exactly the same time as the MVIRI image, by linear interpolation. Given the position of the cloud and both the satellites, the MVIRI and SEVIRI lines of sight can be generated. From the intersection of these it is easy to compute the height.

3.3. ACTH Estimation from the Intersection of Lines of Sight

After the image matching procedure, triples of geographic coordinate pairs were also determined for each pixel. The effects of possible advection of the eruption cloud between the MVIRI and the two SEVIRI data collection times is considered for each pixel triple: the coordinates of a virtual SEVIRI pixel are interpolated from the position of both the actual SEVIRI pixels to the time of the MVIRI acquisition. The conversion from geographic coordinates into geocentric Cartesian coordinates allows the final calculation based on vector algebra. The lines connecting the virtual SEVIRI pixels with the MSG satellite position (SEVIRI lines) and the corresponding lines connecting MVIRI pixels and the Meteosat-7 satellite position can be expressed as parametric equations in the 3D space. The intersection of these line pairs is ideally given by the solution of a linear system.

In practice, the linear system is over determined, and can only be solved by a least-square technique: due to the discrete nature of the datasets the MVIRI and SEVIRI lines never intersect and the solution is not a single point real intersection, but instead is given by the two points at a minimum distance between the lines. The position of one of these two points (or their average) represents the ACTH estimation result. High ACTH estimation accuracy is achieved with this procedure when the distance between the MVIRI and SEVIRI lines is small. For further details see [23].

3.4. ACTH Estimation Error

A description is now provided of the analytical solution of ACTH estimation accuracy from observations collected by two geostationary satellites.

Considering the spatial resolution that meteorological satellites offer today and the height of volcanic clouds, the surface of the Earth is assumed to be flat and not curved.

Given the zenith angle of a satellite (see Figure 2), the relationship between parallax and *ACTH* can be given as:

$$\tan z = \frac{p}{ACTH}$$

Parallax *p* can be observed only if a point is observed from at least two different positions. The parallax can, thus, be defined as the sum of two parallaxes projected to the same vertical plane:

$$p = p_S + p'_M = ACTH \cdot (\tan z_S - \cos \Delta a \cdot \tan z_M)$$

In the first part of the equation above, the sign between both terms is plus, and in the second part it is minus. This is a consequence of the projection of the second parallax to the vertical plane defined by the first parallax. The projection is achieved using the cosine of the difference Δa between azimuths pointing in the directions of both parallaxes. This equation can thus also be used to estimate *ACTH*:

$$ACTH = \frac{p}{\tan z_S - \cos \Delta a \cdot \tan z_M} \tag{1}$$

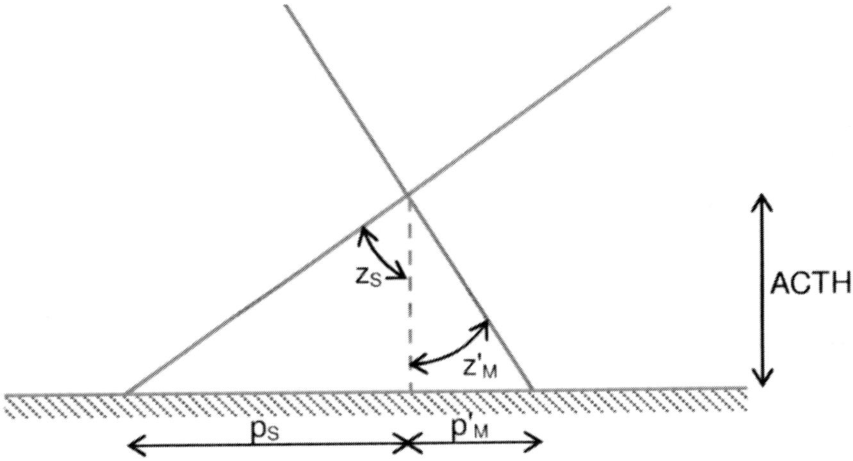

Figure 2. The scheme of observations of a cloud from two directions (subscripts at the parallax p and zenith angle z correspond to S for the blue line directed to SEVIRI and M for the red line directed to MVIRI). The cloud is positioned at the intersection of the coloured lines. The scheme presents the plane defined by the normal to the WGS84 ellipsoid and the direction of the longest parallax (here the blue line). The other line is projected (the apostrophe on the symbol denotes that the parameter has been projected on the same plane by multiplying the parallax by the cosine of the angle between the red and blue lines).

To estimate the analytical accuracy of *ACTH*, it can be assumed that the automatic image matching provides results with errors of no more than a half a pixel. Therefore, for each pixel in the SEVIRI grid it can be estimated under which zenith and azimuth angles the pixel was observed from the SEVIRI and MVIRI positions. Zenith angle is computed from the scalar product between the vector normal to the surface and the vector to the satellite (both unit vectors). Azimuth estimation is even easier. Since the nominal latitude of geostationary satellites is zero, the azimuth a toward the satellites can be computed from:

$$\tan a = \frac{\sin(\lambda - \lambda_0)}{(1 - e^2) \cdot \tan \varphi} \tag{2}$$

where λ is the pixel longitude, λ_0 is the longitude of the geostationary satellite, e is the first eccentricity of the WGS84 ellipsoid, and φ the pixel latitude. Once the azimuth and zenith angles to both satellites are known for each pixel seen from both satellites, the *ACTH*can be estimated for a parallax that measures one single pixel. This value provides the following information: (a) an estimate of the minimum *ACTH* that can be reliably estimated; and (b) half this value equals the accuracy of the *ACTH* estimate, if the image matching provides results with a maximum error of half a pixel.

To evaluate the error on the ACTH estimate it is first necessary to compute the azimuth of the parallax, because it influences the parallax value, the parallax is greater if it is diagonal rather than columnar or linear in direction.

The triangle CSM in Figure 3 is defined by the projection of the cloud true position (point C) to the horizontal plane at sea level, the end point of the parallax from SEVIRI (point S), and the end point of the parallax from MVIRI (point M). Point P is defined by the intersection of a line running North and the parallax (p; line SM). The azimuth of the parallax a can be estimated, once all the angles in the triangle CSP or CPM are known. To estimate angle γ in the triangle CSP, first the parallax p needs to be estimated using the law of cosines (angle MCS is known from the difference of azimuths a_S and a_M, side MC corresponds to tan z_M, and side CS to tan z_S.

$$p^2 = \tan^2 z_S + \tan^2 z_M + 2 \cdot \tan z_S \cdot \tan z_M \cdot \cos \Delta a \quad (3)$$

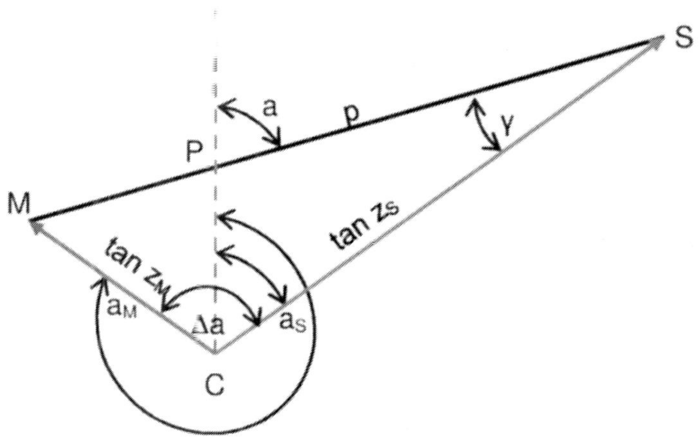

Figure 3. The scheme of the lines of sight (lines connecting points MC and CS) projected to the horizontal plane from two satellites (in colour) forming the corresponding parallax p (MS); both lines of sight are presented as vectors that start in the nadir point of the cloud C and end at the position observed by the satellites (M and S, for MVIRI and SEVIRI, respectively). The zenith and azimuth angles to the satellites are known for both points and the azimuth a of the parallax needs to be estimated.

The law of sines is then used to compute γ:

$$\sin\gamma = \frac{\sin\Delta a}{p}\cdot\tan z_M \tag{4}$$

Finally, azimuth a can be estimated showing the direction of the observed parallax:

$$a = a_S + \gamma \tag{5}$$

After the parallax azimuth has been estimated, it is possible to determine which neighbouring pixel is the closest in this direction. Given the distance between neighbouring pixels in this direction, Equation (1) can be used to estimate the minimal detectable ACTH. Considering the error of one single pixel, the number of possible azimuths of the parallax can be reduced to eight directions (eight neighbor pixels). Thus, one can expect discontinuities in those areas, where the azimuth of the parallax changes from one to another of the eight values. The results for the northern hemisphere of the combination of SEVIRI and MVIRI satellites are shown in Figure 4. The results are given in the grid of the SEVIRI HRV channel situated on 9.5° E. The area presented shows only the pixels for which both instruments have a zenith angle of less than 80°.

However, it is worth noting that this estimation assumed that the spatial resolution of both instruments is almost the same. Over Europe the MVIRI instrument has a significantly lower spatial resolution than SEVIRI. This has a large influence on the automatic image matching, making it difficult to properly match all the SEVIRI and MVIRI pixels.

In addition, the halves of values in Figure 1 can be interpreted as an estimate of accuracy, considering that the image matching provides an accuracy of half a pixel. Again, this is not a totally realistic estimate, since the spatial resolution of MVIRI over Europe is ~5000 m. It is therefore necessary to check all the internal measurements of accuracy first: the suggested values are realistic if correlation between SEVIRI and MVIRI is high (>0.9) for all image pyramids (see [23,43] for details) and if the estimated distance between both lines of sight is not greater than half a SEVIRI pixel.

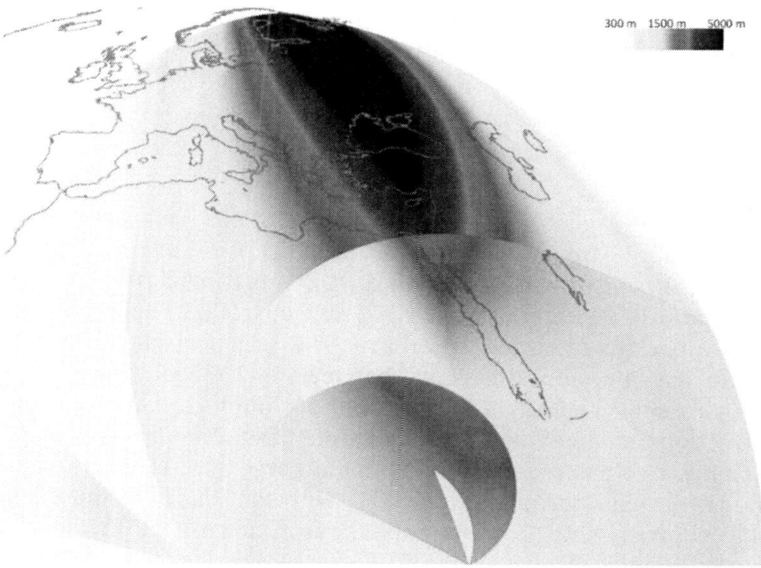

Figure 4. The minimum detectable cloud height considering that a parallax of exactly one pixel is observed. Note the discontinuities—the data depends on the direction of parallax—if it indicates a diagonal neighboring pixel, the results will be greater than when parallax indicates a neighboring pixel along the same line or column.

Note that in the case of a higher cloud, the parallax would be larger than a single pixel (which is assumed in Figure 4), with the result that the discontinues in Figure 4 would be softened in the accuracy image, because the parallax would not be limited to only eight different azimuths.

4. RESULTS

4.1. Etna 23.11.2013 Case Study

On 23 November 2013, starting at 8:00 UTC, the Strombolian activity [44,45] of the new South East Crater (NSEC) of Mt. Etna, which had started the previous day, intensified and around 9:30 UTC evolved into a lava fountain [46,47]. The emission of ash and the eruptive column height increased quickly with lava emissions reaching maximum height around 10:10 UTC. At 10:15 UTC the height of the jets dropped dramatically and at 10:20 UTC activity returned to Strombolian explosions visible until 10:55 UTC. The good weather conditions allowed acquisition of the entire eruptive event by the SEVIRI and MVIRI sensors, and these image sequences form the basis for this study.

4.2. ACTH Results

The images used to estimate the ACTH of the volcanic cloud emitted during the 23 November 2013 Mt Etna eruption are the VIS channel of the MVIRI sensor aboard the MFG (IODC) platform at longitude 57.5° E, and the HRV channel of the SEVIRI sensor in RSS mode platform on the MSG-2 backup satellite located at 9.5° E (see Section 2). The RSS mode offers the minimum difference between the image acquisition times of the two sensors (max 2.5 min), and so minimizes errors due to wind advection by applying the correction described in Section 3.2. Figure 5 shows two sample images extracted from the sequences captured by the two sensors around 11:49 UTC, with the volcanic clouds marked in blue ellipses. Note that the spatial resolution of MVIRI is significantly lower than the resolution of SEVIRI for the Mediterranean area. This is a consequence firstly of lower resolution of MVIRI at nadir and secondly of the viewing geometry—MVIRI observes Europe quite close to the hemisphere's horizon (seen in the upper left corner of Figure 5b). Thus, the island of Sicily looks significantly smaller in the MVIRI data (spatial resolution ~5 km for Sicily) than in SEVIRI data (spatial resolution ~1 km for Sicily). Consequently, it is more difficult to see the position of the ash cloud in a single MVIRI image, while it is much easier to observe it in the SEVIRI data of Figure 5a. The volcanic cloud can easily be observed with both instruments in the animation of the complete sequences of images of the eruption provided in the supplementary data.

Once a volcanic cloud has been correctly identified, it is very important to check the co-registration of the image pairs, since the final ACTH accuracy is highly dependent on the quality of the image matching (see Section 3). A useful visual tool for a quick quality control of the results is shown in Figure 6. Here, the MVIRI image and the two SEVIRI images acquired immediately before and after the MVIRI acquisition at 11:49 UTC have been superimposed into an RGB image, with first SEVIRI as the red component, second SEVIRI as the green component, and MVIRI as the blue component. To make the image more appealing, the colours were inverted—in the original images the clouds appear bright while the sea is dark. The result is a color composite image that highlights the parallax between SEVIRI and MVIRI in yellow. The thicker the yellow belt northeast of the cloud, the greater the parallax, and the greater the cloud's height. The turquoise colour indicates areas influenced by cloud displacement during the 5 min that elapsed between the two subsequent

SEVIRI images: the thicker the turquoise belt, the faster the movement of the clouds. Everything appearing white or grey has a low elevation.

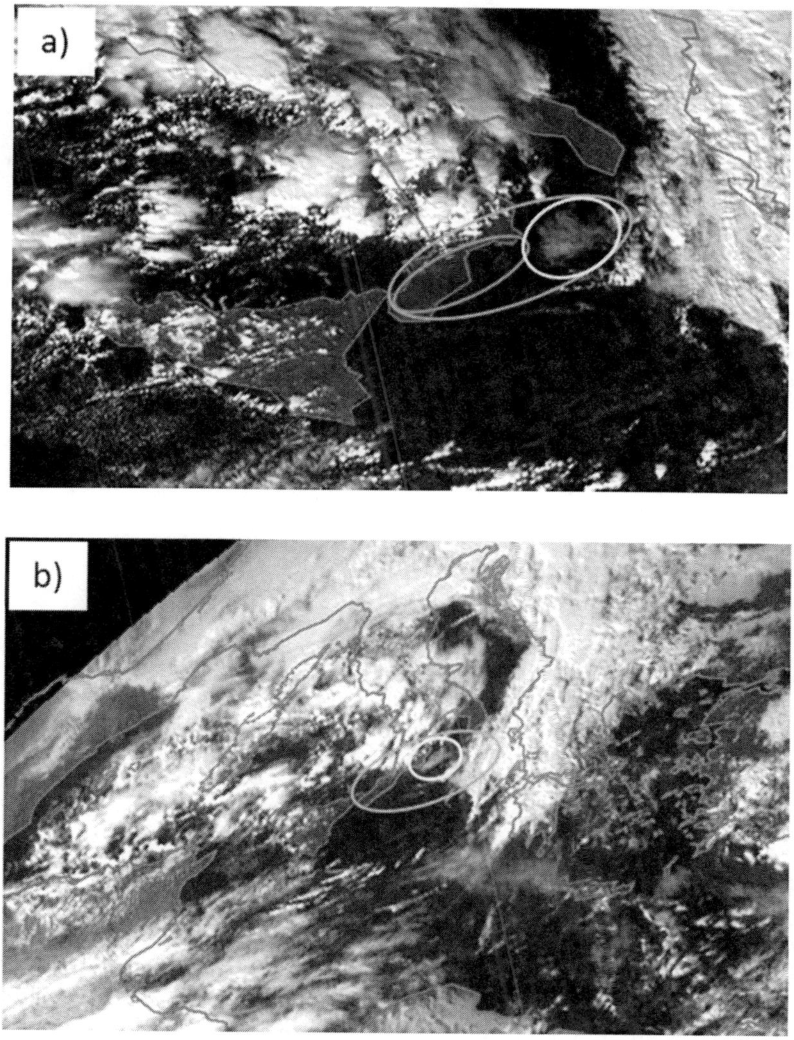

Figure 5. The Etna 23.11.2013 eruption viewed in the images collected around 11:49 UTC from (**a**) SEVIRI (9.5° E); and (**b**) MVIRI (57.5° E). The cloud of volcanic origin is marked with a blue ellipse. The cloud for which ACTH is estimated is marked in yellow. The ash cloud marked in green is not visible in the MVIRI image. The complete sequence of SEVIRI and MVIRI images is provided in the supplementary material.

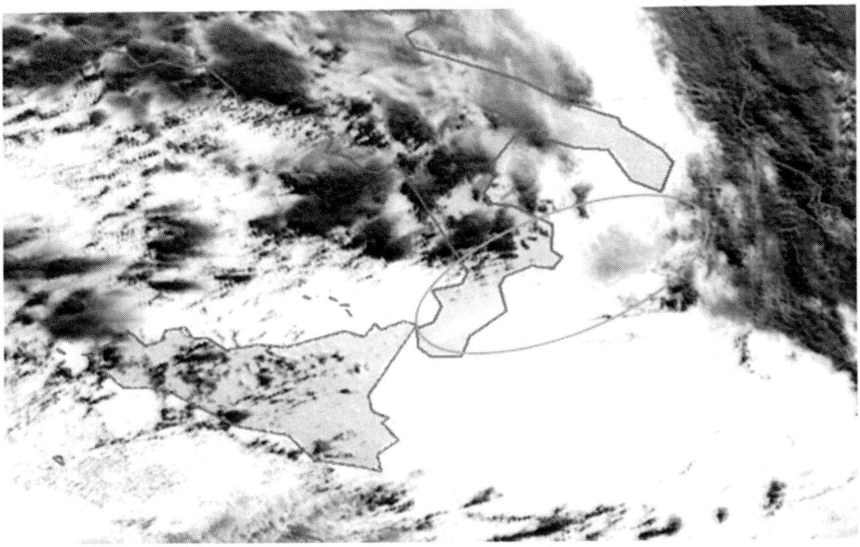

Figure 6. Color composite image representing the comparison of the situation as seen by the two sensors at about 11:49 UTC. This figure visually highlights the parallax between MVIRI and SEVIRI in yellow, and the effects of wind between the two SEVIRI images in turquoise. The volcanic cloud is marked by a blue ellipse.

In this representation, if the images are not well co-registered these colour belts are immediately observed along the coastlines.

As already stated in Section 2, the choice to use visible channels enables estimation of the ACTH with the maximum accuracy achievable using the current operational geostationary instruments and, thus, better demonstrating the potential of the proposed height estimation scheme. Another reason is that ash cloud detection algorithms are based on channels in the thermal infrared range. The simplest effective ash cloud detection is obtained by evaluating the brightness temperature difference (BTD) defined for two channels centred at around 11 and 12 μm [48,49]. Since MVIRI has only one channel in the TIR range, the BTD cannot be evaluated, and meteorological and volcanic ash clouds cannot be discriminated based on MVIRI data. In the HRV SEVIRI image of Figure 5, a dark plume can be seen over the southern tip of the Calabria peninsula (inside the green ellipse) extending into a brighter cloud (marked by the yellow ellipse) over the western part of the Ionian Sea. The dark part of the plume is recognized as ash using BTD and the bright part as a meteorological cloud. Given that both are of volcanic origin, the focus here is only on the bright part of the cloud (which is probably composed of a mixture of ash and ice crystals or water droplets [50]), because the dark part of the plume is too transparent to be clearly detected in the visible bands. This brighter, highly-reflective cloud can be observed with both instruments and is used here to apply the ACTH estimation method.

Figure 7 shows the final results for the same scene presented in Figure 5. Figure 7a shows the correlation between the MVIRI and SEVIRI data prior to the MVIRI observation. It is low for most pixels because of the low level of detail in MVIRI: the spatial resolution of MVIRI is very coarse compared to SEVIRI. Over the bright volcanic cloud, the correlation is only high (>0.95) on the northern and southern edge of the cloud (in white), otherwise it is below 0.8. Figure 7b presents the computed parallax: the bright areas in the blue ellipse represent 30 km, and orange to yellow areas about 20 km parallax. It appears quite "blocky" because there was good correlation over a coarser pyramid level, but the finer pyramids are not highly enough correlated and so the estimated shift in column and line direction is preserved from the lower pyramid.

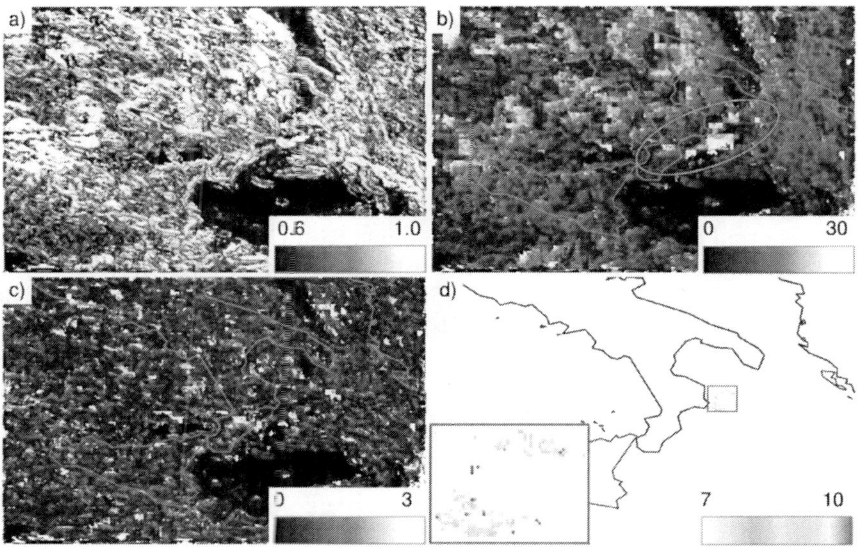

Figure 7. (**a**) Correlation; (**b**) parallax shift (in km); (**c**) distance between intersecting lines (km); and (**d**) final height based on the following pixel-based selection criteria: pixels belonging to the ash cloud, small distances between the intersecting lines, AHCT > 3 km (Etna is 3300 m). The inset in the lower left corner is a zoomed region of the volcanic cloud area.

The distance between the intersecting lines is related to the accuracy of the ACTH estimate (Figure 7c). The lines of sight never intersect because the data used have finite resolution. Thus, the closest points on both lines are determined and the distance between these points is a measure of the uncertainty of the results, which should be as small as possible. Here, only the darkest pixels are used for height estimation since they have distances between the points of less than 600 m (half a SEVIRI HRV pixel in the case study area). Several pixels were discarded because they did not correspond to the volcanic cloud, because the distance between the intersecting lines was too large (>600 m), or because the retrieved height was too low (the cloud height was assumed to be > 3 km, given the meteorological conditions and Etna's altitude of 3300 m.a.s.l.). Moreover, the final ACTH was estimated only for the pixels selected as definitely belonging to the volcanic cloud by visual inspection of the images. The pixels in red in Figure 7d are higher than 10 km, reaching almost 11 km, the purple are lower than 7 km, and the majority of them have values of ~8–9 km (in turquoise and green and yellow).

Following the same procedure described for the two sample images of Figure 5, the ACTH distribution was analyzed for six image pairs between 10:49 and 13:19 UTC. The data collected before this time also contain a clear volcanic plume, but the image matching procedure did not provide reliable results because of the marked heterogeneity of the surface beneath the cloud, which was located over or near land. The results are summarized in the histograms shown in Figure 8, representing the evolution of the ACTH estimated with the parallax method. Not even the initial situation at 10:49 is optimal, but it can still be seen that 50% of the volcanic cloud has its ACTH in the range between 8 and 8.5 km. In the following situation it can be observed that 50% of the volcanic cloud has its ACTH in the range between 8 and 9.5 km. The highest pixels are as high as 11 km. The distribution appears to shift towards lower ACTHs with time.

4.3. Validation

The results obtained for the eruption are also in close agreement with those independently estimated with integrated measurements provided by the Mt. Etna video surveillance camera, by the radar of the Fontanarossa airport of Catania (Sicily), by applying a simpler scheme for parallax height estimation obtained from a comparison of MODIS and SEVIRI satellite data, with the IASI retrieval based on optimal estimation, and with the results obtained from the HYSPLIT model. All these estimates of the ash cloud height released during the Mt. Etna eruption event on 23 November 2013 are in close agreement with the results obtained herein and are thoroughly presented and discussed in a separate article by Corradini *et al.* [50] in this special issue.

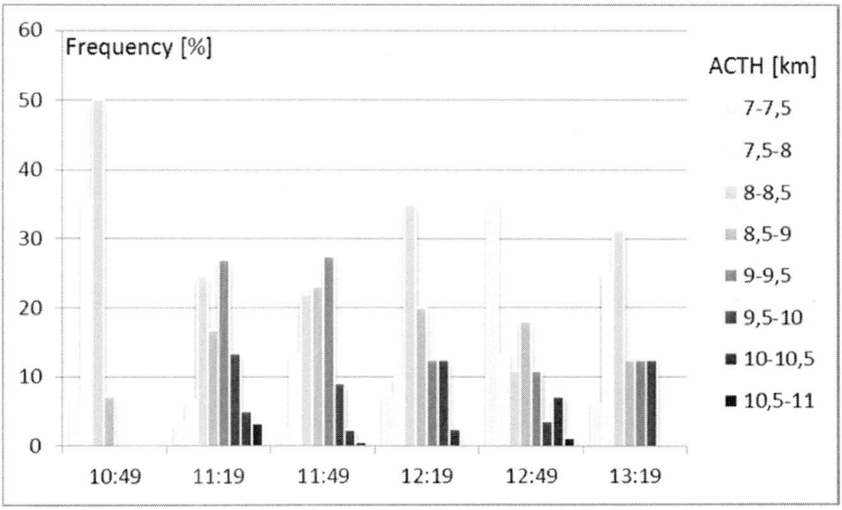

Figure 8. Evolution of ACTH estimated with the parallax method during six MVIRI image acquisitions. The x-axis is UTC time and the y-axis is relative frequency of cloud pixels in ACTH classes binned for each 500 m.

The results of ACTH can be compared with those obtained from radiosonde profiles of temperature and wind speed acquired on 23 November 2013 at 12:00 UTC by the WMO meteorological station of Trapani (Sicily, Italy). From these profiles and the SEVIRI images, two different independent estimates of the cloud height can be easily obtained. Firstly, by estimating the brightness temperature of the cloud and comparing it with the temperature radiosonde vertical profile. This method, also known as darkest pixel method [51], was applied on the six SEVIRI images closest to the MVIRI images from 10:49 to 13:19 UTC with retrieved temperatures between −43.7 and −52.0 °C. The resulting cloud heights range from about less than 9 to 11 km (see Figure 9a). Secondly, by evaluating a wind speed of 38.7 $m \cdot s^{-1}$ from the displacement of similar structures measured on two SEVIRI images divided by the elapsed time between the two acquisitions. This wind speed was then compared to the radiosonde wind speed vertical profile (see Figure 9b). This method is very simple, but prone to error if the reference structures identified in the cloud are not clearly maintained in the two images. In this case it can be assumed as indicative of a good agreement with an acceptable accuracy not better that 2 $m \cdot s^{-1}$.

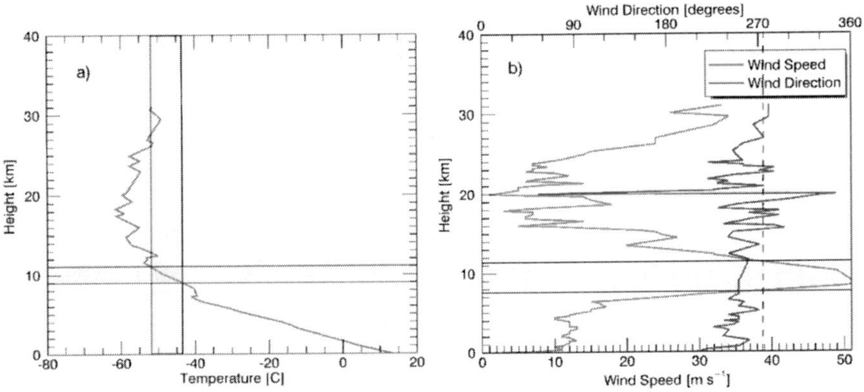

Figure 9. Comparison of the Trapani (Sicily, Italy) WMO station radiosounding vertical profiles of (**a**) the cloud temperature estimated with the darkest pixel method and (**b**) the wind speed and direction collected on 23 November 2013 at 12:00 UTC. The cloud temperature was estimated with the darkest pixel method on six different images, and the wind speed from the cloud displacement measured from two different images.

5. DISCUSSION

There follows a discussion of the most relevant characteristics of the proposed method. Compared to other methods, the one under discussion requires two satellites that can observe the same area nearly simultaneously. Its main drawback is that the accuracy of the results is dependent on the positional accuracy of the input images. It is not particularly important for the absolute geolocation to be accurate, but the co-registration between the images from both satellites must fit perfectly. For this reason it makes sense to check the co-registration accuracy during pre-processing when all datasets are transposed to the same coordinate system. This can be done automatically by adjusting the co-registration along coastlines.

The automatic image matching, which takes the most of the processing time, requires texture or structure in clouds. Without any texture, matching points in different images cannot be found, possibly leaving only some points on the edges of the cloud where ACTH can be estimated. Normally this is not an issue, as volcanoes usually erupt in pulses which are visible in volcanic clouds. Texture might be a problem only if the spatial resolution of the input data is very fine (finer than 50 m) or too coarse (coarser than 5000 m). Considering SEVIRI, volcanic and meteorological clouds usually have enough structure in the HRV channel. Conversely, MVIRI is more problematic. It does provide significant texture close to its nadir, but in the present case study area, the spatial resolution is almost too coarse to provide enough texture. In addition, MVIRI acquires only 8-bit data in the VIS channel, meaning that fine texture details cannot be resolved as well as in SEVIRI (10-bit).

Another significant limitation is cloud transparency. According to Corradini *et al.* [50], the 550 nm AOD of the ash cloud was below 0.5. The ash part of Etna's cloud could not be accurately identified with SEVIRI's HRV channel, nor with MVIRI's VIS channel. However, even if the cloud can be observed with a satellite instrument, it might still be so transparent that the automatic image matching is distracted by the details of the underlying surface visible through the cloud. This is not a problem over homogeneous surfaces like ocean or desert, but it can be over areas with diversity in land cover.

One of the main advantages of cloud photogrammetry is its independence from ancillary datasets or assumptions on the atmosphere state at the moment of the data acquisition. It is not important if the cloud is below the tropopause or above it. Emissivity of the cloud is irrelevant. There is no need for auxiliary data like temperature or humidity profiles.

Moreover, the proposed method is fast. The case study area of 720 by 450 pixels was processed on a notebook with a four-core Intel i7 processor in 30 s. It is also worth noting that the ACTH code was written in IDL, which is a high-level language useful for algorithm development, but probably not the fastest solution for coding real-time applications. In the present example only 10% of the processing time was used for pre-processing the MVIRI data. A similar pre-processing took much longer for the MODIS data [23] because of its finer spatial resolution, but still only about 60 s were required to process a whole 250 m MODIS image.

The most important characteristic of the method is its accuracy, discussed in Section 3. This is a factor of about three times higher than the expected accuracy of operational height estimates based on CO_2 absorption [22], and less than a factor of four times lower than that achievable with space-based LiDARs [18]. For the case study area the analytical estimate of the ACTH accuracy is 700 m.

As mentioned in Section 2, in this study only the visible channels of both instruments were used. This choice allowed exploitation of the much better spatial resolution of these channels compared to those in the thermal infrared range (at nadir, 2.5 and 1 km in the visible, *vs.* 5 and 3 km in the TIR for MVIRI and SEVIRI respectively), thus obtaining the most accurate ACTH estimation results achievable with these two geostationary instruments. Conversely, this choice limits the use of the method to daylight hours. Another drawback is the difficulty of obtaining a reliable detection of a volcanic ash cloud from a single image in the visible range. Fundamental for this purpose are both the fixed observation point and the high repetition cycle of geostationary sensors, which allow acquisition of a complete sequence of images of an eruption. It is, thus, possible to follow the evolution of a cloud following its emission from the volcano vent. Tracking its trajectory in this way provides proof of the volcanic origin of the cloud even at great distances from the source.

Having detected and tracked a volcanic cloud, the next step is to verify its ash composition. To achieve this (although only at reduced resolution), it is possible to use ash inverted absorption in the SEVIRI bands centred at 10.8 and 12 μm. This can be detected using the brightness temperature difference (BTD) [48] which, under certain conditions [49], allows differentiation between meteorological and volcanic clouds. As previously mentioned in Section 4, this verification is not possible with MVIRI since it only has one thermal infrared band centred at 11.5 μm.

It should be noted that the Mt. Etna eruption discussed in this article was chosen to describe the parallax method applied to geostationary satellites using available data because it is a small eruption, difficulty to detect and characterize, and suitable to reveal the problems that this approach can have. The method works well with a huge eruption releasing an extended ash cloud over the ocean, as already demonstrated by Zakšek *et al.* [23] for the 2010 Eyjafjallajökull eruption, although with a different sensor pair on GEO and LEO platforms with higher spatial resolution. Furthermore, the ACTH estimation presented here refers to a cloud for which a visual inspection of both image sequences identifies as being of clear volcanic origin. Conversely, the positive BTD that can be found on the SEVIRI images of the same part of the volcanic cloud used for the ACTH estimate, does not detect it as composed of volcanic ash, but rather of ice or water droplets [50]. Nevertheless, this case study is very useful to demonstrate the potential of the proposed method which, in principle, can be used to estimate the ACTH of any type of cloud using GEO multispectral imagers, regardless of whether they are volcanic ash or SO_2 clouds derived from a detection and retrieval procedure, or meteorological clouds directly identified in the satellite datasets.

Another point is that although volcanic ash detection and retrieval is feasible with several currently operational sensors (*i.e.*, IMAGER on GOES-13 and -15, and MTSAT-2), their spatial resolution is still too coarse to fully exploit the proposed ACTH method. The multi-purpose VIS/IR imagery capability from GEO platforms, as defined by the World Meteorological Organization, is already partly implemented with a

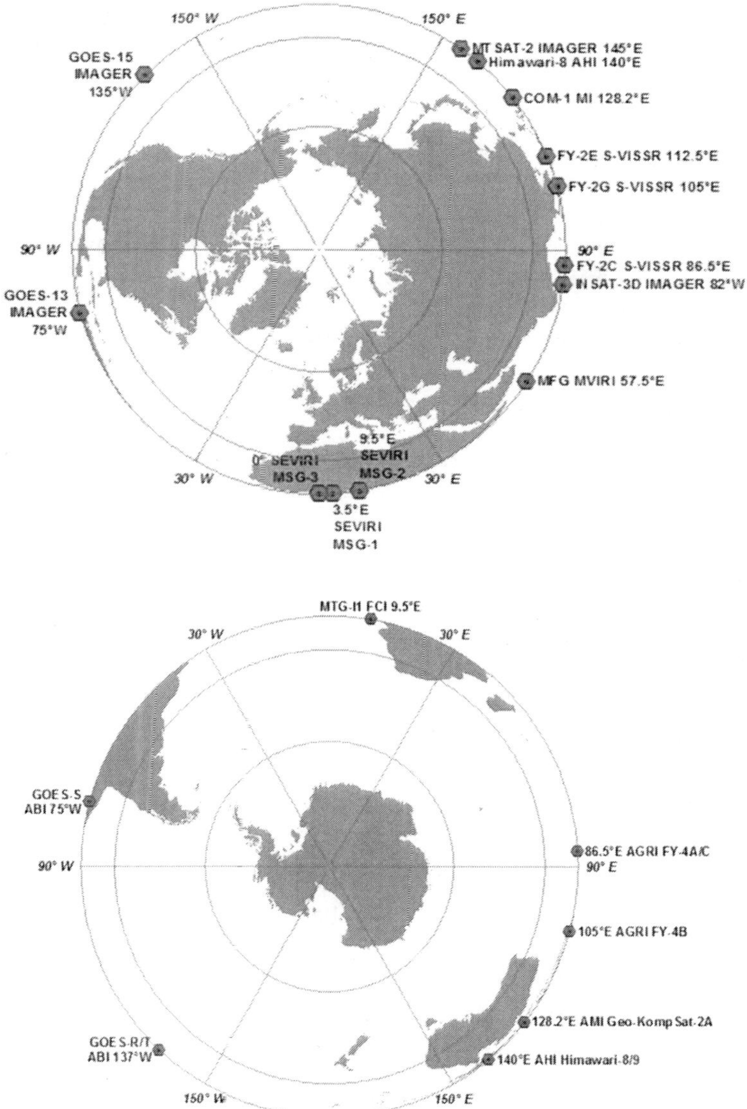

Figure 10. The Northern hemisphere image shows the currently operational geostationary satellites, and the Southern hemisphere the implementation of Earth Observation satellites planned by year 2020 [53].

number of medium-resolution multi-channel radiometers operating in the VIS and IR range in geostationary orbit. The reference observing strategy consists of six evenly spaced 60° wide sectors along the equator with at least one SEVIRI-class instrument per sector and a backup instrument as similar as possible nearby.

However, in the near future the situation will improve significantly. The new generation multispectral sensors—*i.e.*, Flexible Combined Imager on MTG-I, Advanced Baseline Imager, Advanced Himawari Imager, Advanced Geostationary Radiation Imager, Electro-GOMS Imager for Electro M (MSU-GSM)—onboard geostationary platforms will have many more bands and a thermal infrared spatial resolution comparable to that currently achieved by the SEVIRI HRV visible channel [52]. After the year 2020 it is expected that all sectors will be covered with redundancy by third generation imagers, except sector 180° ± 30°, which lacks backup. Nevertheless, even this latter sector will be partially covered by instruments on board the Himawari and GOES-W platforms (see Figure 10).

The GEO VIS/IR imagery will benefit from ever-improving sensor spatial and spectral resolutions, acquisition repetition cycles, and differences in angle of view between the sensors. All these improvements will enable full advantage to be taken of the proposed method both day and night for effective monitoring of volcanic clouds on a global scale.

6. CONCLUSIONS

A volcanic ash cloud-top height estimation method was proposed, based on the apparent shift due to parallax of images collected by the multispectral imagers MVIRI and SEVIRI on board two different geostationary satellites. Previously introduced by Zakšek *et al.*[23] for polar and geostationary image pairs, the method was improved here with detailed discussion of its accuracy, and applied to the case study of the Mt. Etna (Sicily, Italy) eruption that occurred on 23 November 2013. The estimated ACTH for the ash cloud released during this Mt. Etna eruption was presented and discussed. Close agreement was demonstrated with the radiosounding profiles acquired by the Trapani WMO meteorological station. Further sources for the validation of the estimated ACTH can be found in a separate article [50] in this special issue. These include independent ash cloud height assessment based on space-based (SEVIRI, MODIS, IASI) measurements, ground-based (radar and surveillance video camera deployed close to Mt. Etna), and ash cloud trajectories obtained with the HYSPLIT model.

The methodology is straightforward, fast, and can valuably complement other techniques currently used for ACTH estimation. The novel application of ACTH estimates to datasets collected by instruments aboard geostationary platforms opens the way to fully exploit the potential of this methodology. It could also be immediately applied to other combinations of geostationary sensors, for instance the SEVIRI-GOES pair, but is still limited by its dependence on visible data, which provides the finest spatial resolution. However, the forthcoming launch of the next generation of meteorological multispectral imagers into geostationary orbits, offering finer spatial resolutions than current systems including infrared channels with global coverage, will allow application of the proposed method for continuous monitoring of ash clouds. This will not only increase the accuracy of results, but will also increase the applicability and effectiveness of the method. The proposed method will therefore also be applicable to meteorological clouds, other natural aerosol clouds, and even to clouds resulting from industrial accidents.

ACKNOWLEDGMENTS

This research was developed within the framework of the Advanced PRocedures for volcanIc and Seismic Monitoring (APhoRISM) FP7 project and funded under grant agreement n. 606738 by the European Union. It was also supported with a grant from the German Science Foundation (DFG) number ZA659/1-1.

AUTHOR CONTRIBUTIONS

Luca Merucci developed the idea and analysed the case study. Klemen Zakšek wrote the computer code. Elisa Carboni contributed evidence for the possible use of stereo pairs of GEO images for ACTH estimation. Stefano Corradini took part in the discussion of the presented work. All the authors have reviewed the manuscript.

CONFLICTS OF INTEREST

The authors declare no conflict of interest.

REFERENCES

1. Heinold, B.; Tegen, I.; Wolke, R.; Ansmann, A.; Mattis, I.; Minikin, A.; Schumann, U.; Weinzierl, B. Simulations of the 2010 Eyjafjallajökull volcanic ash dispersal over Europe using COSMO-MUSCAT. *Atmos. Environ.* **2012**, *48*, 195–204. [ ] [CrossRef]
2. Wen, S.; Rose, W.I. Retrieval of sizes and total masses of particles in volcanic clouds using AVHRR bands 4 and 5. *J. Geophys. Res. Atmos.* **1994**, *99*, 5421–5431. [ ] [CrossRef]
3. Prata, A.J.; Grant, I.F. Retrieval of microphysical and morphological properties of volcanic ash plumes from satellite data: Application to Mt Ruapehu, New Zealand. *Q. J. Royal Meteorol. Soc.* **2001**, *127*, 2153–2179. [ ] [CrossRef]
4. Corradini, S.; Merucci, L.; Prata, A.J. Retrieval of SO_2 from thermal infrared satellite measurements: Correction procedures for the effects of volcanic ash. *Atmos. Meas. Tech.* **2009**, *2*, 177–191. [ ] [CrossRef]
5. Mastin, L.G.; Guffanti, M.; Servranckx, R.; Webley, P.; Barsotti, S.; Dean, K.; Durant, A.; Ewert, J.W.; Neri, A.; Rose, W.I.; *et al.* A multidisciplinary effort to assign realistic source parameters to models of volcanic ash-cloud transport and dispersion during eruptions. *J. Volcanol. Geotherm. Res.* **2009**, *186*, 10–21. [ ] [CrossRef]
6. Stohl, A.; Prata, A.J.; Eckhardt, S.; Clarisse, L.; Durant, A.; Henne, S.; Kristiansen, N.I.; Minikin, A.; Schumann, U.; Seibert, P.; *et al.* Determination of time- and height-resolved volcanic ash emissions and their use for quantitative ash dispersion modeling: The 2010 Eyjafjallajökull eruption. *Atmos. Chem. Phys.* **2011**, *11*, 4333–4351. [ ] [CrossRef]
7. Rose, W.I.; Kostinski, A.B.; Kelley, L. Real-time C-band radar observations of 1992 eruption clouds from Crater Peak, Mount Spurr volcano, Alaska. *US Geol. Surv. Bull.* **1995**, *2139*, 19–26. [ ]
8. Lacasse, C.; Karlsdóttir, S.; Larsen, G.; Soosalu, H.; Rose, W.I.; Ernst, G.G.J. Weather radar observations of the Hekla 2000 eruption cloud, Iceland. *Bull. Volcanol.* **2003**, *66*, 457–473. [ ] [CrossRef]
9. Marzano, F.S.; Lamantea, M.; Montopoli, M.; Di Fabio, S.; Picciotti, E. The Eyjafjöll explosive volcanic eruption from a microwave weather radar perspective. *Atmos. Chem. Phys.* **2011**, *11*, 9503–9518. [ ] [CrossRef]
10. Scollo, S.; Prestifilippo, M.; Pecora, E.; Corradini, S.; Merucci, L.; Spata, G.; Coltelli, M. Eruption column height estimation of the 2011–2013 Etna lava fountains. *Ann. Geophys.* **2014**, *57*, S0214. [ ]
11. Scollo, S.; Boselli, A.; Coltelli, M.; Leto, G.; Pisani, G.; Spinelli, N.; Wang, X. Monitoring Etna volcanic plumes using a scanning LiDAR. *Bull. Volcanol.* **2012**, *74*, 2383–2395. [ ] [CrossRef]
12. Wang, X.; Boselli, A.; D'Avino, L.; Pisani, G.; Spinelli, N.; Amodeo, A.; Chaikovsky, A.; Wiegner, M.; Nickovic, S.; Papayannis, A.; *et al.* Volcanic dust characterization by EARLINET during Etna's eruptions in 2001–2002. *Atmos. Environ.* **2008**, *42*, 893–905. [ ] [CrossRef]
13. Emeis, S.; Forkel, R.; Junkermann, W.; Schäfer, K.; Flentje, H.; Gilge, S.; Fricke, W.; Wiegner, M.; Freudenthaler, V.; Groβ, S.; *et al.* Measurement and simulation of the 16/17 April 2010 Eyjafjallajökull volcanic ash layer dispersion in the northern Alpine region. *Atmos. Chem. Phys.* **2011**, *11*, 2689–2701. [ ] [CrossRef]
14. Gasteiger, J.; Groß, S.; Freudenthaler, V.; Wiegner, M. Volcanic ash from Iceland over Munich: Mass concentration retrieved from ground-based remote sensing measurements. *Atmos. Chem. Phys.* **2011**, *11*, 2209–2223. [ ] [CrossRef]
15. Hervo, M.; Quennehen, B.; Kristiansen, N.I.; Boulon, J.; Stohl, A.; Fréville, P.; Pichon, J.-M.; Picard, D.; Labazuy, P.; Gouhier, M.; *et al.* Physical and optical properties of 2010 Eyjafjallajökull volcanic eruption aerosol: Ground-based,

Lidar and airborne measurements in France. *Atmos. Chem. Phys.* **2012**, *12*, 1721–1736. [ ] [CrossRef]
16. Merucci, L.; Burton, M.; Corradini, S.; Salerno, G.G. Reconstruction of SO_2 flux emission chronology from space-based measurements. *J. Volcanol. Geotherm. Res.* **2011**, *206*, 80–87. [ ] [CrossRef]
17. Genkova, I.; Seiz, G.; Zuidema, P.; Zhao, G.; Di Girolamo, L. Cloud top height comparisons from ASTER, MISR, and MODIS for trade wind cumuli. *Remote Sens. Environ.* **2007**, *107*, 211–222. [ ] [CrossRef]
18. Winker, D.M.; Liu, Z.; Omar, A.; Tackett, J.; Fairlie, D. CALIOP observations of the transport of ash from the Eyjafjallajökull volcano in April 2010. *J. Geophys. Res.* **2012**, *117*, D00U15. [ ] [CrossRef]
19. Winker, D.M.; Pelon, J.; Coakley, J.A.J.; Ackerman, S.A.; Charlson, R.J.; Colarco, P.R.; Flamant, P.; Fu, Q.; Hoff, R.M.; Kittaka, C.; *et al.* The CALIPSO Mission: A Global 3D View of Aerosols and Clouds. *Bull. Am. Meteorol. Soc.* **2010**, *91*, 1211. [ ] [CrossRef]
20. Wiegner, M.; Madonna, F.; Binietoglou, I.; Forkel, R.; Gasteiger, J.; Geiß, A.; Pappalardo, G.; Schäfer, K.; Thomas, W. What is the benefit of ceilometers for aerosol remote sensing? An answer from EARLINET. *Atmos. Meas. Tech.* **2014**, *7*, 1979–1997. [ ] [CrossRef]
21. Frey, R.A.; Baum, B.A.; Menzel, W.P.; Ackerman, S.A.; Moeller, C.C.; Spinhirne, J.D. A comparison of cloud top heights computed from airborne lidar and MAS radiance data using CO_2 slicing. *J. Geophys. Res.* **1999**, *104*, 24547–24555. [ ] [CrossRef]
22. Holz, R.E.; Ackerman, S.A.; Nagle, F.W.; Frey, R.; Dutcher, S.; Kuehn, R.E.; Vaughan, M.A.; Baum, B. Global Moderate Resolution Imaging Spectroradiometer (MODIS) cloud detection and height evaluation using CALIOP. *J. Geophys. Res.* **2008**, *113*, D00A19. [ ] [CrossRef]
23. Zakšek, K.; Hort, M.; Zaletelj, J.; Langmann, B. Monitoring volcanic ash cloud top height through simultaneous retrieval of optical data from polar orbiting and geostationary satellites. *Atmos. Chem. Phys.* **2013**, *13*, 2589–2606. [ ] [CrossRef]
24. Carn, S.A.; Pallister, J.S.; Lara, L.; Ewert, J.W.; Watt, S.; Prata, A.J.; Thomas, R.J.; Villarosa, G. The Unexpected Awakening of Chaitén Volcano, Chile. *Eos. Trans. AGU* **2009**, *90*, 205–206. [ ] [CrossRef]
25. Kursinski, E.R.; Hajj, G.A.; Schofield, J.T.; Linfield, R.P.; Hardy, K.R. Observing Earth's atmosphere with radio occultation measurements using the Global Positioning System. *J. Geophys. Res.* **1997**, *102*, 23429–23465. [ ] [CrossRef]
26. Solheim, F.S.; Vivekanandan, J.; Ware, R.H.; Rocken, C. Propagation delays induced in GPS signals by dry air, water vapor, hydrometeors, and other particulates. *J. Geophys. Res.* **1999**, *104*, 9663–9670. [ ] [CrossRef]
27. Oppenheimer, C. Re : Volcanological applications of meteorological satellites. *Int. J. Remote Sens.* **1998**, *19*, 2829–2864. [ ] [CrossRef]
28. Eckhardt, S.; Prata, A.J.; Seibert, P.; Stebel, K.; Stohl, A. Estimation of the vertical profile of sulfur dioxide injection into the atmosphere by a volcanic eruption using satellite column measurements and inverse transport modeling. *Atmos. Chem. Phys.* **2008**, *8*, 3881–3897. [ ] [CrossRef]
29. Dubuisson, P.; Frouin, R.; Dessailly, D.; Duforêt, L.; Léon, J.-F.; Voss, K.; Antoine, D. Estimating the altitude of aerosol plumes over the ocean from reflectance ratio measurements in the O_2 A-band. *Remote Sens. Environ.* **2009**, *113*, 1899–1911. [ ] [CrossRef]
30. Richards, M.S. Volcanic Ash Cloud Heights Using the MODIS CO2-Slicing Algorithm. M.Sc. Thesis, University of Wisconsin, Madison, WI, USA, 6 January 2006. [ ]
31. Chang, F.-L.; Minnis, P.; Lin, B.; Khaiyer, M.M.; Palikonda, R.; Spangenberg, D.A. A modified method for inferring upper troposphere cloud top height using the GOES 12 imager 10.7 and 13.3 μm data. *J. Geophys. Res.* **2010**, *115*, D06208. [ ]

32. Glaze, L.S.; Francis, P.W.; Self, S.; Rothery, D.A. The 16 September 1986 eruption of Lascar volcano, North Chile: Satellite investigations. *Bull. Volcanol.* **1989**, *51*, 149–160. [ ] [CrossRef]
33. Hasler, A.F. Stereographic observations from geosynchronous satellites: An important new tool for the atmospheric sciences. *Bull. Amer. Meteor. Soc.* **1981**, *62*, 194–212. [ ] [CrossRef]
34. Prata, A.J.; Turner, P.J. Cloud-top height determination using ATSR data. *Remote Sens. Environ.* **1997**, *59*, 1–13. [ ] [CrossRef]
35. Scollo, S.; Kahn, R.A.; Nelson, D.L.; Coltelli, M.; Diner, D.J.; Garay, M.J.; Realmuto, V.J. MISR observations of Etna volcanic plumes. *J. Geophys. Res.* **2012**, *117*, D06210. [ ] [CrossRef]
36. Virtanen, T.H.; Kolmonen, P.; Rodríguez, E.; Sogacheva, L.; Sundström, A.-M.; de Leeuw, G. Ash plume top height estimation using AATSR. *Atmos. Meas. Tech.* **2014**, *7*, 2437–2456. [ ] [CrossRef]
37. Carboni, E.; Grainger, R.G.; Thomas, G.E.; Poulsen, C.A.; Siddans, R.; Smith, A.; Sayer, A.; Peters, D. *Volcanic Plume Characterization Using Satellite Measurements in the Visible and Thermal Infrared*; USEReST Workshop: Neaples, Italy, 2008. [ ]
38. Francis, P.N.; Cooke, M.C.; Saunders, R.W. Retrieval of physical properties of volcanic ash using Meteosat: A case study from the 2010 Eyjafjallajökull eruption. *J. Geophys. Res.* **2012**, *117*, D00U09. [ ] [CrossRef]
39. Pavolonis, M.J.; Heidinger, A.K.; Sieglaff, J. Automated retrievals of volcanic ash and dust cloud properties from upwelling infrared measurements. *J. Geophys. Res. Atmos.* **2013**, *118*, 1436–1458. [ ] [CrossRef]
40. Poulsen, C.A.; Siddans, R.; Thomas, G.E.; Sayer, A.M.; Grainger, R.G.; Campmany, E.; Dean, S.M.; Arnold, C.; Watts, P.D. Cloud retrievals from satellite data using optimal estimation: Evaluation and application to ATSR. *Atmos. Meas. Tech.* **2012**, *5*, 1889–1910. [ ] [CrossRef]
41. Stoter, A. *Observations of the Karthala Eruption, Mph Report*; University of Oxford: Oxford, UK, 2009. [ ]
42. Scambos, T.A.; Dutkiewicz, M.J.; Wilson, J.C.; Bindschadler, R.A. Application of image cross-correlation to the measurement of glacier velocity using satellite image data. *Remote Sens. Environ.* **1992**, *42*, 177–186. [ ] [CrossRef]
43. Adelson, E.H.; Anderson, C.H.; Bergen, J.R.; Burt, P.J.; Ogden, J.M. Pyramid methods in image processing. *RCA Eng.* **1984**, *29*, 33–41. [ ]
44. Valentine, G.A.; Gregg, T.K.P. Continental basaltic volcanoes—Processes and problems. *J. Volcanol. Geotherm. Res.* **2008**, *177*, 857–873. [ ] [CrossRef]
45. Vergniolle, S.; Ripepe, M. From Strombolian explosions to fire fountains at Etna Volcano (Italy): What do we learn from acoustic measurements? *Geol. Soc. Lond. Spec. Publ.* **2008**, *307*, 103–124. [ ] [CrossRef]
46. UFVG Rapporto UFVG del 28 Novembre 2013: L'attività Parossistica del 23 Novembre 2013 al Nuovo Cratere di SE: Dispersione del Deposito di Caduta e Caratteristiche dei Prodotti Eruttati. Available online: http://www.ct.ingv.it/it/rapporti/vulcanologia.html?view=docman (accessed on 25 February 2016).
47. Bollettino settimanale sul monitoraggio vulcanico, geochimico e sismico del vulcano Etna 26/11/2013. Available online: http://www.ct.ingv.it/it/rapporti/multidisciplinari.html?view=docman&start=210 (accessed on 25 February 2016).
48. Prata, A.J. Observations of volcanic ash clouds in the 10–12 μm window using AVHRR/2 data. *Int. J. Remote Sens.* **1989**, *10*, 751–761. [ ] [CrossRef]
49. Prata, F.; Bluth, G.; Rose, B.; Schneider, D.; Tupper, A. Comments on "Failures in detecting volcanic ash from a satellite-based technique". *Remote Sens. Environ.* **2001**, *78*, 341–346. [ ] [CrossRef]
50. Corradini, S.; Montopoli, M.; Guerrieri, L.; Ricci, M.; Scollo, S.; Merucci, L.; Marzano, F.S.; Pugnaghi, S.; Prestifilippo, M.; Ventress, L.J.; *et al.* A Multi-Sensor

Approach for Volcanic Ash Cloud Retrieval and Eruption Characterization: The 23 November 2013 Etna Lava Fountain. *Remote Sens.* **2016**, *8*, 58. [ ] [CrossRef]
51. Corradini, S.; Merucci, L.; Prata, A.J.; Piscini, A. Volcanic ash and SO_2 in the 2008 Kasatochi eruption: Retrievals comparison from different IR satellite sensors. *J. Geophys. Res.* **2010**, *115*, D00L21. [ ]
52. WMO, OSCAR. Available online: http://www.wmo-sat.info/oscar/ (accessed on 25 February 2016).
53. WMO, OSCAR. Available online: http://www.wmo-sat.info/oscar/observingm ission s/ view/2 (accessed on 25 February 2016).

Index

S

T

U

V